Sustainable Environmental Waste Management Strategies

Series Editors

Alok Prasad Das, Department of Life Sciences
Rama Devi Women's University
Bhubaneswar, Odisha, India

Eric van Hullebusch, Équipe ACE
Institut de Physique du Globe de Paris
PARIS CEDEX 05, France

Sustainable waste management is an absolute imperative in a world that is already dealing with the effects of climate change. This book series seeks to publish an extensive collection of scientific books targeting researchers, scientists, professionals, academics, students, and anyone else involved in the wide, diverse field of sustainable waste management. The series accepts monographs, edited volumes, textbooks, and conference proceedings. Topics covered include but are not limited to environmental pollution and remediation, sources of environmental waste, environmental and human health, sustainable waste management, circular economy, nature-inspired environmental solutions, pollution prevention, and energy-efficient technologies for waste management. The series puts a particular emphasis on developing countries/the majority world but accepts proposals globally as long as the topic is in scope.

Subhadra Rajpoot • Alok Prasad Das
Editors

Innovative Technologies for Waste Management

Strategies, Challenges and Future Directions

Springer

Editors
Subhadra Rajpoot
Amity University
Greater Noida, Uttar Pradesh, India

Alok Prasad Das
Department of Life Sciences
Rama Devi Women's University
Bhubaneswar, Odisha, India

ISSN 3005-1622 ISSN 3005-1630 (electronic)
Sustainable Environmental Waste Management Strategies
ISBN 978-3-031-97183-9 ISBN 978-3-031-97184-6 (eBook)
https://doi.org/10.1007/978-3-031-97184-6

This Springer imprint is published by the registered company Springer Nature Switzerland AG
The registered company address is: Gewerbestrasse 11, 6330 Cham, Switzerland

Preface

The growing complexity and scale of waste generation across the globe have called for a paradigm shift in how we manage and mitigate its impacts. *Innovative Technologies for Waste Management* brings together the latest advancements and forward-thinking solutions in the field, exploring how science, engineering, and sustainability intersect to address one of the most pressing environmental challenges of our time. This book serves as a comprehensive resource for researchers, practitioners, and policymakers seeking to understand and implement cutting-edge approaches, such as waste-to-energy systems, biotechnological treatments, smart waste tracking, and circular economy models. Through interdisciplinary insights and real-world applications, this volume aims to inspire innovation and collaboration in the pursuit of more efficient, sustainable, and resilient waste management systems.

Greater Noida, Uttar Pradesh, India — Subhadra Rajpoot
Bhubaneswar, Odisha, India — Alok Prasad Das

Acknowledgments

The editors would like to extend their deepest gratitude to all the contributing authors whose expertise, dedication, and insightful research have been instrumental in shaping this book. Your commitment to advancing the field of waste management through innovative technologies is not only commendable but also essential to fostering sustainable practices worldwide. We also acknowledge the invaluable support and collaboration of our editorial team, whose meticulous attention to detail and unwavering enthusiasm made this publication possible. This book stands as a testament to the collective effort of a passionate community working toward a cleaner and more sustainable future.

Contents

Managing Waste: Definitions, Categories and Innovative Techniques

Subhadra Rajpoot and Alok Prasad Das

This is a very complex waste that can come from household, commercial, industrial or agricultural activities. The type and volume of solid waste are always growing due to the world's population expansion, permanent improvements in living standards and extensive technological advancements.

The production of municipal solid waste, along with the substantial amount of organic material found in it and its frequently improper disposal, leads to widespread ecological contamination, primarily from the release of greenhouse effect-causing gases like carbon dioxide (CO2) and methane (CH4). To handle the increasing amounts of municipal solid garbage, municipal authorities are presently being asked to create techno-economic and political solutions that are more efficient due to this environmental concern. Biodegradable material, which makes up the majority of municipal (mostly urban) solid waste, contributes significantly to greenhouse gas emissions in modern cities worldwide. The preferred approach to managing this problem, based on present understanding, is integrated solid waste management; nevertheless, these approaches need to be improved to accommodate the increasing organic portions of municipal solid waste. If strategically carried out, this can both help achieve the desired decrease in greenhouse gas emissions and possibly even have positive economic effects. Therefore, systems for managing municipal solid waste sustainably are desirable and promising subjects of research to evaluate present consumption patterns in various parts of the world and to safeguard the environment. This leads to unsanitary diseases and thus increases health problems. Separately, the co-disruption of hazardous waste in industrial/residence

S. Rajpoot (✉)
Amity University Greater Noida, Greater Noida, India
e-mail: srajpoot@gn.amity.edu

A. P. Das
Department of Life Sciences, RDW University, India

S. Rajpoot, A. P. Das (eds.), *Innovative Technologies for Waste Management*, Sustainable Environmental Waste Management Strategies,
https://doi.org/10.1007/978-3-031-97184-6_1

buildings with urban waste can put people at chemical and radioactive risks. Period-free fixed waste can also interfere with the drainage of rainwater, which can lead to the formation of stagnant water, which is a breeding ground for disease. It has the inevitable impact of the entire contaminated water supply and water chain, along with the streets, river benches, abandoned quarries, seas and lakes.

As the importance of decreasing or eliminating environmental pollution grows, so do international attempts to create innovative methods for progressively lowering the amounts of biodegradable municipal solid wastes in landfills: (a) source-separated collection of the organic fraction of municipal solid waste, which is then converted into compost; (b) incineration of organic waste for energy production; and (c) mechanical/biological processing to produce compostable material are the steps in the process to reduce organic pollution.

The concepts of municipal solid waste management, including landfilling and recycling technologies, are introduced to the reader in this introductory chapter. Additionally, the makeup of the various categories of municipal solid waste will be discussed. To give a strong scientific foundation for these processes that are in place or under development, as well as the elements required to evaluate the sustainability of these processes critically and simply using cutting-edge sustainable assessment tools, the most practical, promising and realistic scenarios for municipal solid waste management are presented.

Emergence and Generation of Solid Waste Rapid urbanization, industrialization, and population increase have made managing municipal solid waste (MSW) a crucial environmental and socioeconomic concern. Developing sustainable waste management methods requires a thorough understanding of its emergence and development. This review of the literature looks at important research on the causes of SW creation, its origins and patterns in trash volume and composition. Because of increased consumption and waste production brought about by the industrial revolution, MSW has become a critical issue. The mainstays of early waste management techniques were crude collecting methods and open dumping. However, waste management regulations became more regulated as urban centres grew in the twentieth century. According to recent research, changing consumption patterns, technological developments and economic expansion have all had a substantial impact on the volume and composition of SW (Hoornweg & Bhada-Tata, 2012). Inadequate infrastructure and policy enforcement have made waste problems in developing countries worse. MSW has become an important issue because of the industrial revolution's increased consumption and waste creation. Open dumping and rudimentary gathering procedures were the cornerstones of early waste management strategies. Nonetheless, as urban centres expanded over the twentieth century, waste management laws became increasingly stringent. Recent studies indicate that the amount and makeup of MSW have been significantly impacted by shifting consumer habits, technological advancements and economic growth (Hoornweg & Bhada-Tata, 2012). Waste issues in developing nations have gotten worse due to inadequate infrastructure and lax implementation of policies.

Numerous socioeconomic and policy-related factors impact the emergence and production of waste. To address the growing waste challenge, an increasing amount of research highlights the necessity of integrated waste management systems, innovative legislation and public participation. Future research should concentrate on how digitization, behavioural therapies and the concepts of the circular economy might lower the production of MSW.

1 Types of Solid Wastes

Waste is typically produced by the actions of both humans and animals. A rise in production and consumption processes brought about by urbanization and the quick development of industry has resulted in the production of trash from a variety of sources, including commercial, residential, industrial, institutional, social and communal activities. These wastes build up over time and can have significant effects on the environment and human health. Therefore, understanding the types, classifications and sources of garbage is essential for effective waste management. For safe disposal, solid wastes must be categorized into classes that present comparable environmental and health threats. Modern waste management systems classify solid wastes according to their type, source, characteristics and impact on the environment and human health.

As a result, this subject provides us with a general overview of wastes and their classification.

Waste Classification Based on Physical Waste: Wastes are often categorized as solid, liquid, or gaseous depending on their physical state.

Solid Waste: Discarded materials produced by commercial, industrial, household, and agricultural operations are referred to as solid waste. It comprises hazardous materials, building waste, medical waste, and home trash. In order to avoid resource depletion, health risks, and environmental contamination, solid waste management must be done correctly. Landfilling, incineration, composting, and recycling are common waste management techniques. To reduce waste production and improve resource recovery, sustainable techniques like waste reduction, circular economy strategies, and technological advancements are being embraced more and more. To guarantee a cleaner and healthier environment, governments, businesses, and communities must work together to manage trash effectively.

Liquid Wastes: Any trash produced in liquid form by homes, businesses or enterprises is referred to as liquid waste. Wastewater, sewage, industrial effluents and dangerous liquids like oils and chemicals are all included. Liquid waste must be properly treated and disposed of in order to preserve environmental sustainability, prevent water pollution and safeguard public health. Sedimentation, filtration, biological processes and chemical treatments are examples of common treatment techniques. Reusing and recycling treated wastewater can help cut down on pollution and save water. To maintain a cleaner and healthier environment, effective liquid waste management necessitates stringent rules, cutting-edge treatment technologies and public awareness.

Gaseous Wastes: Unwanted gases generated from domestic sources, automobile emissions, agricultural operations and industrial processes make up gaseous waste. Common examples that contribute to air pollution, climate change and respiratory illnesses are carbon dioxide (CO_2), methane (CH_4), sulphur dioxide (SO_2) and nitrogen oxides (NO_x). Cleaner energy sources, sustainable industrial practices and emission control technology including scrubbers, filters and catalytic converters are all necessary for the efficient management of gaseous waste. Reducing the negative effects of gaseous waste on the environment and human health requires regulations, air quality monitoring and public awareness campaigns.

2 Classification on the Basis of Source

Municipal Solid Waste (MSW): Commonly known as trash or garbage, municipal solid waste includes things like dead animals, market waste, abandoned cars, domestic trash, building and demolition debris, packaging materials, trade refuges and more. They are gathered from homes, marketplaces, streets and other locations, primarily in cities, and disposed of by local governments. Depending on the local eating habits, lifestyle, standard of living and level of commercial and industrial activity, the percentage of various constituents in municipal garbage varies from location to location and season to season. The table below shows the contents and source of municipal trash. Based on their physical, chemical and biological characteristics, municipal solid wastes are further divided into different categories.

Industrial Wastes: Chemical solvents, colours, sandpaper, paper goods, industrial, metals and radioactive waste are several types of industrial waste. Hazardous and nonhazardous wastes are further classifications for industrial solid wastes. Certain industrial solid wastes are hazardous, requiring handling and disposal techniques. If not categorized as hazardous garbage, electrical and electronic waste might be included under industrial waste. The district's industrial practices have a significant impact on the makeup of industrial waste. When it comes to packaging, plastics, paper and metallic materials, a lot of industrial trash is comparable to commercial and household waste. However, a percentage of industrial trash—often referred to as "hazardous waste", "special waste" or "radioactive wastes"—comes from chemical processes and uses. Hazardous and nonhazardous wastes are disposed of in different ways. Depending on the waste's chemical makeup, the processes for hazardous wastes may also vary to lessen the toxicity.

Institutional/Commercial Wastes: Agency and commercial waste are associated with waste produced by businesses, offices, schools, hospitals, hotels and other agencies that residents do not have. These wastes usually include dangerous materials such as paper, cardboard, plastic, food waste, e-waste and medical and chemical waste. The composition of such waste varies depending on the type of agency and business. This will create large amounts of paper waste, while restaurants and hotels will generate large amounts of food waste. Suitable strategies for waste management, such as waste reduction, recycling and responsible disposal, are important to minimize environmental impacts. Implementing sustainable practices such as the implementation of digital solutions to reduce organic waste composting and paper use can help contribute to the round economy (Khan and Kaneesamkandi 2013).

Agricultural Wastes: Natural (organic) and nonnatural wastes produced by farming operations are both considered agricultural wastes. Litter, pesticides, fertilizers, rotten food grains, vegetables and animal and plant wastes are a few examples of agricultural waste. Industries that process agricultural goods, such as sugarcane factories, tobacco processing facilities, slaughterhouses, animals, poultry, etc., generate additional agricultural wastes. While the majority of agricultural waste is biodegradable, some pollutants, such as fertilizers and pesticides, are harmful.

Medical trash, often referred to as clinical waste, typically refers to waste materials generated from healthcare facilities, including clinics, hospitals, doctor's offices, labs and assisted living facilities. Due to its harmful effects, medical waste cannot be recycled. The largest benefit to waste management for medical waste will come from source separation. The costliest treatment option is incineration, which is typically used for them. Typically, infectious trash originates from a healthcare centre or similar establishment. Most of the infectious trash can be prepared for autoclaving or incineration. You can dispose of the residue in a landfill that has been permitted. Blood wastes can be released into a municipal sanitary sewer under certain circumstances as long as secondary treatment is used; liquids can be chemically disinfected; and pathological wastes can be incinerated or buried, if allowed. Additionally, shredding, gas-vapour treatment and disinfection (sodium hypochlorite) can make infectious trash harmless. Medical waste includes more than just infectious waste (Tchobanoglous, 2002).

3 Solid Waste Generation

According to the Ninth Malaysian Plan from 2006 to 2010, the amount of fixed waste in the Malaysian Peninsula increased from 16,200 tonnes per day in 2001 to 19,100 tonnes per day in 2005. On average, Malaysians generate 0.8 kg of fixed waste per day at the time. Ashes in industrial plants and power plants are often disposed of by city and filling. Large-scale manufacturers of industrial waste are integrated iron and steel factories that produce coal ash, slag and steel melting, integrated iron and steel factories that produce non-egg industries such as aluminium, zinc and copper. It is a thermal plant that produces red mud and tails and creates publications for the sugar industry. Industrial waste management is not the responsibility of local organizations. Industry generating solid waste must manage it themselves and obtain prior approval from their respective property management authorities in accordance with the relevant regulations.

4 The Main Element in MSW Management

The processes involved in the oversight of urban solid waste, spanning from its inception to ultimate disposal, can be categorized into six essential components: (a) waste production; (b) waste management and classification, storage and treatment at the origin; (c) collection; (d) sorting, treatment and conversion; (e) transfer and transportation; and (f) disposal.

5 Waste Generation

The act of generating waste encompasses activities where materials are deemed unworthy (in their existing form) and are either thrown away or gathered for disposal. At present, waste generation is a largely uncontrollable activity. Nonetheless, it is expected that stricter regulations will be enforced on waste creation in the future. Minimizing waste at the source, although not overseen by waste management authorities, is now part of system evaluations as a tactic to reduce the amount of waste produced.

6 Waste Handling, Sorting, Storage and Processing at the Source

Waste handling, sorting, storage and processing at the source is the second of the solid waste management system's six functional components. The tasks involved in managing wastes until they are put in storage containers for collection are referred to as waste handling and sorting. Moving loaded containers to the collecting location is also included in handling. One crucial stage in the treatment and storage of solid waste at the source is the sorting of waste components. For instance, at the point of generation, waste materials are best separated for recycling and reuse. Families are starting to understand how important it is to separate ferrous and non-ferrous materials, kitchen garbage, bottles and glass and newspapers and cardboard (Batool and Chaudhry 2009).

Due to aesthetic considerations and public health concerns, on-site storage is crucial. Many residential and business sites frequently have unsightly improvised containers and even open ground storage, both of which are unattractive. In the case of individuals, the household or the management of commercial and industrial assets often bears the cost of providing storage for solid wastes at the source. Composting backyard waste is one example of processing at the source.

7 Sorting, Processing and Transformation of Solid Waste

Waste transformation is the process of reducing waste's size, weight, volume or toxicity without recovering any resources. Numerous mechanical (like shredding), thermal (like burning without energy recovery) or chemical (like encapsulation) methods can be used to transform. The fourth functional element is the sorting, processing and transformation of solid waste materials. This functional component includes the recovery of sorted elements, solid waste processing and solid waste transformation that mostly takes place in areas distant from the waste-producing source. Combustion facilities, transfer stations, disposal sites and materials recovery facilities (MRFs) are often where commingled (mixed) wastes are sorted (Rajpoot 2019). Sorting frequently involves separating bulky goods, separating ferrous and nonferrous metals, separating waste components by size using screens and manually separating trash components. The purpose of waste processing is to recover energy and conversion products. Numerous biological and thermal activities can change the organic portion of municipal solid waste (MSW). Aerobic composting is the most used biological transformation method. Incineration is the most common method of thermal change.

8 Transfer and Transport

Two steps make up the functional part of transfer and transport:

1. Moving the waste from the smaller collection vehicle to the bigger transport equipment.
2. Moving the waste to a processing or disposal location, typically over significant distances. Usually, a transfer station is where the transfer happens.

9 Disposal

Disposal is the last functional component of the solid waste management system. These days, the final fate of all solid wastes—whether they are collected and transported straight to a landfill, leftover materials from materials recovery facilities (MRFs), residue from burning solid waste, composting rejects or other materials from various solid waste-processing facilities—is landfilling or uncontrolled dumping. An engineering facility for disposing of solid waste on land or in the Earth's mantle is called a municipal solid waste landfill plant. Since it is the most efficient method of lowering waste production, handling costs and environmental effects, reduction at the source is ranked first in the hierarchy.

The Process of Handling of MSW Solid waste management is becoming a major issue in every municipality as the population and economic activity grow. Common urban phenomena include air pollution, odour disturbance, hazardous gas releases and public health. Disposing of MSW calls for proper environmental control throughout the trash collection, disposal and, lastly, routine site monitoring. Although certain municipalities or city halls (like Kuala Lumpur City Hall) have contracted out the solid waste collection service to private companies, the local government is still in charge of this function in most Malaysian municipalities. However, they are still in charge of overseeing the management of MSW (Rajpoot 2020). Malaysia's MSW management condition is comparable to that of other Asian nations. All three levels of government—federal, state and local—are involved in the management of these processes, either directly or indirectly; each department has a specific responsibility.

10 Role of the Local Authorities

Local governments play an important role in waste management by ensuring proper investigation, transport, treatment and disposal of waste in their jurisdictions. They develop and implement regulations, promote recycling initiatives and implement waste reduction programmes to minimize environmental impacts. Additionally, it provides waste collection services, maintains landfill locations and monitors waste disposal plants. Campaigns for public sympathy and activities in municipalities are also organized to promote responsibility—resident awareness of waste disposal practices. Additionally, local governments are working with private waste management companies, non-governmental organizations and other stakeholders to improve the efficiency and sustainability of waste management. Through these efforts, they contribute to environmental protection, public health and sustainable development (Rajpoot 2021).

11 The MSW Collection Method

Solid waste is gathered during collection and put into empty containers with distinct sections for recyclables. Before being disposed of at the disposal sites, the collecting vehicles then manually gather the waste surrounding the disposal centres. City employees (municipal collection—city council, district council, city hall, etc.), collectors from private companies that have contracts with the city government (colonial collection contract—Alam Flora Sdn. Bhd.) or collectors from private companies that have contracts with private citizens (private collection) are among the parties responsible for managing this aspect of MSW management. Curb side or alley pickup, set-out and set-back collection and backyard pickup or tote barrel method are the three fundamental approaches.

12 The MSW Collection Method

The three basic methods are:

1. **Curb side or alley pickup:** The most cost-effective and rapid point of collection, which uses a standard container that the local council has designed. Compared to backyard collecting, it is only half as expensive. At the same time, the crews gather from both sides of the street. Municipal ordinances or administrative regulations typically specify when the containers must be placed at the curb or in the alley for pickup, as well as how long they may stay there after pickup. Typically, they leave by 7 am and return by 7 pm. Work moves quickly when solid waste is loaded into curbs or alleyways. Two collectors plus a driver make up the atypical crew. In addition to the method's financial advantages, it also removes the need for collectors to enter private property, and each homeowner receives a comparatively consistent level of treatment.
2. **Set-out, set-back pickup collection:** Although the collector must enter private property, this type of approach removes the majority of the curb method's drawbacks. The following operations make up this method: (1) Collectors must enter the property; (2) before the collection vehicle arrives, the set-out crew transports full containers from the resident storage facility to the curb or alley; (3) collection crews put their garbage into the vehicle; and (4) set-back crews return the containers to the storage place. The homeowner may demand that one of the steps be completed, or any member of the team may be asked to perform multiple tasks. There is no evidence that this approach is more cost-effective or beneficial than the backyard approach.
3. **Backyard pickup or tote barrel method:** Pickup in the backyard is typically done with tote barrels. Using this procedure, the collector goes onto the resident's property, fills a tote barrel, carries the container to the truck and then dumps it. Before going back to the vehicle to dump, the collector could pick up trash from multiple homes. The homeowner's convenience is this system's main benefit. The expense is the biggest drawback. A lot of homeowners are against the collectors entering their homes.
4. **Disposal methods:** Selecting the appropriate disposal method is essential since improper and unscientific methods used for MSW disposal are socially and economically unacceptable. Two important criteria that should be considered while designing an effective, economical and ecologically friendly disposal system are the quantity and characteristics of the MSW.

One can choose the appropriate disposal method which is generally categorized as follows:

13 For Large-Scale Disposal

13.1 Open Dumps

Open dumping is the simplest and least expensive way to dispose of MSW. It involves depositing the waste in low-lying areas on the outskirts of the city and occasionally levelling it using bulldozers. It is not scientific to dispose of rubbish in an open manner. Open dumps are unprotected locations where trash is disposed of without environmental safeguards.

The trash is not separated, left exposed or processed. Despite being straightforward to implement, this traditional trash management system has proven very expensive, especially for large cities. Open, uncontrolled dumping is not a good idea. Open landfills are vulnerable to rats and flies. It also produces an unpleasant odour and an ugly appearance. The movement of loose trash is distributed via wind. The WHO Expert Committee (1967) condemned dumping as "a most unsanitary method that creates public health hazards, a nuisance, and severe pollution of the environment. Dumping should be outlawed and replaced by sound procedures".

13.2 Landfill

Burying rubbish is the process of disposing of it in a landfill, and this is still the norm in many nations. Landfills are typically found in cities where a lot of rubbish is produced and needs to be disposed of in one location. The necessary equipment is reasonably priced, and it may be utilized for other municipal tasks. Additionally, it avoids the major health risks to the community that come with open dumping or burning. Landfills were frequently built in borrow pits, mining voids and unusable or closed quarries. It is a pit that is excavated in the ground as opposed to an open dump. After the waste is disposed of, the pit is equally covered in layers of dirt and debris at the disposal site. Every day, a layer of dirt is dumped on top of it, and the trash is compressed using a device—typically an earthmoving machine—to create a cell. As a result, trash is deposited daily and turns into a cell (Rajpoot 2024).

14 Sanitary Landfills

Large amounts of municipal solid waste (MSW) can be diverted from disposal through source reduction, reuse, recycling, recovery and composting, but some garbage still needs to be dumped in landfills. In accordance with rules, modern landfills are well-engineered structures that are situated, planned, run, observed, closed, maintained and managed after closure. Every regulation was put in place to safeguard the environment and public health. The primary causes of trash are the growing urban population and garbage production, the lack of adequate and restricted resources for solid waste management and the low professional standing of waste management personnel. To guarantee a high-quality, sustainable operation, waste management is a crucial municipal function that calls for exceptional managers to make difficult, decisive decisions.

The widely recognized, scientific or popular meaning of disposing of waste in accordance with engineering procedures is "sanitary landfill". It involves keeping wastes out of the environment until they are rendered harmless by physical, chemical and biological degrading processes in a landfill. Since sanitary landfilling is a controlled operation, there is no risk to the public's health. To reduce environmental contamination, solid wastes are essentially disposed of daily by spreading them out in thin layers, compacting them to the smallest possible volume and covering them with soil cover. Sanitary landfills, which are more hygienic and constructed methodically, are an alternative to landfills or modern landfills that partially address the issue of leaching. "Landfill" refers to a waste disposal site where residual solid waste is deposited in a facility that has been designed with safeguards against air, surface and groundwater pollution, fugitive dust, wind-blown litter, foul odours, fire hazards, bird threats, rodents or pests, greenhouse gas (methane) emissions, slope instability and erosion. These are constructed on impermeable soil and lined with impermeable materials like clay and polymers. To improve property values, deposited garbage is typically covered to keep rodents (such mice or rats) away and compacted to boost its density and stability. All garbage, including infectious hospital waste, usually ends up at the disposal site because there is very little waste segregation at the source. Industrial hazardous trash is frequently dumped at landfills intended for household waste. In most cases, the waste that is dumped at the site is not regularly compacted or dispersed. Additionally, there is no inert material covering it. As a result, the disposal sites are extremely unsanitary. The conditions in which trash workers operate are often unsanitary and unhealthy. If leach is not adequately managed, it seeps into the ground and contaminates the groundwater if it is not stopped.

Considering the entire waste management system, including resource recovery and waste prevention, and looking for a solution that best fits the environment, economy and society in question. These days, there are many examples of technologies that were unsustainable in the particular economy, society and environment due to an overemphasis on technological features (Klundert, 1999). Sanitary landfill design is crucial from an operational standpoint as well. Climate factors including temperature, wind and rain have a direct impact on the project's nature and cost. For instance, windy sites require someone to clean up the environment at the end of the day and trash gates at the operating area. When the soil dries, these locations can also become extremely dusty, which might annoy those who live or work nearby. In order to keep dust and debris contained within a sanitary landfill, trees are planted around its perimeter. Along haul highways made of gravel or crushed stone, water sprinkling or other dust palliatives are frequently required. Rain affects the breakdown of solid waste by seeping into the sanitary landfill.

Incineration Incineration is the process of burning garbage at high temperatures in big furnaces. One form of disposal that involves burning waste is incineration. "Thermal treatment" is sometimes used to refer to incineration and other high-temperature waste treatment methods. Both small-scale incineration by individuals and large-scale incineration by industry are practised. Solid, liquid and gaseous wastes are all disposed of with it. In general, incinerators do not need as much space as landfills. Ash is all that remains after the procedure is finished. Incinerator combustion is not always flawless, and gaseous emissions from incinerator stacks have raised concerns over micropollutants. Some extremely persistent organics, including dioxins, which could be produced in the incinerator, have drawn special attention. There is an issue with how to dispose of this ash (Fang & Farghali (2023). The method is capital-intensive due to the cost of the incinerator and additional expenditure on pollution control equipment. Due to their capital-intensive nature and the low calorific value of the available city rubbish, large-scale incinerators are not economically feasible in India.

15 For Small-Scale Disposal

15.1 Composting

Since the beginning of life on this planet, solid organic waste has been decomposing and stabilizing naturally. The controlled decomposition and stabilization of organic matter is known as composting. Plant matter, food scraps and paper products are examples of waste materials that can be recycled by biological composting and digestion processes that break down the organic matter. It is a biological process wherein microorganisms, primarily bacteria and fungi, transform biodegradable organic waste into a material that resembles humus. Furthermore, industrial waste gases, such methane, can be recovered and converted into electricity.

Simple home compost piles to large-scale enclosed vessel digestion of mixed household waste are just a few of the many different composting and digesting techniques and technologies available. Anaerobic and aerobic methods are the two categories into which biological decomposition techniques are divided, while there are also hybrids of the two. In India, between 35% and 40% of the municipal solid trash produced is made up of organic stuff. Composting, one of the earliest methods of disposal, can be used to recycle this material. Composting may drastically cut down on the quantity of disposable trash in addition to being safe, affordable and clean. Of organic fertilizer, 200–300 kg can be produced from one MT of wet waste. It improves the soil's capacity to retain water and facilitates cultivation. At the community level, vermi-composting is incredibly successful, but it has not yet reached the commercial stage (Batool and Chaudhry 2009). In smaller urban areas, composting is done by hand. Even though towns were equipped with mechanical composting units, very few of them are still in use today. The technique is not economically viable due to several issues, including the high cost of mechanical composting plants and the non-utilization of by-products. The segregation operation is the most important step in the composting process. The compost plant's manual rubbish sorting process is costly and unhygienic.

16 The Concepts of 4R's

When it comes to trash management, recycling, waste reduction and waste reuse are the best choices. The application of these techniques has numerous positive effects on the environment. In the hierarchy of solid waste management, waste reduction ought to come first. Prior to garbage entering the municipal waste stream, it lowers the quantity and toxicity of waste. Less waste is produced, and less natural resources are consumed when items are designed, manufactured, bought or used in methods that lessen the quantity or toxicity of rubbish produced. The approach to cutting and reusing waste is waste avoidance, sometimes known as "source reduction". Businesses and consumers may reduce waste management and protect natural resources. Waste management costs by reducing. Malaysians are adopting a "use and throw" consumption pattern, whereby they consume a growing number of disposable goods, in tandem with their rising standard of living. To lessen the amount of garbage that enters the waste stream, this consumption pattern should be reversed.

17 Saves Natural Resources

Not all waste is produced by consumers discarding things. Every phase of a product's life cycle—raw material extraction, shipping, processing, manufacturing facilities and use—produces waste. Waste is significantly reduced by reusing products or creating them with less material. In the end, less materials will require landfilling, recycling or combustion for energy.

18 Reduces Toxicity of Waste

Another crucial element of source reduction is choosing less hazardous or non-toxic products. Waste toxicity can be decreased by utilizing less hazardous substitutes for some products (such as insecticides and cleaning supplies), sharing products that contain hazardous chemicals rather than discarding leftovers, carefully reading label instructions and using the least amount required.

19 Reduces Costs

Reducing dependency on alternative waste disposal methods is only one advantage of waste prevention. Additionally, preventing waste can save money for businesses, organizations, communities and individual customers.

Reuse: In order to stop waste at its source, reuse frequently involves avoiding or postponing a material's entry into the trash collection and disposal system. Energy and resources can be saved by reusing the product in the same way that it was designed. Anyone using an object should be aware of its utilization value. For instance, a polystyrene cup may be used multiple times before being thrown away, and a plastic bag can be used to bring items home from the market (Akmal and Jamil 2021). In order for commodities to be used according to their utilization value, private sector involvement should be promoted, and repair facilities should be made available. Large manufacturers of electronics, gadgets and other products ought to set up a collection centre where broken goods can be fixed and repurposed.

Recovery: Materials recovery is the process of taking municipal solid waste out of the waste stream so that it can be recycled. Source reduction and recovery are different in that the former keeps garbage out of the waste stream, while the latter extracts recyclables from it.

Recycling: Recycling creates useful resources out of items that would otherwise end up as waste. Gathering used newspapers, cans and bottles and bringing them to the curb or a collection centre is only the first stage in a process that produces numerous social, economic and environmental benefits. Both locally and internationally, some of these advantages are realized. Recycling offers numerous environmental, economic and social benefits. It reduces waste sent to landfills and incinerators, helping to conserve valuable land and reduce pollution. This reduction in energy consumption also lowers greenhouse gas emissions, contributing to the fight against climate change. Economically, recycling creates jobs in collection, processing and manufacturing industries, boosting local economies. Furthermore, it promotes a culture of environmental responsibility and sustainability, encouraging individuals and businesses to adopt more eco-friendly practices. Overall, recycling plays a crucial role in protecting the planet while fostering economic and social development.

20 Solutions and Recommendations

Population expansion is mostly to blame for the rise in the rate of municipal solid trash output. But, on average, each person is also producing more waste every day. Waste discharge is a reflection of people's social activities and lifestyles. The demand for packaged goods has grown as more people rely on others for necessities like food and services and as people value convenience and leisure time. Older durable things have become obsolete more quickly as technology has advanced. As a result, both the kind and volume of municipal solid trash have undergone significant changes throughout time.

Until source reduction begins, the amount of municipal solid trash generated will keep rising. Waste discharge differs from district to district and season to season. The degree of urbanization and industrialization, per capita income, social conventions and climate all have an impact on the characteristics of municipal solid garbage. Generally speaking, cities generate more waste than rural places. Communities with higher economic status produce less ash, ceramics, organic material, etc. Without a doubt, municipal solid trash has grown in importance as a global social and environmental issue. Reports of both immediate and long-term environmental impacts from the previous solid waste management practices have frightened the public in many nations. According to economists, it is less expensive to discard any material than to repurpose it. Environmental scientists disagree with this notion considering public health and finite natural resources. We must therefore come up with appropriate, secure and cost-effective solutions. Because of our everyday lives and industrial operations, waste disposal is unavoidable. In terms of efficiency, it is also crucial to analyse present waste management systems to evaluate their sustainability. What steps may be taken to transform present waste management systems into sustainable ones? Technical, financial, social, regulatory and environmental factors should all be considered when posing and evaluating this topic. The system will benefit from this.

21 Future Research Directions

Waste products from industrial or public operations can either directly or indirectly endanger public health. Uncontrolled or uncollected garbage can lead to respiratory infections. Where burial is conducted, it is quite seldom and does not accomplish its intended goal, and the major communities dispose of their refuse in a crude manner by dumping. The selection of sites is typically predicated on a little knowledge of the possible impacts on groundwater and surface water resources. Some dumping has an impact on domestic animals' and people's health in addition to the condition of the water. Surface streams continue to provide drinking water for a large portion of the rural population. Since scavengers frequently scavenge for food or other useful materials, unregulated dumping puts rural, urban and peri-urban regions at risk. There are clear issues with pollution, habitat loss and aesthetic appeal. The present solid waste management infrastructure will face growing demands from the metropolitan population. The fact that a youngster passes away from a water-related illness every 8 s is very tragic for humanity. Over five million people perish each year from diseases associated with contaminated drinking water, dirty homes and inappropriate excreta disposal. As a result, this issue has a connection to public health in addition to environmental issues (Hettiarachchi et al. 2018).

Waste affects social life, the environment and corporate life in a financial and legal way. Accurate waste definitions are necessary, but so are consistent reporting procedures. Compiling precise data on waste arising is a crucial component of the plan. This makes it possible to set recycling goals, track results and divert various waste types from disposal, among other things. It will be feasible to select the appropriate technology for investment at the appropriate time and scale if there is only reliable data.

22 Conclusion

Environmental issues including pollution, solid waste and wastewater treatment are always very expensive to deal with. But because of technology advancements, environmental challenges can now be benefited from with less expense and trouble. Additional technological advancements will result in a waste management system that is both economically viable and sustainable. The same wastes can now be treated using a variety of methods, which can be readily assessed based on the circumstances and the waste's type, quantity and requirements. According to White et al. (1995) and Warmer Bulletin 49 (1996), environmental sustainability refers to the creation of a waste management system that minimizes the overall environmental impacts of waste management, such as energy consumption; pollution of the air, water and land; and loss of amenity (Khan and Kaneesamkandi 2013). Overall, the technology offers a variety of treatment system options. Particular focus has been paid to low-cost, environmentally friendly alternatives that can be recycled or used for low-pollution energy recovery. These methods include anaerobic digestion, composting, gasification, pyrolysis and combined pyrolysis and gasification. Pyrolysis, gasification and combined pyrolysis/gasification systems are three thermal treatment technologies that are gaining popularity as practical, cost-effective and environmentally friendly alternatives for waste processing. Compared to traditional garbage incineration or landfilling, these methods provide a variety of benefits.

References

Abdul-Aziz H. M. et al. (2007). A. study of baseline data regarding solid waste management in the holy city of Makkah during Hajj. The Custodian of the Two Holy Mosques Institute of the Hajj Research; Medina, Saudi Arabia: . Unpunished Report.

Abubakar, I. R. (2017). Household response to inadequate sewerage and garbage collection services in Abuja, Nigeria. *Journal of Environmental and Public Health, 2017*, 5314840. https://doi.org/10.1155/2017/5314840

Abubakar, I. R., & Aina, Y. A. (2016). *Population growth and rapid urbanization in the developing world.* IGI Global. Achieving sustainable cities in Saudi Arabia: Juggling the competing urbanization challenges; pp. 234–255.

Abul, S. (2010). Environmental and health impact of solid waste disposal at Mangwaneni dumpsite in Manzini: Swaziland. *Journal of Sustainable Development in Africa, 12*, 64–78.

Addo, I. B., Adei, D., & Acheampong, E. O. (2015). Solid waste management and its health implications on the dwellers of Kumasi Metropolis, Ghana. *Current Research Journal of Social Sciences, 7*, 81–93. https://doi.org/10.19026/crjss.7.5225

Akmal, T., & Jamil, F. (2021). Health impact of solid waste management practices on household: The case of metropolitans of Islamabad-Rawalpindi, Pakistan. *Heliyon*. https://doi.org/10.1016/j.heliyon.2021.e07327

Balogun, A. L., et al. (2022). Digitalization for transformative urbanization, climate change adaptation, and sustainable farming in Africa: Trend, opportunities, and challenges. *Journal of Integrative Environmental Sciences, 19*, 17–37. https://doi.org/10.1080/1943815X.2022.2033791

Batool, S. A., & Chaudhry, M. N. (2009). The impact of municipal solid waste treatment methods on greenhouse gas emissions in Lahore, Pakistan. *Waste Management, 29*, 63–69. https://doi.org/10.1016/j.wasman.2008.01.013

Chappell, P. (1995). *Drawing board to reality*. Energy Technology Support Unit (ETSU), Harwell.

European Commission. (2003). *Waste generated and treated in Europe*. European Commission, Office for Official Publications of the European Communities.

Fang, B., Yu, J., & Farghali, M. (2023). Artificial intelligence for waste management in smart cities: A review. *Environmental Chemistry Letters, 21*, 1959–1989.

Gazzeh, K., Abubakar, I. R., & Hammad, E. (2022). Impacts of COVID-19 pandemic on the global flows of people and goods: Implications on the dynamics of urban systems. *Land, 11*, 429. https://doi.org/10.3390/land11030429

Giusti, L. (2009). A review of waste management practices and their impact on human health. *Waste Management, 29*, 2227–2239. https://doi.org/10.1016/j.wasman.2009.03.028

Goren, S. (2005). *Sanitary landfill*. Fatih University Press.

Guerrero, L. A., Maas, G., & Hogland, W. (2012). Solid waste management challenges for cities in developing countries. *Waste Management, 33*, 220–232. https://doi.org/10.1016/j.wasman.2012.09.008

Hettiarachchi, H., Meegoda, J., & Ryu, S. (2018). Organic waste buyback as a viable method to enhance sustainable municipal solid waste management in developing countries. *International Journal of Environmental Research and Public Health, 15*, 2483. https://doi.org/10.3390/ijerph15112483

Hoorn, W. D., & Giannelli, N. (2007). *Managing municipal solid waste in Latin America and the Caribbean: Integrating the private sector, harnessing incentives*. World Bank.

Hoornweg, D., & Bhada-Tata, P. (2012). *What a waste: A global review of solid waste management. Urban development series*. World Bank. Knowledge Papers No. 15.

Kabera, T., Nishimwe, H. (2019). Systems analysis of municipal solid waste management and recycling system in East Africa: Benchmarking Performance in Kigali City, Rwanda. EDP Sciences; Les Ulis, France.

Khan, M. S. M., & Kaneesamkandi, Z. (2013). Biodegradable waste to biogas: Renewable energy option for the Kingdom of Saudi Arabia. *International Journal of Innovation and Applied Studies, 4*, 101–113.

Klundert, A. V. (1999). Integrated sustainable waste management, CEDARE/IETC inter-regional workshop on technologies for sustainable waste management, held 13–15 July 1999 in Alexandria, Egypt.

Manaf, L. A., Samah, M. A. A., & Zukki, N. I. M. (2009). Municipal solid waste management in Malaysia: Practices and challenges. *Waste Management, 29*, 2902–2906. https://doi.org/10.1016/j.wasman.2008.07.015

McBean, E. A., Rovers, F. A., Farquhar G. J. (1994). Solid Waste landfill engineering and design, Prentice Hall PTR, ISBN 0–13–079187-3.

McDougall, F. R., White, P. R., Marina, F., & Peter, H. (2001). *Integrated solid waste management: A life cycle inventory* (2nd ed., p. 24). Blackwell Publishing Ltd.

Menikpura, S. N. M., Gheewala, S. H., & Bonnet, S. (2012). Sustainability assessment of municipal solid waste management in Sri Lanka: Problems and prospects. *Journal of Material Cycles and Waste Management, 14*, 181–192. https://doi.org/10.1007/s10163-012-0055-z

Miezah, K., Obiri-Danso, K., Kádár, Z., Fei-Baffoe, B., & Mensah, M. Y. (2015). Municipal solid waste characterization and quantification as a measure towards effective waste management Ghana. *Waste Management, 46*, 15–27. https://doi.org/10.1016/j.wasman.2015.09.009

Ojok, J. (2013). Rate and quantities of household solid waste generated in Kampala City, Uganda. *Science Journal of Environmental Engineering Research, 2013*. https://doi.org/10.7237/sjeer/237

Pokhrel, D., & Viraraghavan, T. (2005). Municipal solid waste management in Nepal: Practices and challenges. *Waste Management, 25*, 555–562. https://doi.org/10.1016/j.wasman.2005.01.020

Rajpoot, S. (2019). Current technologies for waste water treatment in chemical industries in India (May 4, 2019). *International Journal for Modern Trends in Science and Technology, 05*(05). https://doi.org/10.2139/ssrn.3390363

Rajpoot, S. (2020). A review on ways to manage biomedical waste at different locations in Faizabad. *International Journal for Modern Trends in Science and Technology, 06*(01), 33–36.

Rajpoot, S. (2021). Impact of nanotechnology on environment—A review. *International Journal for Modern Trends in Science and Technology, 7*, 159–164. https://doi.org/10.46501/IJMTST0710026

Rajpoot, S. (2024). Eco-friendly solutions for sustainable waste management: An approach towards a cleaner and greener future. *Waste Management and Life Cycle Assessment for Sustainable Business Practice*, pp. 43–64.

Ramachandra, T. V., Bharath, H. A., Kulkarni, G., & Han, S. S. (2018). Municipal solid waste: Generation, composition and GHG emissions in Bangalore, India. *Renewable and Sustainable Energy Reviews, 82*, 1122–1136. https://doi.org/10.1016/j.rser.2017.09.085

Saeed, M. O., & Hassan, M. N. (2009). Assessment of municipal solid waste generation and recyclable materials potential in Kuala Lumpur, Malaysia. *Waste Management, 29*, 2209–2213. https://doi.org/10.1016/j.wasman.2009.02.017

Scarlat, N., Motola, V., Dallemand, J. F., Monforti-Ferrario, F., & Mofor, L. (2015). Evaluation of energy potential of municipal solid waste from African urban areas. *Renewable and Sustainable Energy Reviews, 50*, 1269–1286. https://doi.org/10.1016/j.rser.2015.05.067

Szász, L., Csíki, O., & Rácz, B.-G. (2021). Sustainability management in the global automotive industry: A theoretical model and survey study. *International Journal of Production Economics, 235*.

Tchobanoglous (2002). Chapter 5—Solid waste management, TCHOBANOGLOUS GEORGE, Davis, California, Salvato Joseph A., Nemerow Nelson, Agardy Frank, Environmental Engineering,

White P. So, Warmer Bulletin 49, What is integrated waste management? (1996), Tonbridge, Kent, Journal of the World Resource Foundation.

White, P., Franke, M., & Hindle, P. (1995). *Integrated solid waste management: A lifecycle inventory*. Blackie Academic and Professional.

Williams, P. T. (2005). *Waste treatment and disposal* (2nd ed.). England John Wiley & Sons, Ltd.

Yang, Q., Fu, L., Liu, X., & Cheng, M. (2018). Evaluating the efficiency of municipal solid waste Management in China. *International Journal of Environmental Research and Public Health, 15*, 2448. https://doi.org/10.3390/ijerph15112448

Zafar, S. (2015). *Solid waste management in Saudi Arabia*. EcoMENA. Retrieved February 24, 2021, from http://www.ecomena.org/tag/dammam/

Zhang, D. Q., Tan, S. K., & Gersberg, R. M. (2010). Municipal solid waste management in China: Status, problems and challenges. *Journal of Environmental Management, 91*, 1623–1633. https://doi.org/10.1016/j.jenvman.2010.03.012

E-Waste: An Overview of Sources, Properties, and Environmental Consequences

Noorie Shama and Aditi Nayak

Abstract Electronic waste, commonly known as e-waste, is a significant concern due to its impacts on the environment and human health. This chapter discussed in detail the different e-waste sources, their characteristics, and environmental impacts. As technology advances, electronic waste sources like phones and computers, which are outdated, and discarded appliances like refrigerators and TVs get replaced very frequently, leading to the growing pile of e-waste. The main concern of e-waste is its composition, containing hazardous elements like lead, mercury, cadmium, and organic pollutants like flame retardants, PAHs, PCBs, and dioxin-related compounds, which need to be handled properly. Outdated or broken medical devices, such as diagnostic machines, patient monitors, and other healthcare electronics, generate e-waste. When this e-waste is dumped in landfills or burned, it causes soil and air pollution. If this e-waste is not disposed of adequately, it leads to the release of harmful chemicals into the environment, causing adverse health effects. In many developing countries, e-waste is often processed in informal settings where workers are exposed to dangerous chemicals without adequate protection. This can result in severe health issues, including respiratory problems and neurological damage. Through a detailed review of present data and research, this chapter aims to provide a clear understanding of the e-waste implications for sustainable development. It is important to discuss in detail the e-waste so that its usage can be reduced and the new trend of replacing them frequently is not followed, to reduce e-waste generation.

Keywords E-waste · Hazardous elements · Flame retardants · Dioxin-related compounds

---

N. Shama · A. Nayak (✉)
Department of Life Science, Guru Nanak Institute of Pharmaceutical Science and Technology, Panihati, Kolkata, West Bengal, India
e-mail: aditi.nayak@gnipst.ac.in

S. Rajpoot, A. P. Das (eds.), *Innovative Technologies for Waste Management*, Sustainable Environmental Waste Management Strategies,
https://doi.org/10.1007/978-3-031-97184-6_2

1 Introduction

E-waste is the discarded, unwanted electronic items operating on electromagnetic fields and electronic currents that are no longer needed. When we look at the material composition of e-waste, its significant portion consists of metals (60%), a smaller portion of plastic (15%), and others (CRT and LCD monitors 12%, cables 2%, PCBs 2%, and pollutants 3%) (Dhir et al., 2021). Major sources of e-waste are domestic appliances like televisions, refrigerators, mixer grinders, AC, mobile, computers, etc., and also medical devices from hospitals like blood pressure monitors, lab analyzers, electrocardiograms, sphygmomanometers, spectrophotometers, MRI machines, etc. Material design of electric and electronic equipment contains 69 elements from periodic tables, of which there are precious metals (gold, silver, iridium, platinum, copper, osmium, and ruthenium), critical raw materials (bismuth, indium, palladium, germanium, cobalt, and antimony), and noncritical materials (iron and aluminum) (Mazrouaa et al., 2019).

Ghana, Nigeria, India, Pakistan, and China are identified as hotspots for the disposal of e-waste; the global e-waste recorded in 2019 was 53.6 MT (Deng et al., 2023). India was reported to be among the top four countries for the e-waste generation as of 2020 (Kuang et al., 2023). E-waste and its components have many toxic properties due to heavy metal content like lead, mercury, and cadmium (Prabhakar et al., 2019; Raj & Das, 2023). There is a significant environmental concern regarding the large volume of e-waste and its hazardous components which has led to the development of inappropriate waste management practices, dangerous landfills, and improper home storage. In many developing nations, particularly in low-income countries, there are significantly unsafe landfill practices employed which pose serious health issues to the public. Disorders among children have been reported in China due to e-waste recycling. This is also most prevalent in Ghana, Pakistan, India, and Vietnam. Wires are burnt to extract copper leading to respiratory disorders, and chemicals such as acids are used for the extraction of valuable metals like gold and palladium; these works are carried out without any safety measures for the workers, posing health hazards (Dhir et al., 2021).

Due to increased demand for electronic products and devices from domestic, medical, and other settings, the electronic market is rapidly increasing which in turn is elevating the level of hazardous e-waste generated each year. Improper handling not only impacts human health but also leads to soil erosion, environmental degradation, and economic concerns due to improper dumping and management (Sari et al., 2021). Contaminants such as pollutants, furans, dioxins, and persistent organic pollutants accumulate in the atmosphere causing respiratory diseases, organ damage, reproductive failure, etc. The pollutants find their way into the food chain when microorganisms ingest particles of e-waste and pass it on to other larger organisms when they feed upon them. The components of the e-waste contaminate atmosphere, aquatic system, and terrestrial land animals which poses serious health implications. There is a need for public awareness on this issue as there is limited public awareness of the management of e-waste; usually, society focuses on municipal solid waste management, but priority must be given to managing electronic waste.

Estimation reports say that global e-waste generation can double by the year 2045; therefore, strategies need to be employed to identify the hazards, their sources, and also how to manage them. Present management practices lag behind scientific methodologies necessary for safe disposal practices (Brindhadevi et al., 2023).

2 Sources of E-Waste

There are different types of e-waste generated in the environment such as discarded domestic equipment. Some sources of e-waste include our home appliances, automobiles, and the medical sector equipments. They include televisions, microwave ovens, refrigerators, smartphones, and hospital equipment such as therapy machines, pulse generators, and electric nerve field generators (Fig. 1). Harmful organic/inorganic components of e-waste pose severe health hazards.

2.1 Discarded Electronics

Items that are replaced or discarded due to errors in functioning, reduced efficiency, and technology advancements are known as discarded electronics. These include laptops, smartphones, computers, tablets, etc., which make up a significant portion

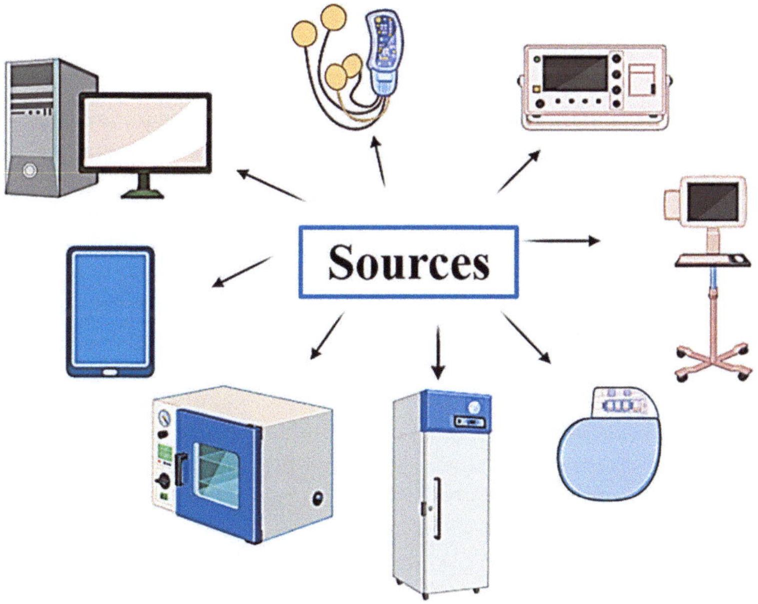

Fig. 1 Sources of e-waste

of e-waste worldwide (Mishra et al., 2023). Many of these items have valuable metals that can be recovered making recycling necessary; much of the e-waste is dumped in landfills and has been piling for years. Because of this unordered way of management, the valuable metals are lost, and the bulk of e-waste remains to accumulate in a large debris (Das et al., 2011; Priyadarsini & Das, 2025). The increased trend and demand for the latest gadgets exacerbates this problem, highlighting the need for sustainable management particles along with effective awareness programs to make people aware of the dangers and reduce the demand for new gadgets before the previous is exhausted (Ghosh et al., 2016; Mohanty et al., 2016; Rene et al., 2021).

2.2 *Medical and Laboratory Equipment*

Medical and laboratory e-waste consists of old and nonfunctioning devices such as X-ray machines, MRI scanners, centrifuges, infusion pumps, etc. These instruments become e-waste either when they are not functioning properly or when more advanced technologies have been upgraded. These devices contain heavy metals, radioactive components, and chemicals. For example, X-ray machines contain mercury and lead which need to be handled properly; also some machines produce hazardous waste which needs to be disinfected and sterilized before being handled for recycling and disposal. As research and health services are increasing, there is a need to address e-waste for supportive care and sustainability (Ahirwar & Tripathi, 2021).

2.3 *Outdated Appliances*

When appliances reach the end of the working period, they are removed or replaced which are known as outdated appliances. This includes refrigerators, washing machines, dishwashers, air-conditioning, microwave ovens, etc. These appliances are recycled to extract their metals, plastics, and steel which can reduce waste generation, but as some underdeveloped regions lack proper infrastructure for e-waste management, these ultimately are dumped in landfills which contaminate air, water, and land (Das et al., 2024; Dey et al., 2023a; Mishra et al., 2019). These appliances release high amounts of chlorofluorocarbon (CFCs) which can contribute to ozone depletion (Ankit et al., 2021).

2.4 *Industrial Waste Generation*

This includes the waste generated during the production process of electronic devices in the industry, such as defective items, damaged parts, raw materials, and packaging waste. Additionally, the transition to new product lines often leaves obsolete inventory that contributes to e-waste (Brindhadevi et al., 2023).

2.5 Entertainment and Leisure Devices

It includes television, playstation gaming consoles, music players, and home theater systems. These devices are usually replaced when new technologies upgrade and new models come that offer better features and efficient performance. This result in the generation of a significant amount of e-waste. These need to be recycled to obtain copper and gold which are major portions of circuits reducing the bulk of the waste. Also, buyback scheme needs to be employed to reduce waste generation and environmental impacts (Shahabuddin et al., 2023).

2.6 Network Infrastructure Upgrade

The waste in this category includes routers, servers, modems, and network cables which are forms of obsolete hardware. Due to the rapid expansion of internet, Wi-Fi, and telecommunication services, providers often replace outdated equipment with newer models to enhance speed and meet the increasing demand. Also, the current acceleration into 5G created a high amount of e-waste. These leave behind large amounts of e-waste like plastics, cables, metals, and electronic circuits; also, hazardous materials like heavy metals, lead, chromium (Das & Mishra, 2008; Paul et al., 2023), and cadmium make waste processing and management challenging. Refurbishment technology needs to be adopted by all people to reduce e-waste generated. Further policies that can encourage manufacturers to form hardware that can easily be recycled and cause less harm to the environment need to be addressed (Abalansa et al., 2021).

2.7 Battery Disposal

Batteries are used in a maximum of small-sized electronic devices like smartphones, laptops, power systems, etc. They can be either charged and used or are disposed of after single use; these single-use batteries make up a high portion of e-waste; also when batteries reach the end of their use, they are not properly discarded which can contaminate soil, water, and air posing severe long-term environmental impacts as they are composed of cadmium, lead, nickel, and lithium. There is a rise in demand for batteries along with renewable resources and electronic devices. Batteries should be thrown in discarded bins made for them without puncturing them as this can cause fire and release harmful gases that can be toxic; reduced use and proper disposal can mitigate the harmful impacts of battery-related e-waste (Wu et al., 2022).

3 Properties of E-Waste

3.1 Physical Properties

E-waste consists of discarded electronic devices of many kinds; its physical properties are crucial to understanding in terms of its reuse or recycling process. Various components are recycled and disposed of in different ways including:

1. Fibers: They may be natural or synthetic and are very often used for manufacturing electronic equipment; they are carbon fiber, ultrahigh molecular weight polyethylene with a high tensile strength, stiffness, and resilience. There are natural fibers, such as vegetable fiber, wood fiber, animal fiber, and mineral fiber, and synthetic fibers, such as fibers made from polyester, nylons, acrylics, and pure chemical fibers (Tsydenova & Bengtsson, 2011).
2. Research indicates that the mechanical properties of e-waste can be improved by the use of natural fibers that can increase their shelf life. Composites with printed circuit boards combined with natural fibers exhibit stiffness and strong tensile strength. It has been reported that tensile strength reaches up to 25.43 MPa and flexural strength 48.30 MPa when natural fibers are used.
3. Different components of e-waste have different properties: metallic, plastic, glass, etc. There is a variation of size and shape for each component, e.g., irregular shapes for plastic, long thin for wires, etc.
4. Density and weight vary for each component with varied fragmentation behavior when disposed of.

3.2 Chemical Properties

1. Metallic properties: E-waste contains electric circuits and batteries made of highly conductive and durable metals like copper, gold, and silver as well as hazardous metals like lead, lithium, cadmium, and nickel (Priyadarsini & Das, 2024). Neodymium and yttrium are some rare elements that are used for marking smartphones; the extraction of these metals requires proper recycling to reduce bulk waste. Some improper recycling methods like acid stripping and burning release harmful gases and substances which cause environmental impacts. Many developing countries are adopting proper recycling methods like metal smelting and bioleaching for sustainability (Das & Ghosh, 2023; Dey et al., 2023b; Kumar et al., 2024).
2. Plastic e-waste: Insulation materials, circuit boards, and wire coverings are made up of plastic polymers PS (polystyrene), PC (polycarbonate), and PP (polypropylene); these make electronics lightweight and increase their shelf life, but recycling them is challenging due to pigments, stabilizers, and retardants. The mixture of types of plastics used makes recycling challenging; innovative tech-

niques like chemical recycling can be adopted to reduce environmental impacts (Dey et al., 2024; Priyadarshini et al., 2024).

3. Halogenated Compounds: Major sources of halogenated components are flame retardants like brominated and polychlorinated flame retardants, basically used to make them resistant to catching fire. When e-waste containing these compounds is burned, it releases dioxins and furans which pose serious health risks. Alternative to flame retardants is being explored to reduce dependence on halogenated compounds while keeping fire safety (Kaya, 2016).

3.3 Electric Properties

The electronic properties of e-waste are determined by the type of material used in it like copper, aluminum, nickel, and various precious metals like gold and silver. These metals contribute to the overall conductivity of the e-waste making it essential for recycling. The ability to recover them is important to reduce the need for fresh raw materials and to reuse the extracted ones. There is a significant difference between the materials of old devices and new ones. Undertaking the electric properties helps in identifying the recycling method and also highlights the importance of safe processing to ensure the least release of hazardous substances that can have a long-term impact on the environment (Lin et al., 2022).

3.4 Toxicological Properties

E-waste often contains flame retardants of different kinds and heavy metals like lead, cadmium, and mercury which are toxic when exposed to the environment. When these leach into the soil, they cause soil degradation and groundwater contamination. Lead from cathode-ray tubes and mercury from LCD screens are highly toxic and accumulate in the aquatic system leading to life destruction and food chain contamination. Workers face long-term neurological disorders due to a lack of equipment to protect the working staff from these hazards; effective recycling protocols with minimal risk to workers and the environment and stringent regulations are crucial to protect public health and the environment (Perkins et al., 2014).

3.5 Hazardous Properties

E-waste is mainly processed in small work areas that lack proper facilities to protect the workers from hazardous chemicals and gases released during the recycling process. Workers are exposed to heavy metals leading to serious health implications, leaking metals, exhaust gases, emissions, emission of volatile compounds, and chlorinated and brominated compounds like dioxins (Eze & Vinken, 2024). When

Table 1 List of different types of e-waste components, sources, and their health hazards

E-waste component (metals/acids/compounds)	Sources	Health hazard	References
Cadmium	Mechanical destruction of plastic components and circuit boards	Acts as a carcinogen that affects kidneys, softens bones, and causes itai-itai disease	Annamalai (2015)
Mercury/beryllium	Breaking and burning of circuit boards	Affects the CNS and immune system and gets ingested by newborns through the mother's milk	Rawat et al. (2020)
Lead	CRT breaking and removal of solder from microchip	It is a neurotoxin affecting the kidney, brain, and reproductive system	Andeobu et al. (2021)
Plastics	Circuits boards, cabinets, and cables are made up of plastics	Carcinogenic activities, reproductive damage, and immune dysfunction	Wirtu and Tucho (2022)
Acids	Used to separate metals from discarded electronics	Causes respiratory disorders and throat infections and is corrosive to the eyes and skin	Ma et al. (2021)
PCB (polychlorinated biphenyl)	User as an additive, heat transfer fluid, and dielectric fluid in the plastic component of electronics	Affects the immune system, reproductive system, nervous system, and endocrine system and causes tumors and cancer	J. J. Li et al. (2021)
Dioxin and furan	Released during the burning of e-waste	Bioaccumulation in the body affects pregnancy and fetus growth, decreases immune responses, and decreases reproductive health	Luo et al. (2023)
BFRs (brominated flame retards)	Flame retards are used in making flame-resistant electronics and are released due to the combustion of e-waste at a lower temperature	Hormonal imbalance and several hormonal disorders	Kalamaras et al. (2021)

no recycling can be performed, end-of-life processes are done on e-waste which poses serious occupational risks and also environmental risks, e.g., exposure to dust and acid fumes, Ba-oxide from broken CRTs, and toner dust. Different components of e-waste and their presence in location along with their possible health implications are listed in Table 1.

4 Environmental Contamination by E-Waste

Global e-waste has been reported to increase by 30% per annum, making it a subject of concern needing immediate attention and new finer remediation technologies to be addressed. Environmental impacts due to e-waste have increased much faster due

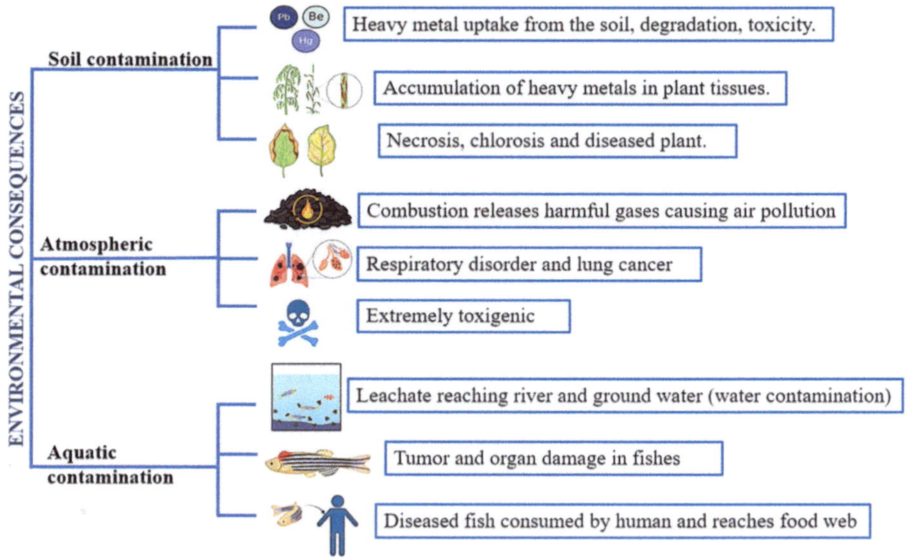

Fig. 2 Environmental consequences of e-waste contamination of soil, atmosphere, and aquatic life

to the increased demands by consumers and the application of informal, inappropriate recycling techniques that expose workers and the natural environment to toxic gaseous exhaust which has affected the soil, air, and aquatic systems leading to consequences affecting life in different forms (Fig. 2). It has been reported that copper (16,000 kg), silver (350 kg), gold (34 kg), and palladium (15 kg) have been found in one million cellphones (Houessionon et al., 2021).

4.1 Contamination of Soil

In underdeveloped regions, e-waste is dumped recklessly in nearby areas and open fields, seeping into the soil and causing soil erosion. Studies have reported that e-waste contamination in soil has been found in various sectors like sales, disposal, improper recycling, etc. Chemicals leak out of old devices when discarded and punched which contaminate the soil and have been seen in soils of processing areas and also nearby locations which means the contaminated soil is spread by different means in far distant regions (Arya et al., 2021).

These pollutants' mobility through different compartments of nature depends on different parameters like soil degradation process, pH, organic composition of the soil, fertility, temperature, etc. In times past, there have been studies indicating the different metal contents of lead, cadmium, mercury, Bi, and Sb at different concentrations in the soil which is increasing manifold yearly. Studies on an informal recycling site of e-waste in Metro Manila and six more areas reported showed a high level of Zn, Cu, Mn, Ni, Co, Pb, and Cd. When e-waste is dumped in landfills, toxic chemicals not only affect the immediate area but also seep into groundwater, posing a threat to aquatic life and contaminating drinking water (Li et al., 2022).

The pollutants alter the chemical composition of the soil, disrupting the normal microflora of the soil affecting microbial accumulation and nutrient cycling processes which are regulated by the microbes present in soil. The presence of heavy metals increases the soil pH level and overall fertility and health of the soil by affecting the organisms that contribute to soil viability.

Additionally, the physical structure and texture of the soil is also altered when the items are dumped and burned, which introduces course particles into the soil decreasing aeration and water retention capacity; these cause soil to compact and disrupt flora leading to long-term impact even when the e-waste is removed and remains contaminated for years complicating land management efforts.

4.2 Contamination of Atmosphere

E-waste contains toxic substances, heavy metals, and chemicals which when burned in an incinerator release high amounts of toxic gases that mix into the atmosphere and cause air pollution, causing ozone depletion by the release of chlorofluorocarbon (CFC). It has been reported that air concentration of PBDEs has increased up to 16,575 pg/m3 (Hu et al., 2023). Some consequences of e-waste contamination in atmospheric air are:

1. Toxic pollutants released when e-waste is burned releasing dioxins, furans, heavy metal particles, and particulate matter. When these accumulate in the air, they cause respiratory disorders, e.g., dioxins persisting in e-waste-contaminated areas causing immune system damage and cancer. Toxic fumes cause respiratory disorders such as bronchitis, asthma, and lung cancer. Workers in e-waste recycling areas are more prone and vulnerable to respiratory disorders as they have high exposure to these hazardous chemicals.
2. E-waste pollutants are exposed to large areas and transported to longer distances from their source before settling to the ground or entering the aquatic system. Even though e-waste processing sites are located far away, they still have hazardous impacts as they mix in the atmospheric air and travel to reach long distances causing air pollution in a widespread environment leading to degradation and health risks beyond the immediate vicinity of e-waste disposal (Yue et al., 2022).
3. Due to emissions by e-waste burning, there is a change in the climate due to temperature elevation and accumulation of greenhouse gases and other climate-altering substances. The release of black carbon or soot from the burning of plastics from the e-waste causes a rise in atmospheric temperature and global warming. There are vulnerable effects on seasonal patterns, i.e., weather changes, sea levels, and ecosystems due to global warming.
4. E-waste and its pollutants due to waste burning cause a lowering of the air quality index. As pollutants mix with atmospheric air, they enter the food chain, disrupting the biological balance and leading to higher disease cases and an

increase in cancer patients. This affects vital ecosystem components that are essential for human survival, also causing a loss of biodiversity.

4.3 Contamination of Aquatic System

E-waste in the dumping sites or processing sites can leach into nearby water bodies like ponds or streams contaminating not only the nearby water body but also seeping to the deeper ground contaminating the water table. Informal processing systems such as hydrometallurgical processes lead to acid disposal into the soil and water bodies (Andeobu et al., 2021). Lagoon sediments of e-waste disposal and burning areas were found to have elevated levels of copper, lead, cadmium, and zinc. Many studies from areas like Accra, Vietnam, India, Japan, Nigeria, Lagos, Alaba, etc., reported elevated levels of heavy metals and PCB (Andrade et al., 2022). Also, bioaccumulation of PCDD/Fs in many river banks like Lianjing and Elbe rivers has been documented (Wirtu & Tucho, 2022).

1. E-waste runoff from improper disposal sites leads to heavy metal accumulation in the aquatic ecosystem, leading to toxic effects, disrupting biological functions, and threatening life.
2. The fine contaminants or pollutants bioaccumulate in the tissues of the organisms as they are in micro- and nanosizes causing increased toxicity in their bodies. This leads to higher concentrations of toxins in the predators which is known as biomagnification affecting the entire food web and aquatic biodiversity.
3. The ingestion of particles by aquatic organisms can cause intestinal blockage, hinder metabolic functions, decrease fertility, and lead to abnormal growth patterns within populations. These effects contribute to a loss of biodiversity, resulting in declining species populations and eventual extinction (Kalamaras et al., 2021).
4. E-waste in the aquatic system can cause an alternation of food chains; as algae are affected very much by pollutants, it can reduce their population and in turn reduce the food available for the herbivore fishes subsequently reducing their population which in turn will affect the carnivorous species which depend on those fishes for their nutrition. The persistent nature of these contaminants has caused harmful effects on the aquatic system for decades. This long-term ecological damage makes this issue a challenge to restore the natural habitats of the aquatic ecosystem (Ullah et al., 2021).

5 E-Waste Contamination and Toxicity in Plants

E-waste-contaminated areas with plants show bioaccumulation of toxins in their tissues. Constituents of e-waste such as PBDEs (polybrominated diphenyl ethers) and PCBs (polychlorinated biphenyl) have been found in tissues of plants in the e-waste processing sites of Guiyu. This translocation of contaminants from the contaminated soil to the roots of the plants and then to higher plants like leaves and tissues results in toxicity in plants. The toxin accumulation in plants leads to stunted growth, diseased leaves, and loss of biodiversity; nutrients and vitamin content of plants are lowered by the toxic accumulation of pollutants in the plant tissues. As herbivores depend on plants for their nutrition, the e-waste enters herbivores when they feed on plants and goes to carnivorous species when they feed on them. In this way, e-waste pollutants find ways to enter the food chain and cause disastrous effects (Luo et al., 2023).

Researchers have shown from reports that crops like rice when grown in e-waste-contaminated fields show lower germination rates, cytotoxic effects, genetic disorders, impaired cellular functions, accumulation of heavy metals, and damaged plant health (Li et al., 2021).

6 E-Waste Contamination and Toxicity in Animals

E-waste exposure to the environment causes significant consequences to both aquatic and terrestrial animals. Many animals from the nearby water body of an e-waste processing site such as water snakes, peregrine falcons, and fishes have been found to have bioaccumulation of e-waste pollutants. Studies have shown a high level of Polybrominated Diphenyl Ethers (PBDEs) in fish, snails, and chicken in Guiyu, China raising concern for public health as these are a very common food consumed by humans. Leachates from e-waste dumping grounds have high toxicity of heavy metals in them and have been reported to have been linked with chromosomal abnormalities and reduced fertility (active sperm count). Also, marine animals affected with e-waste pollutants (PBDEs) like crucian carp and *Clarias gariepinus* showed significant genotoxic effects such as oxidative stress, DNA damage, and liver damage (Alabi et al., 2021).

When electric devices are improperly disposed of, hazardous substances such as heavy metals, chemicals, and flame retardants seep into the environment, leading to exposure to wildlife. Animals consume these with food and water, resulting in serious health risks. The contaminants accumulate and disrupt the predator-prey relationship by the release of toxins in the local ecosystem and imbalance of the species population. The pollutants from aquatic systems move to fishes and marine life which in turn affects terrestrial animals like birds and mammals when they rely on these species for food. Air pollution due to e-waste fumes causes respiratory issues and can damage internal organs. Bioaccumulation and biomagnification of toxins from e-waste to lower organisms, to higher organisms, and finally to the food chain leads to neurological damage and reproductive failure. Improper handling and disposal of e-waste destroys habitat. This degradation makes the area uninhabitable forcing them to move to a new environment that can further strain ecosystems. Animals that live in e-waste dump sites can get injured by sharp edges or chemicals released due to puncture; they may get trapped in discarded, broken electronic equipment. Additionally, to toxic effects of these fumes, chemicals, toxins, and other components of hazardous materials elevate their risk to dangerous diseases and mortality (Frazzoli et al., 2022).

7 Human Exposure and Public Health Effects of E-Waste

The main routes through which e-waste hazardous contaminants could enter the human body are dermal contact, inhalation, and ingestion. Human exposure to e-waste contaminants can also occur through dust, air, water, soil, and food; e-waste recycling site workers, old people, pregnant women, disabled people, and young children have additional exposure risk. Children are more sensitive to these exposures due to their low rate of toxin elimination and frequent food uptake with very low hygiene and very frequent hand-mouth activities. Pregnant women in exposure sites have been found to have more burdens of lipophilic pollutants present in e-waste due to their role in lactational and transplacental transfer to the baby (Rajesh et al., 2022). In humans, mainly children and pregnant women, exposure to e-waste toxic substances like lead and mercury causes severe neurological damage, cognitive impairments, abnormalities, and increased risk of attention deficit hyperactivity disorder (ADHD). Long-term exposure of pregnant women to e-waste contaminants causes a higher risk of stillbirth, premature birth, and lower body weight with long-term effects on the health of the newborn. Inhalation of burning fumes from e-waste incinerators or waste recycling plants leads to chronic respiratory issues with breathing issues, asthma, lung damage, decreased lung function, and also lung cancer due to accumulation of carcinogenic substances in the body leading to tumor formation and serious health conditions (Abubakar et al., 2022).

E-waste contaminants lead to nutritional loss and hormonal imbalance, affecting reproductive health, reducing fertility and active sperm count, and reducing metabolic rate leading to health conditions and reproductive failure. Exposure to toxic chemicals and particles from e-waste can weaken the immune system, making the individual sensitive to infections and diseases; a weak immune system can have significant long-term health implications. Awareness programs for the public need to be conducted to make people aware of the health risks of e-waste and the possible ways to reduce waste generation and to make proper infrastructure for the dumping, processing, disposal, and recycling of e-waste to reduce their environmental consequences for sustainability (Ma et al., 2021).

8 Conclusion

E-waste has emerged as a highly rising environmental challenge posing significant health risks to the environment and human health. The rising demand for e-waste coupled with complex composition, inadequate disposal, and recycling practices has led to the accumulation of contaminants of e-waste like chemical components, plastics, halogenated compounds, toxins, and heavy metals in the soil, air, and water systems. Different properties like physical, chemical, density, and toxicity make e-waste challenging to be removed from the ecosystem. One of the most concerning aspects is the contamination of soil destroys soil flora and soil health which in turn affects the plants growing in them through the upward translocation of contaminants by the tissue system. Herbivores depend on plants for their nutrition, and as they feed upon contaminated tissues of plants, the e-waste finds its way into the food web directly as the animals are in turn consumed by carnivores and humans leading to severe health implications. There are growth effects, disorders of plants, and disorders of the animals by the ingestion of infected plants causing respiratory disorders. Additionally, they mix in the atmospheric air causing respiratory disorders and cancers. The persistence of heavy metals and toxins of e-waste in the aquatic ecosystem causes health hazards such as reproductive failure affecting biodiversity population. There is a need for awareness among the public to enhance the technologies for e-waste recycling and to reduce the use of e-waste for the betterment of the environment and a sustainable ecosystem.

References

Abalansa, S., El Mahrad, B., Icely, J., & Newton, A. (2021). Electronic waste, an environmental problem exported to developing countries: The GOOD, the BAD and the UGLY. *Sustainability, 13*(9), 5302. https://doi.org/10.3390/su13095302

Abubakar, A., Zangina, A. S., Maigari, A. I., Badamasi, M. M., Ishak, M. Y., Abdullahi, A. S., & Haruna, J. A. (2022). Pollution of heavy metal threat posed by e-waste burning and its assessment of human health risk. *Environmental Science and Pollution Research, 29*(40), 61065–61079. https://doi.org/10.1007/s11356-022-19974-6

Ahirwar, R., & Tripathi, A. K. (2021). E-waste management: A review of recycling process, environmental and occupational health hazards, and potential solutions. *Environmental Nanotechnology, Monitoring & Management, 15*, 100409. https://doi.org/10.1016/j.enmm.2020.100409

Alabi, O. A., Adeoluwa, Y. M., Huo, X., Xu, X., & Bakare, A. A. (2021). Environmental contamination and public health effects of electronic waste: An overview. *Journal of Environmental Health Science and Engineering, 19*(1), 1209–1227. https://doi.org/10.1007/s40201-021-00654-5

Andeobu, L., Wibowo, S., & Grandhi, S. (2021). An assessment of e-waste generation and environmental management of selected countries in Africa, Europe and North America: A systematic review. *Science of the Total Environment, 792*, 148078. https://doi.org/10.1016/j.scitotenv.2021.148078

Andrade, M., Soares, A. M. V. M., Solé, M., Pereira, E., & Freitas, R. (2022). Will climate changes enhance the impacts of e-waste in aquatic systems? *Chemosphere, 288*, 132264. https://doi.org/10.1016/j.chemosphere.2021.132264

Ankit, Saha, L., Kumar, V., Tiwari, J., Sweta, Rawat, S., Singh, J., & Bauddh, K. (2021). Electronic waste and their leachates impact on human health and environment: Global ecological threat and management. *Environmental Technology & Innovation, 24*, 102049. https://doi.org/10.1016/j.eti.2021.102049

Annamalai, J. (2015). Occupational health hazards related to informal recycling of E-waste in India: An overview. *Indian Journal Of Occupational And Environmental Medicine, 19*(1), 61. https://doi.org/10.4103/0019-5278.157013

Arya, S., Rautela, R., Chavan, D., & Kumar, S. (2021). Evaluation of soil contamination due to crude E-waste recycling activities in the capital city of India. *Process Safety and Environmental Protection, 152*, 641–653. https://doi.org/10.1016/j.psep.2021.07.001

Brindhadevi, K., Barceló, D., Lan Chi, N. T., & Rene, E. R. (2023). E-waste management, treatment options and the impact of heavy metal extraction from e-waste on human health: Scenario in Vietnam and other countries. *Environmental Research, 217*, 114926. https://doi.org/10.1016/j.envres.2022.114926

Das, A. P., & Ghosh, S. (2023). Mixed culture bioleaching: An insight into manganese biomining process and efficacy. *Manganese Mining Microorganisms*, 129–141. https://doi.org/10.1016/B978-0-12-822147-1.00001-7

Das, A. P., & Mishra, S. (2008). Hexavalent chromium (VI): Health hazards and environmental pollutant. *Journal of Environmental Research and Development, 2*(3), 386–392.

Das, A. P., Sukla, L. B., Pradhan, N., & Nayak, S. (2011). Manganese biomining: A review. *Bioresource Technology, 102*(16), 7381–7387. https://doi.org/10.1016/J.BIORTECH.2011.05.018

Das, A., Ray, S. K., Mohanty, M., Mohanty, J., Dey, S., & Das, A. P. (2024). Ecotoxicity of microplastic wastes and their sustainable management: A review. *Environmental Chemistry and Ecotoxicology, 6*, 144–152. https://doi.org/10.1016/J.ENCECO.2024.05.003

Deng, S., Xiao, Z., Zhang, W., Noble, A., Das, S., Yih, Y., & Sutherland, J. W. (2023). Economic analysis of precious metal recovery from electronic waste through gas-assisted microflow extraction. *Resources, Conservation and Recycling, 190*, 106810. https://doi.org/10.1016/j.resconrec.2022.106810

Dey, S., Das, A., Mallick, K., Sahu, A., & Das, A. P. (2023a). Environmental Petroleum Waste: Pollution. *Toxicity, Sustainable Remediation, 159–175*. https://doi.org/10.1007/978-3-031-48220-5_7

Dey, S., Tripathy, B., Kumar, M. S., & Das, A. P. (2023b). Ecotoxicological consequences of manganese mining pollutants and their biological remediation. *Environmental Chemistry and Ecotoxicology*. https://doi.org/10.1016/J.ENCECO.2023.01.001

Dey, S., Dash, N., Ray, S. K., Priyadarsini, S., Chhotaray, C., & Das, A. P. (2024). Occurrence and quantification of tire wear plastics pollutants from highway city dust of Bhubaneswar City of Odisha. *Sustainable Microbial Technology for Synthetic and Cellulosic Microfiber Bioremediation, 77–89*. https://doi.org/10.1007/978-3-031-62600-5_5

Dhir, A., Koshta, N., Goyal, R. K., Sakashita, M., & Almotairi, M. (2021). Behavioral reasoning theory (BRT) perspectives on E-waste recycling and management. *Journal of Cleaner Production, 280*, 124269. https://doi.org/10.1016/j.jclepro.2020.124269

Eze, C., & Vinken, M. (2024). E-waste: Mechanisms of toxicity and safety testing. *FEBS Open Bio, 14*(9), 1420–1440. https://doi.org/10.1002/2211-5463.13863

Frazzoli, C., Ruggieri, F., Battistini, B., Orisakwe, O. E., Igbo, J. K., & Bocca, B. (2022). E-WASTE threatens health: The scientific solution adopts the one health strategy. *Environmental Research, 212*, 113227. https://doi.org/10.1016/j.envres.2022.113227

Ghosh, S., Mohanty, S., Nayak, S., Sukla, L. B., & Das, A. P. (2016). Molecular identification of indigenous manganese solubilising bacterial biodiversity from manganese mining deposits. *Journal of Basic Microbiology, 56*(3), 254–262. https://doi.org/10.1002/JOBM.201500477

Houessionon, M. G. K., Ouendo, E.-M. D., Bouland, C., Takyi, S. A., Kedote, N. M., Fayomi, B., Fobil, J. N., & Basu, N. (2021). Environmental heavy metal contamination from electronic waste (E-waste) recycling activities worldwide: A systematic review from 2005 to 2017. *International Journal of Environmental Research and Public Health, 18*(7), 3517. https://doi.org/10.3390/ijerph18073517

Hu, H., Zeng, X., Zheng, K., Zeng, Z., Dai, C., & Huo, X. (2023). Risk assessment and partitioning behavior of PFASs in environmental matrices from an e-waste recycling area. *Science of the Total Environment, 905*, 167707. https://doi.org/10.1016/j.scitotenv.2023.167707

Kalamaras, G., Kloukinioti, M., Antonopoulou, M., Ntaikou, I., Vlastos, D., Eleftherianos, A., & Dailianis, S. (2021). The potential risk of electronic waste disposal into aquatic media: The case of personal computer motherboards. *Toxics, 9*(7), 166. https://doi.org/10.3390/toxics9070166

Kaya, M. (2016). Recovery of metals and nonmetals from electronic waste by physical and chemical recycling processes. *Waste Management, 57*, 64–90. https://doi.org/10.1016/j.wasman.2016.08.004

Kuang, H.-X., Li, M.-Y., Li, L.-Z., Li, Z.-C., Wang, C.-H., Xiang, M.-D., & Yu, Y.-J. (2023). Co-exposure levels of volatile organic compounds and metals/metalloids in children: Implications for E-waste recycling activity prediction. *Science of the Total Environment, 863*, 160911. https://doi.org/10.1016/j.scitotenv.2022.160911

Kumar, P., Singh, S., Gacem, A., Yadav, K. K., Bhutto, J. K., Alreshidi, M. A., Kumar, M., Kumar, A., Yadav, V. K., Soni, S., Kumar, R., Qasim, M. T., Tariq, M., & Alam, M. W. (2024). A review on e-waste contamination, toxicity, and sustainable clean-up approaches for its management. *Toxicology, 508*, 153904. https://doi.org/10.1016/j.tox.2024.153904

Li, J. J., Zhao, X. Q., Wang, J. L., & Shen, R. F. (2021). Strategies of cadmium and copper uptake and translocation in different plant species growing near an E-waste dismantling site at Wenling, China. *Environmental Science and Pollution Research, 28*(44), 62562–62571. https://doi.org/10.1007/s11356-021-15072-1

Li, X., Wu, Y., & Tan, Z. (2022). An overview on bioremediation technologies for soil pollution in E-waste dismantling areas. *Journal of Environmental Chemical Engineering, 10*(3), 107839. https://doi.org/10.1016/j.jece.2022.107839

Lin, S., Ali, M. U., Zheng, C., Cai, Z., & Wong, M. H. (2022). Toxic chemicals from uncontrolled e-waste recycling: Exposure, body burden, health impact. *Journal of Hazardous Materials, 426*, 127792. https://doi.org/10.1016/j.jhazmat.2021.127792

Luo, K., Qiao, Z., Liang, W., Lu, C., Fu, M., Zhou, S., Han, Y., Peng, C., & Zhang, W. (2023). Contamination characteristics and potential health risk of brominated flame retardants in paddy soils and rice plants around a typical e-waste recycling site in South China. *Environmental Pollution, 334*, 122160. https://doi.org/10.1016/j.envpol.2023.122160

Ma, Y., Stubbings, W. A., Cline-Cole, R., & Harrad, S. (2021). Human exposure to halogenated and organophosphate flame retardants through informal e-waste handling activities—A critical review. *Environmental Pollution, 268*, 115727. https://doi.org/10.1016/j.envpol.2020.115727

Mazrouaa, A. M., Mansour, N. A., Abed, M. Y., Youssif, M. A., Shenashen, M. A., & Awual, M. R. (2019). Nano-composite multi-wall carbon nanotubes using poly(p-phenylene terephthalamide) for enhanced electric conductivity. *Journal of Environmental Chemical Engineering, 7*(2), 103002. https://doi.org/10.1016/j.jece.2019.103002

Mishra, S., Rout, P. K., & Das, A. P. (2019). Solar photovoltaic panels as next generation waste: A review. *Biointerface Research in Applied Chemistry, 9*, 6.

Mishra, S., Ghosh, S., van Hullebusch, E. D., Singh, S., & Das, A. P. (2023). A critical review on the recovery of base and critical elements from electronic waste-contaminated streams using microbial biotechnology. *Applied Biochemistry and Biotechnology, 2023*, 1–30. https://doi.org/10.1007/S12010-023-04440-X

Mohanty, S., Ghosh, S., Nayak, S., & Das, A. P. (2016). Isolation, identification and screening of manganese solubilizing fungi from low-grade manganese ore deposits. *Geomicrobiological Journal, 34*(4), 309–316. https://doi.org/10.1080/01490451.2016.1189016

Paul, A., Dey, S., Ram, D. K., & Das, A. P. (2023). Hexavalent chromium pollution and its sustainable management through bioremediation. *Geomicrobiology Journal, 1–11*. https://doi.org/10.1080/01490451.2023.2218377

Perkins, N., Brune Drisse, M.-N. D., Nxele, T., & Sly, P. D. (2014). E-Waste: A global hazard. *Annals of Global Health, 80*(4), 286. https://doi.org/10.1016/j.aogh.2014.10.001

Prabhakar, A., Mishra, S., & Das, A. P. (2019). Isolation and identification of Lead (Pb) solubilizing bacteria from automobile waste and its potential for recovery of lead from end of life waste batteries. *Geomicrobiological Journal, 36*(10), 894–903. https://doi.org/10.1080/01490451.2019.1654044

Priyadarshini, B., Santhosh Kumar, M., & Das, A. P. (2024). Microbial degradation of synthetic plastics released from COVID-19 generated single use disposable face masks. *Environmental Science and Engineering, Part, F2827*, 1–16. https://doi.org/10.1007/978-3-031-50840-0_1

Priyadarsini, S., & Das, A. P. (2024). Analytical and structural characterization of waste lithium-ion batteries for their effective recycling strategy. *Environmental Chemistry and Ecotoxicology*. https://doi.org/10.1016/j.enceco.2024.12.004

Priyadarsini, S., & Das, A. P. (2025). Lithium bioleaching: A review on microbial-assisted sustainable technology for lithium bio-circularity. *Journal of Water Process Engineering, 69*, 106744. https://doi.org/10.1016/j.jwpe.2024.106744

Raj, K., & Das, A. P. (2023). Lead pollution: Impact on environment and human health and approach for a sustainable solution. *Environmental Chemistry and Ecotoxicology, 5*, 79–85. https://doi.org/10.1016/J.ENCECO.2023.02.001

Rajesh, R., Kanakadhurga, D., & Prabaharan, N. (2022). Electronic waste: A critical assessment on the unimaginable growing pollutant, legislations and environmental impacts. *Environmental Challenges, 7*, 100507. https://doi.org/10.1016/j.envc.2022.100507

Rawat, S., Verma, L., & Singh, J. (2020). Environmental hazards and management of E-waste. In *Environmental concerns and sustainable development* (pp. 381–398). Springer Singapore. https://doi.org/10.1007/978-981-13-6358-0_16

Rene, E. R., Sethurajan, M., Kumar Ponnusamy, V., Kumar, G., Bao Dung, T. N., Brindhadevi, K., & Pugazhendhi, A. (2021). Electronic waste generation, recycling and resource recovery: Technological perspectives and trends. *Journal of Hazardous Materials, 416*, 125664. https://doi.org/10.1016/j.jhazmat.2021.125664

Sari, D. P., Masruroh, N. A., & Asih, A. M. S. (2021). Consumer intention to participate in E-waste collection programs: A study of smartphone waste in Indonesia. *Sustainability, 13*(5), 2759. https://doi.org/10.3390/su13052759

Shahabuddin, M., Uddin, M. N., Chowdhury, J. I., Ahmed, S. F., Uddin, M. N., Mofijur, M., & Uddin, M. A. (2023). A review of the recent development, challenges, and opportunities of electronic waste (e-waste). *International journal of Environmental Science and Technology, 20*(4), 4513–4520. https://doi.org/10.1007/s13762-022-04274-w

Tsydenova, O., & Bengtsson, M. (2011). Chemical hazards associated with treatment of waste electrical and electronic equipment. *Waste Management, 31*(1), 45–58. https://doi.org/10.1016/j.wasman.2010.08.014

Ullah, Z., Qureshi, M. I., Ahmad, A., Khan, S. U., & Javaid, M. F. (2021). An experimental study on the mechanical and durability properties assessment of E-waste concrete. *Journal of Building Engineering, 38*, 102177. https://doi.org/10.1016/j.jobe.2021.102177

Wirtu, Y. D., & Tucho, G. T. (2022). E-waste: Growing environmental and health problems and its management alternatives in developing countries. *Environmental Reviews, 30*(4), 524–536. https://doi.org/10.1139/er-2021-0120

Wu, Y., Li, G., & An, T. (2022). Toxic metals in particulate matter and health risks in an E-waste Dismantling Park and its surrounding areas: Analysis of three PM size groups. *International Journal of Environmental Research and Public Health, 19*(22), 15383. https://doi.org/10.3390/ijerph192215383

Yue, C., Ma, S., Liu, R., Yang, Y., Li, G., Yu, Y., & An, T. (2022). Pollution profiles and human health risk assessment of atmospheric organophosphorus esters in an e-waste dismantling park and its surrounding area. *Science of the Total Environment, 806*, 151206. https://doi.org/10.1016/j.scitotenv.2021.151206

From Contaminants to Cleanup: The Power of Microorganisms in Bioremediation

Shafeeq Ur Rehman, Minal Hussain, Humera Nazir, Faizan Ali, Ghanwa Maryam, Furqan Shafqat, and Maria Nazir

Abstract Bioremediation is a process that entails using microorganisms to convert hazardous waste into safe waste; therefore, this is a suitable, cost-effective, and eco-friendly technique for waste management and restoration of the natural balance of the environment. It is necessary because of increasing contamination with petroleum hydrocarbons, heavy metals, and synthetic organic chemicals for large-scale habitat restoration. Pollutants can be chemical, biological, and even physical entities that cause an adverse effect on air, water, soil, or living organisms. Heavy metals as chemical pollutants, pesticides, hydrocarbons, and industrial chemicals are harmful; they are nonbiodegradable and have adverse effects on both humans and animals. The most common class of organisms, and the most widely studied for bioremediation, are bacteria because of their versatile nature and their ability to metabolically degrade a variety of xenobiotics. The number of bacterial species varies according to the degree of oxygen availability in an ecosystem such as *Pseudomonas* species that can degrade hydrocarbons and PCBs. *Geobacter* spp. can be used to remediate anaerobic conditions with inorganic ions of nitrate, sulfate, and other heavy metals. Genetically engineered bioremediation bacteria can target and degrade environmental pollutants due to bioremediation advances in genetic engineering.

S. U. Rehman (✉) · M. Hussain · F. Ali · M. Nazir
Department of Microbiology, Cholistan University of Veterinary and Animal Sciences, Bahawalpur, Pakistan

H. Nazir
Department of Microbiology and Molecular Genetics, Bahauddin Zakariya University, Multan, Pakistan

G. Maryam
Department of Zoology Wildlife and Fisheries, University of Agriculture Faisalabad, Faisalabad, Pakistan

F. Shafqat
Department of Medical Biotechnology, University of Wollongong, Wollongong, NSW, Australia

S. Rajpoot, A. P. Das (eds.), *Innovative Technologies for Waste Management*, Sustainable Environmental Waste Management Strategies,
https://doi.org/10.1007/978-3-031-97184-6_3

Keywords Bioremediation · Contaminants · Pollutants · Hydrocarbon degradation · Mycoremediation

1 Introduction to Bioremediation

Bioremediation is a new method that taps into the beneficial properties of microorganisms for pollution sites. Bioremediation involves the use of natural and artificially engineered bacteria and fungi to safely break down or decompose potential dangers into harmless by-products. It was developed as a technique for controlling various environmental pollution sources and hence promotes waste management and restoration of ecosystems in a more cost-efficient, effective, and eco-friendly way (Kuppan et al., 2024). For example, bioremediation can involve cleaning polluted soil, water, or air by adding microbes or other biological agents that can neutralize or eliminate toxic pollutants. It is a branch of environmental biotechnology that uses microorganism's genomic resources to affect the biosynthesis or bioremoval of a range of pollutants including hydrocarbons, heavy metals, and complex contaminants like polychlorinated biphenyls (PCBs) and pesticides (Jing et al., 2018; Visvanathan et al., 2024). The fundamental concept in bioremediation is that microorganisms feed on pollutants like food and metabolize these toxic elements into harmless products.

Bioremediation is not merely about cleaning up contaminated sites but also reconnecting soil food webs dislocated by industrial, agricultural, and domestic spoilage. Conventional decontamination techniques often entail secondary pollution in terms of emissions of smoke, gas, and chemicals to the atmosphere or are expensive to implement. Similarly, bioremediation is cheaper, less environmentally disruptive, and healthier than most conventional waste disposal methods (Azubuike et al., 2020). Bioremediation is a process that deploys microorganisms to neutralize contaminated areas of the environment. There are two broad categories of bioremediation: in situ and ex situ. In situ, bioremediation occurs in the polluted zone without requiring the removal of the contaminated material. This method comprises bioventing, biosparging, and phytoremediation in which plants and microbes are used (Sharma, 2020). In contrast, the ex situ bioremediation process seeks to transport the contaminated material to another location for treatment through bioreactors or carefully designed conditions (Hussain et al., 2022).

Flexibility is one of the critical considerations in bioremediation. Bioremediation is a very flexible approach since microorganisms can be adapted concerning various substances to remove. For example, some microbes can break down spilled crude oil, and others can metabolize chlorinated solvents and heavy metals (Sharma et al., 2021). Bioremediation is generally practical based on the originality and balance of factors like pH, temperature, dissolved oxygen (DO), and nutrients that elicit microbial action, which can be manipulated for maximum efficiency. Bioremediation must be regarded more seriously as a global approach to large-scale habitat restoration. Soil and water pollution, especially by petroleum hydrocarbons, heavy metals, and synthetic organic chemicals, have rapidly risen due to enhanced industrialization and urbanization worldwide (Swain, 2024). These contaminants are a real menace to wildlife and human health, hence the need to find efficient cleaning methods.

Another advantage of bioremediation is that it isolates the polluters and prevents other polluters from causing a lot of harm to the environment. Cleaning can also be less expensive than conventional cleaning; hence, less power input is used during the process. Moreover, bioremediation methods can extend to different aquatic and terrestrial habitats, making it an effective weapon against pollution (Aliko et al., 2022). Bioremediation not only treats the efficiency of removing contaminants but is also beneficial for the continuous improvement of the ecosystem. The process contributes to the recovery of impaired biota and the remediation of soil and water body productivity. In addition, bioremediation is essential in combating climate change since it lowers greenhouse gases linked to other more carbon-consuming remediation approaches (Aparicio et al., 2022; Skinder et al., 2020).

2 History and Evolution

Bioremediation is nothing new; it has been known since the beginning of the twentieth century when researchers stumbled upon an enigmatic property of microorganisms that allowed them to break down organic compounds independently. However, the idea of employing microbes for the remediation of the environment spread a small propaganda only in the 1970s and 1980s due to increased consciousness regarding the environment and the progression in the microbiological and biotechnological fields (Hazen, 2018). Diverse bioremediation techniques are shown in Fig. 1.

In 1972, American microbiologist George Robinson developed the first oil-eating bacterium, which was applied to oil spillage (DeWitt & Melvin, 1975; Wess, 2023). Exxon Valdez in Alaska in 1989 was the landmark event in the application of bioremediation since it exemplified a more extensive scale application of the technology through nutrient application for boosting microbial action (Amundsen, 2022; Smith, 1995). This resulted in this event drawing a positive shift in how people view the concept of bioremediation in disaster management. As with most other pollution control forms, bioremediation is rooted in genetic engineering and molecular biology. Science has occasioned the bioremediation of microorganisms, which have been used to modify their capability to break down certain pollutants like PCBs, dioxins, and heavy metals (Omondi et al., 2022).

Bioremediation technology has later developed to recognize techniques such as bioaugmentation and biostimulation in the recent past. Bioaugmentation includes the addition of selected microorganisms, which have a higher degradative capability for specific pollutants (Adams et al., 2015). At the same time, biostimulation improves the indigenous microorganisms by either the addition of nutrients or modification in client conditions (Hazen, 2018). Plants that can enhance microbial degradation have also attracted much focus under phytoremediation, making it a

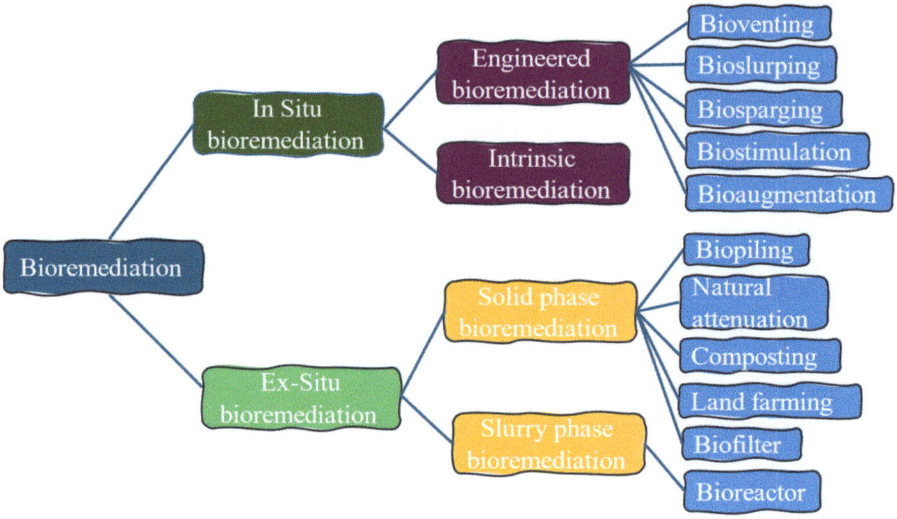

Fig. 1 Diverse bioremediation techniques

visually attractive and viable method in land management and restoration (Rocha et al., 2022).

Several key milestones have marked the evolution of bioremediation as a vital technology for environmental cleanup:

1. **1972, historical background:** Bioremediation as an engineered solution for the birth of engineered microbes could be attributed to the patenting of the first oil-degrading bacterium in 1982 by George Robinson (Pete et al., 2021; Yenn, 2015).
2. **1989, Exxon Valdez oil spill:** After the Exxon Valdez spillage, one of the most considerable bioremediation procedures was applied that more than proved the efficacy of nutrient supplements to boost innovative natural microbes to break down the oil (Venosa, 2004; Youvan, 2024).
3. **1990s, genetically modified microbes:** In the 1990s, more advanced genetically engineered microorganisms were developed for the degradation of more resistant compounds like PCBs and dioxins (Saibu et al., 2020; Unterman, 1996).
4. **2000s, new molecular tools for microbial ecologists**: Molecular biology tools made researchers aware of the microbial consortia and their metabolisms more effectively, which helped in designing efficient bioremediation measures (Brune & Bayer, 2012; Kumar et al., 2023).
5. **2010s, combination of bioremediation with phytoremediation:** Scientists started combining microbial and plant-aided remedial strategies, adding to updated effective models for site rehabilitation (Seshadri et al., 2015).
6. **2020s, interest in carbon storage:** The emerging focus on climate change and sustainable development has renewed interest in bioremediation as an approach to carbon sequestration and emission reduction during remedial actions (Babaniyi et al., 2023).

3 Understanding Contaminants

Pollution is a growing global issue that threatens the bio-environment, human health, and the planet's overall balance. Contaminants, which cause pollution, enter the environment from various sources and exist in different forms, making their management a complex and multifaceted challenge. These pollutants can be chemical, biological, or physical and have diverse impacts on air, water, soil, and living organisms. Industrial activities release harmful substances like heavy metals and toxic chemicals (Singh et al., 2022; Weldeslassie et al., 2018). At the same time, agricultural practices contribute to pollution through excessive use of fertilizers and pesticides, which eventually leach into nearby water bodies. With their high population densities and activities, urban areas generate large amounts of waste and emissions, further adding to the contamination load (Craswell, 2021). Different hazardous organic and inorganic pollutants and their degrading microorganisms are given in Table 1.

Chemical contaminants, such as heavy metals, pesticides, hydrocarbons, and industrial chemicals, are among the most persistent and harmful pollutants. For instance, lead and mercury from industrial waste can bioaccumulate organisms, leading to severe health problems in humans and wildlife (Filote et al., 2021). Similarly, pesticides and herbicides widely used in agriculture contaminate water and soil, causing long-term damage to ecosystems. Biological contaminants, such as harmful bacteria, viruses, and fungi, are another primary concern. These microorganisms can spread diseases through contaminated water or food, posing direct

Table 1 Examples of naturally and non-naturally occurring contaminants with their health effects

Contaminants	Source	Health effects
Non-naturally occurring contaminants		
Pharmaceuticals	Human and animal use	Endocrine disruption, antibiotic resistance
Plastics	Manufacturing, waste disposal	Environmental pollution, potential health effects
Industrial chemicals (e.g., PCBs, dioxins)	Industrial processes	Cancer, immune system suppression
Pesticides	Agricultural practices	Neurological effects and endocrine disruption
Herbicides	Weed control	Cancer, reproductive problems
Naturally occurring contaminants		
Heavy metals (e.g., mercury, lead)	Volcanic activity, mineral deposits	Neurological damage, developmental problems
Radon	Radioactive decay of uranium in rocks and soil	Lung cancer
Certain microbes (e.g., harmful bacteria, viruses)	Natural presence in water and soil	Infectious diseases
Arsenic	Natural deposits in groundwater	Skin lesions, cancer, cardiovascular disease

risks to public health (Garvey, 2019). Physical contaminants, including plastics and microplastics, disrupt ecosystems by clogging waterways, harming marine life, and entering the food chain, causing potential health risks to humans (Yuan et al., 2022). The sources of these contaminants can be categorized as point sources and non-point sources. Point sources are specific and identifiable, such as a factory discharging wastewater into a river or an oil spill from a tanker (Lai et al., 2011). On the other hand, non-point sources are diffuse and more complicated to pinpoint, such as agricultural runoff containing fertilizers and pesticides or urban stormwater carrying multiple pollutants (Hou et al., 2022).

Air pollution, caused by emissions of gases like carbon monoxide and sulfur dioxide, leads to respiratory illnesses, smog formation, and climate change (Manisalidis et al., 2020). Water pollution, resulting from oil spills, industrial discharges, and agricultural runoff, depletes oxygen levels in water bodies, harms aquatic organisms, and contaminates drinking water sources (Singh, 2024). Soil pollution from industrial chemicals and agricultural practices reduces fertility, affects plant growth, and contaminates crops, posing risks to food security and human health (Singh & Singh, 2020). Moreover, contaminants lead to biodiversity loss by degrading habitats, poisoning species, and disrupting food chains, affecting ecosystem services like pollination and nutrient cycling. Traditional cleanup methods, such as chemical treatments and physical removal, are often expensive and can sometimes cause further environmental harm. For example, burning waste releases harmful gases into the atmosphere, while chemical neutralization can generate secondary pollutants. This complexity has led to the exploration of more sustainable and environmentally friendly methods, such as bioremediation (Ahmed et al., 2022; Sharma et al., 2018). The significant factors affecting bioremediation are shown in Fig. 2.

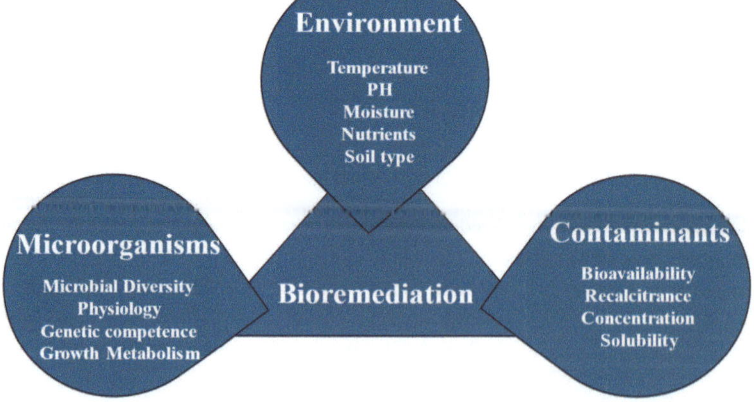

Fig. 2 This triad represents the significant contributing factors of bioremediation

3.1 Naturally Occurring Vs Non-naturally Occurring Contaminants

3.1.1 Organic Contaminants

Organic contaminants contain carbon in them, originating from biological matters or human activities. They contain hydrocarbons, pesticides, herbicides, and other industrial chemicals. These are comparatively large molecules; thus, they are more resistant to the environment and may need particularly favorable microorganisms for their breakdown (Saha et al., 2017; Weldeslassie et al., 2018). For instance, hydrocarbons sourced from petroleum products are frequently categorized as organic pollutants and, when present in the soil and water, result in severe environmental impacts (Mohammadi et al., 2020).

Pesticides present the other common organic pollutants, which find wide use in modern agriculture as agents for controlling hot pests and ensuring increased crop productivity. However, their improper application causes soil and water contamination through non-target species and even the food chain. Most pesticides are persistent organic pollutants (POPs) that can take years and even decades to degrade in the environment, hence the need for bioremediation (Akhtar et al., 2021; Boudh & Singh, 2019).

3.1.2 Inorganic Contaminants

Inorganic contaminants consist of compounds formed of elements that are not carbon. These include lead, mercury, cadmium, arsenic, nitrates, phosphates, and salts. In contrast with other organic pollutants, inorganic pollutants cannot be metabolized or degraded into different substances (Dhiman et al., 2023). Nevertheless, bioremediation processes convert some recalcitrant compounds into less hazardous forms or fix them to reduce their mobility. For instance, *Pseudomonas* and *Bacillus* can also reduce soluble hexavalent chromium [Cr (VI)] to insoluble trivalent chromium [Cr (III)], making it easier to sequester and treat (Sharma et al., 2022b). Microbial remediation techniques of inorganic pollutants involve biodegradation, biosorption, or bioventing processes by which certain organisms concentrate or remove heavy metals from the medium. Examples of naturally and non-naturally occurring contaminants with their health effects are shown in Table 1.

3.1.3 Common Pollutants

- **Hydrocarbons**: The sources of hydrocarbons include oil and petroleum products, making them some of the most common organic pollutants. They are comprised of benzene, toluene, ethylbenzene, xylene (BTEX), and polycyclic aromatic hydrocarbons (PAHs) (Kuppusamy et al., 2020). Hydrocarbon accidents happen during drilling, oil transportation, and processing, contaminating soils and water sources. These contaminants include; pollutants that pollute the water and water resources, affect the aquatic and terrestrial life, and even cause water pollution, making water an unhealthy resource to be used for drinking (Kumar, 2022).

- **Heavy Metals**: Heavy metals can be defined as nondegradable inorganic elements, which, most of the time, spread within the environment in the form of dust or solutions through industrial processes and use, mining, and dumping of wastes (Cimboláková et al., 2019). Some of the most toxic and bioaccumulative elements are metals such as arsenic, lead, mercury, cadmium, and chromium. They have been known to contribute to various ailments, including cancer, organ failure, and development complications, more so when they find their way into the food cycle through water or crops (Singh et al., 2024).
- **Pesticides:** These refer to insecticides, herbicides, and fungicides, all critical in modern farming yet dangerous to the environment. As the name implies, these chemicals can find their way to the ground and water sources when improperly used, polluting beneficial insects, fishes, and even ourselves. The fact that some of these pesticides remain relatively long in the environment compounds the problem of their removal: for example, there are the organochlorine remnant pesticide types (Hossain et al., 2022; Rani et al., 2021). Several organic and inorganic pollutants and their degrading microorganisms are given in Table 2.

4 Effects of Contaminants on the Environment and Health

Biological treatments, notably bioremediation, have been deemed necessary to manage environmental pollutant impacts. Therefore, considering these impacts, we should discover effective bioremediation techniques.

4.1 Environmental Impacts

Environmental pollutants are hazardous because they can drastically change the Earth's soil, water, and air composition. For example, hydrocarbon, dispersed in oil spills, can smother species and harm coastal structures. At the same time, a buildup of heavy metals in soil can hinder plant growth and therefore inevitably decrease the species dwelling in the ecosystem and their interaction during food webs (Shah & Soni, 2024). The two angled pesticides can hinder the growth and survival of helpful insects, such as bees and butterflies, necessary for pollination, blunting crop yield, and any balanced societal ecosystem (Shetty et al., 2023). Pollutants also affect occurrences such as eutrophication through high nutrient pollution to algal blooms that reduce oxygen consumed by water body organisms, killing the water organisms. These environmental disturbances may be long-lasting, so rehabilitating these affected sites may be difficult and expensive (Mishra, 2023; Vantarakis, 2021).

Table 2 Different hazardous organic and inorganic pollutants and their degrading microorganisms (Bala et al., 2022)

Substrate	Compound	Microorganisms
Organic substrate	Diesel oil	*B. coagulans, Serratia ficaria, B. cereus, and P. cepacia*
	N,N-dimethyl-p-phenylenediamine	*Klebsiella pneumonia (RS-13)*
	Endosulfan	*Bacillus, Staphylococcus*
	Chlorobenzenes	*P. putida*
	Remazol black B	*Kluyveromyces marxianus*
	Oil-based paints	*B. subtilis strain NAP2, NAP4*, and *NAP1*
	Sulfonate benzene	*A. radiobacter (S2)*
	Crude oil	*Candida krusei, C. glabrata*, and *Aspergillus niger*
	Aromatic hydrocarbons	*Microbacterium* sp., *Acinetobacter* sp., *and Ralstonia* sp.
	4,4'-Dibromodiphenyl ether	*Phanerochaete chrysosporium*
	Methyl parathion and chlorpyrifos	*Enterobacter* sp., *Photobacterium* sp., *Acinetobacter* sp., and *Pseudomonas* sp.
	Phenol	*Corynebacterium propinquum, P. aeruginosa*, and *B. subtilis*
	Polycyclic aromatic hydrocarbons	*Myceliophthora thermophila* and *Burkholderia* sp.
Inorganic substrate	Cadmium	*A. fumigatus, Paecilomyces* sp., *A. versicolor, Paecilomyces* sp., *Cladosporium* sp., and *Trichoderma* sp.
	Chromium, copper, nickel, and uranium	*Aeromonas* sp. *and P. aeruginosa*
	Hg^{2+}	*Pseudochlorococcum typicum, Cyclotella cryptica, Spirogyra hyaline*, and *Chlamydomonas reinhardtii*
	Lead and mercury nickel	*Cunninghamella elegans* and *Saccharomyces cerevisiae*
	Cr^{6+}	*Pseudomonas putida*
	Chromium, cobalt, lead, and copper	*Lysinibacillus sphaericus*
	$Cr_2O_7^{22}$	*Spirulina* sp. and *Chlorella* spp.
	Cr^{51}	*Ulothrix tenuissima, C. reinhardtii*, and *Spirulina* sp.

4.2 *Health Effects*

The effects of substances on human health will, therefore, depend on the kind of substance, its levels of concentration in the water, and the period of exposure. Hydrocarbon and pesticides are common organic contaminants because they can be carcinogenic or lead to respiratory and neurological diseases. Long-term effects of ingesting heavy metals like lead and mercury contain dialog hepatopathy, cerebral abasement, and retardation in children (Mallah et al., 2022). Pollution of drinking water sources is a severe reality because it can cause diseases and long-term illnesses for the inhabitants. For instance, nitrate leaching from agricultural activities has been proven to cause methemoglobinemia, or blue baby syndrome, in underage children. It is a long-term hazard because contaminants tend to remain in the environment and can accumulate in bio-magnets, living organisms that finally get into the human food chain, causing compound effects on human health (Nieder & Benbi, 2022).

5 Microorganisms: The Invisible Workforce

Microorganisms are also central to the definition of bioremediation since they act as unknown workers transforming or degrading contaminants. Different microorganisms used in bioremediation are given in Table 3. Recognizing that these pollutant substances are organic compounds, bioremediation uses bacteria, fungi, and algae and their inherent metabolic functions to clean up the environment. Different groups of microorganisms play a role in degrading organic and inorganic pollutants, and each group has its metabolic pathways responsible for bioremediation (Gaur et al., 2018). Bioremediation involves using an array of microorganisms, each of which possesses specific metabolic reactions that can degrade or immobilize contaminants. The overview of bioremediation is given in Fig. 3, and the mechanism of bioremediation is given in Table 5.

5.1 Bacteria

The most common and investigated microorganisms in bioremediation are bacteria, owing to their versatility and ability to break down diverse pollutants. Bioremediation using bacteria can be aerobic or anaerobic, depending on oxygen availability within the ecosystem (Abatenh et al., 2017).

- **Aerobic Bacteria**: These bacteria need oxygen to filter contaminants, and their standard action involves decomposing complex organic substances such as hydrocarbons, pesticides, and solvents. Some of the most common types include *Pseudomonas* species; these bacteria are well known for cleaning up organic pollutants such as hydrocarbons and polychlorinated biphenyls (PCBs) (Elangovan et al., 2019; Nzila, 2013).
- **Anaerobic Bacteria**: Anaerobic bacteria are used in anaerobic conditions, where they accumulate or convert inorganic compounds, including nitrates, sulfates, or several heavy metal ions. For example, the species from the genus *Desulfovibrio* are described as having the capacity to resolute sulfate materials. In this capacity, the dangerous heavy metal chemical chromium is converted into a much less

Table 3 Different microorganisms used in bioremediation (Ayilara & Babalola, 2023)

Organisms	Microorganism's species	Contaminant remediated
Bacteria	*Bacillus sonorensis and Bacillus licheniformis*	Naphthalene
	E. coli	Hexavalent chromium
	Rhodovibrio sp., *Bacillus* sp., *and Formosa* sp. *Rhodopirellula* sp.	Hydrocarbon
	Spirulina sp. *and Chlorella* sp.	Nickel, lead, and dichromate
	Pseudomonas sp., *Bacillus* sp., *Shewanella* sp., *Thermus* sp., and *Enterobacter* sp.	Chromium
	Bacillus licheniformis	Dyes
	Staphylococcus sp. and *Bacillus* sp.	Endosulfans
	Alcaligenes odorans, P. aeruginosa, Fusarium sp., *and Corynebacterium propinquum*	Oils
	Cunninghamella elegans and *Saccharomyces cerevisiae*	Heavy metals and mercury
	Pseudomonas aeruginosa	Crude oil
	Spirogyra hyaline and *Cyclotella cryptica*	Mercury
	A. fumigatus, Cladosporium sp., and *Trichoderma* sp.	Cadmium
	Corynebacterium propinquum, P. aeruginosa, and *Alcaligenes odorans*	Phenol
	B. cereus, P. cepacia, and *B. coagulans*	Diesel oil
Fungi	*Aspergillus* sp.	N-hexadecane
	Ganoderma lucidum	Pyrene
	Phanerochaete chrysosporium	4,4'-Dibromodiphenyl ether
	Coprinus comatus	4-Hydroxy-3,5-dichlorobiphenyl
	Phanerochaete chrysosporium	Toluene, benzene, N-heterocyclic explosives, and xylene
	Penicillium sp. and *Aspergillus* sp.	Polycyclic aromatic, hydrocarbons, chlorophenols, and aliphatic hydrocarbons
Algae	*Chlorococcum humicola*	Iron
	F. vesiculosus	Nickel, lead, cadmium, and chromium
	Microcystis aeruginosa	Cadmium
	Chlorococcum humicola	Cobalt
	Microcystis aeruginosa	Arsenic
	Fucus vesiculosus	Zinc
	Scenedesmus protuberans	Cadmium

hazardous element (Kumar et al., 2018). A few bacterial examples used for bioremediation are given below:

- ***Pseudomonas* spp.**: *Pseudomonas* species are well known for their alkane, polycyclic aromatic hydrocarbons, PCBs, and other organic contaminants. For instance, *Pseudomonas putida* has the toluene dioxygenase, which is involved in the oxidation of the compound toluene, a common solvent (Kahlon, 2016).

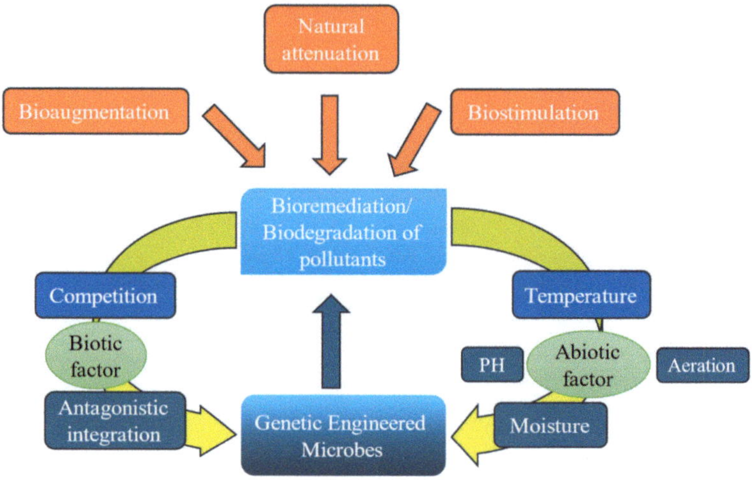

Fig. 3 Overview of bioremediation

Table 5 Mechanisms of bioremediation (Ayilara & Babalola, 2023)

Microorganisms	Pollutant remediated	Mechanisms of remediation
Lysinibacillus sp., *Rhodococcus* sp., and *Bacillus* sp.	Cadmium, copper, lead, and Aluminum	Bioaugmentation
Pseudomonas sp.	Copper and lead	Bioattenuation
Bacillus sp.	Nickel	Biosorption
Desulfovibrio desulfuricans	Copper, zinc, and cadmium	Extracellular sequestration
Pseudomonas aeruginosa	Cadmium	Biosorption
Lysinibacillus sphaericus	Azo dyes	Enzymatic reductase
Cupriavidus sp.	Cadmium	Bioprecipitation
Bacillus cereus and *Pseudomonas aeruginosa*	Cadmium and lead	Bioaugmentation
Oudemansiella canarii	Congo red dye	Enzymatic reduction

- ***Geobacter* spp.**: This group of anaerobic bacteria is highly competent in metal reduction, particularly in contaminated environments with heavy metals such as uranium and chromium. Patel and Sharma identify *Geobacter metallireducens* as a microorganism that can transform poisonous Cr (VI) into less hazardous Cr (III); these microorganisms can act as a bioremediation tool in the removal of heavy metals (Rawat et al., 2023).
- ***Dehalococcoides* spp.**: Such anaerobic bacteria are selective for reducing chlorinated organic contaminants like trichloroethylene (TCE) and vinyl chloride. They are usually applied to remove polluted groundwater by industrial solvents (Dutta et al., 2022).

5.2 Fungi

Due to their extracellular enzyme systems, white-rot fungi are efficient degraders of the target contaminants. The nematodes also lack some enzymes that enable the fungi to degrade complex pollutants that the bacteria cannot break down (Temporiti, 2023).

- **Ligninolytic Enzymes**: Ligninolytic fungi like *Phanerochaete chrysosporium* biosynthesize ligninolytic enzymes like lignin peroxidase, manganese peroxidase, and laccase that are capable of metabolizing various complex organic pollutants and compounds: polycyclic aromatic hydrocarbons (PAHs), dyes, and some pesticides. These enzymes can degrade the hardened chemical bonds in the pollutants to produce simpler, nonhazardous compounds (Baker et al., 2019; Wang et al., 2018).
- **Mycoremediation**: Bioremediation involving fungi is known as mycoremediation and can be used on soils heavily polluted with hard-to-degrade organic pollutants such as dioxins, pesticides, and industrial dyes (Vickram et al., 2024). A few examples of fungi used for bioremediation are given below:
- ***Phanerochaete chrysosporium***: This white-rot fungus is one of the most investigated fungi of bioremediation because it produces such enzymes as lignin peroxidase and manganese peroxidase. They can remove many types of pollutants, such as polycyclic aromatic hydrocarbons (PAHs), dyes, and pesticides (Ellouze & Sayadi, 2016).
- ***Aspergillus niger***: This fungus has compelling profiles to biosorb heavy metals from water bodies that have been contaminated. They are composed of polysaccharides in the cell walls that entrap metal ions such as mercury, cadmium, and lead making it easy to adsorb them from water (Kanamarlapudi et al., 2018).

5.3 Algae

In recent years, algae have been established to be effective in bioremediation, mostly in aquatic habitats. Biosorption, bioaccumulation, and biodegradation are the primary abilities of algae to remove contaminants (Mustafa et al., 2021).

- **Biosorption and Bioaccumulation**: Algae can adsorb heavy metals and many other inorganic impurities dissolved in water by the uptake of these materials in their cells. For instance, *Chlorella* and *Spirulina* can remove lead, cadmium, and mercury metals from industrial effluents since they are known bioindicators of metal accumulation (Aghababai Beni et al., 2024).
- **Phytoremediation of Nutrients**: Algae are used in water purification or general phytoremediation of nutrients, especially nitrogen and phosphorus, in wastewater. Such precautions restrain water eutrophication while, at the same time, water purification is enhanced (Bhalla et al., 2022). A few examples of algae used for bioremediation are given below:

- ***Chlorella vulgaris***: *Chlorella* is a microalgae that has been shown to have the ability to absorb heavy metals as a biosorbent and bioremediation agent. The treatment frequency of industrial wastewater is high because it demonstrates a high ability to absorb metals (Alharbi et al., 2023).
- ***Spirulina platensis***: Despite its limitations, this cyanobacterium is proficient in the uptake of metals, such as arsenic and cadmium, from polluted water and hence could be helpful in aquatic bioremediation (Sardar et al., 2018).

6 Biochemical Mechanisms of Biodegradation of Pollutants by Microorganisms

Microorganisms have specific metabolism phenomena that break down or convert contaminants. These processes are generally classified into oxidative and nonoxidative energy utilization pathways.

6.1 Aerobic Metabolism

Besides providing the energy for the reaction, oxygen is oxidized into the contaminant molecules during aerobic degradation by enzymes such as monooxygenases and dioxygenases. These enzymes tie the pollutant to water, which enhances its metabolism because it makes it easier for aquatic life to metabolize it (Sardar et al., 2018).

- **Hydrocarbon Degradation:** Facultative anaerobic bacteria, including *Pseudomonas* species, first use enzymes, including monooxygenases and dioxygenases, to activate hydrocarbons. These enzymes find the hydrocarbon molecules and incorporate oxygen molecules into them, making them more soluble in water and easily degraded (Abbasian et al., 2015; Dyes, 2023).
- **Enzyme Production:** Most aerobic fungi secrete enzymes that oxidize pollutants to make them vulnerable to other microbes. For instance, lignin peroxidase and laccase enzymes originating from the fungi *Phanerochaete chrysosporium* can bioremediate a host of organic pollutants (Ali et al., 2019; Baker et al., 2019). Table 4 represents the biodegradation method by aerobic with their enzymes.

6.2 Anaerobic Metabolism

Anaerobic metabolism is work done by microorganisms that perform their metabolism without oxygen and use other substances for electron acceptors, for example, nitrates, sulfates, or metals. Reductive processes have been reported frequently in

Table 4 Aerobic biodegradation mechanisms with enzymes

Mechanism	Description	Enzymes	Example
Hydrolysis	Water is used to break down complex molecules	Hydrolases	Degradation of pesticides
Ring cleavage	Opening of aromatic rings in complex organic compounds	Dioxygenases	Degradation of aromatic hydrocarbons
Oxidation	Microorganisms use oxygen to break down pollutants	Oxygenases, dehydrogenases	Degradation of petroleum hydrocarbons
Dehalogenation	Removal of halogen atoms from organic pollutants	Dehalogenases	Degradation of chlorinated solvents

anaerobic degradation in which halogenated compounds are dehalogenation, sulfur-containing compounds desulfurized, or toxicants are detoxified through the transfer of electrons to other acceptors (Tiwari et al., 2017).

- **Reductive Dechlorination**: *Dehalococcoides* is one of the anaerobic bacteria that can perform an activity called reductive dechlorination, where bacteria chop chlorine atoms off chlorinated organic compounds such as PCBs or TCE. These changes make the pollutants less hazardous and friendlier to other degradation processes (Greenwood, 2010; Meuser & Meuser, 2013).
- **Metal Reduction**: For example, removing chromium, uranium, and arsenic, which are lethal metals, reduces them to less soluble phases that are not dangerous. For example, *Geobacter* species can reduce Cr (VI), a powerful oxidizer and highly toxic metal, to Cr (III), a metal that is less toxic and more stable (Pastorella, 2014).

6.3 Bioaccumulation and Biosorption

Algae also show high biosorption levels since their cell walls are naturally used for the biosorption of heavy metals and other inorganic pollutants. *Chlorella vulgaris* and *Aspergillus niger* algae and fungi, respectively, have been discovered to be bioremediate heavy metal-polluted water bodies, thus providing effaced methods of treating wastewater (Mustafa et al., 2021; Silva et al., 2019).

7 Bioaugmentation and Biostimulation

The fundamental means of improving such conditions are bioaugmentation and biostimulation, making bioremediation more efficient.

7.1 Bioaugmentation

Bioaugmentation is a process where specific strains are added to the contaminated area to worsen the pollutant at a higher rate. Acidity control is applied when the native microbial population is suboptimal or does not possess the desired metabolic functions required to eliminate certain pollutants (Petrovskis et al., 2013).

- **Mechanism**: In bioaugmentation, a source of the specific microbial strains is chosen with a view that such microbes possess the ability to decompose the particular pollutant. These strains can be isolated from nature, supplemented from comparable environments, or genetically modified to provide even higher yields (Bala et al., 2022).
- **Applications**: Bioaugmentation is widely applied in industrial water treatment, oil spillage, and rehabilitation of soil polluted by hardy substances such as PCBs and dioxins. This approach is useful when applied to bioreactors because the microbial activity can be adjusted and standardized (Shrivastava, 2020).

7.2 Biostimulation

Biostimulation mainly uses amendments, which include the addition of nutrients, oxygen, or any other material, which increases the rate of native microorganisms at the contaminated site (Goswami et al., 2018).

- **Mechanism:** So biostimulation involves the installation of nutrients such as nitrogen, phosphorus, or electron donors, preferentially lactate for anaerobic processes, to stimulate the activity of microbes. This approach provides a suitable environment in which microorganisms can break down pollutants more effectively (Colville, 2019).
- **Applications:** In situ and ex situ bioremediation programs can be achieved through biostimulation. Similarly, the augmentation of fertilizers or the aeration of contaminated soil will increase the rate of biodegradation of hydrocarbon from oil in water, while the addition of nutrients in water rich in nitrate and chlorinated solvents will reduce their concentration (Perez-Vazquez et al., 2024).

8 Enzyme Mechanisms

Contaminants are broken down by microorganisms by way of enzymatic reactions and pathways. These are genetic factors that impact the rate of synthesis of enzymes and other cellular constituents, the ability of cells to take up the substrate, and the metabolic rate (Kumari & Das, 2023).

8.1 Biological Categories

Biological categories are the pathways used by microorganisms to degrade or metabolize contaminants, and they are of two types. The first type involves enzymatic pathways. Such enzymes can either be attached to the cells (intracellular) or free-moving in the extracellular environment (ectolysins) (Dangi et al., 2019).

- **Oxidative Enzymes**: In aerobic degradation, monooxygenases, dioxygenases, and cytochrome P450 oxidases are significant players that have been reported. These enzymes introduce oxygen atoms to organic pollutants, increase their reactivity, and can be easily degraded (Liang et al., 2018; Narayanan et al., 2023).
- **Reductive Enzymes**: In anaerobic degradation, reductases, hydrogenases, and dehalogenases are used in reductive processes where contaminants are changed into less noxious substances by donating electrons to other acceptors (Kameswaran et al., 2021).

8.2 Genetic Control

The biodegradation process is controlled genetically; the particular genes code for the enzymes that decompose the contaminants. These genes can be present on the main chromosome or on the plasmids that are mobile genetic components that can be distributed between microorganisms (Khan & Rao, 2019) (Table 5).

- **Horizontal Gene Transfer (HGT)**: The ability to dispense biodegradation duties is mobile and relies primarily on lateral gene transfer in microbial populations. It enables microorganisms to upgrade their capacity to metabolize numerous pollutants by obtaining fresh metabolic genes. For instance, plasmids with markers specifying hydrocarbon or chlorinated solvent degradation can be transferred across bacteria to enhance more effective bioremediation (Ganguly et al., 2023).
- **Genetic Engineering**: Newer techniques have enabled scientists to bioengineer microorganisms by boosting biodegradation features. For example, specific *Pseudomonas* and *Escherichia coli* have been bioengineered as highly efficient pollutant degraders compared to their unaltered selves (Sharma et al., 2022a).

9 Approaches to Bioremediation: Methods and Practices

Bioremediation techniques can only succeed by using a mixture of different approaches and steps that depend on the actual type of pollutant, the state of the environment, and available options (Prasad & Aranda, 2018). Here are some standard techniques and strategies:

9.1 Bioreactors

Bioreactors are systems used to mix specially cultivated microorganisms with contaminated materials and substances, such as soil or water. These reactors provide the choice of environmental conditions like PH, temperature, and nutrient level, which in turn assist the correct microbial break (Srivastava et al., 2022).

9.2 Bioventing and Biosparging

- **Bioventing**: This in situ technique involves the injection of air or oxygen within the contaminated soil to enhance the aerobic degradation. It is usually applied to treat hydrocarbons in the subsurface soil (Kuppusamy et al., 2016).
- **Biosparging**: In biosparging, air is flown into the water table to facilitate the biological degradation of soluble pollutants by using dissolved oxygen. This technique applies to removing petroleum hydrocarbons and volatile organic compounds (VOCs) in aquifers (Riser-Roberts, 2020; Wu, 2000).

9.3 Constructed Wetlands

Constructed wetlands, also known as artificial wetlands, are man-made structures used to help in water treatment. The microbial decomposition and nutrient biosorption make the wetland suitable for industrial effluents, agricultural drainages, and domestic sewage (Khan et al., 2020).

9.4 Microbial-Mediated Plant-Assisted Remediation

Phytoremediation is an innovative process of treating contaminated sites using plants either individually or in association with microorganisms. For example, plant roots contain mutualistic bacteria and fungi that improve the biodegradation of such pollutants as heavy metals, hydrocarbons, and pesticides (Kumar & Saxena, 2020; Maurya et al., 2023).

9.5 Genetically Modified Microorganisms (GMMs)

GMMs are advanced to exhibit greater biodegradation efficiencies. For instance, genetically modified *E. coli* and *Pseudomonas* have been proven to degrade contamination faster and a range of pollutants. Even though GMMs have remarkable

SPs to deliver, they entail specific pertinent regulatory and ethical concerns (Sharma et al., 2022a).

10 Recent Advancements and Future Trends in Bioremediation

Bioremediation technologies remain current, with the frontiers in biotechnology, nanotechnology, genetic engineering, and other developing facets of biotechnology being improved. All these developments are improving this bioremediation solution's efficiency, rate, and range in dealing with different contamination issues. Recent advancements in bioremediation are given in Table 6.

10.1 Nanotechnology in Bioremediation

Nanotechnology has emerged as a promising field in bioremediation, offering innovative solutions for contaminant detection, degradation, and removal.

- **Nanoparticles for Enhanced Degradation**: Nanoparticles, such as iron oxide and titanium dioxide, accelerate the breakdown of organic and inorganic contaminants. These nanoparticles can be functionalized with enzymes or microorganisms to enhance degradation. For example, iron nanoparticles combined with *Pseudomonas* spp. have been used to treat chlorinated solvents in groundwater more effectively (Soleimani et al., 2021).

Table 6 Recent advancements in bioremediation

Advancement	Description	Benefits
Bioelectrochemical systems	Integration of biological and electrochemical processes for enhanced bioremediation	Treatment of a wider range of contaminants
Bioaugmentation with specific microbial consortia	Tailored microbial communities for specific contaminants and environmental conditions	Enhanced degradation efficiency and predictability
Nanotechnology	Use of nanoparticles for enhanced bioremediation	Increased surface area, improved delivery of nutrients and microorganisms
Metagenomics	Study of microbial communities and their genetic potential for bioremediation	Identification of novel biodegradative pathways and enzymes
Genetic engineering	Modification of microorganisms to enhance their biodegradative capabilities	Improved efficiency and effectiveness of bioremediation

- **Nanobiosensors for Contaminant Detection**: Nanobiosensors equipped with microbial enzymes or DNA sequences are being developed to detect specific contaminants in real time. These sensors enable better monitoring of bioremediation processes, providing timely feedback on contamination levels and the effectiveness of remediation strategies (Salouti & Khadivi Derakhshan, 2020; Thakur & Kumar, 2022).
- **Carbon Nanotubes for Microbial Support**: Carbon nanotubes (CNTs) are being used to support microbial growth in contaminated environments. It prepares a suitable environment for the microorganisms to degrade some pollutants such as hydrocarbons and pesticides. CNTs have been incorporated into bioreactors to improve microbial activity and bioremediation rate (Sarkar et al., 2018).

10.2 Genetic Engineering in Bioremediation

Through genetic engineering, the potential of bioremediation has been revolutionized to develop microorganisms with a superior capacity for degrading particularly contaminants. Genetic engineering has been used to genetically modify bacteria for the removal of genetically engineered bacteria for specific contaminants, developing bacteria that can degrade recalcitrant pollutants such as PCBs, dioxins, and heavy metals. Genetically modified strains of *Escherichia coli* and *Pseudomonas putida* expressing such enzymes more efficiently degrade complex organic compounds (Malik et al., 2021).

Artificial microbial consortia designed by synthetic biology are capable of simultaneously targeting multiple contaminants. Genetically engineered microbes in these consortia possess complementary metabolic pathways in which these microbes can degrade hydrocarbons, heavy metals, and other different types of pollutants in mixed contamination scenarios (Che & Men, 2019; Jaiswal & Shukla, 2020). CRISPR-Cas9 gene editing technology can be used in bioremediation that is tried to increase the efficiency of biodegradation of microorganisms. Specific genes for enhancements in the rate of degradation enzymes, improvement in tolerance to toxic surrounds, and supply of microorganisms with the capacity to decontaminate materials formerly not biodegradable (Sahoo et al., 2022; Sharma et al., 2024). The advantages and disadvantages of bioremediation are shown in Fig. 4.

11 Conclusion

In the case of bioremediation, microorganisms are used to break down hazardous materials into harmless products to make this a safe, proven, and cheap method of hazardous waste disposal and of restoring the ecosystem. It is a process of treating polluted soil, water, and air with microbes or other living forms to remove or detoxify hazardous substances. More recently, over time, techniques of bioaugmentation

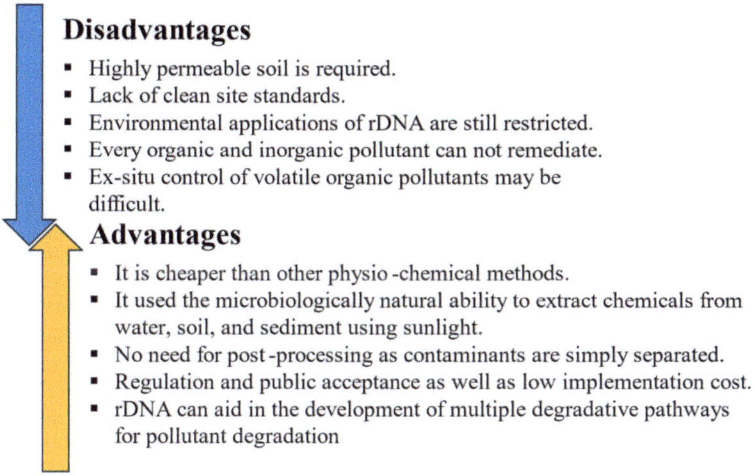

Fig. 4 Advantages and disadvantages of bioremediation

as well as biostimulation have been developed, with key milestones in the development of engineered microbes for oil-degrading bacteria in 1982, tanker Exxon Valdez oil spill in 1989, genetically modified microorganisms in the 1990s, and combined bioremediation and phytoremediation in 2010. Bioremediation is using bacteria, fungi, and algae to break down organic and inorganic pollutants. Microorganisms including bacteria are the most investigated microorganisms employed in bioremediation as they have diversity to occupy in between plants and can break down a large variety of pollutants. Aerobic or anaerobic, they depend on the availability or the lack of oxygen in the ecosystem. Fungi including white-rot fungi, ligninolytic enzymes, and mycoremediation are powerful "creatures of degradation" of target contaminants. Fungi can be used in mycoremediation on soils very polluted with hard degrading organic pollutants like dioxins, pesticides, or industrial dyes. Many of the metabolism phenomena for microorganisms have unique metabolic processes that remove or change contaminants, such as oxidative and nonoxidative energy utilization. Bioaugmentation and biostimulation of bioremediation are essential methods for the improvement of these conditions.

References

Abatenh, E., Gizaw, B., Tsegaye, Z., & Wassie, M. (2017). The role of microorganisms in bioremediation—A review. *Open Journal of Environmental Biology, 2*(1), 038–046.

Abbasian, F., Lockington, R., Mallavarapu, M., & Naidu, R. (2015). A comprehensive review of aliphatic hydrocarbon biodegradation by bacteria. *Applied Biochemistry and Biotechnology, 176*, 670–699.

Adams, G. O., Fufeyin, P. T., Okoro, S. E., & Ehinomen, I. (2015). Bioremediation, biostimulation and bioaugmentation: A review. *International Journal of Environmental Bioremediation & Biodegradation, 3*(1), 28–39.

Aghababai Beni, A., Haghmohammadi, M., Delnabi Asl, S., Hakimzadeh, S. M., & Nezarat, A. (2024). Algae application for treating wastewater contaminated with heavy metal ions. In *Algae as a natural solution for challenges in water-food-energy nexus: Toward carbon neutrality* (pp. 297–322). Springer.

Ahmed, M., Mavukkandy, M. O., Giwa, A., Elektorowicz, M., Katsou, E., Khelifi, O., et al. (2022). Recent developments in hazardous pollutants removal from wastewater and water reuse within a circular economy. *npj Clean Water, 5*(1), 1–25.

Akhtar, A. B. T., Naseem, S., Yasar, A., & Naseem, Z. (2021). Persistent organic pollutants (POPs): Sources, types, impacts, and their remediation. *Environmental Pollution and Remediation*, 213–246.

Alharbi, R. M., Sholkamy, E. N., Alsamhary, K. I., Abdel-Raouf, N., & Ibraheem, I. B. M. (2023). Optimization study of the capacity of Chlorella vulgaris as a potential bio-Remediator for the bio-adsorption of arsenic (III) from aquatic environments. *Toxics, 11*(5), 439.

Ali, M., Husain, Q., & Ishqi, H. M. (2019). Fungal peroxidases mediated bioremediation of industrial pollutants. In *Fungal bioremediation* (pp. 22–61). CRC Press.

Aliko, V., Multisanti, C. R., Turani, B., & Faggio, C. (2022). Get rid of marine pollution: Bioremediation an innovative, attractive, and successful cleaning strategy. *Sustainability, 14*(18), 11784.

Amundsen, T. (2022). The use and environmental impacts of dispersants for marine oil spill cleanup focusing on Corexit: a literature review. (Doctoral Dissertation)

Aparicio, J. D., Raimondo, E. E., Saez, J. M., Costa-Gutierrez, S. B., Alvarez, A., Benimeli, C. S., & Polti, M. A. (2022). The current approach to soil remediation: A review of physicochemical and biological technologies, and the potential of their strategic combination. *Journal of Environmental Chemical Engineering, 10*(2), 107141.

Ayilara, M. S., & Babalola, O. O. (2023). Bioremediation of environmental wastes: The role of microorganisms. *Frontiers in Agronomy, 5*, 1183691.

Azubuike, C. C., Chikere, C. B., & Okpokwasili, G. C. (2020). Bioremediation: An eco-friendly sustainable technology for environmental management. *Bioremediation of Industrial Waste for Environmental Safety: Volume I: Industrial Waste and Its Management*, 19–39.

Babaniyi, B. R., Ogundele, O. D., Bisi-Omotosho, A., Babaniyi, E. E., & Aransiola, S. A. (2023). Remediation approaches in environmental sustainability. In *Microbiology for cleaner production and environmental sustainability* (pp. 321–346). CRC Press.

Baker, P., Tiroumalechetty, A., & Mohan, R. (2019). Fungal enzymes for bioremediation of xenobiotic compounds *Recent advancement in white biotechnology through fungi: Volume 3: Perspective for sustainable environments*, 463–489.

Bala, S., Garg, D., Thirumalesh, B. V., Sharma, M., Sridhar, K., Inbaraj, B. S., & Tripathi, M. (2022). Recent strategies for bioremediation of emerging pollutants: A review for a green and sustainable environment. *Toxics, 10*(8), 484.

Bhalla, G., Bhalla, B., Kumar, V., & Sharma, A. (2022). Bioremediation and phytoremediation of pesticides residues from contaminated water: A novel approach. In *Pesticides remediation technologies from water and wastewater* (pp. 339–363). Elsevier.

Boudh, S., & Singh, J. S. (2019). Pesticide contamination: environmental problems and remediation strategies. *Emerging and eco-friendly approaches for waste management*, 245–269.

Brune, K. D., & Bayer, T. S. (2012). Engineering microbial consortia to enhance biomining and bioremediation. *Frontiers in Microbiology, 3*, 203.

Che, S., & Men, Y. (2019). Synthetic microbial consortia for biosynthesis and biodegradation: Promises and challenges. *Journal of Industrial Microbiology and Biotechnology, 46*(9–10), 1343–1358.

Cimboláková, I., Uher, I., Laktičová, K. V., Vargová, M., Kimáková, T., & Papajová, I. (2019). Heavy metals and the environment. *Environmental factors affecting human health,* 1, 29.

Colville, S. D. (2019). *Enhanced petroleum hydrocarbon remediation by biostimulation: Effects on groundwater microbial communities, geochemistry, and mineralogy*. University of Saskatchewan.

Craswell, E. (2021). Fertilizers and nitrate pollution of surface and ground water: An increasingly pervasive global problem. *SN Applied Sciences, 3*(4), 518.

Dangi, A. K., Sharma, B., Hill, R. T., & Shukla, P. (2019). Bioremediation through microbes: Systems biology and metabolic engineering approach. *Critical Reviews in Biotechnology, 39*(1), 79–98.

DeWitt, F. A., & Melvin, P. (1975). *Oil spill and oil pollution reports: November 1974–February 1975* (Vol. 1). National Environmental Research Center.

Dhiman, S., Kumar, S., Kumar, M., & Kumar, G. (2023). Origin and Management of Inorganic and Organic Contaminants. In *Management and mitigation of emerging pollutants* (pp. 21–67). Springer.

Dutta, N., Usman, M., Ashraf, M. A., Luo, G., & Zhang, S. (2022). A critical review of recent advances in the bio-remediation of chlorinated substances by microbial dechlorinators. *Chemical Engineering Journal Advances, 12*, 100359.

Dyes, A. (2023). Microbial degradation of pollutants. In *Environmental microbiology* (pp. 161–285). Springer.

Elangovan, S., Pandian, S. B. S., SJ, Geetha, & Joshi, S. J. (2019). Polychlorinated biphenyls (PCBs): Environmental fate, challenges and bioremediation. *Microbial metabolism of xenobiotic compounds*, 165–188.

Ellouze, M., & Sayadi, S. (2016). White-rot fungi and their enzymes as a biotechnological tool for xenobiotic bioremediation. *Management of hazardous wastes*, 103–120.

Filote, C., Roşca, M., Hlihor, R. M., Cozma, P., Simion, I. M., Apostol, M., & Gavrilescu, M. (2021). Sustainable application of biosorption and bioaccumulation of persistent pollutants in wastewater treatment: Current practice. *PRO, 9*(10), 1696.

Ganguly, D., Prasanna, K., Neelapu, S., & Goswami, G. (2023). Role of microbes in bioremediation. In *Industrial microbiology and biotechnology: Emerging concepts in microbial technology* (pp. 549–584). Springer.

Garvey, M. (2019). Food pollution: A comprehensive review of chemical and biological sources of food contamination and impact on human health. *Nutrire, 44*(1), 1.

Gaur, N., Narasimhulu, K., & PydiSetty, Y. (2018). Recent advances in the bio-remediation of persistent organic pollutants and its effect on environment. *Journal of Cleaner Production, 198*, 1602–1631.

Goswami, M., Chakraborty, P., Mukherjee, K., Mitra, G., Bhattacharyya, P., Dey, S., & Tribedi, P. (2018). Bioaugmentation and biostimulation: A potential strategy for environmental remediation. *Journal of Microbiology and Experimentation, 6*(5), 223–231.

Greenwood, S. J. (2010). *Ì This thesis may be used within the Department of National Defence, but copyright for open publication remains the property of the author*. Royal Military College of Canada.

Hazen, T. C. (2018). In situ: groundwater bioremediation. In Consequences of Microbial Interactions with Hydrocarbons, Oils, and Lipids: Biodegradation and Bioremediation (pp. 1-18). Springer, Cham.

Hazen, T. C. (2018). Bioremediation. In *Microbiology of the terrestrial deep subsurface* (pp. 247–266). CRC Press.

Hossain, M. E., Shahrukh, S., & Hossain, S. A. (2022). Chemical fertilizers and pesticides: Impacts on soil degradation, groundwater, and human health in Bangladesh. In *Environmental degradation: Challenges and strategies for mitigation* (pp. 63–92). Springer.

Hou, L., Zhou, Z., Wang, R., Li, J., Dong, F., & Liu, J. (2022). Research on the non-point source pollution characteristics of important drinking water sources. *Water, 14*(2), 211.

Hussain, A., Rehman, F., Rafeeq, H., Waqas, M., Asghar, A., Afsheen, N., et al. (2022). In-situ, *ex-situ*, and nano-remediation strategies to treat polluted soil, water, and air–A review. *Chemosphere, 289*, 133252.

Jaiswal, S., & Shukla, P. (2020). Alternative strategies for microbial remediation of pollutants via synthetic biology. *Frontiers in Microbiology, 11*, 808.

Jing, R., Fusi, S., & Kjellerup, B. V. (2018). Remediation of polychlorinated biphenyls (PCBs) in contaminated soils and sediment: State of knowledge and perspectives. *Frontiers in Environmental Science, 6*, 79.

Kahlon, R. S. (2016). Biodegradation and bioremediation of organic chemical pollutants by Pseudomonas. In *Pseudomonas: Molecular and applied biology* (pp. 343–417). Springer.

Kameswaran, S., Ramesh, B., Pitchika, G. K., Chandra, M. S., & Srinivasulu, M. (2021). Microbial reductive dehalogenation and its role in bioremediation. *Innovations in biotechnology for a sustainable future*, 205–226.

Kanamarlapudi, S., Chintalpudi, V. K., & Muddada, S. (2018). Application of biosorption for removal of heavy metals from wastewater. *Bios, 18*(69), 70–116.

Khan, A., & Rao, T. S. (2019). Molecular evolution of xenobiotic degrading genes and mobile DNA elements in soil bacteria. In *Microbial diversity in the genomic era* (pp. 657–678). Elsevier.

Khan, S., Nawab, J., & Waqas, M. (2020). Constructed wetlands: a clean-green technology for degradation and detoxification of industrial wastewaters. *Bioremediation of industrial waste for environmental safety: Volume II: Biological agents and methods for industrial waste management*, 127–163.

Kumar, N. (2022). An overview of water pollution and its multiple causes. *Applied Science and Biotechnology Journal for Advanced Research, 1*(2), 7–12.

Kumar, V., & Saxena, G. (2020). *Microbe-assisted phytoremediation of environmental pollutants: Recent advances and challenges*. AK Peters/CRC Press.

Kumar, V., Shahi, S., & Singh, S. (2018). Bioremediation: an eco-sustainable approach for restoration of contaminated sites. *Microbial bioprospecting for sustainable development*, 115–136.

Kumar, V., Bilal, M., Ferreira, L. F. R., & Iqbal, H. M. (Eds.). (2023). Genomics approach to bioremediation: principles, tools, and emerging technologies.

Kumari, S., & Das, S. (2023). Bacterial enzymatic degradation of recalcitrant organic pollutants: Catabolic pathways and genetic regulations. *Environmental Science and Pollution Research, 30*(33), 79676–79705.

Kuppan, N., Padman, M., Mahadeva, M., Srinivasan, S., & Devarajan, R. (2024). A comprehensive review of sustainable bioremediation techniques: eco friendly solutions for waste and pollution management. *Waste Management Bulletin*.

Kuppusamy, S., Palanisami, T., Megharaj, M., Venkateswarlu, K., & Naidu, R. (2016). *In-situ* remediation approaches for the management of contaminated sites: A comprehensive overview. *Reviews of Environmental Contamination and Toxicology, 236*, 1–115.

Kuppusamy, S., Maddela, N. R., Megharaj, M., Venkateswarlu, K., Kuppusamy, S., Maddela, N. R., … Venkateswarlu, K. (2020). An overview of total petroleum hydrocarbons. *Total Petroleum Hydrocarbons: Environmental Fate, Toxicity, and Remediation*, 1–27.

Lai, Y., Yang, C., Hsieh, C., Wu, C., & Kao, C. (2011). Evaluation of non-point source pollution and river water quality using a multimedia two-model system. *Journal of Hydrology, 409*(3–4), 583–595.

Liang, Y., Wei, J., Qiu, X., & Jiao, N. (2018). Homogeneous oxygenase catalysis. *Chemical Reviews, 118*(10), 4912–4945.

Malik, G., Arora, R., Chaturvedi, R., & Paul, M. S. (2021). Implementation of genetic engineering and novel omics approaches to enhance bioremediation: a focused review. *Bulletin of environmental contamination and toxicology*, 1–8.

Mallah, M. A., Changxing, L., Mallah, M. A., Noreen, S., Liu, Y., Saeed, M., et al. (2022). Polycyclic aromatic hydrocarbon and its effects on human health: An overview. *Chemosphere, 296*, 133948.

Manisalidis, I., Stavropoulou, E., Stavropoulos, A., & Bezirtzoglou, E. (2020). Environmental and health impacts of air pollution: A review. *Frontiers in Public Health, 8*, 14.

Maurya, A., Sharma, D., Partap, M., Kumar, R., & Bhargava, B. (2023). Microbially-assisted phytoremediation toward air pollutants: Current trends and future directions. *Environmental Technology & Innovation, 31*, 103140.

Meuser, H., & Meuser, H. (2013). Groundwater, soil vapour and surface water treatment. *Soil remediation and rehabilitation: treatment of contaminated and disturbed land*, 279–346.

Mishra, R. (2023). The effect of eutrophication on drinking water. *British Journal of Multidisciplinary and Advanced Studies, 4*(1), 7–20.

Mohammadi, L., Rahdar, A., Bazrafshan, E., Dahmardeh, H., Susan, M. A. B. H., & Kyzas, G. Z. (2020). Petroleum hydrocarbon removal from wastewaters: A review. *PRO, 8*(4), 447.

Mustafa, S., Bhatti, H. N., Maqbool, M., & Iqbal, M. (2021). Microalgae biosorption, bioaccumulation and biodegradation efficiency for the remediation of wastewater and carbon dioxide mitigation: Prospects, challenges and opportunities. *Journal of Water Process Engineering, 41*, 102009.

Narayanan, M., Ali, S. S., & El-Sheekh, M. (2023). A comprehensive review on the potential of microbial enzymes in multipollutant bioremediation: Mechanisms, challenges, and future prospects. *Journal of Environmental Management, 334*, 117532.

Nieder, R., & Benbi, D. K. (2022). Reactive nitrogen compounds and their influence on human health: An overview. *Reviews on Environmental Health, 37*(2), 229–246.

Nzila, A. (2013). Update on the cometabolism of organic pollutants by bacteria. *Environmental Pollution, 178*, 474–482.

Omondi, E. A., Ndiba, P. K., & Koech, G. C. (2022). Complexity of E-waste and its management challenges in developing countries–A review. *International Journal of Environmental Sciences and Natural Resources, 31*(2), 1–17.

Pastorella, G. (2014). *Investigation of the electrochemical activity of chromium tolerant mutants of Geobacter metallireducens*. Dublin City University.

Perez-Vazquez, A., Barciela, P., & Prieto, M. (2024). In situ and ex situ bioremediation of different persistent soil pollutants as agroecology tool. *PRO, 12*(10), 2223.

Pete, A. J., Bharti, B., & Benton, M. G. (2021). Nano-enhanced bioremediation for oil spills: A review. *ACS ES&T Engineering, 1*(6), 928–946.

Petrovskis, E. A., Amber, W. R., & Walker, C. B. (2013). Microbial monitoring during bioaugmentation with Dehalococcoides. *Bioaugmentation for groundwater remediation*, 171–197.

Prasad, R., & Aranda, E. (2018). *Approaches in bioremediation*. Springer.

Rani, L., Thapa, K., Kanojia, N., Sharma, N., Singh, S., Grewal, A. S., . . . Kaushal, J. (2021). An extensive review on the consequences of chemical pesticides on human health and environment. Journal of Cleaner Production, 283, 124657.

Rawat, C. D., Phian, S., Gupta, R., Verma, H., Kumar, M., Kaur, J., & Rawat, V. S. (2023). Microbial bioprocesses in remediation of contaminated environments and resource recovery. In *Microbial bioprocesses* (pp. 225–274). Elsevier.

Riser-Roberts, E. (2020). *Remediation of petroleum contaminated soils: Biological, physical, and chemical processes*. CRC press.

Rocha, C. S., Rocha, D. C., Kochi, L. Y., Carneiro, D. N. M., Dos Reis, M. V., & Gomes, M. P. (2022). Phytoremediation by ornamental plants: A beautiful and ecological alternative. *Environmental Science and Pollution Research*, 1–19.

Saha, J. K., Selladurai, R., Coumar, M. V., Dotaniya, M., Kundu, S., Patra, A. K., . . . Dotaniya, M. (2017). Organic pollutants. Soil Pollution-An Emerging Threat to Agriculture, 105–135.

Sahoo, S., Routray, S. P., Lenka, S., Bhuyan, R., & Mohanty, J. N. (2022). CRISPR/Cas-mediated functional gene editing for improvement in bioremediation: An emerging strategy. In *Omics insights in environmental bioremediation* (pp. 635–664). Springer.

Saibu, S., Adebusoye, S. A., & Oyetibo, G. O. (2020). Aerobic bacterial transformation and biodegradation of dioxins: A review. *Bioresources and Bioprocessing, 7*(1), 7.

Salouti, M., & Khadivi Derakhshan, F. (2020). Biosensors and nanobiosensors in environmental applications. *Biogenic nano-particles and their use in agro-ecosystems*, 515–591.

Sardar, U. R., Bhargavi, E., Devi, I., Bhunia, B., & Tiwari, O. N. (2018). Advances in exopolysaccharides based bioremediation of heavy metals in soil and water: A critical review. *Carbohydrate Polymers, 199*, 353–364.

Sarkar, B., Mandal, S., Tsang, Y. F., Kumar, P., Kim, K.-H., & Ok, Y. S. (2018). Designer carbon nanotubes for contaminant removal in water and wastewater: A critical review. *Science of the Total Environment, 612*, 561–581.

Seshadri, B., Bolan, N., & Naidu, R. (2015). Rhizosphere-induced heavy metal (loid) transformation in relation to bioavailability and remediation. *Journal of Soil Science and Plant Nutrition, 15*(2), 524–548.

Shah, G., & Soni, V. (2024). Comprehensive insights into the impact of oil pollution on the environment. *Regional Studies in Marine Science, 103516.*

Sharma, I. (2020). Bioremediation techniques for polluted environment: Concept, advantages, limitations, and prospects. In *Trace metals in the environment-new approaches and recent advances*. IntechOpen.

Sharma, S., Tiwari, S., Hasan, A., Saxena, V., & Pandey, L. M. (2018). Recent advances in conventional and contemporary methods for remediation of heavy metal-contaminated soils. *3 Biotech, 8*, 1–18.

Sharma, J., Goutam, J., Dhuriya, Y. K., & Sharma, D. (2021). Bioremediation of industrial pollutants. *Microbial Rejuvenation of Polluted Environment, 2*, 1–31.

Sharma, P., Bano, A., Singh, S. P., Sharma, S., Xia, C., Nadda, A. K., et al. (2022a). Engineered microbes as effective tools for the remediation of polyaromatic aromatic hydrocarbons and heavy metals. *Chemosphere, 306*, 135538.

Sharma, P., Singh, S. P., Parakh, S. K., & Tong, Y. W. (2022b). Health hazards of hexavalent chromium (Cr (VI)) and its microbial reduction. *Bioengineered, 13*(3), 4923–4938.

Sharma, V., Sehgal, R., Angra, V., & Gupta, R. (2024). Applications of genome editing in bioremediation. In *Genome editing in bacteria (part 2)* (pp. 150–183). Bentham Science Publishers.

Shetty, S. S., Deepthi, D., Harshitha, S., Sonkusare, S., Naik, P. B., & Madhyastha, H. (2023). Environmental pollutants and their effects on human health. *Heliyon, 9*, 9.

Shrivastava, N. G. (2020). Bioremediation of solid waste management. *Handbook of solid waste management: Sustainability through circular economy*, 1–51.

Silva, A., Delerue-Matos, C., Figueiredo, S. A., & Freitas, O. M. (2019). The use of algae and fungi for removal of pharmaceuticals by bioremediation and biosorption processes: A review. *Water, 11*(8), 1555.

Singh, V. (2024). Water pollution. In *Textbook of environment and ecology* (pp. 253–266). Springer.

Singh, S., & Singh, M. (2020). Soil pollution and human health. *Plant responses to soil pollution*, 205–220.

Singh, A., Sharma, A., Verma, R. K., Chopade, R. L., Pandit, P. P., Nagar, V., et al. (2022). Heavy metal contamination of water and their toxic effect on living organisms. In *The toxicity of environmental pollutants*. IntechOpen.

Singh, D., Yadav, B., & Choudhary, S. (2024). Heavy metals in the environment and their toxicological consequences on animal health. Life Science Informatics Publications, 10(2), 23.

Skinder, B. M., Uqab, B., & Ganai, B. A. (2020). Bioremediation: A sustainable and emerging tool for restoration of polluted aquatic ecosystem. *Fresh Water Pollution Dynamics and Remediation*, 143–165.

Smith, G. (1995). *Biotechnology in industrial waste treatment and bioremediation*. CRC Press.

Soleimani, M., Ghasemi, J. B., Ziarani, G. M., Karimi-Maleh, H., & Badiei, A. (2021). Photocatalytic degradation of organic pollutants, viral and bacterial pathogens using titania nanoparticles. *Inorganic Chemistry Communications, 130*, 108688.

Srivastava, V., Puri, M., Srivastava, T., Nidheesh, P., & Kumar, M. S. (2022). Integrated soil washing and bioreactor systems for the treatment of hexachlorocyclohexane contaminated soil: A review on enhanced degradation mechanisms, and factors affecting soil washing and bioreactor performances. *Environmental Research, 208*, 112752.

Swain, C. K. (2024). Environmental pollution indices: A review on concentration of heavy metals in air, water, and soil near industrialization and urbanisation. *Discover Environment, 2*(1), 5.

Temporiti, M. E. E. (2023). Fungi degrading plastics: selection and evaluation of their activity. (Doctoral Thesis).

Thakur, A., & Kumar, A. (2022). Recent advances on rapid detection and remediation of environmental pollutants utilizing nanomaterials-based (bio) sensors. *Science of the Total Environment, 834*, 155219.

Tiwari, S., Tripathi, A., & Gaur, R. (2017). Bioremediation of plant refuges and xenobiotics. *Principles and applications of environmental biotechnology for a sustainable future*, 85–142.

Unterman, R. (1996). A history of PCB biodegradation. *Biotechnology Research Series, 6*, 209–253.

Vantarakis, A. (2021). Eutrophication and public health. *Chemical lake restoration: Technologies, innovations and economic perspectives*, 23–47.

Venosa, A. D. (2004). Literature review on the use of commercial bioremediation agents for cleanup of oil-contaminated estuarine environments. *National risk management research laboratory, office of research and development, US environmental protection agency, Cincinnati, OH*, 1–56.

Vickram, A., Shofia, S. I., Palanivelu, J., Karishma, S., Saravanan, A., & Yaashikaa, P. (2024). A comprehensive analysis and exploration of the recent developments in the utilization of genetically modified microorganisms for the remediation of hazardous dye pollutants. *Groundwater for Sustainable Development, 101315.*

Visvanathan, P., Kirubakaran, D., Selvam, K., Manimegalai, P., Rajkumar, M., & Vasantharaj, K. (2024). Recent strategies for natural bioremediation of emerging pollutants: Development of a green and sustainable environment. *Bioremediation of Emerging Contaminants in Water, 2*, 1–19.

Wang, X., Yao, B., & Su, X. (2018). Linking enzymatic oxidative degradation of lignin to organics detoxification. *International Journal of Molecular Sciences, 19*(11), 3373.

Weldeslassie, T., Naz, H., Singh, B., & Oves, M. (2018). Chemical contaminants for soil, air and aquatic ecosystem. *Modern age environmental problems and their remediation*, 1–22.

Wess, L. (2023). *The curious world of bacteria*. Greystone Books Ltd.

Wu, Y. (2000). *A biosparging system with pulsed air injection for the remediation of petroleum-contaminated sites*. Faculty of Graduate Studies and Research, University of Regina.

Yenn, R. (2015). *A study on problems associated with crude oil contamination in Assam and its control using biological methods*: Lulu. com.

Youvan, D. C. (2024). Bioremediation in oil spill response: Case studies from Exxon Valdez to modern environmental remediation.

Yuan, Z., Nag, R., & Cummins, E. (2022). Human health concerns regarding microplastics in the aquatic environment-from marine to food systems. *Science of the Total Environment, 823*, 153730.

Microorganisms' Significance in Bioremediation

Amiya Ojha, Deeplina Das, and Tarun Kanti Bandyopadhyay

Abstract Bioremediation is a biological process that involves the breakdown of contaminants through the utilisation of microorganisms. Bioremediation, a sustainable and eco-friendly approach to environmental restoration, leverages the innate metabolic capabilities of microorganisms to degrade, detoxify, or transform pollutants in various environments. To overcome this obstacle, microorganisms are the most effective choice. Because of the metabolic activity that they employ, microorganisms are able to thrive in any and all of the harsh environments that exist on Earth. Two categories of elements influence the rate of degradation of biological and abiotic environments. Various strategies and procedures are employed globally. Common examples include biostimulation, bioaccumulation, bioventilation, biopiles, and bioattenuation. Case studies of successful bioremediation efforts in oil spills, contaminated soils, and industrial wastewater are examined to demonstrate the practical applications of microbial interventions.

This chapter explores the pivotal role microorganisms play in bioremediation, highlighting their significance in addressing diverse pollutants such as hydrocarbons, heavy metals, and pesticides. Microbial species, including bacteria, fungi, and archaea, possess the unique ability to metabolise harmful compounds, either by using them as energy sources or by converting them into less toxic forms. This degradation occurs through various mechanisms, including aerobic and anaerobic respiration, co-metabolism, and enzymatic pathways. Furthermore, the chapter addresses the challenges and limitations of microbial bioremediation, including factors affecting microbial efficacy, such as environmental conditions, genetic adaptability, and the presence of competing organisms.

A. Ojha · D. Das (✉)
Department of Bio Engineering, National Institute of Technology Agartala,
Jirania, Tripura, India
e-mail: deeplina.bio@nita.ac.in

T. K. Bandyopadhyay
Department of Chemical Engineering, National Institute of Technology Agartala,
Jirania, Tripura, India

S. Rajpoot, A. P. Das (eds.), *Innovative Technologies for Waste Management*, Sustainable Environmental Waste Management Strategies,
https://doi.org/10.1007/978-3-031-97184-6_4

Keywords Bioremediation · Bioaugmentation · Biostimulation · Microbial metabolism

1 Introduction

Bioremediation, the utilisation of biological agents to decompose or neutralise environmental pollutants, constitutes one of the most promising and sustainable methods for environmental restoration. This method utilises the intrinsic metabolic properties of microorganisms—such as bacteria, fungus, and algae to convert hazardous contaminants into less dangerous or inert compounds. As global enterprises develop, the quantity and diversity of pollutants released into ecosystems have risen, posing substantial problems for environmental management. Microorganisms provide an effective and environmentally sustainable solution to these challenges, capable of degrading complex contaminants that resist physical or chemical cleansing procedures. Conventional pollution management techniques, such as chemical treatments, incineration, and physical removal, can entail substantial operating expenses and adverse ecological consequences, including secondary contamination (Ajibade et al., 2021). Conversely, bioremediation utilises naturally existing microorganisms to decompose harmful compounds, offering a sustainable option that harmonises with natural ecosystems. Increasing industrial production, particularly in developing areas, is resulting in a greater prevalence of pollutants such as hydrocarbons, heavy metals (HM), pesticides, and pharmaceutical waste contaminating soil and water systems (Barinova et al., 2020). These creatures have adapted to thrive in extreme conditions, including oil spills and metal-contaminated mining sites, where they metabolise toxic substances as a survival strategy. The efficacy of bioremediation depends on microbial variety and adaptation, enabling the conversion of complex chemical compounds, such as polycyclic aromatic hydrocarbons (PAHs) and polychlorinated biphenyls (PCBs), into non-toxic forms. Bacterial species like *Pseudomonas*, *Bacillus*, and *Mycobacterium* and fungi such as *Aspergillus* and *Phanerochaete* have shown significant capacity for degrading pollutants, making them ideal candidates for bioremediation applications (Kour et al., 2022). Under aerobic circumstances, microorganisms employ oxygen to decompose contaminants, proving especially efficient for organic pollutants like hydrocarbons. Anaerobic processes are advantageous in oxygen-depleted conditions, when bacteria utilise alternate electron acceptors to decompose pollutants. The efficacy of microbial bioremediation is contingent upon several parameters, including nutrient availability, environmental conditions (such as pH, temperature, and moisture), and the bioavailability of pollutants. Nutrient supplements, including nitrogen and phosphorus, are frequently necessary to promote microbial proliferation and improve their degradation abilities. Environmental parameters must be optimised to maximise efficiency, since excessive pH or temperature might impede microbial activity. Moreover, the physical and chemical properties of the pollutant affect its availability to microbes; hydrophobic substances, for instance, may necessitate emulsifiers or surfactants to enhance their solubility, hence promoting microbial breakdown (Cameotra et al., 2010).

This chapter sets the stage for a comprehensive exploration of microorganisms' role in bioremediation, emphasising their metabolic diversity, ecological adaptability, and potential applications in various contaminated environments. Chapters will explore microbiological processes, technological innovations, and case studies that illustrate effective bioremediation applications. Additionally, analyse the environmental, economic, and societal ramifications of bioremediation, along with prospective avenues for study and application.

2 Types of Bioremediations

Depending on the environmental conditions and type of contaminant, bioremediation can occur naturally (natural attenuation), or it can be enhanced through biostimulation or bioaugmentation.

Ex Situ *Bioremediation*: These procedures mostly include the excavation of samples, with treatment conducted distant from polluted locations, leading to elevated costs and damage of natural areas. Contaminants are removed from the site and treated elsewhere. A prevalent instance of ex situ bioremediation is composting including land farming, bioreactor, and windrows (Fig. 1), which involves the decomposition of organic waste materials by microbes, often at high temperatures 55–65 °C (Pavel et al., 2008).

In Situ *Bioremediation*: Treatment occurs at the site of contamination without excavation. In situ bioremediation refers to the implementation of a field-based biological treatment to eradicate hazardous substances in the environment. It includes techniques like bioventing and biosparging for soil and groundwater remediation (Maitra, 2018) (Fig. 1).

Biosparging: This technique elevates groundwater oxygen levels to accelerate the biological breakdown and volatilisation of pollutants, facilitated by naturally existing microorganisms (Johnson et al., 2001). ***Biostimulation***: This method involves the injection of certain nutrients into the soil or groundwater to enhance the activity of naturally occurring microbes. It is an indigenous or naturally occurring bacteria. It emphasises the stimulation of the fungal community—initially, by identifying fertilisers, growth enhancers, and minerals and, secondly, by supplying additional environmental parameters such as pH, temperature, and aeration to enhance their metabolic rate and pathways. Trace levels of pollutants can serve as catalysts for the activation of opioids within biological enzymes (Tyagi et al., 2011). This strategic approach often involves the supplementation of nutrients and oxygen to support indigenous bacteria. These nutrients enable the fundamental building blocks and microbes of life to fulfil essential requirements, such as energy production, cellular biology, and enzymatic functions to mitigate pollution. ***Bioventing***: It facilitates the spontaneous in situ biodegrada-

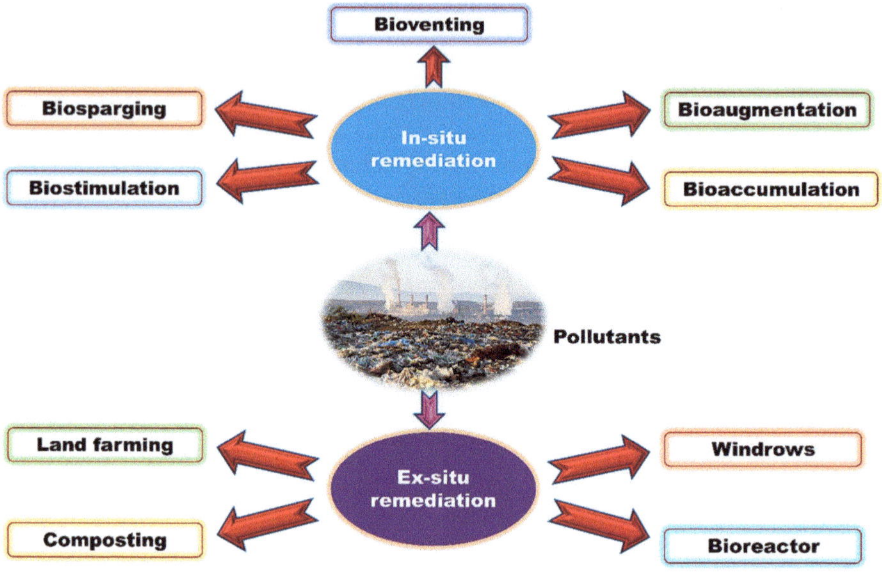

Fig. 1 Microbial bioremediation for the restoration of the contaminated sites via in situ and ex situ remediation

tion of substances that can be broken down aerobically by providing oxygen to soil microbes that already present with a modest airflow rate, ensuring sufficient oxygen to sustain microbial activity. Oxygen is directly injected into residual soil contamination via wells (Lee & Swindoll, 1993). Consequently, the fuel residues adsorbed on soil particles, along with associated contaminants, undergo biodegradation processes. Simultaneously, volatile organic compounds (VOCs) are biodegraded as they gradually migrate through biologically active soil layers. ***Bioaccumulation***: It refers to the incorporation of embedded microorganisms into a system that functions as aphoridometers to efficiently and thoroughly eliminate complicated pollutants. Moreover, genetically engineered microorganisms exhibit the capability to improve the degrading efficiency of several environmental contaminants. Due to its distinct metabolic profile, it is transformed into less complicated and innocuous end products. Natural species lack the capacity to decompose certain compounds, necessitating genetic modification through DNA manipulation; genetically engineered microorganisms degrade pollutants significantly more rapidly than their natural counterparts and exhibit high competitiveness with indigenous species, predators, and various abiotic factors (Wang, 2016). Genetically modified bacteria have shown potential for the biogenesis of soil, groundwater, and activated sludge, exhibiting improved degradation capacities for various chemical and physical contaminants. ***Biopiling***: It is a large-scale method that involves heaping excavated soils into heaps for bioremediation under optimal aeration conditions. The contaminants are then decomposed into CO_2 and water (Zhu et al., 2024). A fundamental biopile system consists of a treatment bed, an aeration mechanism, an irrigation network, and a leachate collection system.

3 Microbes Used in Bioremediation

Microorganisms may transform poisonous elements into water, carbon dioxide, and other less harmful molecules, which are then decomposed by more bacteria in a process known as mineralisation. Bioremediation can utilise bacteria, fungus, algae, and other microorganisms. Microbes are omnipresent in nature and employ many substrates as carbon sources; hence, they inhabit atypical situations where they can assimilate various contaminants. Their capacity to endure in unusual settings enhances their efficiency. ***Aerobic bioremediation***: Aerobic reactions are very efficient in degrading hydrocarbons, including those present in oil spills. Bacteria such as *Pseudomonas*, *Alcaligenes*, and *Acinetobacter* oxidise hydrocarbons, transforming them into carbon dioxide, water, and cellular biomass (Fragkou et al., 2021). Essential enzymes, such as oxygenases, are pivotal in integrating oxygen atoms into pollutant molecules, so aiding their degradation. ***Anaerobic bioremediation***: In oxygen-deficient settings, bacteria engage in anaerobic respiration, using contaminants as electron acceptors instead of oxygen. This process effectively reduces hazardous chemicals, such as chlorinated hydrocarbons, to innocuous end products. Anaerobic bacteria such as *Dehalococcoides* play a crucial role in the remediation of chlorinated solvents often present in groundwater (Fathepure & Tiedje, 1999).

3.1 Bacteria

Bacteria are essential for bioremediation, serving as vital biosorbents for the cleanup of polluted environments due to their capacity to proliferate under regulated settings and withstand harsh environmental conditions. They demonstrate effective biosorption capabilities for heavy metals in contaminated settings. Several bacterial species, including *Bacillus subtilis*, *Bacillus cereus*, and *Pseudomonas putida*, can effectively facilitate the removal of Cr (VI), while *Kocuria flava*, *Pseudomonas veronii*, and *Sporosarcina ginsengisoli* possess the capability to eliminate As, Zn, Cd, and Cu from contaminated environments (Ahmad et al., 2024; Hao et al., 2024; Lata et al., 2019; Sable et al., 2024; Sharma et al., 2021) (Table 1). Consortia of bacterial strains demonstrate greater efficacy than single-strain cultures in the bioremediation of heavy metals.

Table 1 Various microbes are utilised in bioremediation

Microorganisms	Species name	References
Bacteria	*Bacillus subtilis, Bacillus firmus, Pseudomonas putida, Pseudomonas aeruginosa N6P6, Dechloromonas aromatica, Alcaligenes aquatilis BU33N, Oleispira antarctica RB-8, Alcanivorax borkumensis, Streptomyces rimosus, Aphanothece halophytica, Thiobacillus ferrooxidans, Ganoderma applanatum, Stereum hirsutum, Ochrobactrum anthropi, Desulfovibrio fructosovorans, Corynebacterium glutamicum, Geobacter metallireducens, Desulfovibrio vulgaris, Sporosarcina ginsengisoli, Desulfuromonas acetoxidans, Zoogloea ramigera, Desulfovibrio norvegicus*	Dell'Anno et al. (2021); Elshafei and Mansour (2024); Marzan et al. (2017); Mondal et al. (2019); Pal et al. (2020)
Fungi	*Aspergillus niger, Aspergillus oryzae, Aspergillus terreus, Exophiala xenobiotica, Penicillium cyclopium, Aspergillus flavus NIOSN-SK56S22, Coriolopsis gallica, Pleurotus ostreatus, Aspergillus sydowii NIOSN-SK56C42, Ganoderma lucidum, Penicillium polonicum, Curvularia senegalensis, Pullularia pullulans, Aspergillus terreus MF12, Polyporus versicolor, Phanerochaete chrysosporium, Aspergillus flavus, Penicillium citrinum NIOSN-M126, Fusarium solani, Mucor rouxii, Phanerochaete chrysosporium, Phanerochaete chrysosporium, Pleurotus ostreatus, Acremonium sclerotigenum NIOSN-M109, Penicillium mononematosum, Mucor racemosus CBMAI 847, Aspergillus sclerotiorum CBMAI 849, Candida parapsilosis*	Dell'Anno et al. (2021); Elshafei and Mansour (2024); Jeyakumar et al. (2023); Mondal et al. (2019); Pal et al. (2020)
Algae	*Chlorella vulgaris, Volvox aureus, Nannochloropsis oculata, Chlamydomonas reinhardtii, Cladophora glomerata, Oscillatoria rubescens, Chroococcus minutus, Phormidium ceylanicum, Nostoc linckia, Gloeocapsa pleurocapsoides, Prototheca zopfii, Anabaena variabilis, Oedogonium westi, Oscillatoria quadripunctulata, Ulva lactuca, Sarcodia suiae, Phormidium ambiguum, Chlorococcum humicola, Oedogonium westi, Enteromorpha intestinalis, Pseudochlorella pringsheimii, Durvillaea antarctica, Chlorococcum humicola, Selenastrum capricornutum, Parachlorella kessleri, Chlorella sorokiniana*	Bhunia et al. (2022); Chugh et al. (2022); Gupta et al. (2019); Jeyakumar et al. (2023); Rath (2012); Touliabah et al. (2022)

3.2 Fungi

As strong biocatalysts for heavy metal bioremediation, fungi can take heavy metals up into their mycelium and spores. How they get rid of metal ions in living things is by collecting them, precipitating them inside and outside of cells, and changing their charge. The only living things on Earth that can break down wood are definitely fungus. Toxins and acids that can break down lignin and cellulose are released by fungi's mycelium. The strain of *Rhizopus oryzae* CDBB-H-1877 has the capability of biosorbing pentachlorophenol via methylation and dechlorination processes (Pal et al., 2020). Numerous fungi from the class *Zygomycetes* and *Aspergillus* species have been confirmed to possess the ability to decolorise and detoxify textile wastewater. Several prominent fungi, including *Fusarium solani*, *Penicillium chrysogenum*, *Penicillium digitatum*, and *Scedosporium apiospermum*, are documented for their capacity to degrade polychlorinated biphenyls (Dell'Anno et al., 2021; Elshafei & Mansour, 2024; Jeyakumar et al., 2023) (Table 1).

3.3 *Yeasts*

Yeasts provide a very prevalent alternative for the bioremediation of HM from contaminated environments due to their capacity to endure a broad spectrum of environmental conditions. Yeast cell walls possess a negative charge and may bond with HM cations, perhaps due to electrostatic interactions. *Saccharomyces cerevisiae* is the predominant yeast species utilised for the biosorption of heavy metals such as Cr, Pb, Zn, Hg, Cd, and Ni (Wang & Chen, 2006). In addition to *S. cerevisiae*, *Cyberlindnera fabianii*, *Candida utilis*, and *Candida tropicalis* have also been documented for their role in the bioremediation of chromium from contaminated habitats (Ayele & Godeto, 2021; Mousavi et al., 2021; Sable et al., 2024).

3.4 *Algae*

Algae has demonstrated significant efficacy in the bioaccumulation of HM and the breakdown of xenobiotics. The algae bioremediation technique has garnered significant attention from academics due to its essential function in carbon dioxide fixing and its beneficial biomass for biofuel generation. Species such as *Monoraphidium braunii* and *Selenastrum capricornutum* possess the capability to eliminate naphthalene, toluene, benzene, and bisphenol from contaminated environments. The bioadsorption of organic and pharmacological pollutants by microalgae has been extensively documented. Species of algae such as *Nannochloris* sp., *Chlamydomonas mexicana*, and *Chlorella sorokiniana* are essential for the bioremediation of diclofenac (Chugh et al., 2022; Gupta et al., 2019; Jeyakumar et al., 2023; Touliabah et al., 2022) (Table 1).

4 Mechanisms of Microbial Bioremediation

Microorganisms engaged in bioremediation employ several metabolic routes to decompose pollutants, which may be categorised as aerobic and anaerobic degradation processes. Every microbe has a distinct enzymatic repertoire that enables it to target particular contaminants and use them as sources of energy, biomass, or metabolic substrates. Immobilisation can be executed by in situ and ex situ technologies. The ex situ procedure entails the extraction of contaminated soils from the pollution site to another location, where a microbiological process is employed to immobilise the metal ions responsible for the contamination. The in situ approach involves treating pollutants on-site. Microorganisms including *B. cereus* and *E. asburiae* have been documented to participate in the immobilisation of HM that contaminate the environment (Wróbel et al., 2023). In bioremediation, microorganisms safeguard themselves against hazardous substances by developing hydrophobic or solvent efflux pumps that shield the cell's outer membrane.

4.1 Enzymatic Bioremediation

Enzymes are classified as biocatalysts that accelerate reaction rates, reduce activation energy, and break down or detoxify hazardous xenobiotic substances. Progress in enzymatic technologies, enzyme-substrate specificity, rapid mobility, reduced size, improved bioavailability, swift action, and diminished mass-transfer limitations have led to the emergence of enzyme-based pollutant and contaminant remediation, termed enzymatic bioremediation (Alaidaroos, 2023). Enzymatic oxidation refers to the biochemical process wherein hazardous compounds are reduced from a higher to a lower oxidation state. During this process, HM undergo electron loss, resulting in a decrease in their overall toxicity. This procedure employs an enzyme (oxidoreductase) secreted by the participating microorganisms. This approach is very efficacious in the cleanup of dyes, phenols, and other contaminants that are not readily degradable by bacteria. Oxidative enzymes generate radicals that can be decomposed into various fractions, ultimately resulting in molecules with elevated molecular weight. Microbes have a harder time dealing with tough environmental conditions like pH, temperature, contamination levels, and salt than enzymes do. The functionality of enzymes is independent of nutritional availability, whereas the microbial population is contingent upon nutrient delivery. It is a more environmentally sustainable technology with several advantages compared to microbial treatment. For the purpose of degrading a wide variety of pollutants, a number of enzymes, including cellulases, phosphotriesterases, laccases, carboxylesterases, peroxidases, haloalkane dehalogenases, lipases, and oxygenases, among others, play important roles (Bhandari et al., 2021). These enzymes have the ability to convert a wide variety of molecules, including aromatic hydrocarbons, phenols, chlorinated and nonchlorinated chemicals, recalcitrants, and others, into substances that are either simpler or conjugated and less reactive (Srinivasulu et al., 2022).

4.2 Bioleaching

The method of bioleaching involves using acidophilic microorganisms to facilitate the solubilisation of heavy metals in solid form derived from the sediment matrix. This procedure is especially helpful for removing contaminants that contain iron or sulphur. As a result, bacteria capable of oxidising iron or sulphur are mostly recruited for this process. Some examples of such organisms are *Aspergillus* sp., *Mucor* sp., *Rhizopus* sp., *Cladosporium* sp., *A. thiooxidans*, and *Penicillium* sp. Heavy metals that are in an immobilised condition are dissolved into an aqueous solution by these bacteria, which also produce an acidic environment inside the environment (Medfu Tarekegn et al., 2020). Microbial metabolism can generate secretions, including low molecular weight organic acids, capable of solubilising HM and soil particles containing HM minerals. Microbes may efficiently utilise resources and energy to excrete organic acids and facilitate the leaching of cadmium under favourable conditions. The leaching rate without nutrients was 9%; however, the inclusion of glucose and other nutrients increased it to 36% (Gopikrishnan et al., 2020).

4.3 Genetically Engineered Microorganisms (GEMs)

Genetically engineered microbes, also known as GEMs, have shown that they have the potential to be used in biological applications in settings such as soil, groundwater, and activated sludge. Furthermore, they have exhibited superior degrading capabilities when it comes to a wide variety of chemical pollutants. Numerous potentials exist to enhance suboptimal performance using genetic engineering approaches. For instance, genetically altering rate-controlling stages in established metabolic pathways can enhance degradation rates, or whole novel metabolic pathways can be integrated into bacterial strains for the breakdown of previously modified substances. Mostly, biomolecular engineering methods like rational designing and guided evolution have been used to change the genes of bacteria and their enzymes so that they can break down persistent organic pollution (POPs) like PCBs, pesticides, and PAHs (Mishra et al., 2021). Numerous advancements in recombinant DNA technology, including the creation of "suicidal-GEMs" (S-GEMs), have been undertaken to facilitate safe and effective bioremediation of polluted locations (Adetunji et al., 2024).

4.4 Nano-Bioremediation

Nano-bioremediation integrates advanced nanotechnology with specialised degradative microbial strains to form a cohesive remediation system. This is a newly developed integrated approach of sustainable remediation that aids in decreasing pollutant concentrations in the environment. This would exemplify a comprehensive predictive solution for effective environmental decontamination. Nano-bioremediation utilises the advantages of nanotechnology in conjunction with an environmentally sustainable bioremediation method. Nanotechnology synthesises and employs various nanomaterials for the detection and elimination of environmental pollutants (Sethi et al., 2024). Depending on their size, shape, structure, and composition, many nanomaterials are used, including nano-tubes, nano-fibres, nano-shells, nano-clusters, and nanocomposites. This technology's greatest strength is its adaptability; it may be used for cleanup of hazardous contaminants either in situ or off-site. Because of this, bioremediation using nanotechnology will be a greener, faster, and more efficient alternative.

5 Factors Affecting Microbial Bioremediation

The efficiency of microbial bioremediation of heavy metals is modulated by a range of factors, including the overall concentration of metal ions, redox conditions, metal speciation, microbial interactions and competition, environmental pH, temperature variations, soil texture and structure, oxygen levels, moisture availability, soil composition, and the aqueous solubility of the metals.

5.1 Ambient Temperature

Temperature is a critical determinant in the function of both external and intracellular microbial enzymes. Microbial enzymes possess several optimal conditions for the degradation of environmental pollutants. Enzymes, being proteins, are vulnerable to elevated temperatures that might denature or alter their properties. Temperatures between 20 and 40 °C were determined to be optimal for the breakdown of synthetic alizarin cyanine green dye by *Stropharia aeruginosa*, since the Lac enzymes become inactive or cleaved beyond these temperatures (Saravanan et al., 2021). Nevertheless, the optimal temperatures often fall between 25 and 35 °C. The absorption of Cd^{2+} by *Bacillus jeotgali* U3 peaks at 35 °C, whereas the optimal temperature for Zn^{2+} is 30 °C (Kushwaha et al., 2013). *Bacillus licheniformis* has comparable results; however, the impact of temperature on it is less pronounced. The fact that *Pseudomonas aeruginosa* was able to degrade polyethylene at temperatures lower than 23 °C raises the possibility that the enzymes responsible for the biodegradation process, including depolymerase and hydrolases, were functioning well at such temperatures (Facey, 2023).

5.2 pH

The pH is very important for microbial biosorption, and different bacteria often need a different pH. Microbes cannot grow well in environments with the wrong pH for a number of reasons. First, pH changes the way enzymes work in bacteria, which changes how fast microorganisms break down heavy metals. The pH has an effect on the surface charge of the bacterium, which in turn has an effect on the microorganism's ability to adsorb heavy metal ions. Additionally, the hydration and mobility of a great number of metal ions in the soil are influenced by pH. The ideal pH for diverse enzymes produced by distinct microorganisms differs. An inappropriate pH value results in diminished enzyme activity. The majority of enzymes operate inside the neutral pH range. Nonetheless, certain enzymes, such alkaline proteases, function at a pH above 8.0. MnP and LiP are examples of fungal lignocellulolytic enzymes that are being investigated for their potential to biodegrade synthetic colours (Kumar & Chandra, 2022). Between 7 and 8 is the pH range that has been proven to be favourable for the biodegradation of crude enzymes during the process. It is important to note that in the polluted site, microorganisms exhibit tolerance to the environmental circumstances and may metabolise polycyclic aromatic hydrocarbons at elevated pH levels. A neutral pH is conducive to the mineralisation of petroleum hydrocarbons. At acidic pH, heavy metals generate free ionic species, resulting in an increased availability of protons that saturate the binding sites of the metals (Priya et al., 2022). The pH of an environment influences the composition of pollutants and the survival capacity of microbes; the optimal pH for effective bioremediation ranges from 6.5 to 8.5 (Sekhar et al., 2024).

5.3 Substrate Concentration

Microbes vie for carbon, a finite resource that functions as an energy source for them. Consequently, the intrinsic capability of microorganisms that excel in degrading heavy metal pollutants will influence the rate of biodegradation. The species and abundance of microbes influence the efficiency and effectiveness of a bioremediation process; for example, a specific strain may effectively bioremediate a certain heavy metal in a laboratory setting, but a consortium of microbes is typically required in field applications. The bioremediation potential of specific microorganisms is governed by their molecular characteristics, the activation of genes and enzymes, metabolite production, growth efficiency, and survival rate (Srivastava et al., 2014). The rate of microbial biodegradation is also influenced by factors such as the ionisation state of chemical moieties within the cell wall, the structural morphology of the microbial cell wall, and the specific sorption sites involved. Low concentrations of the contaminant inhibit the induction of bacterial breakdown enzymes. The consequences of toxicity are, however, seen at extremely high concentrations of contaminants (Bilal et al., 2019). The degradation kinetics of catabolic enzymes are accelerated through synergistic interactions involving multiple constituents of a contaminant.

5.4 *Bioavailability*

The term "pollutant bioavailability" refers to the amount of pollutants that may be rapidly absorbed by microorganisms and broken down throughout the degradation process. There is a possibility that the presence of contaminants in the aqueous medium will have an effect on enzyme-mediated bioremediation activities. A new alkaline protease enzyme discovered in *Bacillus cereus* KM201428 was responsible for the biodegradation of the malachite green colour (Wanyonyi et al., 2017). Surfactants are introduced to contaminated soils to enhance the mass transfer of hydrophobic organic pollutants. Chemical surfactants generated from petroleum include SDS, Tween 80, and Triton X-100. Nonetheless, there are food-grade surfactants such as T-MAZ 60, T-MAZ 10, and T-MAZ 28 (Setiadi et al., 2022). For the purpose of producing a plant-based surfactant, the pericarp of *Sapindus mukorossi* and humic acid are utilised. Certain microorganisms are also capable of producing surfactants, which are utilised in the bioremediation process for the purpose of reducing the amount of toxins in the environment (Ugwu, 2019).

5.5 *Nutrient Availability*

Microorganisms require nutrients including C, N, P, K, and Ca for their development. Furthermore, the relative concentrations of the available nutrients can influence the breakdown of the pollutant. An overabundance of N, K, and P adversely affected the biodegradation of hydrocarbons. The rate of biodegradation is influenced by the accessibility of organic contaminants to microorganisms, also known as bioavailability (Olaniran et al., 2013). When mass transfer is a constraining factor, elevated biotransformation is not attained despite significant microbial conversion. Research indicates that, even after 50 years, the explosive contaminants in the soil have not undergone biodegradation; nevertheless, intensive soil mixing and fragmentation of coarse particles have promoted biodegradation (Chakraborty et al., 2022). Ageing or weathering denotes the decline in bioavailability with time. Ageing occurs as a result of many physical and chemical processes.

6 Practical Applications of Microbial Bioremediation

6.1 Heavy Metal Bioremediation

Microbial biomass exhibits diverse biosorptive capacities that vary significantly across different microbial species. The biosorptive efficiency of individual microbial cells depends on factors such as pretreatment processes and the specific experimental conditions employed. Furthermore, microbial cells require acclimatisation to fluctuations in physical, chemical, and bioreactor conditions to optimise biosorption performance (Nwankwegu & Onwosi, 2017). Bacteria serve as significant biosorbents owing to their prevalence, size, capacity for regulated growth, and resistance to environmental conditions. Multiple heavy metals have been evaluated utilising bacterial species such as *Bacillus*, *Pseudomonas*, *Flavobacterium*, *Enterobacter*, and *Micrococcus* sp. (Taran et al., 2019) (Table 2). Their exceptional biosorption capacity arises from their elevated surface-to-volume ratios and the presence of putative active chemisorption sites (teichoic acid) on the cell wall. Microorganisms alter the ionic state of heavy metals, affecting their solubility, bioavailability, and mobility in both soil and aquatic environments. The mobilisation or immobilisation of heavy metals facilitates microbial remediation, followed by oxidation-reduction, chelation, alteration of the metallic complex, and biomethylation (Rahman et al., 2020) (Fig. 3). Microbial enzymatic catalysis reduces metals from higher to lower oxidation states; for example, *Thiobacillus ferrooxidans* and *T. thiooxidans* facilitate the enzymatic oxidation of uranium (Tuovinen, 1972). *Aspergillus* species is utilised for the extraction of chromium from tannery effluent. Chromium removal was 85% at pH 6 in a bioreactor system using synthetic medium, whereas removal from tannery effluent was 65% (Sharma et al., 2011). This may result from the presence of organic contaminants that impede the organism's development. Studies have explored the use of biofilms for the extraction of heavy metals. Biofilm serves as an effective bioremediation instrument and a biological stabilisation agent. Biofilms have a remarkable resistance to harmful inorganic substances, even at deadly doses (Sharma et al., 2024). A research on *Rhodotorula mucilaginosa* has shown that the metal removal effectiveness ranged from 4.79% to 10.25% for planktonic cells and from 91.71% to 95.39% for biofilm cells (Grujić et al., 2017). Algae biomass is utilised for the bioremediation of HM-contaminated wastewater by adsorption or cellular integration (Goswami et al., 2022). Phycoremediation is the use of diverse algae and cyanobacteria to remediate heavy metals through the removal or breakdown of harmful substances. Microorganisms facilitate the movement of heavy metals through membrane-associated transport pathways and convert them into non-toxic forms (Fig. 2). Microorganisms employ mechanisms such as biosorption, bioaccumulation, biotransformation, and bioleaching to survive in metal-contaminated environments.

Table 2 Bioremediation of pollutants used by different microbes

Pollutant	Source	Microorganisms	Mechanisms	References
Nickel	Soil and water systems	*Enterobacter cloacae, Desulfovibrio desulfuricans, Durvillaea antarctica, Oedogonium westi, Rhizopus nigricans*	Biosorption	Taran et al. (2019)
Copper	Mining, electroplating	*Chlorella vulgaris, Bacillus firmus, Sphaerotilus natans, Aspergillus niger, Pseudomonas stutzeri, Daedalea quercina, Pseudomonas aeruginosa, Aspergillus lentulus, Sulfolobus solfataricus, aspergillus versicolor, Sphaerotilus natans*	Intracellular sequestration	Ashokkumar et al. (2017); Marzan et al. (2017); Thakare et al. (2021)
Zinc	Smelting, electroplating	*Streptomyces rimosus, Desulfovibrio desulfuricans, Aphanothece halophytica, Azospirillum lipoferum, Rhizobium leguminosarum, Trametes versicolor*	Extracellular sequestration	Alabssawy and Hashem (2024); Hussain et al. (2022); Thakare et al. (2021)
Cadmium	Waste batteries, paint sludge, incinerations	*Ganoderma applanatum, Stereum hirsutum, Phanerochaete chrysosporium, Pleurotus sapidus, Candida utilis, Bacillus subtilis, Pseudomonas putida, Pseudomonas aeruginosa, Aspergillus niger*	Biosorption, bioprecipitation, bioaugmentation	Ahmad et al. (2024); Hao et al. (2024); Lata et al. (2019); Sable et al. (2024)
Chromium	Mining, industrial coolants	*Aspergillus lentulus, Desulfuromonas acetoxidans, Aspergillus foetidus, Desulfovibrio norvegicus, Rhizopus oligosporus, Lysinibacillus sphaericus* CBAM5, *Ulothrix tenuissima*	Biosorption	Ayele and Godeto (2021); Kamaludeen et al. (2003); Mousavi et al. (2021); Sable et al. (2024)

(continued)

Table 2 (continued)

Pollutant	Source	Microorganisms	Mechanisms	References
Lead	Lead-acid batteries, paints, smelting operations	*Bacillus subtilis*, *Ochrobactrum anthropi*, *Sphingomonas paucimobilis*, *Arthrospira platensis*, *Saccharomyces cerevisiae*, *Bacillus firmus*, *Ganoderma applanatum*, *Rhodopseudomonas palustris*, *Streptomyces longwoodensis*, *Phormidium valderium*, *Spirogyra insignis*	Bioimmobilisation, bioattenuation	Kumar and Singh (2024); Sevak et al. (2021); Tiquia-Arashiro (2018)
Mercury	Electrical appliances, fluorescent lamps	*Cyclotella cryptica*, *Rhizopus arrhizus*, *Vibrio fluvialis*, *Pseudochlorococcum typicum*, *Ganoderma applanatum*, *Spirogyra hyaline*, *Geobacter metallireducens*, *Bacillus licheniformis*, *Chlamydomonas reinhardtii*, *Candida parapsilosis*	Biosorption, bioleaching	Ghosh et al. (2024); Kumar et al. (2024); Sable et al. (2024)
Arsenic	Smelting operations, groundwater, fuel-burning	*Microcystis aeruginosa*, *Sporosarcina ginsengisoli*, *Chlamydomonas reinhardtii*, *Saccharomyces cerevisiae*	Bioaccumulation, immobilisation	Khan et al. (2024); Mitra et al. (2017); Olaya-Abril et al. (2024)
Molybdenum	Spent catalyst, wastewater	*Bacillus amyloliquefaciens*, *Chlorella sorokiniana* TU5, *Enterobacter cloacae*, *Picochlorum oklahomensis*, *Pseudomonas aeruginosa*, *Acinetobacter calcoaceticus*	Bioprecipitation, bioaccumulation, sequestration	Abd Shukor, 2024; Tambat et al. (2023); Yakasai et al. (2021)
Cobalt	Mining, volcanic eruptions	*Colpomenia sinuosa*, *Chlorococcum humicola*, *Ulva fasciata*	–	Salem et al. (2021)

(continued)

Table 2 (continued)

Pollutant	Source	Microorganisms	Mechanisms	References
Uranium	Nuclear radiation plant	*Thiobacillus ferrooxidans*, *T. thiooxidans*	Biostimulation	Tuovinen (1972)
Naphthalene	Cigarette smoke, vehicle exhaust, air fresheners, paints, stains	*Bacillus licheniformis*, *Halomonas pacifica Cnaph3*, *Bacillus sonorensis*	Biosurfactants	Mohapatra et al. (2021)
Phenol	Sewage treatment	*Corynebacterium propinquum*, *Chlorella* sp., *Isochrysis galbana*, *Alcaligenes odorans*, *Phaeodactylum tricornutum*, *Pseudomonas* sp. sp48	Immobilisation, oxidation	Whiteley et al. (2000)
Hydrocarbon	Fossil fuels, coal	*Pseudoalteromonas* sp., *Rhodopirellula* sp., *Agarivorans* sp., *Halomonas* sp.	Biostimulation	Fathepure (2014); Ławniczak et al. (2020)
Azo dyes	Pharmaceutical factories and hospitals	*Lysinibacillus sphaericus*, *Aeromonas hydrophila*, *Citrobacter freundii*	Adsorption, reduction	Ajaz et al. (2020)
Dyes	Textile	*Bacillus licheniformis*, *Stenotrophomonas acidaminiphila*, *Lysinibacillus macroides*	Adsorption	Ihsanullah et al. (2020)
Congo red dye	Wastewater	*Oudemansiella canarii*, *Aspergillus flavus*, *Penicillium chrysogenum*	Adsorption	Talha et al. (2018)
Iron	Soil, groundwater	*Chlorococcum humicola*, *Pseudomonas fluorescence*	Bioleaching, bioaccumulation, biostimulation	Braunschweig et al. (2013); Wang et al. (2025)
Oils	Oil spills during extraction	*Alcaligenes odorans*, *Ralstonia pickettii*, *Corynebacterium propinquum*, *Marinobacter hydrocarbonoclasticus*	Bioaugmentation, biostimulation	Cohen (2002); Saranya et al. (2024)
Pyrene	Wood preservative industry, aluminium smelters	*Pseudomonas aeruginosa* N6P6, *Ganoderma lucidum*, *Pseudomonas pseudoalcaligenes* NP103	Oxidation	Mortazavi Mehrizi et al. (2022); Ostrem Loss and Yu (2018)

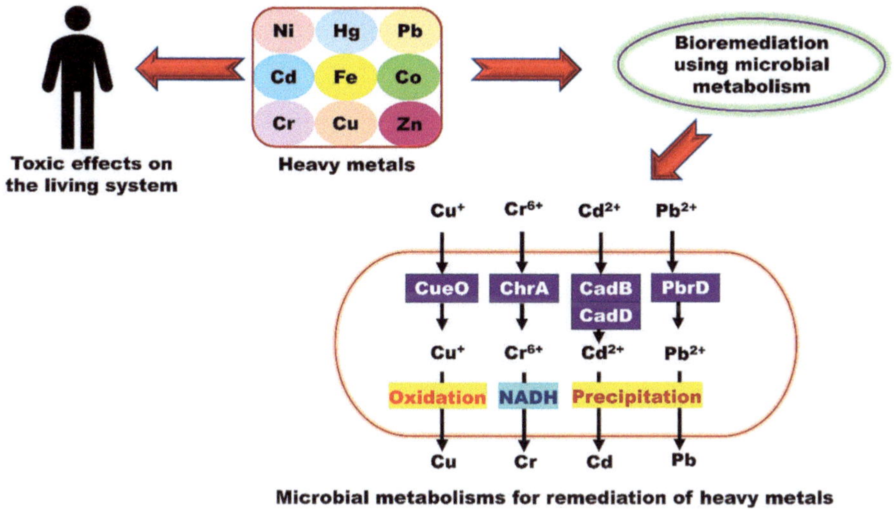

Fig. 2 Microbial metabolic pathways for heavy metal bioremediation

6.2 *Wastewater Treatment*

Environmental protection and energy security have emerged as key concerns for the economy of the entire world as a result of the rise in industrialisation and population growth. The requirements of humans have been satisfied by the utilisation of natural resources such as water and fossil fuels. Industrial operations, when utilised for residential purposes, emit hazardous gases into the environment and generate contaminated water, which is detrimental to ecological recipients. The accumulation of these hazardous substances adversely affects the aquatic ecology and heightens public health issues. Water quality monitoring programmes have become more essential for safeguarding public health in water bodies susceptible to contamination. The wastewater treatment plant is utilised to transform untreated influents into somewhat neutral effluents suitable for disposal. Algae- and bacteria-based treatments are widely used in wastewater management to remove excess nutrients, heavy metals, and pathogens from effluent streams (Mathew et al., 2022) (Fig. 3). Algae can absorb high levels of nitrogen and phosphorus, while bacteria, particularly those in activated sludge systems, help degrade organic pollutants. *Pseudomonas*, *Bacillus*, *Achromobacter*, *Aspergillus*, and *Rhodococcus* are important microorganisms in bioremediation (Fouad et al., 2023). Fungal bioremediation involves the degradation of contaminants by fungus, including *Actinomycetes* and *Aspergillus* species (Dinakarkumar et al., 2024). The decomposition of carbon by cyanobacterial bioremediation involves the use of polyethylene. Textile wastewater comprises several substances detrimental to the environment, such as organic dyes, vibrant colours, elevated COD levels, heavy metals, phosphates, nitrates, and sulphates. Microalgae may assimilate these nutrients to facilitate their growth and transform

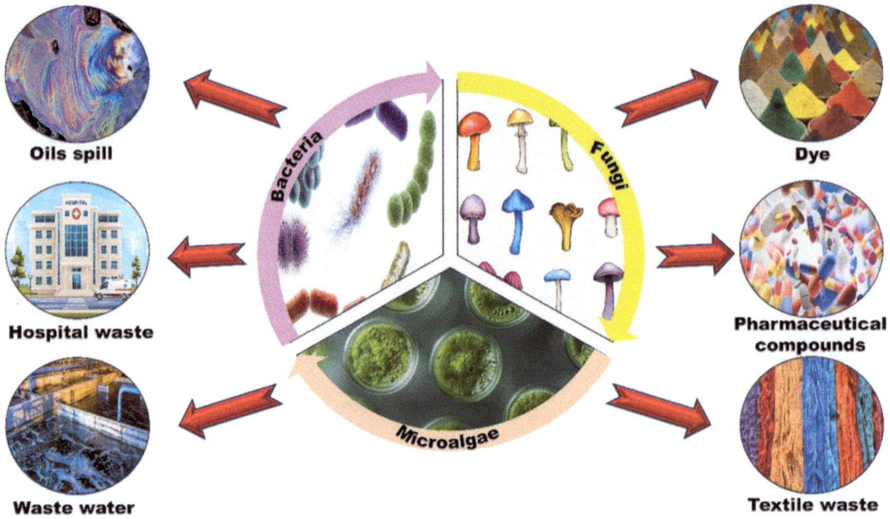

Fig. 3 Field applications of microbial bioremediation

them into valuable biofuels such as biodiesel. HM including arsenic (As), cadmium (Cd), fluoride (F), lead (Pd), and mercury (Hg) are prevalent pollutants in wastewater (Table 2). The paper industry generates substantial quantities of wastewater with elevated concentrations of phenolic, chlorinated, and lignin-derived pollutants, along with sulphonated contaminants (Shukla et al., 2022). Metal contamination in aquatic organisms has been associated with the release of industrial effluents and sewage into water bodies. They effectively attach to obstructive particles, accumulate in the riverbed, and ultimately discharge into the adjacent water, posing a lethal hazard. Microbial bioremediation has significant potential for wastewater treatment due to its versatility in addressing various contaminants and its negligible environmental effect. Innovative methods like microbial mats and fuel cells advance the boundaries of sustainable wastewater management. This approach corresponds with the pressing global demand for cleaner, environmentally sustainable, and economical water treatment techniques.

6.3 *Landfill Leachates Remediation*

Landfilling is a prevalent and widely endorsed method for waste disposal worldwide, hence resulting in the generation of leachate. Landfill leachate (LFL) refers to the effluent that permeates through the decomposing solid waste. It has become a significant environmental concern affecting both groundwater and surface water systems. The remediation of LFLs presents a considerable challenge for contemporary society, for which bioremediation offers a viable alternative due to its reduced operational costs compared to physicochemical approaches. They were able to obtain a removal rate of 90% for ammonia and a removal rate of 60% for nitrate and phosphate from LFL using bacterial strains belonging to the phyla *Firmicutes*, *Actinobacteria*, and *Proteobacteria* (Morris et al., 2018). A substantial reduction of 70–100% in ammonia nitrogen was observed using the algal strain SW15aRL of *Chlamydomonas* sp., while the bioremediation efficacy of indigenous fungi *Phanerochaete sanguinea* MUT 1284 and *Pseudallescheria boydii* MUT 721 resulted in a commendable and rapid decolorisation rate of 60%, accompanied by detoxification (Ganguly et al., 2023; Pal et al., 2020).

6.4 Oil Spill Bioremediation

Oil spills pose considerable ecological concerns, particularly in coastal regions, where they affect species, disrupt ecosystems, and result in enduring pollution. Microorganisms are essential in the natural degradation of oil via bioremediation, a process that utilises the capacity of microorganisms to metabolise hydrocarbons in oil as an energy source, converting them into less toxic chemicals. Diverse bacterial, fungal, and some archaeal species have developed mechanisms to break down hydrocarbons, rendering them efficient agents in oil spill remediation (Table 2). Bioremediation is seen as a cost-efficient, sustainable, and eco-friendly alternative to physical and chemical remediation techniques. Oil spills present a significant environmental hazard. Bacterial consortia, often from the genera *Pseudomonas*, *Alcanivorax*, and *Bacillus*, have shown effectiveness in degrading petroleum hydrocarbons (Ławniczak et al., 2020). These bacteria thrive on hydrocarbons as a primary energy source, converting them into non-toxic byproducts, thus aiding in the restoration of marine and terrestrial ecosystems. The elevated degrading capabilities of bacteria in marine settings are largely attributable to their halophilic characteristics, enabling them to function under varying degrees of salt stress. A halophilic consortium including *Halomonas*, *Dietzia*, and *Arthrobacter* has demonstrated the ability to degrade approximately 40% of diesel oil (Rezaei Somee et al., 2018). *Halomonas* strains appear to possess significant potential for hydrocarbon-degrading capabilities (Fathepure, 2014). The use of surfactant molecules generated by microorganisms offers a viable method to enhance the bioremediation efficacy of contaminated settings. Due to their amphiphilic nature, these chemicals improve bioremediation processes by promoting the partitioning of hydrophobic contaminants into the hydrophobic cores of surfactant micelles. This mechanism enhances the extraction and removal of pollutants from sediment environments. A unique method to accelerate the bioremediation of oil spills in marine environments involves using new materials as potential carriers for hydrocarbonoclastic bacteria. One such potential carrier, composed of puffed chitosan, calcium alginate, and *Panicum miliaceum*, is capable of immobilising oil-degrading bacteria inside its porous framework (Dell'Anno et al., 2021). This construction has demonstrated biodegradability and buoyancy in oil-contaminated waters. Numerous terrestrial fungal species have been identified as significant contributors to hydrocarbon biodegradation; however, only a limited number of research have explored the bioremediation capacity of fungi concerning petroleum hydrocarbons in marine ecosystems. Nonetheless, other studies have demonstrated that fungus found in diverse marine habitats may effectively degrade petroleum hydrocarbons. The fungus species *Penicillium citrinum* may diminish the total crude oil mass by 77% and reduce each n-alkane mass by an average of 95% (Ramdass & Rampersad, 2021). Hydrocarbon-contaminated sediments from the western coast of Algeria demonstrated the capacity of 12 fungal strains from the genera *Cladosporium*, *Aspergillus*, and *Penicillium* to effectively break down crude oil and create biosurfactants (Ebrahimian et al., 2024). Metabolic pathways of fungi that facilitate hydrocarbon degradation under aerobic circumstances may start with oxidation processes mediated by cytochrome P450 monooxygenases and alkane oxygenase enzymes (Abdel-Shafy & Mansour, 2018). The breakdown of polycyclic aromatic hydrocarbons by ligninolytic fungi has been thoroughly investigated in recent years, particularly in white rot fungi. These fungi generate extracellular peroxidases that facilitate the early oxidation of PAHs. *Chlorophyllous* microalgae *Chlorella vulgaris* have shown significant promise for the remediation of crude oil-contaminated waters, with a bioremediation effectiveness of 88–94% (Gul et al., 2022). *Chlorella vulgaris* BS1 can degrade 98% of petroleum hydrocarbons at starting concentrations of 115 mg L^{-1} in water after 14 days (Dell'Anno et al., 2021). Five cyanophytes, specifically *Westiellopsis prolifica*, *Anabaena variabilis*, *Oscillatoria princeps*, *Phormidium mucicola*, and *Lyngbya digueti*, have shown the ability to diminish the quantities of various hydrocarbon compounds in oil refinery wastewater by 24–92% (Cepoi et al., 2016; Chaillan et al., 2006). These studies highlight the adaptability and efficacy of microbial bioremediation techniques in many environmental contexts. Utilising the enzymatic pathways of indigenous microbial populations and enhancing environmental conditions can facilitate the breakdown of oil pollutants, resulting in economical and environmentally sustainable rehabilitation of impacted regions.

7 Emerging Research and Future Directions

Numerous research initiatives have been conducted on the use of microbial enzymes for the bioremediation of waste materials; nonetheless, enhancing the process is crucial to guarantee a safer and more sustainable environment. It is essential to enhance research to discover new microorganisms capable of efficiently and swiftly bioremediating various contaminants, particularly those originating from industrial sources. Gagandeep Singh and associates emphasised the extensive application and constraints of several bioremediation methods in the context of soil contamination. This study emphasises the need of choosing appropriate microbiological methods suited to various contaminants (COMBAT, B. A. A. S. T., 2024). Athwal et al. investigated the amalgamation of microbial populations with nanomaterials to improve bioremediation in agricultural settings. The research highlights the necessity for enhanced comprehension of microbial-nanomaterial interactions to optimise efficacy while reducing environmental hazards (Athwal et al., 2024). Patel and associates investigated the application of synthetic biology to develop microbial strains proficient in the effective removal of heavy metals from polluted settings. The method has opportunities for developing effective bioremediation agents, although it entails ethical and ecological issues that require thorough assessment (Mishra et al., 2024). Y.J. Song et al. examined the dual use of microbial species for bioenergy production and pollution remediation, particularly in microbial fuel cells. The study promotes the ongoing advancement of microbial technologies that utilise natural processes for energy generation and pollution reduction (Song et al., 2024). A thorough investigation by Tang et al. regarding microbial remediation methods highlights the necessity for advancements in genetic and bioengineering methodologies. These innovations seek to optimise microbial strains for the bioremediation of HM, hence improving the efficacy of environmental restoration initiatives (Tang et al., 2024). Mohamed Saad Hellal's study elucidates the biological remediation of polluted water through the application of microbial consortia. Advancements in this domain concentrate on enhancing detoxification rates and microbiological efficacy (Hellal et al., 2025). Ullah et al. examined the function of bacteria in facilitating phytoremediation of harmful substances in plants. Their findings underscore the significance of symbiotic interactions in bioremediation and propose pathways for biotechnological advancements (Yu et al., 2024). A review by Esa and Him examines the advantages of implementing specialised microbial strains for bioaugmentation in polluted areas and outlines future possibilities, including the application of modified microorganisms for expedited and more effective cleanup (Esa & Him, 2024). Krishnan and Pratap investigated microbial nano-bioremediation as a solution for micro- and nanoplastic contamination. Research demonstrates that enzyme technology and nanomaterials can markedly improve the bioremediation of enduring plastic contaminants (Pratap et al., 2024). Hegde and Selvaraj's research on chitosan—a biodegradable biopolymer sourced from environmental waste—demonstrates its application as a bioremediation agent, transforming contaminants into less harmful substances (Hegde et al., 2024). Through genetic engineering, scientists may augment certain metabolic pathways in microorganisms, enabling them to decompose enduring organic contaminants such as plastics. Recombinant DNA technology can establish novel degradation pathways or augment existing ones in microbes, so increasing their efficacy in bioremediation. The future of bioremediation is employing machine learning to engineer microbial consortia with ideal characteristics for certain contaminants and environmental circumstances. These consortia may be customised to leverage synergies among various microbial species, improving degrading efficiency and robustness to environmental variations.

8 Conclusion

Microorganisms are crucial agents in bioremediation owing to their distinctive capacity to metabolise contaminants and acclimatise to diverse settings. With the advancement of bioremediation technology, microbial bioremediation is set to become a fundamental element of sustainable environmental management. By comprehending and utilising the potential of these microbes, humans may tackle significant environmental concerns, converting contaminated areas into rejuvenated ecosystems. Microbial bioremediation serves as a crucial option for alleviating environmental contamination, providing a sustainable alternative to traditional remediation techniques. It possesses significant potential for international environmental management initiatives, particularly as climate change and industrial operations persist in aggravating pollution challenges. Nonetheless, its comprehensive implementation requires a nuanced comprehension of microbial ecology, diligent oversight, and pioneering biotechnological innovations to successfully tackle existing constraints. Ongoing investment and research position microbial bioremediation as essential for preserving and rehabilitating environmental health.

References

Abd Shukor, M. Y. (2024). Bacterial reduction of molybdenum as a tool for its bioremediation. In Eco-restoration of polluted environment (pp. 238–277). CRC Press.

Abdel-Shafy, H. I., & Mansour, M. S. M. (2018). Microbial degradation of hydrocarbons in the environment: an overview. Microbial action on hydrocarbons. Springer, Singapore. 353–386. https://doi.org/10.1007/978-981-13-1840-5_15

Adetunji, C. O., Ahamed, M. I., & Altalhi, T. (2024). *Genetically engineered organisms in bioremediation*. CRC Press.

Ahmad, I. et al. (2024). Microbial strategies for potential cadmium bioremediation: A review. Cadmium Toxicity in Water: Challenges and Solutions. Springer, Cham. 101–137. https://doi.org/10.1007/978-3-031-54005-9_6

Ajaz, M., Shakeel, S., & Rehman, A. (2020). Microbial use for azo dye degradation—a strategy for dye bioremediation, International Microbiology, 23, 149–159. https://doi.org/10.1007/s10123-019-00103-2

Ajibade, F. O. et al. (2021). Environmental pollution and their socioeconomic impacts. In Microbe mediated remediation of environmental contaminants (pp. 321–354). Elsevier. https://doi.org/10.1016/B978-0-12-821199-1.00025-0

Alabssawy, A. N., & Hashem, A. H. (2024). Bioremediation of hazardous heavy metals by marine microorganisms: a recent review, Archives of microbiology, 206(3), 103. https://doi.org/10.1007/s00203-023-03793-5

Alaidaroos, B. A. (2023). Advancing eco-sustainable bioremediation for hydrocarbon contaminants: Challenges and solutions, Processes, 11(10), 3036. https://doi.org/10.3390/pr11103036

Ashokkumar, P., Loashini, V. M., Bhavya, V. (2017). Effect of pH, Temperature and biomass on biosorption of heavy metals by Sphaerotilus natans, International Journal of Microbiology and Mycology, 6(1), 32–38.

Athwal, S., Garg, K., Sudan, S. K., Koul, A., Ray, A., & Magotra, S. (2024). Microbial intervention with nanomaterials: An effective approach for bioremediation in agro-ecosystems. In *Discovery, disruption, and future implications of nanomaterials* (pp. 497–534). IGI Global.

Ayele, A., & Godeto, Y. G. (2021). Bioremediation of chromium by microorganisms and its mechanisms related to functional groups, Journal of Chemistry, 2021(1), 7694157. https://doi.org/10.1155/2021/7694157

Barinova, G. M., Gaeva, D. V., Krasnov, E. V. (2020). Hazardous chemicals and air, water, and soil pollution and contamination. In Good Health and Well-Being, Springer, Cham, 255–266. https://doi.org/10.1007/978-3-319-69627-0_48-1

Bhandari, G., Sharma, M. (2021). Microbial enzymes in the bioremediation of pollutants: Emerging potential and challenges. In Mycoremediation and Environmental Sustainability, Springer, Cham. 75–94. https://doi.org/10.1007/978-3-030-54422-5_3

Bhunia, B., Debnath, C., Muthuraj, M. (2022). Innovations in sustainable treatment technologies for Heavy Metals, D., & Xenobiotics, O. Trends in bioremediation of dyes from wastewater. 140. Bentham Science Publishers, https://doi.org/10.2174/97898150497251220101

Bilal, M., Adeel, M., Rasheed, T., Zhao, Y., & Iqbal, H. M. J. E. I. (2019). Emerging contaminants of high concern and their enzyme-assisted biodegradation–a review, *124*, 336–353. https://doi.org/10.1016/j.envint.2019.01.011

Braunschweig, J., Bosch, J., & Meckenstock, R. U. J. N. B. (2013). Iron oxide nanoparticles in geomicrobiology: from biogeochemistry to bioremediation, *30*(6), 793–802. https://doi.org/10.1016/j.nbt.2013.03.008

Cameotra, S. S., & Makkar, R. S. (2010). Biosurfactant-enhanced bioremediation of hydrophobic pollutants, Pure and Applied Chemistry, 82(1), 97–116.

Cepoi, L., Donţu, N., Şalaru, V., & Şalaru, V. J. C. F. B. O. W. (2016). Removal of organic pollutants from wastewater by cyanobacteria. 27–43. Springer, Cham. https://doi.org/10.1007/978-3-319-26751-7_4

Chaillan, F., Gugger, M., Saliot, A., Coute, A., & Oudot, J. J. C. (2006). Role of cyanobacteria in thc biodegradation of crude oil by a tropical cyanobacterial mat, *62*(10), 1574–1582. https://doi.org/10.1016/j.chemosphere.2005.06.050

Chakraborty, N., Begum, P., & Patel, B. K. (2022). Counterbalancing common explosive pollutants (TNT, RDX, and HMX) in the environment by microbial degradation. In *Development in wastewater Treatment research and processes* (pp. 263–310). Elsevier. https://doi.org/10.1016/B978-0-323-85839-7.00012-8

Chugh, M., Kumar, L., Shah, M. P., & Bharadvaja, N. J. E. N. (2022). Algal bioremediation of heavy metals: An insight into removal mechanisms, recovery of by-products, challenges, and future opportunities, *7*, 100129. https://doi.org/10.1016/j.nexus.2022.100129

Cohen, Y. J. I. M. (2002). Bioremediation of oil by marine microbial mats, *5*, 189–193. https://doi.org/10.1007/s10123-002-0089-5

COMBAT, B. A. A. S. T. (2024). Bioremediation as a strategy to combat soil pollution. Futuristic Trends in Biotechnology, Volume 3, Book 11, Part 4, Chapter 1.

Dell'Anno, F., Rastelli, E., Sansone, C., Brunet, C., Ianora, A., & Dell'Anno, A. J. M. (2021). Bacteria, fungi and microalgae for the bioremediation of marine sediments contaminated by petroleum hydrocarbons in the omics era, *9*(8), 1695. https://doi.org/10.3390/microorganisms9081695

Dinakarkumar, Y., Gnanasekaran, R., Reddy, G. K., Vasu, V., Balamurugan, P., & Murali, G. (2024). Fungal bioremediation: An overview of the mechanisms, applications and future perspectives. Environmental Chemistry and Ecotoxicology, 6, 293-302. https://doi.org/10.1016/j.enceco.2024.07.002

Ebrahimian, A., Kariminik, A., Hassanshahian, M., & Khoshroo, S. M. R. J. G. J. (2024). The roles played by fungi in bioremediation technologies, *41*(8), 808–817. https://doi.org/10.1080/01490451.2024.2395362

Elshafei, A. M., & Mansour, R. J. D. S. (2024). Microbial bioremediation of soils contaminated with petroleum hydrocarbons, *1*(1), 1–13. https://doi.org/10.1007/s44378-024-00004-5

Esa, F. N., & Him, N. R. N. (2024). Bioaugmentation as bioremediation approach for contaminated soil: A review, Jurnal Teknologi (Sciences & Engineering), 86(5), 89–102.

Facey, F. S. B. (2023). Exploring polyethylene degrading capabilities of bacteria using bioprospecting and synthetic biology. (Doctoral dissertation) Macquarie University.

Fathepure, B. Z. J. F. I. M. (2014). Recent studies in microbial degradation of petroleum hydrocarbons in hypersaline environments, *5*, 173. https://doi.org/10.3389/fmicb.2014.00173

Fathepure, B. Z., & Tiedje, J. M. J. B. O. C. S. (1999). Anaerobic bioremediation: microbiology, principles, and applications, *37*, 339–396. https://doi.org/10.2134/agronmonogr37.c14

Fouad, F. A., Youssef, D. G., Shahat, F. M., & Abd El-Ghany, M. N. (2023). Role of microorganisms in biodegradation of pollutants. In Handbook of biodegradable materials (pp. 221-260). Springer.

Fragkou, E., Antoniou, E., Daliakopoulos, I., Manios, T., Theodorakopoulou, M., & Kalogerakis, N. J. (2021). In situ aerobic bioremediation of sediments polluted with petroleum hydrocarbons: a critical review, Journal of Marine Science and Engineering, 9(9), 1003. https://doi.org/10.3390/jmse9091003

Ganguly, D., Prasanna, K., Neelapu, S., & Goswami, G. (2023). Role of microbes in bioremediation. In *Industrial microbiology and Biotechnology: Emerging concepts in microbial Technology* (pp. 549–584). Springer.

Ghosh, S., Sarkar, B., Imboon, T., & Thongmee, S. (2024). Microbe-mediated mercury bioremediation for wastewater treatment. In *Microbial metagenomics in effluent Treatment plant* (pp. 239–255). Elsevier.

Gopikrishnan, V., et al. (2020). Microbial leaching of heavy metals from e-waste: opportunities and challenges. Biovalorisation of wastes to renewable chemicals and biofuels, 189–216. https://doi.org/10.1016/B978-0-12-817951-2.00010-9

Goswami, R. K., Agrawal, K., Shah, M. P., & Verma, P. (2022). Bioremediation of heavy metals from wastewater: a current perspective on microalgae-based future, Letters in Applied Microbiology, 75(4), 701–717.

Grujić, S., Vasić, S., Radojević, I., Čomić, L., & Ostojić, A. (2017). Comparison of the Rhodotorula mucilaginosa biofilm and planktonic culture on heavy metal susceptibility and removal potential, Water, Air, & Soil Pollution, 228, 73. https://doi.org/10.1007/s11270-017-3259-y

Gul, M. Z., Rupula, K., & Beedu, S. R. (2022). Microbial consortia for pollution remediation—Success stories. In *Microbial consortium and biotransformation for pollution decontamination* (pp. 95–121). Elsevier.

Gupta, P. K., Ranjan, S., Gupta, S. K. (2019). Phycoremediation of petroleum hydrocarbon-polluted sites: application, challenges, and future prospects. Application of Microalgae in Wastewater Treatment: Volume 1: Domestic and Industrial Wastewater Treatment, 145–162. Springer, https://doi.org/10.1007/978-3-030-13913-1_8

Hao, X., Zhu, P., Liu, X., Jiang, L., Jiang, H., Liu, H., & Chen, Z. (2024). Bioreactor Expansion Affects Microbial Succession of Mixotrophic Acidophiles and Bioremediation of Cadmium-Contaminated Soils, Toxics, 12(5), 362. https://doi.org/10.3390/toxics12050362

Hegde, S., & Selvaraj, S. (2024). Chitosan: An in-depth analysis of its extraction, applications, constraints, and future prospects, Journal of microbiology, biotechnology and food sciences, 14(2), e10563–e10563.

Hellal, M. S., Ziembińska-Buczyńska, A., El-Liethy, M. A., & Surmacz-Górska, J. (2025). The biological remediation of water and wastewaters using different Treatment techniques. In *Eco-restoration of polluted Environment* (pp. 129–160). CRC Press.

Hussain, S., Khan, M., Sheikh, T. M. M., Mumtaz, M. Z., Chohan, T. A., Shamim, S., & Liu, Y. (2022). Zinc essentiality, toxicity, and its bacterial bioremediation: A comprehensive insight, Frontiers in microbiology, 13, 900740. https://doi.org/10.3389/fmicb.2022.900740

Ihsanullah, I., Jamal, A., Ilyas, M., Zubair, M., Khan, G., & Atieh, M. A. (2020). Bioremediation of dyes: Current status and prospects, Journal of Water Process Engineering, 38, 101680. https://doi.org/10.1016/j.jwpe.2020.101680

Jeyakumar, P., Debnath, C., Vijayaraghavan, R., & Muthuraj, M. (2023). Trends in bioremediation of heavy metal contaminations, Environmental Engineering Research, 28(4), 220631.

Johnson, P. C., Johnson, R. L., Bruce, C. L., & Leeson, A. (2001). Advances in in situ air sparging/biosparging, Bioremediation Journal, 5(4), 251–266. https://doi.org/10.1080/20018891079311

Kamaludeen, S., Arunkumar, K., Avudainayagam, S., & Ramasamy, K. (2003). Bioremediation of chromium contaminated environments, Indian journal of experimental biology, 41(9), 972–985.

Khan, A., et al. (2024). The sustainable approach of microbial bioremediation of arsenic: an updated overview, International Journal of Environmental Science and Technology, 21(11), 7849–7864. https://doi.org/10.1007/s13762-024-05594-9

Kour, D., et al. (2022). Microbe-mediated bioremediation: Current research and future challenges, J Appl Biol Biotechnol, 10(2), 6–24.
Kumar, A., & Chandra, R. (2022). Extremophilic nature of microbial ligninolytic enzymes and their role in biodegradation. In *Microbial Biodegradation and bioremediation* (pp. 571–589). Elsevier.
Kumar, K., & Singh, D. (2024). Toxicity and bioremediation of the lead: a critical review, International Journal of Environmental Health Research, 34(4), 1879–1909. https://doi.org/10.1080/09603123.2023.2165047
Kumar, A., Kumar, V., Chawla, M., Thakur, M., Bhardwaj, R., Wang, J., et al. (2024). Bioremediation of mercury contaminated soil and water: A review, Land Degradation & Development, 35(4), 1261–1283. https://doi.org/10.1002/ldr.4989
Kushwaha, S., & Sudhakar, P. (2013). Sorption mechanism of Cd (II) and Zn (II) onto modified palm shell, Adsorption Science & Technology, 31(6), 503–519.
Lata, S., Kaur, H. P., & Mishra, T. (2019). Cadmium bioremediation: A review, Int. J. Pharm. Sci. Res, 10(9), 4120–4128.
Ławniczak, Ł., Woźniak-Karczewska, M., Loibner, A. P., Heipieper, H. J., & Chrzanowski, Ł. (2020). Microbial degradation of hydrocarbons—basic principles for bioremediation: A review, Molecules, 25(4), 856. https://doi.org/10.3390/molecules25040856
Lee, M. D., & Swindoll, C. M. (1993). Bioventing for in situ remediation, Hydrological sciences journal, 38(4), 273–282.
Maitra, S. (2018). In situ bioremediation—An overview. Responsibility of Life Science Informatics Publications, 4(6), 576-598.
Marzan, L. W., Hossain, M., Mina, S. A., Akter, Y., & Chowdhury, A. M. A. (2017). Isolation and biochemical characterization of heavy-metal resistant bacteria from tannery effluent in Chittagong city, Bangladesh, bioremediation viewpoint, Bioremediation viewpoint. Egyptian Journal of Aquatic Research, 43(1), 65–74. https://doi.org/10.1016/j.ejar.2016.11.002
Mathew, M. M., Khatana, K., Vats, V., Dhanker, R., Kumar, R., Dahms, H.-U., & Hwang, J.-S. (2022). Biological approaches integrating algae and bacteria for the degradation of wastewater contaminants—A review, Frontiers in Microbiology, 12, 801051. https://doi.org/10.3389/fmicb.2021.801051
Medfu Tarekegn, M., Zewdu Salilih, F., & Ishetu, A. I. (2020). Microbes used as a tool for bioremediation of heavy metal from the environment, Cogent Food & Agriculture, 6(1), 1783174. https://doi.org/10.1080/23311932.2020.1783174
Mishra, B., Varjani, S., Kumar, G., Awasthi, M. K., Awasthi, S. K., Sindhu, R., et al. (2021). Microbial approaches for remediation of pollutants: innovations, future outlook, and challenges, Energy & Environment, 32(6), 1029–1058.
Mishra, S., Patel, A., Bhatt, P., Chen, S., & Srivastava, P. K. (2024). Perspective evaluation of synthetic biology approaches for effective mitigation of heavy metal pollution, Reviews of Environmental Contamination and Toxicology, 262(1), 21. https://doi.org/10.1007/s44169-024-00072-2
Mitra, A., Chatterjee, S., & Gupta, D. K. (2017). Potential role of microbes in bioremediation of arsenic. Arsenic contamination in the environment: The issues and solutions, 195–213. Springer, Cham. https://doi.org/10.1007/978-3-319-54356-7_10
Mohapatra, B., & Phale, P. S. (2021). Microbial degradation of naphthalene and substituted naphthalenes: metabolic diversity and genomic insight for bioremediation, Frontiers in Bioengineering and Biotechnology, 9, 602445. https://doi.org/10.3389/fbioe.2021.602445
Mondal, S., & Palit, D. (2019). Effective role of microorganism in waste management and environmental sustainability. Sustainable Agriculture, Forest and Environmental Management. 485–515. Springer, Singapore. https://doi.org/10.1007/978-981-13-6830-1_14
Morris, S., Garcia-Cabellos, G., Enright, D., Ryan, D., & Enright, A.-M. (2018). Bioremediation of landfill leachate using isolated bacterial strains, International Journal of Environmental Bioremediation & Biodegradation, 6(1), 26–35.
Mortazavi Mehrizi, M., Yousefinejad, S., Jafari, S., Baghapour, M. A., Karimi, A., Mahvi, A. H., & Jahangiri, M. (2022). Bioremediation and microbial degradation of benzo [a] pyrene in aquatic environments: A systematic review, International Journal of Environmental Analytical Chemistry, 102(15), 3508–3523. https://doi.org/10.1080/03067319.2020.1770743

Mousavi, S. M., Hashemi, S. A., Iman Moezzi, S. M., Ravan, N., Gholami, A., Lai, C. W., et al. (2021). Recent advances in enzymes for the bioremediation of pollutants, Biochemistry research international, 2021(1), 5599204. https://doi.org/10.1155/2021/5599204

Nwankwegu, A. S., & Onwosi, C. O. (2017). Microbial cell immobilization: a renaissance to bioaugmentation inadequacies—A review, Environmental Technology Reviews, 6(1), 186–198. https://doi.org/10.1080/21622515.2017.1356877

Olaniran, A. O., Balgobind, A., & Pillay, B. (2013). Bioavailability of heavy metals in soil: impact on microbial biodegradation of organic compounds and possible improvement strategies, International journal of molecular sciences, 14(5), 10197–10228. https://doi.org/10.3390/ijms140510197

Olaya-Abril, A., Biełło, K., Rodríguez-Caballero, G., Cabello, P., Sáez, L. P., Moreno-Vivián, C., et al. (2024). Bacterial tolerance and detoxification of cyanide, arsenic and heavy metals: Holistic approaches applied to bioremediation of industrial complex wastes, Microbial Biotechnology, 17(1), e14399. https://doi.org/10.1111/1751-7915.14399

Ostrem Loss, E. M., & Yu, J. H. (2018). Bioremediation and microbial metabolism of benzo (a) pyrene, Molecular microbiology, 109(4), 433–444. https://doi.org/10.1111/mmi.14062

Pal, A. K., Singh, J., Soni, R., Tripathi, P., Kamle, M., Tripathi, V., & Kumar, P. (2020). The role of microorganism in bioremediation for sustainable environment management. In *Bioremediation of pollutants* (pp. 227–249). Elsevier.

Pavel, L. V., & Gavrilescu, M. (2008). Overview of ex situ decontamination techniques for soil cleanup, Environmental Engineering & Management Journal (EEMJ), 7(6), 815.

Pratap, J. & K., Krishnan, K. (2024). Microbial nanobioremediation of micro-nanoplastics: Current strategies, challenges, and future prospects. Management of Micro and Nano-plastics in Soil and Biosolids, 419–445. Springer, Cham. https://doi.org/10.1007/978-3-031-51967-3_17

Priya, A., Gnanasekaran, L., Dutta, K., Rajendran, S., Balakrishnan, D., & Soto-Moscoso, M. (2022). Biosorption of heavy metals by microorganisms: Evaluation of different underlying mechanisms, Chemosphere, 307, 135957. https://doi.org/10.1016/j.chemosphere.2022.135957

Rahman, Z., & Singh, V. P. (2020). Bioremediation of toxic heavy metals (THMs) contaminated sites: Concepts, applications and challenges, Environmental Science and Pollution Research, 27(22), 27563–27581. https://doi.org/10.1007/s11356-020-08903-0

Ramdass, A. C., & Rampersad, S. N. (2021). Biodiversity and biocatalyst activity of culturable hydrocarbonoclastic fungi isolated from Marac–Moruga mud volcano in South Trinidad, Scientific Reports, 11(1), 19466. https://doi.org/10.1038/s41598-021-98979-6

Rath, B. (2012). Microalgal bioremediation: current practices and perspectives, Journal of Biochemical Technology, 3(3), 299–304.

Rezaei Somee, M., Shavandi, M., Dastgheib, S. M. M., & Amoozegar, M. A. (2018). Bioremediation of oil-based drill cuttings by a halophilic consortium isolated from oil-contaminated saline soil, 3 Biotech, 8, 1–13. https://doi.org/10.1007/s13205-018-1261-8

Sable, H., Singh, V., Kumar, V., Roy, A., Pandit, S., Kaur, K., & Malik, S. (2024). Toxicological and bioremediation profiling of nonessential heavy metals (mercury, chromium, cadmium, aluminium) and their impact on human health: A review. Toxicologie Analytique et Clinique, 36(3), 205-234. https://doi.org/10.1016/j.toxac.2024.03.096

Salem, D. M., Moawad, M. N., & El-Sayed, A. A. (2021). Comparative study for bioremediation of cobalt contaminated aqueous solutions by two types of marine macroalgae, Egyptian Journal of Aquatic Research, 47(1), 13–19. https://doi.org/10.1016/j.ejar.2021.02.002

Saranya, S., Thamanna, L., Sreekutty, V., Dhayanithi, S., & Chellapandi, P. (2024). Bioremediation of oil and natural gas industry waste using methanogens: Current status and future perspective to biohythane production. Arabian Journal for Science and Engineering, 1–19. https://doi.org/10.1007/s13369-024-09389-5

Saravanan, A., Kumar, P. S., Vo, D.-V. N., Jeevanantham, S., Karishma, S., & Yaashikaa, P. (2021). A review on catalytic-enzyme degradation of toxic environmental pollutants: Microbial enzymes, Journal of Hazardous Materials, 419, 126451. https://doi.org/10.1016/j.jhazmat.2021.126451

Sekhar, M., Christie, S., Trivedi, A., Agrawal, S., & Jejal, A. (2024). Bioremediation: For sustainable environment management. The Agriculture Magazine, 3(3), 124-129.

Sethi, S., Bharshankh, A., Sahu, R., & Biswas, R. (2024). Nano-bioremediation for Sustainable Treatment of waste: Applications, advancements, and challenges. In *Environmental Engineering and waste management: Recent trends and perspectives* (pp. 539–575). Springer.

Setiadi, T., Harimawan, A., Sumampouw, G. A., & Indarto, A. (2022). The mixture of agricultural pesticides and their impact on populations: Bioremediation strategies. In *Emerging contaminants in the environment* (pp. 511–546). Elsevier.

Sevak, P. I., Pushkar, B. K., & Kapadne, P. N. (2021). Lead pollution and bacterial bioremediation: A review, Environmental Chemistry Letters, 19(6), 4463–4488. https://doi.org/10.1007/s10311-021-01296-7

Sharma, S., Malik, A., Satya, S., & Mishra, A. (2011). Development of a biological system employing Aspergillus lentulus for Cr removal from a small-scale electroplating industry effluent, Asia-Pacific Journal of Chemical Engineering, 6(1), 55–63. https://doi.org/10.1002/apj.449

Sharma, P., Kumar, S., & Pandey, A. (2021). Bioremediated techniques for remediation of metal pollutants using metagenomics approaches: a review, Journal of Environmental Chemical Engineering, 9(4), 105684. https://doi.org/10.1016/j.jece.2021.105684

Sharma, R., Nath, P. C., Debnath, S., Ojha, A., Sarkar, B., Bandyopadhyay, T. K., & Bhunia, B. (2024). Exopolysaccharides for bioremediation. In Microbial exopolysaccharides (pp. 275–303). CRC Press.

Shukla, H., Gautam, S., Bapodariya, H., Maliwad, M. B., Gangawane, A. K., & Kaushal, R. S. (2022). Bioremediation of waste water: In-depth review on current practices and promising perspectives. Journal of Aquaculture in the Tropics, 37(1–4), 117–152.

Song, Y.-J., Zhao, N.-L., Dai, D.-R., & Bao, R. (2024). Prospects of pseudomonas in microbial fuel, bioremediation, and sustainability. ChemSusChem, 18(2), e202401324. https://doi.org/10.1002/cssc.202401324

Srinivasulu, M., Chandra, M. S., Maddela, N. R., Golla, N., & Ramesh, B. (2022). Recent trends in bioremediation of pollutants by enzymatic approaches. In *Cost effective Technologies for Solid Waste and Wastewater Treatment* (pp. 115–134). Elsevier.

Srivastava, J., Naraian, R., Kalra, S., & Chandra, H. (2014). Advances in microbial bioremediation and the factors influencing the process, International Journal of environmental science and technology, 11(6), 1787–1800. https://doi.org/10.1007/s13762-013-0412-z

Talha, M. A., Goswami, M., Giri, B., Sharma, A., Rai, B., & Singh, R. (2018). Bioremediation of Congo red dye in immobilized batch and continuous packed bed bioreactor by Brevibacillus parabrevis using coconut shell bio-char, Bioresource technology, 252, 37–43. https://doi.org/10.1016/j.biortech.2017.12.081

Tambat, V. S., Tseng, Y.-S., Kumar, P., Chen, C.-W., Singhania, R. R., Chang, J.-S., et al. (2023). Effective and sustainable bioremediation of molybdenum pollutants from wastewaters by potential microalgae, Environmental Technology & Innovation, 30, 103091. https://doi.org/10.1016/j.eti.2023.103091

Tang, H., Xiang, G., Xiao, W., Yang, Z., & Zhao, B. (2024). Microbial mediated remediation of heavy metals toxicity: mechanisms and future prospects, Frontiers in Plant Science, 15, 1420408. https://doi.org/10.3389/fpls.2024.1420408

Taran, M., Fateh, R., Rezaei, S., & Gholi, M. K. (2019). Isolation of arsenic accumulating bacteria from garbage leachates for possible application in bioremediation, Iranian Journal of Microbiology, 11(1), 60.

Thakare, M., Sarma, H., Datar, S., Roy, A., Pawar, P., Gupta, K., et al. (2021). Understanding the holistic approach to plant-microbe remediation technologies for removing heavy metals and radionuclides from soil, Current research in biotechnology, 3, 84–98. https://doi.org/10.1016/j.crbiot.2021.02.004

Tiquia-Arashiro, S. M. (2018). Lead absorption mechanisms in bacteria as strategies for lead bioremediation. Applied microbiology and biotechnology, 102(13), 5437–5444. https://doi.org/10.1007/s00253-018-8969-6

Touliabah, H. E.-S., El-Sheekh, M. M., Ismail, M. M., & El-Kassas, H. (2022). A review of microalgae-and cyanobacteria-based biodegradation of organic pollutants, Molecules, 27(3), 1141. https://doi.org/10.3390/molecules27031141

Tuovinen, O. J. (1972). Microbiological aspects in the leaching of uranium by Thiobacillus ferrooxidans, Atomic Energy Review, 10(2), 251–258

Tyagi, M., da Fonseca, M. M. R., & de Carvalho, C. C. (2011). Bioaugmentation and biostimulation strategies to improve the effectiveness of bioremediation processes, Biodegradation, 22, 231–241. https://doi.org/10.1007/s10532-010-9394-4

Ugwu, E. C. (2019). Remediation of contaminated soil in copper mining areas using plant-based surfactants. (Doctoral dissertation) Heriot-Watt University.

Wang, W.-X. (2016). Bioaccumulation and biomonitoring. In *Marine ecotoxicology* (pp. 99–119). Elsevier.

Wang, J., & Chen, C. (2006). Biosorption of heavy metals by Saccharomyces cerevisiae: A review, Biotechnology advances, 24(5), 427–451. https://doi.org/10.1016/j.biotechadv.2006.03.001

Wang, Y., Ning, X., Liang, J., Wang, A., & Qu, J. (2025). Enhancing microbial superoxide generation and conversion to hydroxyl radicals for enhanced bioremediation using iron-binding ligands, Journal of Environmental Sciences, 147, 597–606. https://doi.org/10.1016/j.jes.2023.11.023

Wanyonyi, W. C., Onyari, J. M., Shiundu, P. M., & Mulaa, F. J. (2017). Biodegradation and detoxification of malachite green dye using novel enzymes from bacillus cereus strain KM201428: kinetic and metabolite analysis, Energy Procedia, 119, 38–51. https://doi.org/10.1016/j.egypro.2017.07.044

Whiteley, A. S., & Bailey, M. J. (2000). Bacterial community structure and physiological state within an industrial phenol bioremediation system, Applied and environmental microbiology, 66(6), 2400–2407.

Wróbel, M., Śliwakowski, W., Kowalczyk, P., Kramkowski, K., Dobrzyński, J. (2023). Bioremediation of heavy metals by the genus Bacillus, International Journal of Environmental Research and Public Health, 20(6), 4964. https://doi.org/10.3390/ijerph20064964

Yakasai, H. M., Rahman, M. F., Manogaran, M., Yasid, N. A., Syed, M. A., Shamaan, N. A., et al. (2021). Microbiological reduction of molybdenum to molybdenum blue as a sustainable remediation tool for molybdenum: A comprehensive review, International Journal of Environmental Research and Public Health, 18(11), 5731. https://doi.org/10.3390/ijerph18115731

Yu, G., Ullah, H., Yousaf, B., Pikoń, K., Antoniadis, V., Prasad, M. N. V., et al. (2024). Microbe-assisted phytoremediation of toxic elements in soils: Present knowledge and future prospects, Earth-Science Reviews, 255, 104854. https://doi.org/10.1016/j.earscirev.2024.104854

Zhu, Y., Dong, B., Zhang, H., Liu, Z., Zou, H., Li, S., et al. (2024). Intermittent anaerobic-aerobic biopiling for remediating energetic compound-contaminated soils, Environmental Technology & Innovation, 35, 103728. https://doi.org/10.1016/j.eti.2024.103728

Harnessing Microbial Magic: Bioremediation for a Greener Future

Suman Kumari, Aditi Dhama, and Subhadra Rajpoot

Abstract The expansion of cities and industrialization are the major reasons for environmental pollution, due to the harmful chemicals emitted by industries and urban cities into the atmosphere. Air, water, soil, and other forms of pollution have increased as a result of the expanding urbanization and industrialization tendencies. Pesticides, fertilizers, and pharmaceuticals released by mining and metallurgy, textile and dyeing, pulp and paper, waste management and landfills, and electrical device manufacturers are the main sources of industrial pollutants that pollute the environment. These businesses damage the ecosystem and cause serious health problems for people by releasing hazardous chemicals, waste products, and heavy metals into the air, water, and land. The health issues increase due to bad air quality, spoiled drinking water, and soil pollution. A preventive method to overcome this environmental pollution, a biological technique known as "bioremediation," has come into existence. The technique involves microorganisms that perform degradation and transformation, detoxification, and immobilization of pollutants in soil, water, and other growth media. Microorganisms are the primary and significant agents in the process of bioremediation due to their adapting ability under any environmental circumstances and various metabolic capabilities that degrade different pollutants. Bioremediation involves the most relevant technique known as degradation. Different microbes develop different mechanisms for degradation through biosorption, biotransformation, and bioaccumulation, for bioremediation based on the particular pollutant degradation. The success rate of bioremediation is based on the type of microorganisms employed in the degradation process, the nature of the contaminant and its concentration, and the prevalent environmental circumstances. This chapter gives a definite explanation of the function of microorganisms in bioreme-

S. Kumari (✉)
CDOE, Vivekanand Global University, Jaipur, Rajasthan, India

A. Dhama · S. Rajpoot
Amity University, Greater Noida Campus, Greater Noida, Uttar Pradesh, India

S. Rajpoot, A. P. Das (eds.), *Innovative Technologies for Waste Management*, Sustainable Environmental Waste Management Strategies,
https://doi.org/10.1007/978-3-031-97184-6_5

diation, assessing their significance in various types of pollutant degradation and transformation processes—inorganic or organic—which occur when heavy metals are present.

Keywords Bioremediation · Microorganisms · Environmental pollution · Pollutants · Bioaugmentation · Bioattenuation · Biostimulation

1 Introduction

The international waste problem is essentially at the top of environmental degradation, as over 2 billion tons of municipal solid wastes is generated annually and poor disposal along with nonworking waste management systems has fueled heavy air, water, and soil pollution partly in developing countries (World Bank Group, 2018). Hazardous chemicals, plastics, and untreated industrial waste all accumulate due to rapid urbanization, industrialization, and population growth (Dey et al., 2024b; Priyadarshini et al., 2024). These factors significantly endanger the health of ecosystems and critically jeopardize human health (Kaza et al., 2018). Perhaps the most important issue is methane from inadequately managed organic waste—the potent greenhouse gas that strongly contributes to global warming (UNEP, 2018). In addition, microplastics from uncontrolled waste streams are increasingly found as contaminants of the aquatic system, causing damage to marine life and entering the food chain, thereby affecting terrestrial and aquatic biodiversity (Singh et al., 2016; Mishra et al., 2021; Sahu & Das, 2024). Thus, in such a scenario, bioremediation, which refers to the biocatalysis that living microorganisms use to detoxify and restore polluted environments, comes into perspective as an eco-friendly measure for a manageable solution of the pollution menace (Singh et al., 2016; Das et al., 2015; Paul et al., 2023). Indeed, bioremediation—a technique describing the process of using microorganisms to degrade and neutralize environmental pollutants—has a long history dating back nearly to the very beginning of the past century (Das et al., 2024; Pradhan et al., 2023). Bioremediation was formalized as a method of using microorganisms to break down environmental pollutants in the 1960s by George M. Robinson. The initial vision for bioremediation was stirred from the first revelation that researchers made about the capabilities of microbes in breaking down organic matter into more readily reusable forms. Developments in molecular biology and microbial ecology have made bioremediation strategies more effective: it is now feasible to genetically modify the microbes, or apply a specific combination of different species of microbes, that leads to more effective degradation of pollutants (Singh et al., 2016). Presently, bioremediation is well regarded as an essential part of the practice of sustainable waste management because it began from less environmentally friendly remediation technologies that require energy to modern forms of energy-efficient and environmentally friendly options. As of the present, bioremediation is an evolving technology; emerging research now focuses on optimizing microbial activity and its scale-up application into even broader spectrums of environmental contaminants.

2 Types of Microorganisms in Bioremediation

Microorganisms are involved in bioremediation through the metabolism and degradation of pollutants that will revive degraded habitats. Some groups of microorganisms that uniquely contribute to the field include bacteria, fungi, archaea, algae, and actinomycetes.

2.1 *Bacteria-Bacteria*

Bacteria-bacteria are the most used microorganisms in bioremediation because they are characterized by a high metabolic diversity and easily adapt to situations (Nag et al., 2023). They can be categorized into two broad types: aerobic and anaerobic bacteria.

(a) Aerobic bacteria: These bacteria utilize oxygen to break down pollutants. *Pseudomonas putida* is recognized for the hydrocarbon-degrading ability and is vastly used in bioremediation of oil spills. There have also been observations of *Acinetobacter* species for its capacity to degrade oils and hydrocarbons, especially in marine environments.
(b) Anaerobic bacteria: These bacteria thrive well in the absence of oxygen, and some anaerobes are specifically used in the degradation of chlorinated compounds. *Dehalococcoides* species are highly useful in the process of biodegradation of TCE along with other chlorinated solvents by breaking down into less harmful compounds. Sulfate-reducing bacteria like *Desulfovibrio* spp. reduce heavy metals into their insoluble forms and hence detoxify contaminated areas.

2.2 *Fungi*

Fungi are always effective in the biodegradation of complex organic contaminants due to lignin breakdown, however at a much slower rate than white-rot and brown-rot species.

(a) White-rot fungi: *Phanerochaete chrysosporium* represents one of the most extensively studied fungi due to their ability to degrade recalcitrant pollutants, such as polycyclic aromatic hydrocarbons (PAH) and pesticides through ligninolytic enzymes. A good example of a white-rot fungus is the oyster mushroom, *Pleurotus ostreatus*, reported to degrade PAHs and synthetic dyes, which have various applications in industrial bioremediation.

(b) Brown-rot fungi: Though primarily cellulose degraders, brown-rot fungi like *Gloeophyllum trabeum* can play a role in bioremediation due to their ability to degrade cellulose, thereby rendering complex organic pollutants more bioavailable to other microbes.

2.3 Archaea

Archaea are extremophiles that survive in extreme conditions, such as high salt concentrations or temperature, and thus dominate the environment—ecologically speaking—in anaerobic and saline ecosystems.

(a) Methanogenic archaea: Methanogens, including species of *Methanosarcina*, are facultative anaerobic archaea that break down organic pollutants into methane. These microorganisms are used extensively in the processes of anaerobic digestion for wastewater and organic waste treatment.
(b) Halophilic archaea: *Halobacterium salinarum* species may be found to degrade hydrocarbons in the very saline oil fields contaminated with petroleum. This is an ideal microorganism for remediation of oil spills in the highly saline oil fields.

2.4 Algae

Algae, more precisely microalgae and cyanobacteria, are largely involved in bioremediation because of their high ability to absorb pollutants as well as nutrients, mainly within aquatic environments.

(a) Microalgae: *Chlorella vulgaris* is highly efficient for the removal of heavy metals like cadmium, lead, and mercury that exist in water and converts such contaminated water bodies to become detoxified. Another genus, *Scenedesmus*, removes nitrates and phosphates that prevent the process of eutrophication in the aquatic ecosystems.
(b) Cyanobacteria: Cyanobacteria like *Anabaena* possess nitrogen-fixation ability and exploit nutrients in water to their maximum extent, thereby aiding in upgrading quality of the water present in polluted water bodies. *Spirulina* is another cyanobacterium which has the capacity to absorb heavy metals and other toxic contaminants that are present in wastewater.

2.5 *Actinomycetes*

Actinomycetes are gram-positive bacteria that produce extracellular enzymes capable of breaking down complex organic chemicals. They are involved in the fragmentation of pollutants like pesticides and hydrocarbons.

(a) *Streptomyces*: This genus of actinomycetes, in addition to the production of antibiotics, is also a degrading organism of organic pollutants, hydrocarbons, and pesticides and thus the selected organism for soil bioremediation.
(b) *Nocardia*: The other species of the genus *Nocardia* is used in industrial waste treatment processes because it degrades oil and many other organic compounds in contaminated soil and water.

3 Types of Bioremediations

In situ bioremediation is a type of bioremediation that treats toxins directly at the source of pollution rather than digging the soil or pumping out groundwater (Dey et al., 2024a). This method uses natural microbial processes to break down or neutralize harmful compounds in the environment while minimizing disturbance to the affected area. In situ, bioremediation is regarded more ecologically benign and cost-effective than other remediation procedures since it works inside the natural ecosystem to detoxify the site.

3.1 *Natural Attenuation (Intrinsic Bioremediation)*

Natural attenuation, or intrinsic bioremediation, relies on naturally occurring microbial populations to degrade contaminants without any human intervention. In this method, microorganisms utilize contaminants as a source of energy and nutrients, converting them into less harmful products, typically carbon dioxide, water, and biomass. This process is passive and occurs over a prolonged period.

3.2 *Enhanced In Situ Bioremediation*

Enhanced in situ bioremediation is a human intervention that improves the effectiveness of the natural degradation process by enhancing ambient conditions for microbial activity (Das et al., 2023). This can be accomplished by the addition of nutrients, electron acceptors, or other substrates that promote microbial growth.

- **Biostimulation**: Biostimulation is one of the most important bioremediation techniques, which encourages autochthonous microorganisms to grow and promote their metabolism to increase the breakdown of pollutants. It involves adding nutrients, electron acceptors, or any growth-promoting agent to the polluted environments. Biostimulation is effective when the native populations have the required catabolic capacities but are lacking in sufficient resources for maximum utilization. Biostimulation that is common involves nitrogen and phosphorus compounds, oxygen, and organic substrates. The technique has been applied with some promising results in remediation of different types of contaminants, such as petroleum hydrocarbons, heavy metals, and chlorinated compounds. Success is however dependent on several factors like soil properties, contaminant bioavailability, and environmental conditions. Optimizing the stimulation application by proper monitoring and adjustment would enhance this approach and provide significant effectiveness without negative impacts on the ecosystem.
- **Bioaugmentation**: Bioaugmentation is a promising bioremediation technique where targeted microorganisms or enzymes are added to enhance degradation in a contaminated environment. This approach overcomes the weakness of natural microbial populations since exogenous, specialized strains with superior degradative performance can be added. So far, this technology has been found to work for contaminants like petroleum hydrocarbons, chlorinated solvents, and pesticides. Bioaugmentation success depends on the survival of the inoculated microorganisms, activity, environment, and presence of pollutants. The recent discovery in biotechnology with respect to genetic engineering, like GMOs with enhanced degradative capabilities, opens the opportunity for this kind of application, but that has very significant ecological and regulatory implications. Despite such challenges in the application of bioaugmentation, it still forms a significant practice tool in environmental remediation, especially if combined with other bioremediation practices, like biostimulation. Present research activities are emphasizing optimizing the microbial selection, new inoculation techniques, and sustaining the survival of incorporated organisms over the long term under varied environmental conditions.

3.3 Engineered In Situ Bioremediation Techniques

Several engineered bioremediation methods are applied directly in the subsurface environment to stimulate and control microbial degradation. These techniques include:

- **Bioventing**: Bioventing involves circulating a controlled amount of air or oxygen into the unsaturated zone of soil so as to hasten the growth and activities of natural aerobic microorganisms. After being provided with nutrients and oxygen, these microorganisms can metabolize and mineralize a wide range of organic contaminants that include petroleum hydrocarbons, chlorinated solvents, and other recalcitrant compounds. Thus, improved microbial activity enhances the natural biodegradation process, and contaminant concentrations in the soil will subsequently become acceptable levels. The other benefits of bioventing over the traditional remedy techniques include cost-effectiveness since it does not have to undergo soil excavation and disposal or energy-intensive thermal or chemical treatments. More, bioventing is a nondestructive technology that causes the slightest disturbance to the soil and its microenvironment, with a sense of more environmentally friendly and sustainable remediation of contaminated sites.

- **Biosparging**: Biosparging is a similar technique where air is injected into the saturated zone of contaminated groundwater. The injected air increases the oxygen concentration, enhancing microbial activity and enabling aerobic degradation of contaminants such as gasoline and diesel.
- **Permeable reactive barriers (PRBs)**: PRBs are barriers installed underground that contain reactive materials, such as iron or organic substrates, which facilitate microbial degradation of contaminants as groundwater flows through the barrier. These barriers are particularly effective for treating groundwater contaminated with metals, chlorinated solvents, and other persistent pollutants.

4 Factors Affecting Microbial Bioremediation

Several factors influence the efficacy and efficiency of microbial bioremediation. These can be further divided into environmental restrictions, complexity of pollutants, concentration levels of toxicity, environmental parameters, availability of nutrients, and properties of contaminants (Bhattacharjee et al., 2021). All these factors either support or oppose microbial degradation of these contaminants. It is highly imperative to have adequate information about their activities so that bioremediation strategies can be maximized.

4.1 Environmental Constraints

Such environmental parameters as pH, temperature, and oxygen availability are main indicators of governing microbial bioremediation in terms of its success. Microorganisms are sensitive to these changes; for instance, pH, temperature, and oxygen conditions influence their microbial metabolism, growth, and enzyme activities.

pH: Most of the microorganisms that take part in biodegradation work well at near-neutral pH ranges of 6–8. In extreme pHs, they fail to be well denatured or the cellular system will be disrupted, thus affecting the metabolic actions of the microbes. For instance, some bacterial genera, which include *Pseudomonas* and *Bacillus*, have been reported to have optimal biodegradation activities at near-neutral pH values. But the fungal genera, such as *Aspergillus* and *Penicillium*, tolerate slightly acidic pH values (Chapman & Wang 2011; Ghosh et al. 2011).

Temperature: The basic process of microbial metabolism is temperature-responsive and enzymatic reactions also vary with temperature. Most bioremediating microorganisms reportedly are active between 20 and 35 °C. Lower temperature cannot be avoided, as a result of which the microbial activity reduces. A very high temperature kills microbes due to the denaturation of proteins and the cellular membrane. For instance, the Geobacter species which are associated with the anaerobic biodegradation of hydrocarbons degrade effectively within moderate temperatures but become inhibited at either low or high extreme values.

Availability of Oxygen: The availability of oxygen determines the degree of dominance by either aerobic or anaerobic processes of degradation. Under aerobic conditions, microorganisms utilize oxygen as a terminal electron acceptor, thereby biodegrading contaminants such as hydrocarbons rapidly. Organisms rely on alternative electron acceptors, such as nitrates, sulfates, or ferric ions, in oxygen-limited or anaerobic conditions to degrade the pollutants, and this process results in a slow rate of degradation. For example, oil-related hydrocarbons degrade quickly through aerobic bacteria like *Pseudomonas* spp. On the other hand, anoxic condition necessitates the presence of anaerobic microorganisms such as *Dehalococcoides* that break down chlorinated solvents.

4.2 Diversity of Pollutants

The chemical structure and complexity of the pollutant will be determined by whether it is easy or very difficult to degrade by microorganisms. Simple organic compounds are easily biodegradable, while complex recalcitrant compounds create insurmountable challenges for microbial communities.

Complex and Resistant Contaminants These are highly complex-structured molecules, high-molecular-weight PAHs, chlorinated solvents, and heavy metals. These compounds resist degradation by microorganisms. For example, the PAHs are lipophilic and therefore stable in terms of forming molecular rings; this makes their bioavailability low to microorganisms. On the other hand, chlorinated solvents such as TCE require unique dechlorination pathways, involving some unique microbes specific to reduce them under anaerobic conditions. In most situations, it is a community of different microbial species with complementary metabolic capabilities that cause such complex pollutants' breakdown.

4.3 *Toxicity Levels*

The toxic nature of pollutants may directly affect microbial survival and activity. A high concentration of toxic contaminants, including heavy metals, pesticides, or aromatic hydrocarbons, can inhibit the growth and enzymatic function of microorganisms, thus limiting the effectiveness of bioremediation.

Heavy Metals: These comprise mercury (Hg), cadmium (Cd), lead (Pb), arsenic (As), and many others, and most microbes are inhibited as they affect the essential biochemical pathways, disrupt cell membranes, and inactivate enzymes by binding to their active sites. However, some microbial species have mechanisms that developed to prevent this toxicity caused due to efflux pumps that extrude the heavy metals from the cellular environment and enzymatic processes of detoxification (Das & Mishra, 2010; Mishra et al., 2022). For instance, *Ralstonia metallidurans* and *Pseudomonas aeruginosa* contain metal resistance genes that give them capabilities to outlive under contaminated environments.

Organic Pollutants: High levels of organic pollutants are also toxic to microorganisms. For example, the presence of high levels of BTEX compounds (benzene, toluene, ethylbenzene, and xylene) can disrupt membranes and inhibit microbial metabolism. While these compounds may be metabolized using species-specific oxygenases and dehydrogenases, in many instances, especially where concentration is high, degradation processes may take longer since such concentration is toxic (Das & Chandran, 2011).

4.4 *Environmental Factors*

Other environmental factors that involve the microbial bioremediation process include moisture content salinity and redox potential.

Moisture Content: Sufficient moisture will create an environment that favors microbial growth and the migration of contaminants and nutrients. Low moisture in soil ecosystems can inhibit microbial functions because pollutants become immobilized. High moisture results in anoxia and the subsequent development of anaerobic conditions. High salinity in marine ecosystems can decrease the microbial population and degrade the hydrocarbon breakdown rate.

Redox Potential: The redox potential (Eh) determines the availability of electron acceptors, which become available to microorganisms for use in metabolism. Under aerobic conditions where high redox potentials prevail, microorganisms can accept oxygen directly as an electron acceptor to increase the oxidation rate of the contaminants. Under anaerobic conditions with low redox potential, supplementary electron acceptors such as nitrates, sulfates, or ferric ions are accepted by microorganisms in slower degradation processes.

4.5 Nutrient Availability

Nitrogen, phosphorus, and trace elements are thus some of the most required nutrients for the growth of microbes that enable bioremediation (Braguglia et al., 2018). In most polluted environments, the primary limiting factor in microbial activity is the availability of necessary nutrients, sometimes which can be replenished by adding fertilizers or nutrient amendments.

Nitrogen and Phosphorus: These nutrients serve as a fuel in several cellular functions, including DNA replication, protein synthesis, and energy metabolism. Nitrogen- or phosphorus-rich compounds can contribute significantly to bioremediation in nutrient-poor environments through biostimulation. While the overuse of nutrients supports algal bloom or eutrophication, which is to the detriment of the ecosystem.

Trace Elements: Trace elements such as iron, magnesium, and zinc are also required in trace amounts for functional microbial enzymes related to bioremediation pathways. In some cases, the degradation process may be slowed down or halted without such trace elements.

4.6 Properties of Contaminants

The physical and chemical properties of contaminants, mainly those related to solubility, volatility, and sorption to soil particles, affect their bioavailability and hence biodegradability.

Solubility: The more freely water-soluble pollutants, including benzene and phenols, are highly available to microbial populations for efficient degradation. Conversely, hydrophobic pollutants that include high molecular PAHs have the propensity to adsorb in soil organic matter, thus reducing their bioavailability and increasing their difficulty in degrading.

Volatility: Compounds, for example, volatile organic compounds including toluene and trichloroethylene, are volatile and thus can evaporate from the soil or water surface ahead of time before being attacked by the microbes, hence reducing the action of bioremediation. Bioventing or biosparging methods can enhance oxygen delivery to the subsurface, thereby increasing the bioavailability of such compounds to the microorganisms (Das & Chandran, 2011).

Sorption: Strong adsorption of pollutants to soil particles or organic matter could make them less bioavailable to the microorganisms and results in a slower degradation rate. The environmental contaminants that are long-lived, such as DDT and PCBs, are known to be retained because of their strong sorption to the soil and sediment particles.

5 Applications of Microbial Bioremediation

Below, we discuss two major applications of microbial bioremediation: soil and water treatment.

5.1 *Soil Remediation*

Industrial leaks, farm runoff, and random waste disposal have led to contamination in soils, resulting in severe environmental and health impacts. Among the sustainable ways of rehabilitating contaminated soils is through microbial bioremediation. Microbial bioremediation involves the use of microorganisms that degrade or transform toxic compounds into harmless forms (Atlas, 1981).

Hydrocarbon Degradation The most famous and substantial application of microbial bioremediation in soil is hydrocarbon biodegradation. This involves petroleum and PAHs. Such compounds are used by some bacteria, like *Pseudomonas* and *Mycobacterium*, as a source of carbon and energy for growth, metabolism, and subsequent transformation into water and carbon dioxide (Singh et al., 2020).

For example, *Pseudomonas aeruginosa* has been demonstrated to biodegrade crude oil in contaminated soils through vast devaluation in the contaminant levels overtime (Das & Chandran, 2011).

5.1.1 Heavy Metal Remediation

Microorganisms also facilitate detoxification and heavy metal immobilization, for example, lead, cadmium, lead, lithium, and mercury (Priyadarsini & Das, 2025). For instance, the bacillus and thiobacillus can convert some of the metals into less bioavailable forms or even immobilize them, thus reducing their toxicity while preventing them from entering the food chain (Priyadarsini & Das, 2024).

Pesticide Degradation DDT and atrazine persist in soil, showing bioaccumulation and adverse ecological effects. Microbial degradation is an excellent way of reducing pesticide residues in agricultural soils. For example, *Pseudomonas putida* has been utilized in breaking down organophosphate pesticides, thereby reducing its environmental persistence and toxicity (Singh & Walker, 2006). The above mechanisms help microbial bioremediation in enhancing soil quality, restoring ecological balances, and reducing the long-term environmental impacts of industrial and agricultural pollutants.

5.2 Water Treatment

One of the major concerns related to environmental sustainability has been the pollution of water due to industrial discharges, oil spills, agricultural runoff, and domestic sewage. Purification of such contaminated water was believed to be very efficiently done by microbial bioremediation, as it occurs naturally and at low expense.

Wastewater Treatment: Treatment of wastewater in plants involves the technique of microbial bioremediation. Oxygen- and nitrogen-resenting bacteria deplete organic contaminants, thereby reducing the BOD and COD of wastewater (Mishra et al., 2024; Mishra & Das, 2023). In nitrification, *Nitrosomonas* and *Nitrobacter* convert ammonia into nitrate. Conversely, during denitrification, *Pseudomonas* species change nitrates to nitrogen gas, relieving the problem of nutrient pollution (Tchobanoglous et al., 2014). Advanced treatment processes usually involve specialized microbes that can target difficult-to-degrade organic pollutants, such as pharmaceuticals and industrial chemicals.

Oil Spill Bioremediation: Oil spills are probably the most harmful types of water pollution when considered from an ecological perspective. The rate at which oil is broken down in marine environments can be improved by bioremediation methods such as hydrocarbonoclastic bacteria, such as *Alcanivorax borkumensis* (Yakimov et al., 2007). These bacteria metabolize hydrocarbons present in crude oil and break them into harmless byproducts that allow for the natural recuperation of ecosystems. Examples of notable success include microbial remediation of the environment after the Deepwater Horizon oil spill in the Gulf of Mexico. Heavy metals are removed from water. Heavy metal removal from water bodies forms an important component of bioremediation, especially in the cases of mining and industrial runoff-prone areas. Sulfate-reducing bacteria (SRB) *Desulfovibrio desulfuricans* can oxidize soluble metal ions to convert them into insoluble metal sulfides, thus implying an increase in the immobility and bioavailability of such metals as cadmium and lead (Nicomrat et al., 2006). Such processes are particularly effective in the removal of groundwater and industrial effluents (Rajpoot, 2019; Rajpoot, 2020; Rajpoot, 2021, Rajpoot, 2024; Rajpoot, 2025). It has advantages over the water treatment procedures where costs are much lower, with minimal interference with the environment and the possibility of in situ application. Continuing researches may also advance genetic engineered microbes capable of enhanced degradative capabilities with increased applications for bioremediation technologies in water treatment (Sutherland et al., 2020).

6 Future Prospects and Innovations in Microbial Bioremediation

The future of this field is shaped by several cutting-edge technologies, driving greater efficiency and expanding the scope of pollutants that can be remediated.

6.1 Advancements in Synthetic Biology

The new generation of synthetic biology is changing the face of microbial bioremediation with the design of enhanced capabilities within microorganisms. Microorganisms are engineered with certain pollutant-degrading enzymes or pathways, hence increasing their degradation capabilities for heavy metals, pesticides, and plastics. Examples include *Pseudomonas* species engineered for degradative oil spills and even genetically modified yeast to degradable industrial chemicals. Such approach will be very effective in doing large-scale as well as more economic environment cleaning by being more precise (Atlas,1981).

6.2 Artificial Intelligence and Machine Learning

These have increasingly been applied and utilized presently to optimize microbe bioremediation processes. In this regard, AI is used to predict microbial efficiency as it analyzes vast datasets of the characteristics of microorganisms, their environmental conditions, and profiles of pollutants (Ranjan et al., 2016). AI can predict the interaction of diverse strains of microorganisms with particular pollutants in various media and aid in the selection of efficient agents for particular remediation operations. Such predictive ability can significantly reduce the processes of trial and error to a considerable extent and makes bioremediation considerably faster and cost-effective.

6.3 CRISPR Technology

CRISPR-Cas9 gene-editing technology has advanced appreciably in microbial bioremediation. Scientists have improved its resistance to extreme environmental conditions by directly modifying the microbial genome or even permitting it to break down more complex pollutants (Ghosh et al. 2011). For example, by using CRISPR, scientists have been able to construct microbes that degrade chlorinated hydrocarbons that are otherwise challenging to clean up. Such a high-precision and efficient method is very useful in engineering microbes for the desired capabilities to degrade pollutants.

6.4 Environmental Sustainability

Bioremediation by microbial populations very well fits into global goals toward sustainability, mainly about pollution and resource preservation (Salt et al., 1995). Since bioremediation is free of chemical processes, it saves much more energy and is relatively environmentally friendly. The breakdown by microbes does not deposit any toxic residue. The whole process sustains the principle of a circular economy: transformation of waste into nontoxic products. This is a sustainable approach to reducing the environmental impacts of industrial activities while minimizing the demand for chemical agents. Scaling up microbial bioremediation from small-scale experiments to large, real-world applications is very challenging. The contaminated sites are usually heterogeneous with pollutants not evenly distributed in space. Spatial heterogeneity poses significant challenges to applying uniform bioremediation strategies throughout the site (Liu & Fang, 2003). Moreover, bioremediation at an enhanced scale is expensive as the process may require supplementary resources including engineered microbes, nutrients, and monitoring systems. This is because bioremediation requires time for the process to achieve satisfactory results. This can be too much in cases where remediation has to be fast.

6.5 Legal and Regulatory Barriers

Strict regulations govern the application of GEMs in bioremediation. Most countries regulate the release of GMOs in the environment, including thorough tests of safety and possible ecological risk. The public's aversion to genetically modified organisms adds to the complexity of the process in applying GEM-based bioremediation processes. Liability issues over potential failure in GEM-based bioremediation processes can pose some legal difficulties for their application.

7 Conclusion

Even though the microorganisms are small in size, their impact on the environment is gigantic. These microscopic beings are natural cleanup agents as they break down pollutants in the environment through their metabolic processes. Whether it is heavy metals, hydrocarbons, pesticides, or plastics, microbes can degrade them all and play an important role in keeping our ecosystems healthy. It is a nonpolluting and less expensive alternative to environmental contamination compared to chemical methods, which are not only hazardous but also consume much energy.

The need for innovative solutions is even more urgent with the further growth of the industrialized world. One such innovative solution is microbial bioremediation. While harmful chemicals are a danger to man, microbes offer a softer, more energy-efficient alternative. Moreover, the scientists engineer these microbes to embrace even more severe pollutants such as plastics and pharmaceutical residues. Progress in synthetic biology, genetic engineering, and microbiology is propelling us very fast toward breakthroughs in enhancing the capabilities of microbes in dealing with diversified and challenging contaminants.

Microbial bioremediation into the future is the core of many efforts toward gaining a sustainable future. Such technologies as CRISPR gene editing and artificial intelligence will help fine-tune the performance of microbes into more efficient forms that can better handle pressing issues such as oil spill pollution, plastic pollution, and hazardous waste. This indeed fits the circular economy perfectly as it does not throw away its waste but instead turns the waste into harmless and less toxic materials. Instead of landfill waste, they do not add to additional damage to the environment since the end products become healthy ecosystems. In a nutshell, microbial bioremediation is an effective, environmentally friendly solution to environmental pollution. The future of science will make microorganisms play an increasingly critical role in ecological balance, creating a cleaner, more sustainable future for our planet.

References

Atlas, R. M. (1981). Microbial degradation of petroleum hydrocarbons: An overview. *Microbial Ecology, 7*(3), 273–287.

Bhattacharjee, J., Mishra, S., & Das, A. P. (2021). Recent advances in sensor-based detection of toxic dyes for bioremediation application: A review. *Applied Biochemistry and Biotechnology, 2021*, 1–20. https://doi.org/10.1007/S12010-021-03767-7

Braguglia, C. M., et al. (2018). Bioremediation of oil spills: An overview of methods and their effectiveness. *Environmental Engineering Science, 35*(7), 521–532.

Chapman, P. M., & Wang, F. (2011). *Environmental toxicology and risk assessment: A practical approach*. CRC Press.

Das, N., & Chandran, P. (2011). Microbial degradation of petroleum hydrocarbon contaminants: An overview. Biotechnology Research International, 2011.

Das, A., & Mishra, S. (2010). Biodegradation of the metallic carcinogen hexavalent chromium Cr(VI) by an indigenously isolated bacterial strain. *Journal of Carcinogenesis, 9*. https://doi.org/10.4103/1477-3163.63584

Das, A. P., Ghosh, S., Mohanty, S., & Sukla, L. B. (2015). Advances in manganese pollution and its bioremediation. In L. B. Sukla, N. Pradhan, S. Panda, & B. K. Mishra (Eds.), *Environmental microbe biotechnology* (1st ed., pp. 313–328). Springer. https://doi.org/10.1007/978-3-319-19018-1_16

Das, A. P., Mohanty, A. K., van Hullebusch, E. D., Figueiredo, G., Das, A. P., Mohanty, A. K., van Hullebusch, E. D., & Figueiredo, G. (2023). Bioremediation as an emerging technology for the removal of synthetic microplastic pollutants from marine ecosystem. *MarPB, 194*, 115297. https://doi.org/10.1016/J.MARPOLBUL.2023.115297

Das, A., Dey, S., Bala, K., Gnanasekaran, R., & Das, A. P. (2024). Microbial bioremediation Technology for Sustainable Treatment and Management of synthetic microfiber waste. *Sustainable Microbial Technology for Synthetic and Cellulosic Microfiber Bioremediation, 221–238*. https://doi.org/10.1007/978-3-031-62600-5_12

Dey, S., Dash, A., & Das, A. P. (2024a). Bioremediation as a sustainable and circular solution for mining waste management and bioenergy generation. *Sustainable Management of Mining Waste and Tailings: A Circular Economy Approach, 107–119*. https://doi.org/10.1201/9781003442455-6/BIOREMEDIATION-SUSTAINABLE-CIRCULAR-SOLUTION-MINING-WASTE-MANAGEMENT-BIOENERGY-GENERATION-SUDESHNA-DEY-AKANSHYA-DASH-ALOK-PRASAD-DAS

Dey, S., Dash, N., Ray, S. K., Priyadarsini, S., Chhotaray, C., & Das, A. P. (2024b). Occurrence and quantification of Tire Wear plastics pollutants from Highway City dust of Bhubaneswar City of Odisha. *Sustainable Microbial Technology for Synthetic and Cellulosic Microfiber Bioremediation, 77–89*. https://doi.org/10.1007/978-3-031-62600-5_5

Ghosh, S., et al. (2011). Genetic engineering of microorganisms for bioremediation of environmental pollutants. *Journal of Environmental Management, 92*(3), 2559–2571.

Kaza, S., Yao, L., Bhada-Tata, P., & Van Woerden, F. (2018). Waste management and global health: Emerging challenges in urbanization. *Journal of Environmental Health, 80*(7), 34–39.

Liu, Y., & Fang, H. H. (2003). *Bioaugmentation for bioremediation of contaminated environments*. Springer.

Mishra, S., & Das, A. P. (2023). Microbial remediation of synthetic microfiber contaminated wastewater. In M. P. Shah (Ed.), *Microbial technologies in industrial wastewater treatment* (pp. 337–350). Springer. https://doi.org/10.1007/978-981-99-2435-6_17

Mishra, S., Swain, S., Sahoo, M., Mishra, S., & Das, A. P. (2021). Microbial colonization and degradation of microplastics in aquatic ecosystem: A review. *Geomicrobiology Journal*. https://doi.org/10.1080/01490451.2021.1983670

Mishra, S., Dash, D., Al-Tawaha, A. R. M. S., & Das, A. P. (2022). A review on heavy metal ion adsorption on synthetic microfiber surface in aquatic environments. *Applied Biochemistry and Biotechnology, 194*(10), 4639–4654. https://doi.org/10.1007/S12010-022-04029-W/METRICS

Mishra, S., Dash, D., & Das, A. P. (2024). Isolation and molecular characterization of polyester-degrading bacterial strains from municipal wastewater and their application in microplastic degradation in contaminated wastewater. *Geomicrobiology Journal*. https://doi.org/10.1080/01490451.2024.2376101

Nag, M., Lahiri, D., Ghosh, S., Sarkar, T., Pati, S., Das, A. P., Ram, D. K., Bhattacharya, D., & Ray, R. R. (2023). Application of microorganisms in biotransformation and bioremediation of environmental contaminant: A review. *Geomicrobiology Journal*. https://doi.org/10.1080/01490451.2023.2261443

Nicomrat, D., Dick, W. A., & Tuovinen, O. H. (2006). Microbial populations identified by fluorescence in situ hybridization in a constructed wetland treating acid coal mine drainage. *Water Research, 40*(10), 1977–1986.

Paul, A., Dey, S., Ram, D. K., & Das, A. P. (2023). Hexavalent chromium pollution and its sustainable management through bioremediation. *Geomicrobiology Journal, 1–11*. https://doi.org/10.1080/01490451.2023.2218377

Pradhan, G., Tripathy, B., Ram, D. K., Digal, A. K., & Das, A. P. (2023). Bauxite mining waste pollution and its sustainable management through bioremediation. *Geomicrobiology Journal, 1–10*. https://doi.org/10.1080/01490451.2023.2235353

Priyadarshini, B., Santhosh Kumar, M., & Das, A. P. (2024). Microbial degradation of synthetic plastics released from COVID-19 generated single use disposable face masks. *Environmental Science and Engineering, Part, F2827*, 1–16. https://doi.org/10.1007/978-3-031-50840-0_1

Priyadarsini, S., & Das, A. P. (2024). Analytical and structural characterization of waste lithium-ion batteries for their effective recycling strategy. *Environmental Chemistry and Ecotoxicology*. https://doi.org/10.1016/j.enceco.2024.12.004

Priyadarsini, S., & Das, A. P. (2025). Lithium bioleaching: A review on microbial-assisted sustainable technology for lithium bio-circularity. *Journal of Water Process Engineering, 69*, 106744. https://doi.org/10.1016/j.jwpe.2024.106744

Rajpoot, S. (2019). Current technologies for waste water treatment in chemical industries in India (May 4, 2019). *International Journal for Modern Trends in Science and Technology, 5*(5). https://doi.org/10.2139/ssrn.3390363

Rajpoot, S. (2020). A review on ways to manage biomedical waste at different locations in Faizabad. *International Journal for Modern Trends in Science and Technology, 06*(01), 33–36.

Rajpoot, S. (2021). Impact of nanotechnology on environment—A review. *International Journal for Modern Trends in Science and Technology, 7*, 159–164. https://doi.org/10.46501/IJMTST0710026

Rajpoot, S. (2024). Eco-friendly solutions for sustainable waste management: An approach towards a cleaner and greener future. *Waste Management and Life Cycle Assessment for Sustainable Business Practice*, pp. 43–64.

Rajpoot, S. (2025). Smart and innovative nanotechnology applications for water purification: Exploring technologies for a cleaner future. *Emerging Trends and Technologies in Water*, 87–120.

Ranjan, M., et al. (2016). Nanotechnology for the remediation of environmental contaminants: A review. *Environmental Nanotechnology, Monitoring & Management, 6*, 79–89.

Sahu, S., & Das, A. P. (2024). *Microplastics degradation mechanism by aquatic plastiomes: A review*. 213–226. https://doi.org/10.1007/978-3-031-50840-0_12.

Salt, D. E., et al. (1995). Phytoremediation: A novel strategy for the removal of toxic metals from the environment using plants. *Biotechnology, 13*(5), 466–472.

Singh, J., & Walker, A. (2006). Microbial degradation of organophosphorus compounds. *FEMS Microbiology Reviews, 30*(3), 428–471.

Singh, S., et al. (2016). Bioremediation of contaminated sites: A low-cost nature's approach for clean-up. *Frontiers in Microbiology, 7*, 1–16.

Singh, R. K., Kumari, S., & Mishra, S. (2020). Bioremediation: Mechanisms and application. In P. Bharagava (Ed.), *Recent advances in environmental management*. CRC Press.

Sutherland, I. W., Kennedy, A. F., & Young, S. A. (2020). Genetically modified microorganisms in bioremediation. *Current Opinion in Biotechnology, 62*, 144–152.

Tchobanoglous, G., Burton, F. L., Stensel, H. D., & Tsuchihashi, R. (2014). *Wastewater engineering: Treatment and resource recovery*. McGraw-Hill Education.

United Nations Environment Programme (UNEP) (2018). Single-use plastics: A roadmap for sustainability. UNEP Report.

World Bank Group. (2018). *What a waste 2.0: A global snapshot of solid waste management to 2050*. World Bank Report.

Yakimov, M. M., Timmis, K. N., & Golyshin, P. N. (2007). Obligate oil-degrading marine bacteria. *Current Opinion in Biotechnology, 18*(3), 257–266.

Waste Management and Classification: Repurposing Garbage to Create Nanoparticles for Water Purification

Rishav Garg, Rajni Garg, Ansar Anjum, Nnabuk Okon Eddy, and Subhadra Rajpoot

Abstract The developing discipline of waste material repurposing to synthesize nanoparticles for efficient and environmentally friendly water filtration is an innovative and sustainable approach. The emphasis on turning trash streams into lucrative assets can solve the world water problem. The chapter explores waste sources fit for green synthesis approaches, nanoparticle synthesis, characterization methodologies, water purification applications, and environmental and financial consequences. The chapter also covers the difficulties and prospects of this sector as well as examines the part waste categorization and sorting play in maximizing nanoparticle production. Future studies should investigate the possibilities of technology and solve these difficulties. The study reveals that the development of more effective synthesis techniques, thorough lifetime evaluations, and integration with other water treatment approaches depends on well-defined policies and legal frameworks help to hasten application. No doubt, one bright future route toward a cleaner and more sustainable one is recycling trash to produce nanoparticles for water purification.

R. Garg (✉)
Department of Civil Engineering, Galgotias College of Engineering and Technology, Greater Noida, Uttar Pradesh, India
e-mail: rishavgarg@science.org.in

R. Garg
Department of Applied Science and Humanities, Galgotias College of Engineering & Technology, Knowledge Park-II, Uttar Pradesh, India

A. Anjum
Department of Applied Chemistry, Delhi Technological University, Delhi, India

N. O. Eddy
Department of Pure and Industrial Chemistry, University of Nigeria, Nsukka, Nigeria

S. Rajpoot
Department of Applied Sciences, Amity University, Greater Noida, Uttar Pradesh, India

S. Rajpoot, A. P. Das (eds.), *Innovative Technologies for Waste Management*, Sustainable Environmental Waste Management Strategies,
https://doi.org/10.1007/978-3-031-97184-6_6

1 Introduction

One of the main challenges is that the worldwide water issue demands innovative and ecologically sustainable solutions. Access to safe drinking water remains a major obstacle for billions of people worldwide, leading to health problems and slowing down social development (Chums-ard et al. 2019). Although accepted, traditional water treatment methods might have disadvantages like high costs, energy consumption, dangerous side effects, and ineffective removal of many pollutants (Mishra et al., 2019). We need to move toward more long-lasting answers because the growing world population is putting even more stress on resources and facilities (Ali et al., 2023). That being said, there is also a big environmental risk that comes from releasing more waste. When waste is dumped into the environment, whether on purpose or by accident, it affects environment and people's health (Ghimire & Ariya, 2020; Das et al., 2011). For instance, the oil refineries and utilization industries make a lot of waste, such as oil fuel ash and waste oil, which pollutes the environment if it is not managed. This calls for new ways to deal with waste that will make the best use of resources and minimize damage to the environment (Garg & Anjum, 2024). The huge amount and range of waste around the world is a big problem for the earth. A lot of plastic waste ends up in dumps and rivers, which is bad for people's health and the health of animals. Also, agricultural and domestic wastes are important contributors. Cotton trash, like other fiber trash, is bad for the earth. The complicated mix of valuable and dangerous parts in electronic trash (e-waste) makes it hard to restore resources and protect the environment (Rizal et al., 2021).

Waste management finds shape in the circular economy. It focuses on lowering waste generation, maximizing resource use, and fostering recycling and reuse instead of considering trash as a disposal problem (Bala et al., 2024a; Mishra et al., 2024; Mohanty et al., 2024b; Sahu et al., 2024). This idea is compatible with using waste for the production of nanoparticles (NPs) (Khalid et al., 2023). Turning trash into useful nanomaterials for water filtration generates financial possibilities and helps to lower pollution levels. Several waste products might be possible precursors for the manufacturing of NPs (Chithra et al., 2023). A byproduct of rice fields, rice husk includes silica, synthesizable into silica NPs (Si-NPs) for water filtration (Bansal et al., 2022). The electrochemical production of iron oxide NPs can use discarded rusted iron nails as electrodes: Iron oxide NPs let spent coffee grinds be functionalized. Cellulose nanocrystals may be made from cotton trash. Low-cost membranes may be made from polyethylene terephthalate bottles' waste (Onu et al., 2023). Cost-effective, environmentally friendly, and consistent with circular economy ideas is using waste resources for nanoparticle production. Still, problems exist. Waste material heterogeneity and contaminants might call for pre-treatment processes to guarantee consistent and high quality of NPs (Sagir et al., 2020).

Evaluation of nanoparticle characteristics and suitability for water purification depend on careful characterization. An eco-friendly substitute for conventional chemical processes is provided by green synthesis techniques (Ramasamy et al.,

2022). These techniques decrease and stabilize metal ions into NPs by the use of bio-reductants derived from biological sources or plant extracts. Biological agents simplify synthesis, lower energy usage, make the process ecologically friendly, and cut the requirement for strong chemicals (Sharma et al., 2023). For the green manufacture of copper (Cu) NPs, plant extract is employed, for instance; rice husk is synthesized as a high-purity silica-based nano-adsorbent (Serang et al., 2024). Another such is the fabrication of zinc oxide (ZnO) NPs imprinted on clay tablets devoid of chemical reducing agents. Temperature, pH, precursor concentration, and reaction times are among the several factors affecting nanoparticle production. High yields of NPs with desired qualities depend on the optimal values of these factors. For methylene blue adsorption, for example, ideal pH and adsorbent dose were found for a silica-Cu nanocomposite (Magnabosco et al., 2023).

Rework of waste using NPs for water treatment offers great environmental benefits. Separating garbage from landfills reduces the environmental impact of conventional nanoparticle production by so lessening the pollution (Mujtaba et al., 2023). One helps to preserve resources by utilizing waste items. Its energy consumption is less than more traditional methods. Green synthesis techniques assist in lowering environmental impact yet more. Moreover, appearing excellent is the financial feasibility (Pujeri et al., 2018). This chapter explores how using waste products helps in sustainable waste management with waste valorization. It will provide the present scenario about waste-to-nanoparticle technologies. Emphasizing different waste sources and their opportunities for sustainable nanomaterial production inside a circular economy framework, this chapter methodically investigates the expanding field of waste to nanomaterials (Bala et al., 2024b; Das et al., 2024a; Dey et al., 2024a). The chapter explores the many waste sources, the methods applied for nanomaterial extraction and synthesis, the characteristics and applications of the generated materials, and the related challenges and future directions of this fascinating issue.

2 Waste Streams as a Source of Nanomaterials: A Circular Economy Approach

Waste management's growing worldwide difficulty calls for creative ideas beyond conventional sequential "take-make-disposal" methods. The circular economy offers a convincing substitute with an emphasis on resource efficiency, waste reduction, and the building of a closed-loop system whereby waste products are reused (Khajuria et al., 2022). This paradigm change presents a special chance to convert many waste sources into high-value nanomaterials as it fits exactly the developing discipline of nanotechnology. Nanomaterials, characterized by their special nanoscale qualities, have great promise in many different fields, including environmental remediation, medicine, and energy (Garg et al., 2023). However, the traditional synthesis of many nanomaterials usually entails the use of dangerous chemicals and energy-intensive procedures, therefore posing environmental issues (Dey et al., 2024c; Mohanty et al., 2024a; Ray et al., 2024). Nanomaterial synthesis using waste streams offers a sustainable replacement that solves waste disposal problems as well as lowers environmental impact (Mujtaba et al., 2023).

The environmental and financial feasibility of green nanotechnology depends on the choice of suitable waste resources for the synthesis of NPs. The type of nanoparticle generated, the synthesis process's efficiency, and the general approach's overall sustainability all depend much on the waste material chosen (Wardanu et al., 2024). Table 1 lists the various waste streams as resources, highlighting the applications in environmental remediation.

2.1 Agricultural Waste as a Source of Nanomaterials

Produced as the outcome of several agricultural activities, agricultural waste is a major but generally neglected resource. Various waste products, including leaves (Das et al. 2020), seeds (Rautela et al. 2019), nutshells (Dang et al. 2019), coconut husk (Sebastian et al. 2018), bagasse (Falk et al. 2019), and rice husk (Bansal et al., 2022), have been sustainably used to generate varied nanomaterials. Not only does using agricultural waste to create nanoparticles address waste disposal issues, but it also offers a reasonably affordable and ecologically friendly substitute for more conventional methods of nanomaterial synthesis (Sharma et al., 2024). Among the easy and eco-friendly methods to turn this trash into nanomaterials include chemical treatments, incineration, and pyrolysis (Shen 2017). Caped by numerous functional groups, the resulting bio-based NPs have a greater surface area and can be employed for the restoration of contaminated surroundings (Garg et al., 2022a).

According to literature studies, different NPs in a green manner may be produced from several agricultural wastes. Being renewable, easily obtained, and reusable, agricultural waste is a safe and appealing material to employ. Many reducing and capping agents found in these waste products include phytochemicals, which change the final characteristics of the NPs (Eddy et al., 2024). Strong antibacterial and antioxidant effects have been demonstrated by green synthesis of encapsulated Cu NPs derived from *Moringa oleifera* leaves (Das et al. 2020). Successful green manufacturing of CuO NPs using *Piper nigrum* leaf extracts has been achieved. Strong antibacterial activity of these NPs against *Escherichia coli* and *Bacillus subtilis* suggests that they could find utility as disinfectants and in wastewater treatment (Naaz et al., 2024). Furthermore, biosynthesized from *Brassica oleracea* (cauliflower), silver (Ag) NPs showed outstanding photocatalytic breakdown of methylene blue dye and Hg^{2+} biosensing properties (Ansar et al. 2020). Ag NPs were synthesized in previous studies from *Tectona grandis* (teak) seeds. These nanoparticles were shown to have antimicrobial activity (Rautela et al. 2019). Ag NPs in a low-impact manner have been produced from the leaves of *Lagenaria siceraria* (bottle gourd) and *Pisum sativum* (pea) (Deepa et al. 2022). Ag NPs indicate the

Table 1 Waste streams as resources: waste classification and application fields

Waste material	Source	Key features/ properties	Applications	Environmental benefits	Economic considerations	References
Farm trash	Agriculture (crop residue, fruit peels, animal manures)	Adsorptive capacity, rich in lignocellulosic materials	Direct adsorbents or value-added products (e.g., biochar, activated carbon)	Reduction in waste to landfills with lower reliance on conventional treatment methods	Costs depend on collection, processing, and reuse potential	Mujtaba et al. (2023)
Rice husk	Rice milling	High silica content, porous structure	Adsorption of heavy metals and other contaminants Potential regeneration and reuse	Decreases environmental impact of landfill waste Reduces reliance on chemical adsorbents	Economic feasibility hinges on processing and regeneration costs vs savings	Bansal et al. (2022)
Biochar	Thermochemical conversion of farm waste	Porous, high adsorption capacity	Removal of organic dyes, pesticides, and heavy metals from water	Carbon sequestration potential Mitigation of climate change	Financial viability depends on feedstock costs, pyrolysis method, and market value of treated water	Khalid et al. (2023)
Industrial wastes	Metallurgical processes, water treatment plants, etc.	Varied properties; high adsorptive capacity for pollutants	Smelter slag is used for removing heavy metals, dyes, and pollutants, while alum sludge is sued for soil stabilization in road construction	Reduced need for virgin materials and offers lower waste disposal in landfills	Cost depends on processing and regeneration vs savings from reduced traditional treatment methods	Mohamed et al. (2022); Uzu et al. (2010)

(continued)

Table 1 (continued)

Waste material	Source	Key features/ properties	Applications	Environmental benefits	Economic considerations	References
Fly ash and red mud	Industrial byproducts	High reactivity, adsorptive properties	Adsorbents, coagulants, or composite membrane components for water treatment	Minimizes landfill waste and reduces dependence on virgin adsorbents	Effectiveness and cost vary based on material properties and processing methods	Ma et al. (2018)
Plastic debris	Discarded plastics	Durability, potential to enhance material properties	Incorporated into asphalt concrete for improved durability and performance with reduction in raw material needs	Diverts plastic waste from landfills and reduces greenhouse gas emissions from virgin plastic production	Costs tied to plastic recycling processes and infrastructure integration	Ali et al. (2023)
Textile waste	Discarded polyester, fabrics	Sorbent potential for oil-water separation	Recycled into textile-based sorbents for oil spill remediation	Reduces textile waste in landfills and offers practical solutions for environmental oil contamination cleanup	Financial viability depends on processing costs and efficiency of the sorbent in market applications	Rizal et al. (2021)

potential low-cost and environmentally friendly reducing agent (Damirchi et al. 2020) by the use of *Viscum album* extract.

Numerous studies have shown that agricultural waste may be used to produce many nanomaterials. The utilization of *Citrullus lanatus* (watermelon) waste to generate gold (Au) nanoparticles emphasizes even more the adaptability of agricultural waste. This work synthesized NPs with antimicrobial action against *Escherichia coli* and *Staphylococcus epidermidis* from both the red and green sections of watermelon waste (Chums-ard et al. 2019). Another instance of how agricultural waste may be utilized is the green generation of Au NPs using *Macadamia* nut shells (Dang et al. 2019). De-oiled *Jatropha* waste has great potential to generate Au NPs, which demonstrates how to obtain valuable nanomaterials from various waste streams (Kanchi et al. 2018). Also, the peels of *Ananas comosus* (pineapple) and *Passiflora edulis* (passion fruit) were used to make nontoxic and environmentally friendly Au NPs (Pechyen et al. 2021). Plants that are easily obtained and have been employed in green manufacturing of iron oxide NPs abound in various types of phytochemicals (Sandeep et al. 2023). Golthi and Kommu (2023) have shown by manufacturing iron oxide NPs that the usually expanding weed *Jatropha gossypifolia* might be exploited as a long-term supply of reducing agents. Similarly, iron oxide NPs in a green manner have been produced from discarded *Citrus aurantifolia* peel, which when applied as a coating on steel has high anticorrosive qualities.

Lycopersicon esculentum (tomato) extract showed promise as a nontoxic and reasonably cheap reducing agent when used to synthesize thorium NPs (Pujeri et al., 2018). The resultant NPs showed anticancer activity, therefore extending the possible uses for this method in biological fields. *Melaleuca leucadendra* waste is used in the synthesis of magnesium oxide NPs to show even further the possibility of this method in using agricultural waste for the manufacturing of metal oxide NPs with antioxidant characteristics (Serang et al., 2024). Olive tree pruning debris has been used to obtain cellulose nanocrystals as a sustainable method for plant bacterial disease prevention (Schiavi et al., 2022), while coconut coir has been used to synthesize reduced graphene oxide/$CoFe_2O_4$ nanoflakes with spinel structure showing promise for use in supercapacitors and photocatalysis (Ramasamy et al., 2022). Literature reports several approaches for synthesizing carbon (Wang et al., 2016) and silica-based nanomaterials (Sharma et al., 2024) from rice trash, underlining their uses in biomedical research, environmental remediation, and energy-related applications. Moreover, the manufacturing of nanomaterials from agricultural waste supports a circular economy by lowering dependency on virgin resources and thereby lessening environmental effects. However, scaling up these techniques for mass manufacturing and guaranteeing consistent quality and characteristics of the produced nanomaterials remains difficult (Magnabosco et al., 2023). Factors including the required nanoparticle type, availability, and type of waste material direct the synthesis. The size, form, and stability of the NPs are strongly influenced by the phytochemical composition of the waste (Sharma et al., 2023).

The agricultural waste is easily accessible which makes them ecologically sustainable and financially appealing choices for the manufacturing of NPs. Spent coffee grounds have also been investigated as a source of reducing and stabilizing agents for the production of Au and Ag NPs. The effectiveness of the nanoparticle production process was shown to be impacted by both the overall antioxidant activity and the total caffeine-yielding content of the green extracts (Yust et al., 2022). Another easily obtained waste product used for the manufacture of nano-calcium carbonate is eggshells. The resultant NPs showed a unique spherical form and smooth, crystalline surface, therefore highlighting the possibilities of this waste product for the synthesis of various useful nanomaterials (Gandhi et al., 2021). Table 2 shows the possibilities of waste valorization via green nanoparticle synthesis, therefore turning trash into useful assets.

2.2 E-Waste as a Nanomaterial Source

Considered a fast-expanding worldwide waste stream, e-waste consists of abandoned electronic equipment and components. With estimates showing a notable rise in the next years, the exponential growth of electronic trash (e-waste) poses a major environmental problem. This waste is a good source of precious metals and other components that may be collected and used for the production of nanomaterials in addition to other purposes (Rajahalme et al., 2022). Recovering these resources will have major environmental and financial advantages, hence lowering the need for virgin resources and lessening the environmental effect of e-waste disposal (Das et al., 2024c; Dey et al., 2023; Mishra et al., 2022; Priyadarsini & Das, 2025). The possibilities of e-waste as a source of different nanomaterials have been investigated several times. A major part of e-waste, waste-printed circuit boards include rare earth elements and precious metals such as palladium and Ag in great abundance (Rajahalme et al., 2022). Liou and Jheng, for instance, effectively manufactured premium-ordered mesoporous carbons (CMK-1) from recycled electronic packaging resin waste employing a silica template source. Appropriate for uses like energy storage and catalysis, the obtained mesoporous carbon displayed a large surface area (1715 m^2/g) and excellent purity (99.94% by weight) (Liou & Jheng, 2018).

For the production of NPs, recovering valuable metals from e-waste presents a potential path for both waste management and resource recovery (Priyadarshini et al., 2024; Tripathy et al., 2024). Recovering Au from e-waste for use in electrocatalysis (He et al., 2022) successfully shows the potential of this approach for sustainable energy usage. This recovery makes e-waste a handy instrument for the rehabilitation of some resources. Nevertheless, the presence of hazardous elements in e-waste demands the development of sustainable recycling strategies to reduce risks. Sometimes valuable metal extraction from e-waste calls for complex processes like pyrometallurgy, hydrometallurgy, and biohydrometallurgy (Desmarais et al. 2020). Using these methods helps one optimize environmental friendliness and efficiency, therefore reducing the total environmental impact of the process. Still, the management of e-waste creates difficulties as it includes toxic elements such persistent organic pollutants (Chithra et al., 2023). Careful maintenance and suitable treatment strategies serve to reduce any hazards to human health and the

Table 2 The emergence of waste valorizing green nanotechnology

Aspect	Description	Advantages	Applications	References
Environmental context	Increasing global agricultural waste and e-waste output present serious environmental challenges	Converts waste into valuable NPs (NPs)	Waste management	Magnabosco et al. (2023)
		Reduces harmful environmental impact	Environmental remediation	
Unique properties of NPs	NPs possess high surface area-to-volume ratio, catalytic activity, antimicrobial properties, and unique optical characteristics	Enhanced functionality in diverse fields	Medicine	Serang et al. (2024)
		High reactivity and specificity for targeted applications	Environmental cleanup	
			Catalysis	
Green synthesis overview	Sustainable alternative using natural reducing and capping agents from waste materials	Avoids harmful chemicals	Synthesis of eco-friendly NPs for various domains	Sandeep et al. (2023)
		Reduces energy consumption and harmful byproducts		
Methods of green synthesis				
Bioselection	Bioreduction using biomolecules like polyphenols, flavonoids, and alkaloids in plant extracts or microbial metabolites	Sustainable, low energy requirement	Synthesis of elemental and oxide NPs for industrial and biomedical use	Garg and Anjum (2024)
		Influences reaction kinetics for tailored NP properties		
Microwave-assisted synthesis	Uses microwaves to accelerate reactions between reducing agents and metal ions	Reduces reaction times and energy consumption	Production of tailored NPs from complex waste materials	Anjum et al. (2024)
		Allows precise control of NP size and shape		
Coprecipitation	Concurrent precipitation of metal ions using plant extracts as reducing and stabilizing agents	Suitable for metal oxide NPs	Synthesis of magnetic NPs (e.g., iron oxides) for environmental remediation and medicine	Onu et al. (2023)
		Allows control over NP size and characteristics through reaction variables		

(continued)

Table 2 (continued)

Aspect	Description	Advantages	Applications	References
Applications and future directions	Green-synthesized NPs are widely used due to their environmental and functional advantages	Promotes sustainable waste management Supports research into eco-friendly technologies for large-scale production	Medicine, environmental remediation, catalysis, electronics, and optical devices	Ghimire and Ariya (2020)
Challenges and research directions	Difficulty in scaling up methods Variability in waste composition affecting reproducibility	Overcoming limitations to ensure sustainability and efficiency at industrial levels	Continued development of sustainable methods for large-scale NP production	Chithra et al. (2023)

surroundings. The incorrect approach of disposal of e-waste including nanomaterials aggravates the problem even more as the discharge of these compounds into the environment may have unanticipated effects (Ghimire & Ariya, 2020). Though the diverse compositions of e-waste offer opportunities for the synthesis of several NPs, careful attention to such compounds is necessary to assure environmentally suitable actions. Thus, the realization of sustainable and ecologically friendly e-waste recycling methods will help to fulfill the complete potential of e-waste as a source of valuable nanomaterials.

2.3 *Plastic Waste as a Nanomaterial Source*

One of the main environmental issues facing us globally is the accumulation of plastic trash which requires innovative concepts for sustainable waste management. Upcycling plastic garbage into valuable carbon nanomaterials like graphene, carbon nanotubes, and carbon nanofibers (Onu et al., 2023) might help to solve this. This approach not only reduces environmental damage but also generates financial profit by transforming waste into very valuable goods (Dey et al., 2024b; Jagatee et al., 2024; John & Das, 2024; Mallick et al., 2023). Many methods have been looked at to turn waste made of plastic into carbon nanoparticles. Modern methods comprise stepwise conversion (Chi & Shaikh, 2024), thermal decomposition, flash joule heating, and chemical vapor decomposition. These methods may be modified to control the characteristics of the generated nanomaterials, hence personalizing them for specific usage. Thermal breakdown, for instance, can produce graphene and carbon nanotubes; CVD can be used to synthesize carbon nanofibers (Ikram et al., 2023). The kind of plastic trash, the targeted properties of the nanomaterials, and the process scalability all affect the choice of the optimum approach. Special carbon nanostructures with enormous surface area, outstanding electrical conductivity, and strong mechanical strength that meet a wide range of purposes including energy storage, catalysis, and composites are produced from plastic waste (el Essawy et al. 2017). Still, achieving the high production and purity of the generated nanoparticles as well as addressing the probable toxicity of some plastic additions proves challenging. Furthermore, major focus should be on the environmental impact of the conversion techniques themselves ensuring their sustainability and environmental balance (Larder & Hatton, 2023). The development of efficient and ecologically acceptable methods for the conversion of plastic waste into carbon NPs (Vijayvargiya et al. 2024) would let one realize the entire potential of this strategy inside a circular economy framework. The economic viability of this approach is also determined by the cost of the conversion process, the effective gathering and sorting of plastic waste, and the market demand for the generated nanomaterials including carbon nanotubes and other carbonaceous nanomaterials (Onu et al., 2023).

2.4 Other Waste Streams and their Prospect for Manufacturing Nanomaterials

Many other waste sources, beyond e-waste, agricultural waste, and plastic trash, have great promise for the manufacturing of nanomaterials inside a circular economy framework. These cover municipal solid waste (Part et al., 2018), olive waste (Sawalha et al., 2021), recycled paper sludge (Maslennikov et al., 2022), and industrial waste (Kadier et al., 2021). Using these several waste sources provides a multifarious method of resource recovery and waste management. One sustainable substance with uses in water treatment is nanocellulose, produced from recycled paper sludge. Using bacteria like *Yarrowia lipolytica*, olive mill wastewater (Sarris et al., 2023), a major byproduct of olive oil manufacture, has been used as a carbon source to synthesize important metabolites such as citric acid. Moreover, via microbial reduction with bacteria like *Geobacter sulfurreducens*, naturally abundant Fe(III) minerals may be converted into functional magnetic NPs, therefore providing possible uses in environmental remediation (Joshi et al., 2018).

Utilizing industrial byproducts, nanomaterial production offers a special chance to minimize waste and recover important resources, thereby fostering a more sustainable industrial environment. Though the complexity and variability of municipal solid waste demand rigorous processing and separation techniques, it also contains some components that can be exploited for the synthesis of nanomaterials (Zhuo et al., 2012). Through nanomaterial manufacturing, the valorization of several waste streams not only solves waste disposal problems but also promotes resource economy and economic growth (Rudovica et al., 2021). Nonetheless, it is important to evaluate the possible effects of the processing techniques and the produced nanomaterials on the environment and health, thereby guaranteeing that the whole approach conforms with the ideas of sustainability (Ramesh & Radhakrishnan, 2019). Furthermore necessary for their general acceptance and use is the development of uniform approaches for characterizing the features of nanomaterials from various waste sources (Zhang & Matharu, 2019). This covers a complete awareness of any possible environmental hazards or toxicity related to these materials (Eddy et al., 2024). Also, the economic feasibility of these techniques needs thorough consideration considering the expenses connected with the collecting, processing, and purifying of the nanomaterials (Part et al., 2018).

3 Nanoparticle Applications in Water Purification: Mechanisms and Efficiency

The worldwide need for safe and clean water is an overwhelming concern due to the always-increasing population, industrial expansion, and intense agricultural methods (Eddy et al., 2024). Although conventional water treatment processes may sufficiently sequester certain pollutants, these approaches cannot remove the persistent organic pollutants. Furthermore lacking in the elimination of microbiological infections and addressing other issues such as disinfection byproducts and the emergence of antibiotic resistance are these techniques (Garg & Anjum, 2024). These constraints show how urgently creative and effective water treatment methods are needed. Since it lets one control materials at the nanoscale, nanotechnology offers a practical approach to overcoming these difficulties (Sharma et al. 2023). The special physicochemical characteristics and high surface area-to-volume ratio of NPs give significant benefits over conventional methods with improved adsorption, catalytic activity, and antimicrobial effects (Eddy et al. 2025). Investigating the fundamental ideas, efficiency, and environmental effects of various nanoparticle-based technologies depends on an analysis of the expanding field of nanoparticle uses in water remediation with versatility in types of NPs, their mechanisms of action, and their efficacy in removing a wide range of water contaminants. NPs are helpful in water remedial operations as they may interact with different contaminants via numerous paths. As seen in Fig. 1 and outlined in Table 3, these processes mostly consist of adsorption, photocatalysis, antibacterial activity, and membrane filtering.

Fig. 1 Technologies and methods of treatment using reusable waste

Table 3 Mechanism for water purification

Mechanism	Description	Factors influencing efficiency	Models/processes	Applications/implications	References
Adsorption	NPs adsorb pollutants via van der Waals forces, electrostatic interactions, and chemical bonding	Surface area, surface charge, pore size, pollutant type, pH (influences nanoparticle charge and ionization of pollutants)	Langmuir and Freundlich isotherms to model adsorption behavior	Effective for removing organic and inorganic pollutants from water. Highly dependent on solution pH and nanoparticle properties	Onu et al. (2023)
Photocatalysis	Metal/metal oxide NPs use light to generate electron-hole pairs, producing radicals (•OH, O2•−) that degrade organic pollutants into CO2 and water	Surface area, pollutant concentration, pH, light intensity, and wavelength	Visible light-responsive photocatalysts are being developed for enhanced efficiency under sunlight	Efficient in degrading organic pollutants. Visible light-responsive photocatalysts enable sustainability by leveraging natural sunlight	Sagir et al. (2020)
Antimicrobial agents	Metal NPs (e.g., AgNPs, CuNPs) exhibit antibacterial activity via membrane disruption, ROS generation, and interference with cellular functions	Nanoparticle size, shape, concentration, and type of bacterium	AgNPs bind to thiol groups on membranes, produce ROS, and disrupt cellular activity	Effective in combating microbial contamination. Smaller NPs exhibit higher efficacy due to larger surface area. High concentration enhances activity but may raise toxicity concerns	Garg et al. (2022b)
Membrane filtering	NPs incorporated into membranes enhance filtration by adsorbing contaminants or inhibiting biofilm growth	Operating conditions, nanoparticle type and concentration, membrane pore size	NPs embedded in or deposited on membranes improve adsorption or filtration capacity	Enhances long-term functionality by resisting biofilm growth. Sustainable nanocellulose membranes are being explored for biodegradability	Anjum et al. (2024)

3.1 *Types of NPs Applied for Water Purification*

Various categories of NPs have shown promise in applications for water remediation based on their unique properties and modes of operation. Usually falling under three categories are metal NPs, metal oxide NPs, and carbon-based nanomaterials.

3.1.1 Metal NPs

Metal nanoparticles have drawn a lot of attention especially Ag and Cu NPs due to their significant antibacterial properties (Garg et al., 2021). Variables defining its antibacterial efficacy are particle size, shape, concentration, and the particular type of microorganisms (Garg et al., 2022b). In inactivating a broad range of bacteria, viruses, and fungus, AgNPs have demonstrated very remarkable effectiveness. Environmental persistence of AgNPs and their possible damage to nontarget species (Eddy et al., 2024) should thus be given great attention. CuNPs show quite significant antibacterial activity even if their mode of action might differ greatly from that of AgNPs. Direct interaction with thiol groups in proteins, for example, has demonstrated AgNPs to rupture and leak microbial cell membranes, therefore disturbing their membranes (Vo et al. 2017). By their mechanism, reactive oxygen species (ROS) including hydroxyl radicals aid to further inactivate bacteria. Among other biological components, interacting with proteins and DNA damages and kills cells at last (Anjum et al., 2024). Furthermore, bimetallic nanoparticles usually show synergistic effects improving performance above their monometallic counterparts. Combining metals can produce either better adsorption capacity, catalytic activity, or antibacterial effect (Xie et al., 2014).

3.1.2 Metal Oxide NPs

Metal oxide NPs, including titanium dioxide, iron oxide, and ZnO NPs that create electron-hole pairs upon UV or visible light exposure, which generates highly reactive ROS primarily from hydroxyl radicals, are especially well known for their photocatalytic activity (Sharma et al., 2023). Strong oxidizing agents, these hydroxyl radicals may break down a wide spectrum of organic pollutants. The photocatalytic efficiency of these NPs depends on particle size, crystallinity, surface area, arriving light intensity, and wavelength (Sharma et al. 2023). TiO2 NPs restrict their practical use in natural sunlight since they display quite poor efficiency in the visible light spectrum. Surface functionalizing or doping with another metal will help to raise their photocatalytic activity in the visible light spectrum (Sagir et al., 2020).

Two other metal oxide nanoparticles with environmental remedial capacity are copper oxide (CuO) and cerium oxide (CeO_2). CuO NPs are shown to be effectively adsorbable for several pollutants, including heavy metals and dyes. Their large surface area and availability of surface functional groups permitting interactions with pollutants characterize their adsorption capability (Bekru et al. 2022). Conversely, redox characteristics of CeO_2 NPs allow their use for the breakdown of organic contaminants. Alternating Ce(III) and Ce(IV) allows CeO_2 to participate in redox activities, hence generating oxidation and breakdown of organic molecules (Murugadoss et al. 2021). ZnO nanoparticles have shown success especially for specific pollutants from water. ZnO NPs have successfully removed heavy metal ions from water, therefore addressing one of the major worldwide health issues. Their huge surface area and surface charge allow significant binding interactions with heavy metal ions, hence explaining their inclination to absorb arsenic (Garg et al., 2022a).

3.1.3 Carbon-Based Nanomaterials

Particularly good adsorbents (Xiang et al., 2020) include carbon-based nanomaterials such as carbon nanotubes (CNTs) and graphene oxide (GO) because of their quite huge surface area and unique physicochemical properties. GO shows remarkable adsorption capacity for many organic pollutants and heavy metals using its several oxygen-containing functional groups—epoxy, hydroxyl, and carboxyl groups—which improve their hydrophilicity and offer sites for chemical bonding and electrostatic interactions with pollutants (Ikram et al., 2023). Furthermore, high aspect ratio and tubular structure of CNTs offer a perfect surface area for the adsorption of pollutants. Apart from their great tensile strength and electrical conductivity, which equip them for membrane filtration and sensor development, to guarantee their efficient use in water filtration, however, problems with the aggregation and dispersion of GO and CNTs in aqueous solutions have to be solved (Arora & Attri, 2020). Techniques include surface modification to promote their dispersion and polymers or surfactants inhibit agglomeration. Chemical functionalizing GO and CNTs tailors their surface characteristics to target certain contaminants, thereby improving their adsorption capacity (Rao et al. 2007). For instance, although functionalization with carboxyl groups may increase the adsorption of cationic contaminants, functionalizing with amine groups may increase the adsorption of anionic pollutants (Dubey et al., 2021).

3.2 *Efficacy of Nanoparticle-Based Water Purification*

Maximizing the production and quality of NPs depends critically on effective waste categorization and sorting. Waste stream contaminants can compromise the production of NPs, therefore affecting either decreased yields or unwelcome characteristics. Sometimes pre-treatment actions are required to eliminate undesired elements (Table 4). The purity and uniformity of the starting ingredients determine directly the efficiency of nanoparticle production. The performance of nanoparticle-based water filtration is determined by a complex interaction of various factors including

Table 4 Factors affecting efficacy of nanoparticle-based water purification

Aspect	Key factors	Details	References
Operational factors			
pH	Determines nanoparticle surface charge and contaminant speciation	Affects adsorption and other interactions	Akintelu et al. (2020)
Temperature	Influences kinetics of adsorption and photocatalytic reactions	Higher temperatures may increase reaction rates	Golthi and Kommu (2023)
Contact time	Dictates the equilibrium between pollutants and NPs	Longer contact allows for maximum pollutant removal	Ali et al. (2023)
Nanoparticle concentration	Affects available adsorption sites and pollutant removal rates	Higher concentration improves efficiency but may lead to agglomeration	Anjum et al. (2024)
Water composition	Compounds in water interfere with nanoparticle-pollutant interactions	Some contaminants compete for adsorption sites, reducing efficiency	Dubey et al. (2021)
Environmental considerations			
Toxicity concerns	NPs can harm nontarget organisms	AgNPs cause oxidative stress and cellular disruptions in aquatic species	Eddy et al. (2024)
Toxicity factors	Depends on size, shape, surface coating, and target organism	Requires extensive ecotoxicological assessment	Ghimire and Ariya (2020)
Environmental fate	NPs may persist in the environment	Factors: size, morphology, surface properties, and ecosystem	Ikram et al. (2023)
Bioaccumulation risks	Potential accumulation in the food chain and impact on higher trophic levels	Requires monitoring and mitigation strategies	Mohamed et al. (2022)
Recovery and disposal	Effective recovery methods are essential to reduce environmental damage	Example: magnetic separation and filtration are under development	Larder and Hatton (2023)
Waste management challenges	Need for cost-effective and eco-friendly disposal methods	Calls for sustainable nanoparticle treatment plans	Khajuria et al. (2022)

the type of nanoparticle utilized, the type of target pollutant, operational parameters, and initial water quality (Xie et al., 2014). Target contaminant determines variations in effectiveness among various NPs. Although TiO_2 excels in removing organic pollutants (Sagir et al., 2020), AgNPs are quite effective against bacteria (Garg et al., 2021). Carbon-based nanomaterials like GO and CNTs are mostly helpful as adsorbents for a wide spectrum of organic compounds and heavy metals (Xiang et al., 2020). Metal oxides like ZnO show especially attraction for some contaminants, including arsenic (Garg et al., 2022a). Thus, the selection of the appropriate nanoparticle type determines most importantly the degree of possible remediation.

Efficiency, scalability, and environmental friendliness of green synthesis techniques differ as well. The waste material, intended nanoparticle characteristics, and sustainability objectives also influence the approach used (Sagir et al., 2020). Different approaches can be found most appropriate by the use of a comparison study. Various ways of water remediation are provided by waste-derived NPs. Adsorption is the binding of contaminants to the surface of NPs. High surface area NPs improve their adsorption capability (Naaz et al., 2024). Photocatalysis makes use of the capacity of the NPs to produce reactive oxygen species, thereby destroying organic contaminants (Kadam et al., 2020). Disinfection is the process by which interaction with NPs inactivates bacteria. Many studies show how well waste-derived NPs eliminate contaminants (Pineda-Vásquez et al., 2024). Composites made of ZnO-clay efficiently remove heavy metal ions, while AgNPs remove organic compounds and heavy metals (Garg & Anjum, 2024). Heavy metals and bacterial pollutants are removed by functionalized nanoporous zerovalent iron, while Congo red dye is removed by iron oxide NPs (Sharma et al., 2023). Biochar-supported nanocomposites aid in the removal of Cu(II), Ni(II), and Cr(VI) (Khalid et al., 2023). Size, form, surface area, and surface charge define nanoparticle efficiency. Interaction between NPs and contaminants is influenced by surface charge. Usually, smaller NPs have more adsorption capability (Ramasamy et al., 2022). Important for pragmatic uses are the reusability and regeneration of waste-derived NPs. Reusing NPs for several purification cycles lowers the environmental impact and expense for water remediation (Garg & Anjum, 2024).

3.3 Financial and Environmental Consequences of the Urgent Need for Ecological Water Treatment

Driven by population rise, industrial development, and agricultural intensification, the global demand for freshwater resources is quickly growing (Zhang & Matharu, 2019). Along with this rise in waste material generation, including agricultural leftovers, industrial wastes, and municipal solid trash, there is concern for sustainable growth (Ali et al., 2023). Growing waste volumes and water scarcity provide a double conundrum for which innovative and sustainable water treatment solutions are needed (Arora & Attri, 2020). Conventional water treatment methods might rely on energy-intensive operations and the consumption of virgin materials, therefore worsening environmental harm and economic expenses (Garg & Anjum, 2024). Recycling waste materials for water treatment is a convincing substitute that provides a road toward a circular economy reducing environmental impact and maybe lowering treatment costs (Das et al., 2024b; Dey et al., 2024c). Analyzing the sustainability and efficiency of several strategies across many waste sources, researchers must investigate closely the financial and environmental consequences of using recycled waste in water treatment technologies (Khajuria et al., 2022). The assessment will highlight knowledge gaps and topics for more study as well as studies unequivocally evaluating the sustainability of some methods covering both financial and environmental elements (Mujtaba et al., 2023).

4 Environmental Effects of Waste-Based Water Treatment and Future Prospective

Evaluating the environmental effects of water treatment strategies depends critically on life cycle assessment. From raw material extraction and processing to manufacturing operation and end-of-life disposal, this assessment evaluates the whole life cycle of a process, therefore measuring energy consumption, greenhouse gas emissions, water use, and other environmental consequences (Ghimire & Ariya, 2020). Studies contrasting waste-based water treatment techniques with conventional systems have found that waste-based treatments can sometimes have major environmental benefits (Chi & Shaikh, 2024). Studies on employing waste plastics in other mixtures, for example, show lower greenhouse gas emissions and less dependency on virgin materials, even if some recycling chain operations might have particularly detrimental effects, including ozone depletion (Ali et al., 2023). Comparatively, another study assessing the effects of nutrient removal and resource recovery in a small-scale water resource recovery plant found a notable decrease in effluent impacts and greenhouse gas emissions (He et al., 2022). Still, environmental performance of waste-based solutions will depend on the kind of waste material, processing techniques, and particular treatment technology used (Part et al., 2018).

Generally speaking, economic evaluations including capital, operational, maintenance, and value of the treated water guide to evaluate the viability of waste-based water treatment methods (Vijayvargiya et al. 2024). Studies have demonstrated that waste-based systems may usually be more reasonably priced than conventional choices especially when considering the cost reductions linked with waste disposal and reduced reliance on virgin resources (Garg et al., 2023). Still, the financial viability is affected by the availability and cost of gathering and processing waste products, the viability of the resultant treatment materials, and the size of the operation (Onu et al., 2023). Careful financial research is needed to ensure that waste-based water treatment technologies are competitive with traditional treatments and financially feasible (Eddy et al. 2025). Support from laws and conventions will help ensure broader recognition and adoption of waste-based water treatment technologies (Nko Okina Solomon et al. 2024). These rules should eliminate any legal obstacles, assure the quality of the generated treated water, and encourage the usage of waste products (Musee 2011). Lack of a clear regulatory framework might make new technologies less welcomed.

Policies should also promote research and development in this field, therefore facilitating the development of very affordable waste-based water treatment systems with ecologically friendly results. Clear rules and procedures help to guarantee the quality and safety of the treated water, therefore fostering public confidence and acceptance of these novel ideas (Sarris et al., 2023).

Evaluating the whole environmental effect of waste-derived NPs depends on their lifetime evaluation as well including the expense of energy and resources for their production, usage, and disposal (Hristova et al., 2023). The evaluation should also consider the leaking of NPs and the effects of waste management strategies on the surroundings. Maintaining the long-term viability of this strategy depends on this thorough study (Komkova & Habert, 2023). There are still difficulties even if waste-derived NPs show promise for water filtration. Some NPs, especially at large dosages, have toxicity that raises issues. Long-term environmental effects call more research (Eddy et al., 2024). Crucially important are scalability and standardizing of synthesis and application techniques. Development and commercialization might be hampered by regulatory obstacles as well as by the absence of uniform procedures (Parthasarathy, 2021). The emphasis of further studies should be on tackling these difficulties. More ecologically friendly and efficient synthesis techniques are still to be developed. Enhancement of purification efficiency can result from optimization of nanoparticle characteristics for particular contaminants (Sharma et al., 2023). Synergistic results can come from looking at how waste-derived nanoparticle technologies could be used with other water treatment approaches (Partow et al., 2022). Development of suitable policy and legal structures is absolutely essential. There is need for well-defined rules on safe production, processing, and disposal of waste-derived NPs (Rudovica et al., 2021).

5 Conclusion

Sustainable water management is a feasible way toward not only managing water shortage but also the growing problem of garbage disposal by means of waste material repurposing for water treatment. Many studies have revealed the capabilities of some waste sources to provide relatively low and ecologically acceptable water treatment solutions. Adsorption, coagulation-flocculation, bioremediation, and membrane filtration are among the treatment methods able to effectively employ recovered waste materials. Life cycle assessments and economic studies are highly important for establishing the sustainability of these solutions considering their complete life cycle impacts and cost-effectiveness. Widespread acceptance of waste-based water treatment depends on supporting legislation and regulations guaranteeing the environmental safety and economic viability of these innovative technologies. More research is needed to utilize present technologies, look at new waste sources, and create strong regulatory frameworks to assist the change toward a more circular and sustainable water management system.

References

Akintelu, S. A., Folorunso, A. S., Folorunso, F. A., & Oyebamiji, A. K. (2020). Green synthesis of copper oxide nanoparticles for biomedical application and environmental remediation. *Heliyon, 6*(7), e04508. https://doi.org/10.1016/j.heliyon.2020.e04508

Ali, B. T. I., Kusumawati, Y., Jaafar, J., Sulistiono, D. O., & Widiastuti, N. (2023). Low-cost membrane from polyethylene terephthalate bottle waste for water purification and chromium removal: Modification and application. *RSC Advances, 13*(13), 8985–8995. https://doi.org/10.1039/d3ra00827d

Anjum, A., Gupta, D., Singh, B., Garg, R., Pani, B., Kashif, M., & Jain, S. (2024). Clay–polymer nanocomposites for effective water treatment: Opportunities, challenges, and future prospects. *Environmental Monitoring and Assessment, 196*(7), 666. https://doi.org/10.1007/s10661-024-12823-8

Ansar, S., Tabassum, H., Aladwan, N. S. M., Naiman Ali, M., Almaarik, B., AlMahrouqi, S., Abudawood, M., Banu, N., & Alsubki, R. (2020). Eco friendly silver nanoparticles synthesis by Brassica oleracea and its antibacterial, anticancer and antioxidant properties. Scientific Reports, 10(1), 18564. https://doi.org/10.1038/s41598-020-74371-8

Arora, B., & Attri, P. (2020). Carbon nanotubes (CNTs): A potential nanomaterial for water purification. *Journal Of Composites Science, 4*(3), 135. https://doi.org/10.3390/jcs4030135

Bala, K., Anshul, V., Dogra, S., Das, A., Kumar, A. P., & Nandini, R. (2024a). Sustainable Management for Urban Plastic Waste Generation. *Sustainable Management of Urban Plastic Waste Through Circular Economic Approaches, 211–236.* https://doi.org/10.1201/9781003465447-13

Bala, K., Dogra, A., Lata, P., & Das, A. P. (2024b). Discussing sources and biological-chemical remediation approaches for microplastic pollution. *Sustainable Microbial Technology for Synthetic and Cellulosic Microfiber Bioremediation, 37–59.* https://doi.org/10.1007/978-3-031-62600-5_3

Bansal, M., Garg, R., Garg, V. K., Garg, R., & Singh, D. (2022). Sequestration of heavy metal ions from multi-metal simulated wastewater systems using processed agricultural biomass. *Chemosphere, 296*(December 2021), 133966. https://doi.org/10.1016/j.chemosphere.2022.133966

Bekru, A. G., Tufa, L. T., Zelekew, O. A., Goddati, M., Lee, J., & Sabir, F. K. (2022). Green Synthesis of a CuOZnO Nanocomposite for Efficient Photodegradation of Methylene Blue and Reduction of 4-Nitrophenol. *ACS Omega, 7*(35), 30908–30919. https://doi.org/10.1021/acsomega.2c02687

Chi, L., & Shaikh, M. O. (2024). Upcycling plastic waste into valuable carbon nanomaterials. *ChemNanoMat.* https://doi.org/10.1002/cnma.202400409

Chithra, R., Kishorekumar, P., Manojkumar, M., & Praveen, S. S. (2023). Sustainable waste management using block chain techniques for Smart City. In *3rd international conference on innovative mechanisms for industry applications, ICIMIA 2023—Proceedings* (pp. 372–376). https://doi.org/10.1109/ICIMIA60377.2023.10426053

Chums-ard, W., Fawcett, D., Fung, C. C., & Poinern, G. E. J. (2019). Biogenic synthesis of gold nanoparticles from waste watermelon and their antibacterial activity against Escherichia coli and Staphylococcus epidermidis. *International Journal of Research in Medical Sciences, 7*(7), 2499. https://doi.org/10.18203/2320-6012.ijrms20192874

Dang, H., Fawcett, D., & Poinern, G. E. J. (2019). Green synthesis of gold nanoparticles from waste macadamia nut shells and their antimicrobial activity against Escherichia coli and Staphylococcus epidermis. *International Journal of Research in Medical Sciences, 7*(4), 1171. https://doi.org/10.18203/2320-6012.ijrms20191320

Damirchi, B. T. Z., Charati, F. R., Akbari, R., & Daneshvar, A. (2020). Green synthesis of silver nanoparticles using the aqueous extract of Viscum album Fruit. *Nanochemistry Research, 5*(1), 104–110. https://doi.org/10.22036/ncr.2020.01.010

Das, A. P., Sukla, L. B., Pradhan, N., & Nayak, S. (2011). Manganese biomining: A review. *Bioresource Technology, 102*(16), 7381–7387. https://doi.org/10.1016/J.BIORTECH.2011.05.018

Das, P. E., Abu-Yousef, I. A., Majdalawieh, A. F., Narasimhan, S., & Poltronieri, P. (2020). Green Synthesis of Encapsulated Copper Nanoparticles Using a Hydroalcoholic Extract of Moringa oleifera Leaves and Assessment of Their Antioxidant and Antimicrobial Activities. *Molecules, 25*(3). https://doi.org/10.3390/molecules25030555

Das, A., Dey, S., Bala, K., Gnanasekaran, R., & Das, A. P. (2024a). Microbial bioremediation Technology for Sustainable Treatment and Management of synthetic microfiber waste. *Sustainable Microbial Technology for Synthetic and Cellulosic Microfiber Bioremediation, 221–238.* https://doi.org/10.1007/978-3-031-62600-5_12

Das, A., Dey, S., & Das, A. P. (2024b). Ecotoxicological impacts of synthetic microfiber pollutants and development of sustainable mitigation strategies. *Environmental Chemistry and Ecotoxicology.* https://doi.org/10.1016/J.ENCECO.2024.12.003

Das, A., Ray, S. K., Mohanty, M., Mohanty, J., Dey, S., & Das, A. P. (2024c). Ecotoxicity of microplastic wastes and their sustainable management: A review. *Environmental Chemistry and Ecotoxicology, 6*, 144–152. https://doi.org/10.1016/J.ENCECO.2024.05.003

Deepa, Ameen, F., Amirul Islam, M., & Dhanker, R. (2022). Green synthesis of silver nanoparticles from vegetable waste of pea Pisum sativum and bottle gourd Lagenaria siceraria: Characterization and antibacterial properties. *Frontiers in Environmental Science*, 10. https://doi.org/10.3389/fenvs.2022.941554

Dey, S., Tripathy, B., Kumar, M. S., & Das, A. P. (2023). Ecotoxicological consequences of manganese mining pollutants and their biological remediation. *Environmental Chemistry and Ecotoxicology.* https://doi.org/10.1016/J.ENCECO.2023.01.001

Dey, S., Bissoyi, A., Koteswara Reddy, G., & Das, A. P. (2024a). Synthetic microfibers: An urban emerging pollutant demanding for sustainable environmental solutions. *Sustainable Microbial Technology for Synthetic and Cellulosic Microfiber Bioremediation, 107–121.* https://doi.org/10.1007/978-3-031-62600-5_7

Dey, S., Dash, N., Ray, S. K., Priyadarsini, S., Chhotaray, C., & Das, A. P. (2024b). Occurrence and quantification of Tire Wear plastics pollutants from Highway City dust of Bhubaneswar City of Odisha. *Sustainable Microbial Technology for Synthetic and Cellulosic Microfiber Bioremediation, 77–89.* https://doi.org/10.1007/978-3-031-62600-5_5

Dey, S., Dash, A., & Das, A. P. (2024c). Bioremediation as a sustainable and circular solution for mining waste management and Bioenergy generation. *Sustainable Management of Mining Waste and Tailings: A Circular Economy Approach, 107–119.* https://doi.org/10.1201/9781003442455-6/BIOREMEDIATION-SUSTAINABLE-CIRCULAR-SOLUTION-MINING-WASTE-MANAGEMENT-BIOENERGY-GENERATION-SUDESHNA-DEY-AKANSHYA-DASH-ALOK-PRASAD-DAS

Desmarais, M., Pirade, F., Zhang, J., & Rene, E. R. (2020). Biohydrometallurgical processes for the recovery of precious and base metals from waste electrical and electronic equipments: Current trends and perspectives. *Bioresource Technology Reports, 11*, 100526. https://doi.org/10.1016/j.biteb.2020.100526

Dubey, R., Dutta, D., Sarkar, A., & Chattopadhyay, P. (2021). Functionalized carbon nanotubes: Synthesis, properties and applications in water purification, drug delivery, and material and biomedical sciences. *Nanoscale Advances, 3*(20), 5722–5744. https://doi.org/10.1039/D1NA00293G

Eddy, N. O., Garg, R., Ukpe, R. A., Akpanudo, N. W., Abugu, H., Garg, R., Anjum, A., & Anand, B. (2024). Ethical and environmental implications associated with the application of nanotechnology. In *Smart and sustainable applications of nanocomposites* (pp. 274–299). https://doi.org/10.4018/979-8-3693-1094-6.ch010

Eddy, N. O., Ukpe, R. A., Garg, R., Garg, R., Odiongenyi, A., Ameh, P., & Nyambi Akpet, I. (2025). Enhancing water purification efficiency through adsorption and photocatalysis: models, applications, and challenges. *International Journal of Environmental Analytical Chemistry, 105*(7), 1577–1594. https://doi.org/10.1080/03067319.2023.2295934

el Essawy, N. A., Konsowa, A. H., Elnouby, M., & Farag, H. A. (2017). A novel one-step synthesis for carbon-based nanomaterials from polyethylene terephthalate (PET) bottles waste. *Journal of the Air and Waste Management Association, 67*(3), 358–370. https://doi.org/10.1080/10962247.2016.1242517

Falk, G., Shinhe, G. P., Teixeira, L. B., Moraes, E. G., & de Oliveira, A. P. N. (2019). Synthesis of silica nanoparticles from sugarcane bagasse ash and nano-silicon via magnesiothermic reactions. *Ceramics International, 45*(17), 21618–21624. https://doi.org/10.1016/j.ceramint.2019.07.157

Gandhi, C. P., Garg, R., & Eddy, N. O. (2021). Application of biosynthesized nano-catalyst for biodiesel synthesis and impact assessment of factors influencing the yield. *Nanosystems: Physics, Chemistry, Mathematics, 12*(6), 808–817. https://doi.org/10.17586/2220-8054-2021-12-6-808-817

Garg, R., & Anjum, A. (2024). Smart and sustainable applications of nanocomposites. In R. Garg & A. Anjum (Eds.), *Smart and sustainable applications of nanocomposites: Vol. i.* IGI Global. https://doi.org/10.4018/9798369310946

Garg, R., Rani, P., Garg, R., & Eddy, N. O. (2021). Study on potential applications and toxicity analysis of green synthesized nanoparticles. *Turkish Journal of Chemistry, 45*(6), 1690–1706. https://doi.org/10.3906/kim-2106-59

Garg, R., Garg, R., Okon Eddy, N., Ibrahim Almohana, A., Fahad Almojil, S., Amir Khan, M., & Ho Hong, S. (2022a). Biosynthesized silica-based zinc oxide nanocomposites for the sequestration of heavy metal ions from aqueous solutions. *Journal of King Saud University - Science, 34*(4), 101996. https://doi.org/10.1016/j.jksus.2022.101996

Garg, R., Rani, P., Garg, R., Khan, M. A., Khan, N. A., Khan, A. H., & Américo-Pinheiro, J. H. P. (2022b). Biomedical and catalytic applications of agri-based biosynthesized silver nanoparticles. *Environmental Pollution, 310*(July), 119830. https://doi.org/10.1016/j.envpol.2022.119830

Garg, R., Garg, R., Sharma, R., Anjum, A., & Eddy, N. O. (2023). Introduction to nanotechnology and nanomaterials. *Nanotechnology for Sustainable Agriculture, Food and Environment, 1–15.* https://doi.org/10.1201/9781003397861-1

Ghimire, H., & Ariya, P. A. (2020). E-wastes: Bridging the knowledge gaps in global production budgets, composition, recycling and sustainability implications. *Sustainable Chemistry, 1*(2), 154–182. https://doi.org/10.3390/suschem1020012

Golthi, V., & Kommu, J. (2023). An eco-friendly and sustainable method for producing Fe3O4 nanoparticles using Jatropha podagrica leaf extract for efficient dye degradation and antibacterial uses. *Hybrid Advances, 4*(1), 100110. https://doi.org/10.1016/j.hybadv.2023.100110

He, Y., Wang, M., Mao, X., Zhang, M., Feng, X., Ji, Z., Li, Y., Xiong, Z., Xing, Z., Hu, J., & Wu, G. (2022). Selective recovery of gold from e-waste with 3D hierarchical porous amidoximated fabrics and its application in the reduction of 4-nitrophenol. *Radiation Physics and Chemistry, 194*, 110006. https://doi.org/10.1016/j.radphyschem.2022.110006

Hristova, T., Savov, N., & Yanev, N. (2023). Municipal solid waste and non-hazardous waste processing for sustainable circular economy through blockchain. *Journal Of Chemical Technology And Metallurgy, 58*(3), 513–521. https://doi.org/10.59957/jctm.v58i3.81

Ikram, R., Mohamed Jan, B., Nagy, P. B., & Szabo, T. (2023). Recycling waste sources into nanocomposites of graphene materials: Overview from an energy-focused perspective. *Nanotechnology Reviews, 12*, 1. https://doi.org/10.1515/ntrev-2022-0512

Jagatee, S., Priyadarshini, S., Rath, C. C., & Das, A. P. (2024). Synthetic microfiber: An enduring environmental problem linked to sustainable development. *Environmental Science and Engineering, Part, F2333*, 93–112. https://doi.org/10.1007/978-3-031-51792-1_6

John, A., & Das, A. P. (2024). Government strategies for achieving circular economy for plastic waste management in Southeast Asia. *Sustainable Management of Urban Plastic Waste Through Circular Economic Approaches, 72–82.* https://doi.org/10.1201/9781003465447-6/GOVERNMENT-STRATEGIES-ACHIEVING-CIRCULAR-ECONOMY-PLASTIC-WASTE-MANAGEMENT-SOUTHEAST-ASIA-AKBAR-JOHN-ALOK-PRASAD-DAS

Joshi, N., Filip, J., Coker, V. S., Sadhukhan, J., Safarik, I., Bagshaw, H., & Lloyd, J. R. (2018). Microbial reduction of natural Fe(III) minerals; toward the sustainable production of functional magnetic nanoparticles. *Frontiers in Environmental Science, 6*(NOV). https://doi.org/10.3389/fenvs.2018.00127

Kadam, J., Dhawal, P., Barve, S., & Kakodkar, S. (2020). Green synthesis of silver nanoparticles using cauliflower waste and their multifaceted applications in photocatalytic degradation of methylene blue dye and Hg2+ biosensing. *SN. Applied Sciences, 2*, 4. https://doi.org/10.1007/s42452-020-2543-4

Kadier, A., Ilyas, R. A., Huzaifah, M. R. M., Harihastuti, N., Sapuan, S. M., Harussani, M. M., Azlin, M. N. M., Yuliasni, R., Ibrahim, R., Atikah, M. S. N., Wang, J., Chandrasekhar, K., Amirulislam, M., Sharma, S., Punia, S., Rajasekar, A., Asyraf, M. R. M., & Ishak, M. R. (2021). Use of industrial wastes as sustainable nutrient sources for bacterial cellulose (BC) production: Mechanism, advances, and future perspectives. *Polymers, 13*(19). https://doi.org/10.3390/polym13193365

Kanchi, S., Kumar, G., Lo, A. Y., Tseng, C. M., Chen, S. K., Lin, C. Y., & Chin, T. S. (2018). Exploitation of de-oiled jatropha waste for gold nanoparticles synthesis: A green approach. *Arabian Journal of Chemistry, 11*(2), 247–255. https://doi.org/10.1016/j.arabjc.2014.08.006

Khajuria, A., Atienza, V. A., Chavanich, S., Henning, W., Islam, I., Kral, U., Liu, M., Liu, X., Murthy, I. K., Oyedotun, T. D. T., Verma, P., Xu, G., Zeng, X., & Li, J. (2022). Accelerating circular economy solutions to achieve the 2030 agenda for sustainable development goals. *Circular Economy, 1*(1), 100001. https://doi.org/10.1016/j.cec.2022.100001

Khalid, S., Chaudhary, M. N., Nazir, R., Ahmad, S. R., Hussain, N., Ayub, Y., & Ibrar, M. (2023). Biochar supported metallo-inorganic nanocomposite: A green approach for decontamination of heavy metals from water. *PLoS One, 18*(9 September). https://doi.org/10.1371/journal.pone.0289069

Komkova, A., & Habert, G. (2023). The urban-industrial metabolism: Contribution of waste recycling to the circular economy objectives within the construction sector. *Journal of Physics: Conference Series, 2600*(17). https://doi.org/10.1088/1742-6596/2600/17/172002

Larder, R. R., & Hatton, F. L. (2023). Enabling the polymer circular economy: Innovations in photoluminescent labeling of plastic waste for enhanced sorting. *ACS Polymers Au, 3*(2), 182–201. https://doi.org/10.1021/acspolymersau.2c00040

Liou, T. H., & Jheng, J. Y. (2018). Synthesis of high-quality ordered mesoporous carbons using a sustainable way from recycling of E-waste as a silica template source. *ACS Sustainable Chemistry and Engineering, 6*(5), 6507–6517. https://doi.org/10.1021/acssuschemeng.8b00310

Ma, L., Wei, Q., Chen, Y., Song, Q., Sun, C., Wang, Z., & Wu, G. (2018). Removal of cadmium from aqueous solutions using industrial coal fly ash-nZVI. *Royal Society Open Science, 5*(2). https://doi.org/10.1098/rsos.171051

Magnabosco, P., Masi, A., Shukla, R., Bansal, V., & Carletti, P. (2023). Advancing the impact of plant biostimulants to sustainable agriculture through nanotechnologies. *Chemical and Biological Technologies in Agriculture, 10*(1). https://doi.org/10.1186/s40538-023-00491-8

Mallick, K., Sahu, A., Dubey, N. K., & Das, A. P. (2023). Harvesting marine plastic pollutants-derived renewable energy: A comprehensive review on applied energy and sustainable approach. *Journal of Environmental Management, 348*, 119371. https://doi.org/10.1016/J.JENVMAN.2023.119371

Maslennikov, A., Peretz, R., Vadivel, V. K., & Mamane, H. (2022). Recycled paper sludge (RPS)-derived nanocellulose: Production, detection and water treatment application. *Applied Sciences (Switzerland), 12*(6). https://doi.org/10.3390/app12063077

Mishra, S., Charan Rath, C., & Das, A. P. (2019). Marine microfiber pollution: A review on present status and future challenges. *Marine Pollution Bulletin, 140*, 188–197. https://doi.org/10.1016/J.MARPOLBUL.2019.01.039

Mishra, S., Dash, D., Al-Tawaha, A. R. M. S., & Das, A. P. (2022). A review on heavy metal ion adsorption on synthetic microfiber surface in aquatic environments. *Applied Biochemistry and Biotechnology, 194*(10), 4639–4654. https://doi.org/10.1007/S12010-022-04029-W/METRICS

Mishra, S., Dash, D., & Das, A. P. (2024). Microbial degradation of polyester microfibers using indigenously isolated bacterial strain Exiguobacterium Sp. *CLEAN—Soil, Air, Water, 52*(12), e202300343. https://doi.org/10.1002/CLEN.202300343

Mohamed, O. A., Farghali, A. A., Eessaa, A. K., & El-Shamy, A. M. (2022). Cost-effective and green additives of pozzolanic material derived from the waste of alum sludge for successful replacement of Portland cement. *Scientific Reports, 12*(1), 1–22. https://doi.org/10.1038/s41598-022-25246-7

Mohanty, J., Dey, S., Bala, K., & Das, A. P. (2024a). Recent advances in microplastic pollution for its sustainable management. *Sustainable Management of Urban Plastic Waste Through Circular Economic Approaches, 27–48.* https://doi.org/10.1201/9781003465447-3/RECENT-ADVANCES-MICROPLASTIC-POLLUTION-SUSTAINABLE-MANAGEMENT-JYOTIRMAYEE-MOHANTY-SUDESHNA-DEY-KIRAN-BALA-ALOK-PRASAD-DAS

Mohanty, M., Tripathy, B., Koteswara Reddy, G., & Das, A. P. (2024b). Greener technologies for establishing restraint over microplastic pollution. *Sustainable Microbial Technology for Synthetic and Cellulosic Microfiber Bioremediation, 239–258.* https://doi.org/10.1007/978-3-031-62600-5_13

Mujtaba, M., Fernandes Fraceto, L., Fazeli, M., Mukherjee, S., Savassa, S. M., Araujo de Medeiros, G., do Espírito Santo Pereira, A., Mancini, S. D., Lipponen, J., & Vilaplana, F. (2023). Lignocellulosic biomass from agricultural waste to the circular economy: A review with focus on biofuels, biocomposites and bioplastics. *Journal of Cleaner Production, 402*, 136815. https://doi.org/10.1016/j.jclepro.2023.136815

Murugadoss, G., Kumar, D. D., Kumar, M. R., Venkatesh, N., & Sakthivel, P. (2021). Silver decorated CeO2 nanoparticles for rapid photocatalytic degradation of textile rose bengal dye. *Scientific Reports, 11*(1), 1–13. https://doi.org/10.1038/s41598-020-79993-6

Naaz, S., Shet, V. B., & Mubarak, N. M. (2024). Green synthesis of copper oxide nanoparticles: Characterization and applications for environmental and biomedical fields. *Canadian Journal of Chemical Engineering, 102*(4), 1454–1465. https://doi.org/10.1002/cjce.25142

Musee, N. (2011). Nanotechnology risk assessment from a waste management perspective: Are the current tools adequate? *Human and Experimental Toxicology, 30*(8), 820–835. https://doi.org/10.1177/0960327110384525

Nko Okina Solomon, Peter Simpa, Olubunmi Adeolu Adenekan, & Scholar Chinenye Obasi. (2024). Sustainable nanomaterials' role in green supply chains and environmental sustainability. *Engineering Science & Technology Journal, 5*(5), 1678–1694. https://doi.org/10.51594/estj.v5i5.1136

Onu, M. A., Ayeleru, O. O., Modekwe, H. U., & Olubambi, P. A. (2023). Valorization of plastic wastes for the development of adsorbent designed for the removal of emerging contaminants in wastewater. *Advances in Environmental and Engineering Research, 04*(04), 1–26. https://doi.org/10.21926/aeer.2304055

Part, F., Berge, N., Baran, P., Stringfellow, A., Sun, W., Bartelt-Hunt, S., Mitrano, D., Li, L., Hennebert, P., Quicker, P., Bolyard, S. C., & Huber-Humer, M. (2018). A review of the fate of engineered nanomaterials in municipal solid waste streams. *Waste Management, 75*, 427–449. https://doi.org/10.1016/j.wasman.2018.02.012

Parthasarathy, M. (2021). Challenges and emerging trends in toner waste recycling: A review. *Recycling, 6*(3). https://doi.org/10.3390/recycling6030057

Partow, A. J., Meng, S., Wong, A. J., Savin, D. A., & Tong, Z. (2022). Recyclable & highly porous organo-aerogel adsorbents from biowaste for organic contaminants' removal. *Science of the Total Environment, 827*. https://doi.org/10.1016/j.scitotenv.2022.154051

Pechyen, C., Ponsanti, K., Tangnorawich, B., & Ngernyuang, N. (2021). Waste fruit peel – Mediated green synthesis of biocompatible gold nanoparticles. *Journal of Materials Research and Technology, 14*, 2982–2991. https://doi.org/10.1016/j.jmrt.2021.08.111

Pineda-Vásquez, T., Rendón-Castrillón, L., Ramírez-Carmona, M., & Ocampo-López, C. (2024). From E-waste to high-value materials: Sustainable synthesis of metal, metal oxide, and MOF nanoparticles from waste printed circuit boards. *Nanomaterials, 14*(1). https://doi.org/10.3390/nano14010069

Priyadarshini, B., Santhosh Kumar, M., & Das, A. P. (2024). Microbial degradation of synthetic plastics released from COVID-19 generated single use disposable face masks. *Environmental Science and Engineering, Part, F2827*, 1–16. https://doi.org/10.1007/978-3-031-50840-0_1

Priyadarsini, S., & Das, A. P. (2025). Lithium bioleaching: A review on microbial-assisted sustainable technology for lithium bio-circularity. *Journal of Water Process Engineering, 69*, 106744. https://doi.org/10.1016/j.jwpe.2024.106744

Pujeri, U. S., Pujar, A. S., & Yadawe, M. (2018). Study on green synthesis, characterization and anticancer activity of thorium nanoparticles using tomato (Lycopersicon Esculentum) extract. *IOSR Journal of Applied Chemistry (IOSR-JAC), 11*(7), 25–29. https://doi.org/10.9790/5736-1107032529

Rajahalme, J., Perämäki, S., & Väisänen, A. (2022). Separation of palladium and silver from E-waste leachate: Effect of nitric acid concentration on adsorption to thiol scavenger. *Chemical Engineering Journal Advances, 10*(February). https://doi.org/10.1016/j.ceja.2022.100280

Ramasamy, T., Satheesh, L. G., Selvaraj, V., Bazaka, O., Levchenko, I., Bazaka, K., & Mandhakini, M. (2022). Spinel CoFe2O4 nanoflakes: A path to enhance energy generation and environmental remediation potential of waste-derived rGO. *Nanomaterials, 12*(21). https://doi.org/10.3390/nano12213822

Ramesh, S., & Radhakrishnan, P. (2019). Cellulose nanoparticles from agro-industrial waste for the development of active packaging. *Applied Surface Science, 484*, 1274–1281. https://doi.org/10.1016/j.apsusc.2019.04.003

Rao, G. P., Lu, C., & Su, F. (2007). Sorption of divalent metal ions from aqueous solution by carbon nanotubes: A review. *Separation and Purification Technology, 58*(1), 224–231. https://doi.org/10.1016/j.seppur.2006.12.006

Ray, S. K., Dey, S., Das, A. P., & John, A. (2024). Microplastic waste management through greener technologies : A review. *Sustainable Management of Urban Plastic Waste Through Circular Economic Approaches, 237–261*. https://doi.org/10.1201/9781003465447-14

Rautela, A., Rani, J., & Debnath (Das), M. (2019). Green synthesis of silver nanoparticles from Tectona grandis seeds extract: characterization and mechanism of antimicrobial action on different microorganisms. *Journal of Analytical Science and Technology, 10*(1), 5. https://doi.org/10.1186/s40543-018-0163-z

Rizal, S., Abdul Khalil, H. P. S., Oyekanmi, A. A., Olaiya, N. G., Abdullah, C. K., Yahya, E. B., Alfatah, T., Sabaruddin, F. A., & Rahman, A. A. (2021). Cotton wastes functionalized biomaterials from micro to nano: A cleaner approach for a sustainable environmental application. *Polymers, 13*(7), 1006. https://doi.org/10.3390/polym13071006

Rudovica, V., Rotter, A., Gaudêncio, S. P., Novoveská, L., Akgül, F., Akslen-Hoel, L. K., Alexandrino, D. A. M., Anne, O., Arbidans, L., Atanassova, M., Bełdowska, M., Bełdowski, J., Bhatnagar, A., Bikovens, O., Bisters, V., Carvalho, M. F., Catalá, T. S., Dubnika, A., Erdoğan, A., et al. (2021). Valorization of marine waste: Use of industrial by-products and beach wrack towards the production of high added-value products. *Frontiers in Marine Science, 8*. https://doi.org/10.3389/fmars.2021.723333

Sagir, M., Tahir, M. B., Akram, J., Tahir, M. S., & Waheed, U. (2020). Nanoparticles and significance of photocatalytic nanoparticles in wastewater treatment: A review. *Current Analytical Chemistry, 17*(1), 38–48. https://doi.org/10.2174/1573411016999200607172553

Sahu, A., Dey, S., Mallick, K., Priyadarsini, S., Behera, I. D., & Das, A. P. (2024). A review on advanced technology for sustainable management of synthetic microplastic waste. *Sustainable Management of Urban Plastic Waste Through Circular Economic Approaches, 262–279*. https://doi.org/10.1201/9781003465447-15

Sandeep, S., Santosh Kumar, S., & Krisn Pratap, M. (2023). Synthesis and Characterization of Iron Oxide Nanoparticles by Murraya Koenigii leaves. *REST Journal on Advances in Mechanical Engineering, 2*(2), 08–11. https://doi.org/10.46632/jame/2/2/2

Sarris, D., Tsouko, E., Kothri, M., Anagnostou, M., Karageorgiou, E., & Papanikolaou, S. (2023). Upgrading major waste streams derived from the biodiesel industry and olive Mills via microbial bioprocessing with non-conventional Yarrowia lipolytica strains. *Fermentation, 9*(3). https://doi.org/10.3390/fermentation9030251

Sawalha, S., Moulaee, K., Nocito, G., Silvestri, A., Petralia, S., Prato, M., Bettini, S., Valli, L., Conoci, S., & Neri, G. (2021). Carbon-dots conductometric sensor for high performance gas sensing. *Carbon Trends, 5*. https://doi.org/10.1016/j.cartre.2021.100105

Schiavi, D., Francesconi, S., Taddei, A. R., Fortunati, E., & Balestra, G. M. (2022). Exploring cellulose nanocrystals obtained from olive tree wastes as sustainable crop protection tool against bacterial diseases. *Scientific Reports, 12*(1), 1–14. https://doi.org/10.1038/s41598-022-10225-9

Sebastian, A., Nangia, A., & Prasad, M. N. V. (2018). A green synthetic route to phenolics fabricated magnetite nanoparticles from coconut husk extract: Implications to treat metal contaminated water and heavy metal stress in Oryza sativa L. *Journal of Cleaner Production, 174*, 355–366. https://doi.org/10.1016/j.jclepro.2017.10.343

Serang, Y., Chen, C., & Lee, M. J. (2024). Green synthesis of magnesium oxide nanoparticles using melaleuca Leucadendra waste. *Materials Science Forum, 1123*, 109–118. https://doi.org/10.4028/p-NSjf2n

Sharma, R., Garg, R., Bali, M., & Eddy, N. O. (2023). Potential applications of green-synthesized iron oxide NPs for environmental remediation. *Environmental Monitoring and Assessment, 195*(11). https://doi.org/10.1007/s10661-023-12035-6

Sharma, R., Garg, R., Bali, M. et al. (2023). Biogenic synthesis of iron oxide nanoparticles using leaf extract of Spilanthes acmella: antioxidation potential and adsorptive removal of heavy metal ions. *Environ Monit Assess 195*, 1345. https://doi.org/10.1007/s10661-023-11860-z

Sharma, R., Lata, S., & Garg, R. (2024). Valorisation of agricultural waste and their role in green synthesis of value-added nanoparticles. *Environmental Technology Reviews, 13*(1), 40–59. https://doi.org/10.1080/21622515.2023.2283412

Shen, Y. (2017). Rice Husk Silica-Derived Nanomaterials for Battery Applications: A Literature Review. *Journal of Agricultural and Food Chemistry, 65*(5), 995–1004. https://doi.org/10.1021/acs.jafc.6b04777

Tripathy, B., Dey, S., Dubey, N. K., Lo, W.-C., & Das, A. P. (2024). *An omics-based approach for managing the risk of COVID-19-generated wastes* (pp. 51–66). https://doi.org/10.1007/978-3-031-50840-0_4

Uzu, G., Sobanska, S., Sarret, G., Muñoz, M., & Dumat, C. (2010). Foliar Lead uptake by lettuce exposed to atmospheric fallouts. *Environmental Science and Technology, 44*(3), 1036–1042. https://doi.org/10.1021/es902190u

Vijayvargiya, R., Sahu, A. K., & Joshi, A. (2024). Sustainable Approach: Utilizing Plastic Waste-Derived rGO for Multifunctional Thermoplastic Nanocomposites. *Journal of Environmental Nanotechnology, 13*(2), 385–390. https://doi.org/10.13074/jent.2024.06.242609

Vo, D. T., Sabrina, S., & Lee, C. K. (2017). Silver deposited carboxymethyl chitosan-grafted magnetic nanoparticles as dual action deliverable antimicrobial materials. *Materials Science and Engineering C, 73*, 544–551. https://doi.org/10.1016/j.msec.2016.12.066

Wang, Z., Yu, J., Zhang, X., Li, N., Liu, B., Li, Y., Wang, Y., Wang, W., Li, Y., Zhang, L., Dissanayake, S., Suib, S. L., & Sun, L. (2016). Large-scale and controllable synthesis of graphene quantum dots from rice husk biomass: A comprehensive utilization strategy. *ACS Applied Materials and Interfaces, 8*(2), 1434–1439. https://doi.org/10.1021/acsami.5b10660

Wardanu, A. P., Indrasti, N. S., & Suprihatin. (2024). Green synthesis of silica nanoparticles (Si-NPs) from palm oil fuel ash (POFA) and its application to purification water: A review. *IOP Conference Series: Earth and Environmental Science, 1358*(1), 012015. https://doi.org/10.1088/1755-1315/1358/1/012015

Xiang, S., Cheng, W., Chi, F., Nie, X., Hayat, T., & Alharbi, N. S. (2020). Photocatalytic removal of U(VI) from wastewater via synergistic carbon-supported zero-valent iron nanoparticles and S. Putrefaciens. *ACS Applied Nano Materials, 3*(2), 1131–1138. https://doi.org/10.1021/acsanm.9b01581

Xie, Y., Fang, Z., Qiu, X., Tsang, E. P., & Liang, B. (2014). Comparisons of the reactivity, reusability and stability of four different zero-valent iron-based nanoparticles. *Chemosphere, 108*, 433–436. https://doi.org/10.1016/j.chemosphere.2014.01.076

Yust, B. G., Rao, N. Z., Schwarzmann, E. T., & Peoples, M. H. (2022). Quantification of spent coffee ground extracts by roast and brew method, and their utility in a green synthesis of gold and silver nanoparticles. *Molecules, 27*(16). https://doi.org/10.3390/molecules27165124

Zhang, Z., & Matharu, A. S. (2019). CHAPTER 9: Adhesives derived from biomass waste streams. In *RSC green chemistry, 2019-January* (Vol. 60, pp. 205–234). https://doi.org/10.1039/9781788012997-00205

Zhuo, C., Alves, J. O., Tenorio, J. A. S., & Levendis, Y. A. (2012). Synthesis of carbon nanomaterials through up-cycling agricultural and municipal solid wastes. *Industrial & Engineering Chemistry Research, 51*(7), 2922–2930. https://doi.org/10.1021/ie202711h

The Impact of Nanotechnology in Water Quality Control and Treatment of Wastewater

Ram Raj Meena

Abstract Water contamination is a critical global health issue, and ensuring access to clean drinking water through affordable and efficient technologies is vital for sustainable economic development. Nanotechnology has emerged as a transformative tool in water quality control and wastewater treatment, offering innovative solutions for pollutant removal. Nanoscale materials, with their exceptional adsorption characteristics, enhanced photocatalytic activity, and high reactivity, have been extensively studied in pilot-scale, laboratory, and in situ applications, showing remarkable potential in improving water treatment processes. This chapter examines the impact of nanotechnology on water quality control and wastewater treatment, focusing on advanced nanomaterials such as carbon nanotubes, graphene, metal nanoparticles (e.g., iron, magnesium zinc, and silver), as well as magnetic-core composites including cobalt, nickel, and iron, metal and, oxide nanoparticles. These materials are compared with conventional treatment methods with respect to cost-efficiency and capability for the removal of organic and inorganic pollutants. Furthermore, this review addresses the challenges, limitations, and future perspectives of employing engineered nanomaterials in the treatment of wastewater and drinking water, highlighting their potential to revolutionize global water quality management.

Keywords Nanotechnology · Wastewater · Nanoparticles

R. R. Meena (✉)
Motilal Nehru National Institute of Technology Allahabad, Prayagraj, India
e-mail: ram.2020rce10@mnnit.ac.in

S. Rajpoot, A. P. Das (eds.), *Innovative Technologies for Waste Management*, Sustainable Environmental Waste Management Strategies,
https://doi.org/10.1007/978-3-031-97184-6_7

1 Introduction

Water is essential for sustaining life and a cornerstone of human civilization. Nevertheless, ensuring universal access to safe, clean, and accessible water remains among the most crucial challenges of the twenty-first century. All over the world, approximately 780 million humans still lack access to better-quality drinking water sources, emphasizing the critical need for effective water treatment solutions, particularly in developing regions with limited or no water and wastewater infrastructure (WHO, 2012). Existing water supply systems are under increasing strain due to longstanding problems and emerging threats undermining their reliability and sustainability. In addition, human activities contribute to water scarcity globally by polluting freshwater resources, impacting communities in both developing and industrialized nations. Also, evolving regulatory frameworks with rigorous water quality standards and emerging novel contaminants demand innovative and adaptive approaches to protect water resources and ensure long-term sustainability (Meena et al., 2023, .John & Das, 2024; Mohanty et al., 2024a, 2024b, 2024c).

Rapid population growth, rising living standards, and global climate change supplementary strain water resources destabilizing freshwater availability and exacerbating the inadequate distribution of water (Paul et al., 2024). To address these challenges, unconventional water sources like stormwater, wastewater, brackish water, contaminated freshwater, and seawater are critical to water management strategies. Nevertheless, traditional wastewater and water treatment methods and infrastructure struggle to meet these demands. Nanotechnology offers transformative opportunities to resolve the boundaries of current water and wastewater treatment systems (Mishra & Das, 2021, 2023). Conventional centralized systems rely heavily on extensive infrastructure and are increasingly unsustainable. In contrast, nanotechnology enables modular, multifunctional, highly efficient processes that promise high-performance and cost-effective water and wastewater treatment solutions. These advancements address the limitations of existing technologies and unlock the potential for economically viable use of unconventional water sources, significantly expanding global water supplies (Ali et al., 2019).

This study explores recent advancements in nanotechnology for improving water quality and wastewater treatment. It provides a detailed analysis of the applications of nanomaterials in various treatment processes while addressing the challenges that limit their widespread adoption. Furthermore, it outlines the research priorities needed to overcome these barriers. While the latent effects of nanomaterials on public health, ecosystems, and treatment processes are critical, these aspects fall outside the scope of this discussion and are not detailed herein.

2 Treatment Applications for Wastewater and Water

Nanomaterials are characterized by having at least one dimension measuring less than 100 nanometers and exhibit distinct physicochemical properties at the nanoscale, which diverge notably from their huge number of counterparts. These properties, from high surface-to-volume ratios and quantum mechanical effects, make nanomaterials highly promising for advanced wastewater and drinking water treatment applications. Applications leveraging nanomaterials can be broadly categorized based on their functional mechanisms. Materials with high specific surface areas, like nano-zero-valent iron (nZVI) and graphene oxide, demonstrate enhanced adsorption capacities, rapid dissolution kinetics, and increased chemical reactivity. These attributes permit the competent removal of contaminants, like heavy metals, organic pollutants, and emerging impurities, through adsorption and catalytic processes (Ahmed & Haider, 2018). In contrast, certain nanomaterials exploit unique discontinuous properties. For instance, superparamagnetic nanoparticles facilitate rapid separation processes under applied magnetic fields, while materials exhibiting localized surface plasmon resonance (e.g., metallic nanoparticles) are utilized in advanced sensing and photothermal applications. Similarly, quantum confinement effects in materials like quantum dots enable precise sensing and photodegradation of pollutants. While most of these applications remain at the experimental stage, several have transitioned to pilot-scale or field-testing phases, demonstrating their feasibility for integration into existing water treatment systems. These advancements highlight the potential of nanotechnology to overcome current limitations in drinking water and wastewater management, offering modular, effective, and scalable solutions.

2.1 Adsorption

Adsorption is widely utilized as an enhancing stage in wastewater and drinking-water treatment processes to remove organic and inorganic impurities. However, the efficiency of traditional adsorbents is often constrained by their limited surface area, a low density of active sites, poor choosiness, and slow adsorption kinetics. Nano-adsorbents, in contrast, offer substantial advantages, including an exceptionally high specific surface area, a greater density of sorption locations, a customizable size of pores, reduced intraparticle diffusion distances, and surface chemistries. These properties make nano-adsorbents a promising alternative for enhancing adsorption performance and overcoming the limitations of conventional adsorbents.

2.1.1 Nanostructured Carbon Adsorbents

2.1.1.1 Organic Removal Using Carbon Nanotubes (CNTs)

Carbon nanotubes (CNTs) exhibit superior adsorption efficiency for a wide range of organic chemicals compared to traditional adsorbents like activated carbon. This enhanced performance is attributed to their exceptionally large specific surface area and the varied interaction mechanisms they enable between contaminants and their surfaces. Primarily the adsorption occurs on the outer surface of CNTs. However, in aqueous environments, the hydrophobic nature of CNTs' graphitic surface often causes them to form loose bundles or aggregates, reducing the effective surface area available for adsorption, although carbon nanotubes (CNTs) tend to aggregate (Alayan et al., 2021; Mashkoor & Nasar, 2020). In comparison, activated carbon, despite having a similar specific surface area, is often limited by its abundance of micropores, which are too small to accommodate larger organic molecules like pharmaceuticals and antibiotics (Bodzek et al., 2020). CNTs, on the other hand, feature larger interstitial pores and more accessible adsorption sites, providing a superior capacity for capturing bulky organic compounds. Moreover, CNTs have a distinct advantage over activated carbon due to their strong attraction for adsorbing low-molecular-weight polar organic complexes. This affinity arises from the variety of interaction mechanisms CNTs enable, including hydrogen bonding, hydrophobic effects, covalent bonding, electrostatic interactions, and π–π interactions (Mashkoor & Nasar, 2020). The unique properties of carbon nanotubes make them highly effective adsorbents for a wide range of organic contaminants. Their graphitic surface, rich in π-electrons, promotes strong π–π bonds with organic molecules containing double bonds (C=C) or aromatic rings, like polar aromatic compounds and polycyclic aromatic hydrocarbons (PAHs) (Kong et al., 2023). Furthermore, functional groups on organic compounds, such as –COOH, –OH, and $-NH_2$, develop hydrogen bonds with the electron-sacrificial surface of CNTs. Under suitable pH conditions, electrostatic interactions also enhance adsorption by binding positively charged organic molecules, including certain antibiotics (Saadi et al., 2019). These versatile adsorption mechanisms allow CNTs to efficiently remove challenging organic pollutants that are not easily addressed with traditional materials.

2.1.1.2 Mitigation of Heavy Metal

Oxidized carbon nanotubes (CNTs) exhibit outstanding metal ion adsorption capabilities, marked by rapid kinetics and high efficiency. This superior performance is primarily due to functional groups on their surface, including carboxyl, hydroxyl, and phenolic groups, which act as important adsorption sites. These groups enable metal ion removal through chemical bonding and electrostatic attraction (Khaled et al., 2019). Oxidation of the CNT surface greatly enhances their adsorption properties, making them highly effective for removing heavy metals. Studies consistently show that CNTs surpass activated carbon in adsorbing metals such as Pb^{2+}, Cd^{2+}, Zn^{2+}, and Cu^{2+} (P. Zhang et al., 2014). This advantage stems from CNTs' extremely controllable adsorption spots and small intraparticle diffusion distances, enabling faster and more efficient adsorption processes. However, CNTs are not suited to replace activated carbon as broad-spectrum adsorbents. Instead of highly tunable surface chemistry, it provides an exceptional edge for targeted applications. These include specialized roles such as polishing steps to remove insistent contaminants in advanced water treatment systems and pre-concentration of trace organic pollutants for investigative determinations (Das et al., 2024b). The limited material quantities required for these applications mitigate the impact of CNTs' higher costs, making them viable in specific, high-value scenarios. In contrast, graphite oxide offers a low-cost alternative for broader applications. Graphite oxide, created through the exfoliation of graphite using strong acids and oxidizing agents, is recognized as an affordable and effective adsorbent (Ali et al., 2019). Studies have demonstrated that sand granules coated with graphite oxide remove heavy metals, such as Hg^{2+}, and large organic pollutants like Rhodamine B dye. This makes graphite oxide-coated sand a promising material for water treatment applications. Its performance has been comparable to commercial activated carbon, suggesting significant potential for large-scale use (Ra et al., 2019). While oxidized CNTs excel in contaminant-specific removal, high-efficiency graphite oxide provides a cost-effective option for more comprehensive applications. Together, these materials offer complementary solutions, proceeding with the capabilities of adsorption technologies in water and wastewater treatment.

2.1.1.3 Recovery and Reimplementation

The ability to regenerate adsorbents is a critical factor influencing their cost-effectiveness and sustainability. C. Lu et al. (2006) stated that carbon nanotubes (CNTs) demonstrate exceptional regeneration performance in metal ion adsorption, largely due to their ability to desorb metal ions by simply reducing the solution pH. At pH below 2, metal ion recovery rates typically exceed 90% and often approach 100%. CNTs maintain their adsorption capacity well over multiple regeneration cycles, showcasing their durability and efficiency. The Zn^{2+} adsorption capacity of single-walled carbon nanotubes (SWCNTs) and multi-walled carbon nanotubes (MWCNTs) decreased by less than 25% even after ten cycles of regeneration and reuse. In stark contrast, the adsorption capacity of activated carbon for Zn^{2+} decreased by 50% or more after just one regeneration cycle, underscoring the superior reusability and stability of CNTs. Further, regression-based statistical analysis indicates that CNT nano-adsorbents can undergo several hundred regeneration and reuse cycles for Zn^{2+} removal while maintaining effective adsorption efficiency, highlighting their durability and suitability for long-term applications. These findings establish CNTs as highly durable and cost-effective adsorbents capable of reducing operational costs and minimizing material replacement in wastewater and drinking-water treatment systems (Lu et al., 2007).

2.1.2 Metallic Nanoparticle Adsorbents

Metal oxides such as titanium dioxide, alumina, and iron oxide are widely recognized for their cost-effectiveness and efficiency as adsorbents in removing heavy metals and radionuclides. The adsorption process is primarily governed by the interaction between dissolved metal ions and oxygen-containing functional groups on the surface of these oxides (Singh et al., 2023). This process occurs in two phases: an initial rapid adsorption phase where ions adhere to the external surface and a slower phase driven by diffusion-limited migration into the micropores. Nanoscale metal oxides demonstrate superior performance to their bulk counterparts, attributed to their increased specific surface area, shorter diffusion pathways, and higher density of reactive sites, including edges, corners, and vacancies. For instance, reducing the particle size of magnetite from 300 nm to 11 nm has enhanced its arsenic adsorption capacity by over 100-fold, underscoring the potential of nanoscale materials in water treatment applications (NABI et al., 2009). Interestingly, while magnetite particles of 300 and 20 nm exhibit similar surface-area-normalized arsenic adsorption capacities (~6 mmol/m^2 or 3.6 atoms/nm^2), further reduction to 11 nm significantly boosts the capacity to approximately 18 mmol/m^2 or 11 atoms/nm^2.This dramatic improvement, known as the "nanoscale effect," is attributed to structural changes on the magnetite surface that create new, high-energy adsorption sites (Jacukowicz-Sobala et al., 2015). Additionally, due to their superior adsorption capacities, certain nanoscale iron oxides, such as nano-maghemite and nano-magnetite, possess superparamagnetic properties. The magnetic behavior of these materials depends on their size: when particle size decreases below a critical threshold (~40 nm), magnetic domains transition from multiple to single domains, increasing magnetic susceptibility (Zhang et al., 2020). Further reduction in size leads to superparamagnetic behavior, wherein particles lose their permanent magnetic moments but remain responsive to external magnetic fields. This property permits efficient separation and recovery of nanoparticles using low-gradient magnetic fields, making them specifically suitable for scalable applications (Fig. 1). Metal oxide nanocrystals have the added benefit of being compressible into porous pellets without significantly reducing surface area when subjected to moderate pressure (Runnerstrom et al., 2016). Adjusting the consolidation pressure allows for precise control over pore size and volume, enabling these materials to function effectively as fine powders or porous pellets, making them highly adaptable for industrial applications.

Metal oxide nanomaterials have shown exceptional capability in removing various heavy metals, with mercury, arsenic, copper, cadmium, lead, nickel, and chromium, outperforming traditional adsorbents such as activated carbon (Adhikari et al., 2023; Sable et al., 2024). Arsenic removal has garnered considerable attention

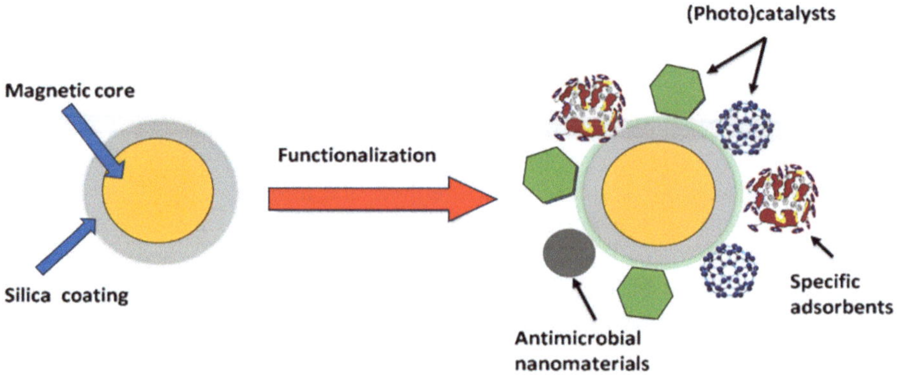

Fig. 1 Functions of nanoparticles

due to activated carbon's relatively low adsorption capacity, particularly for arsenic in its As(V) form (Zhao et al., 2010). In contrast, nanoscale metal oxides such as magnetite and titanium dioxide exhibit significantly higher arsenic adsorption performance than activated carbon. Moreover, metal hydro-oxide nanoparticles can be incorporated into porous materials like activated carbon, allowing for the simultaneous removal of arsenic and organic co-contaminants (Mohanty et al., 2024b). This multifunctional capability makes them especially suitable for point-of-use water treatment systems (Hristovski & Markovski, 2017). In summary, metal oxide nanomaterials offer remarkable water and wastewater treatment advantages, including enhanced adsorption efficiency, tunable functionality, and scalability. Their flexibility and ability to address various pollutants position them as transformative solutions for advanced adsorption technologies.

2.1.3 Nano-polymer Adsorptive Materials

Dendrimers are highly specialized adsorbents with a unique, branched structure that makes them effective for removing organic pollutants and heavy metals from aqueous solutions. The interior shells can be designed to exhibit hydrophobic properties, optimizing them for organic compound adsorption, while the exterior branches function with groups like hydroxyl or amine to target heavy metal ions specifically (Mohammedali et al., 2022). These versatile adsorbents operate through a combination of mechanisms, including electrostatic interactions, complexation, hydrogen bonding, and hydrophobic effects, allowing them to address various contaminants. Ultrafiltration systems utilizing dendrimers have proven highly effective for recuperating metal ions from aqueous solutions. For example, a solution with an initial Cu^{2+} concentration of 10 ppm achieved nearly complete removal by maintaining a Cu^{2+} to PAMAM dendrimer-NH_2 ratio 0.2. Once the dendrimers bound the metal ions, ultrafiltration separated them efficiently. The system's reusability was ensured by regenerating it through pH adjustment to 4, maintaining consistent performance over multiple cycles (Lazar et al., 2023). This process permitted the dendrimers to be recycled multiple times without significant loss in adsorption performance. The ability of dendrimers to selectively target and remove miscellaneous contaminants, combined with their efficient recovery and regeneration, highlights their potential as advanced materials for sustainable water treatment (Das et al., 2024a; Dey et al., 2024a, b; Mohanty et al., 2024a; Ray et al., 2024). Their personalized adsorption capabilities and reusability make dendrimers a promising solution for complex environmental remediation challenges.

2.1.4 Integration of Nano-adsorbents into Water Treatment Systems

Nano-adsorbents are highly efficient materials that can be incorporated into current water treatment systems through adsorbers or slurry reactors. In slurry reactors, nano-adsorbents in powder form maximize efficiency by utilizing their entire surface area, with mixing further enhancing mass transfer. However, this approach necessitates a supplementary separation unit to recuperate nanoparticles after treatment. Alternatively, Nano-adsorbents are commonly functional in fluidized or fixed adsorbers, often in porous granules, beads, or pellets. Fixed-bed reactors provide the advantage of eliminating the need for nanoparticle recovery; however, they can encounter challenges such as head loss and mass transfer limitations. In the case of arsenic removal, the nano-adsorbents have proven to be both effective and cost-efficient in commercial applications (Sable et al., 2024). Examples include ArsenX^{np}, a hybrid ion-exchange material incorporating iron oxide nanoparticles within a polymer matrix, and ADSORBSIA, a nanocrystalline titanium dioxide adsorbent available as beads ranging from 0.25 to 1.2 mm. Both products have demonstrated excellent arsenic removal performance, with ArsenX^{np} requiring minimal backwashing, making these technologies practical for water treatment applications (L. Das et al., 2020). Cost analysis from pilot-scale studies highlights the economic feasibility of these materials. ArsenXnp and ADSORBSIATM have been successfully deployed in small- to medium-scale drinking water purification units, proving efficient and cost-effective with conventional adsorbents. These findings highlight the transformative potential of nano-adsorbents in water treatment applications. The nano-adsorbant's ability to efficiently remove persistent contaminants like arsenic and their scalability and cost-effectiveness position nano-adsorbents as a promising solution for advanced water purification methods.

2.2 Enhanced Membrane Methods for Advanced Treatment of Water

Water treatment aims to eliminate undesirable constituents, ensuring water quality for diverse applications. Membrane technologies play a critical role by serving as physical barriers that separate contaminants dependent on size, enabling the utilization of alternative water sources. These systems are fundamental to modern water treatment and reuse due to their high automation, minimal land and chemical requirements, and modular designs, which provide scalability and flexibility (Ezugbe & Rathilal, 2020). Despite their advantages, membrane arrangements face several challenges. One primary limitation is the integral tradeoff between permeability and selectivity. Pressure-driven membrane processes are energy-intensive, which hinders their widespread adoption. Additionally, membrane fouling—caused by the deposit of contaminants on the membrane's outer surface—increases energy consumption, complicates system operation, and decreases the lifespan of modules and membranes. These factors collectively impact membrane systems' efficiency, reliability, and cost-effectiveness. The efficiency of membrane systems is mainly determined by the materials used. Integrating functional nanomaterials into membrane structures has emerged as a promising solution to address these limitations. Nanomaterial-enhanced membranes offer improved permeability, enhanced fouling resistance, and greater mechanical and thermal stability (Bera et al., 2022). Also, they can be engineered to introduce advanced functionalities such as contaminant degradation, antimicrobial effects, and self-cleaning properties (Mishra & Das, 2021). These innovations pave the way for more sustainable, efficient, and cost-effective water treatment methods, positioning nanomaterial-integrated membranes as a transformative solution for future applications.

2.2.1 Electrospinning for Fabricating Advanced Nanofiber Membranes

Electrospinning is a versatile, cost-effective technique for producing ultra-fine fibers from diverse materials, like polymers, ceramics, and metals. Nanofibers created through this process include unique features like high specific surface area, significant porosity, and intricate pore structures in nanofiber mats. The technique allows precise control over critical properties, including morphology, spatial alignment, composition, fiber diameter, and secondary structure, making it adaptable to various applications. While nanofiber membranes are extensively used in the filtration of air, their specialization in water remediation is still underutilized. These membranes efficiently eliminate micron-sized particles from water with high refusal rates and minimal fouling (Shi et al., 2022). Such capabilities make them ideal for pretreatment layers in advanced filtration systems like ultrafiltration and reverse osmosis (RO). Integrating functional nanomaterials into electro-spun nanofibers improves their functionality and efficiency. These materials can be utilized in the spinning solution or synthesized in situ on the nanofiber surface during fabrication (Yalcinkaya, 2019). This process enables the creation of multifunctional membranes tailored for specific tasks. Electro-spun nanofibers offer an exceptional platform for developing next-generation filtration technologies. Intrinsically multifunctional materials, like titanium dioxide (TiO_2), can be used directly, or functional agents can be added to the nanofiber scaffold to address specific contaminants. For example, ceramic nanoparticles or specialized capture agents integrated into nanofibers enable the development of affinity membranes capable of removing organic pollutants and heavy metals during filtration (Shi et al., 2022). Electro-spun nanofiber membranes are a transformative solution for advanced water treatment applications with outstanding properties, adaptability, and efficiency.

2.2.2 Nanocomposite-Enhanced Membrane Technology: Developments and Functions

Significant advancements in membrane nanotechnology have been accomplished by integrating several nanomaterials into polymeric and inorganic membranes, improving their performance and empowering multifunctional capabilities. Common nanomaterials used to contain hydrophilic metal oxide nanoparticles (such as zeolites, Al_2O_3, and TiO_2), antimicrobial agents such as nano-silver and carbon nanotubes (CNTs), and (photo)catalytic materials like TiO_2 and bimetallic nanoparticles. Hydrophilic metal oxide nanoparticles are primarily added to enhance surface hydrophilicity, thereby mitigating membrane fouling. Research indicates that nanoparticles such as alumina, silica, zeolite, and TiO_2 improve hydrophilicity and increase water permeability and resistance to fouling. These materials also enhance polymeric membranes' mechanical and thermal stability, minimizing the negative effects of compaction and heat on their overall performance (Mladin et al., 2022).

Antimicrobial nanomaterials, including CNTs and nano-Ag, are utilized to combat biofouling. When incapacitated or surface-grafted onto membranes, Nano-Ag inhibits attachment of bacteria and biofilm formation and inactivates viruses. However, its long-term effectiveness remains unclear, and strategies for replenishing nano-Ag in practical applications must be developed. CNTs, in contrast, provide antimicrobial activity through direct contact with bacteria, achieving high inactivation rates. CNTs are insoluble and not consumed during filtration, eliminating the requirement for replenishment (Tiwari et al., 2022). However, their antimicrobial efficiency could be affected by fouling, requiring further investigation in long-term applications. Oxidized multi-walled CNTs (MWNTs) also enhance the hydrophilicity and permeability of poly-sulfone membranes when added in small concentrations. Incorporating (photo)catalytic nanoparticles into membranes has led to the development of reactive membranes that combine physical separation with catalytic degradation of contaminants. Photocatalytic membranes using nano-TiO_2 or modified forms of TiO_2 have been widely employed for degrading pollutants under light exposure. Similarly, metallic and bimetallic nanoparticles, such as nano zero-valent iron (nZVI) and noble metals supported on nZVI, are used for the reductive degradation of chlorinated compounds. nZVI is an electron donor in these systems, while noble metals act as catalysts to accelerate the reaction (Salehi & Wang, 2022). Integrating nanomaterials into membranes addresses critical limitations of traditional technologies, offering enhanced fouling resistance, improved antimicrobial properties, and catalytic degradation functionalities. These multifunctional membranes represent a transformative approach to tackling complex water treatment challenges, providing effective, sustainable, and adaptable solutions for modern environmental needs (Tiwari et al., 2022).

2.2.3 Bioengineered Membrane Systems

Drawing inspiration from nature, biologically inspired membranes offer exceptional selectivity and permeability, making them a promising foundation for advanced water treatment technologies. Aquaporins, protein channels facilitating water transport across cell membranes, are particularly notable for their extraordinary water permeability and selectivity. Aligned carbon nanotubes (CNTs) present another innovative approach to high-performance membranes. According to studies, water transport through aligned carbon nanotubes (CNTs) occurs at speeds far exceeding those predicted by the Hagen-Poiseuille equation. This unusually fast flow rate is due to the atomically smooth inner walls of the CNTs, which reduce friction, and the single-file organization of water molecules within the constricted nanotube channels (Salehi & Wang, 2022). However, producing CNTs with uniform sub-nanometer diameters is a significant challenge, limiting their ability to achieve high salt and small molecule rejection. Functional group modifications at CNT openings have been explored to enhance selectivity. For instance, attaching carboxyl groups onto sub-2-nm openings of CNT has achieved 98% denial of $Fe(CN)_6^{3-}$ ions at small ionic strength via Donnan exclusion (Natarajan et al., 2023). However, the denial of smaller salts, such as KCl, was only 50% at 0.3 mM and dropped to near zero at 10 mM. While attaching bulky functional clusters at CNT openings could improve salt rejection by steric exclusion (Prakashkumar et al., 2022), it significantly reduces water permeability. Achieving reliable desalination with CNT membranes requires uniform diameters below 0.8 nm.

Scaling up the production and fabrication of aquaporin and aligned CNT membranes remains a key challenge. Large-scale generation and decontamination of aquaporins are particularly difficult. For CNTs, chemical vapor deposition (CVD) remains the primary production method. Recent advancements, such as high-yield CVD prototypes for vertically aligned CNTs, have paved the path for higher-scale generation (De Villoria et al., 2011). Additionally, alignment techniques utilizing magnetic fields have been established after manufacturing to enhance fabrication precision, as Wychowaniec and Brougham (2022) stated. These advancements hold particular promise for nanocomposite and thin-film nanocomposite (TFN) membranes, which represent scalable and efficient alternatives in membrane technology. These membranes can be manufactured using existing industrial processes and provide high water permeability, reducing operational pressure and membrane area requirements. This leads to lower energy consumption and operational costs, particularly for low-osmotic-pressure water treatment. However, their impact on seawater RO is limited, as these systems are already near their thermodynamic energy minimum (Wang et al., 2017). A current review of membrane nanotechnologies ranked their potential for performance improvement and commercial scalability, emphasizing the transformative potential of these innovations. While biologically inspired and nanotechnology-enhanced membranes hold immense promise for advancing water treatment, addressing scalability, cost, and technical challenges is critical for widespread adoption and implementation.

2.2.4 Osmotic Gradient-Driven Separation Process (Forward Osmosis)

Forward osmosis (FO) is an innovative water treatment method that relies on osmotic gradients to transfer water from a solution with low osmotic pressure to one with higher osmotic pressure, known as the draw solution. The diluted draw solution is then processed, often through reverse osmosis (RO) or thermal methods, to recover clean water. FO offers several advantages over pressure-driven reverse osmosis, such as operating without high pressures and reducing membrane fouling, making it a more energy-efficient and dependable alternative. The effectiveness of FO largely depends on the choice of draw solute, which must exhibit high osmolality and allow for efficient separation from water. Common draw solutes like sodium chloride (NaCl) and ammonia bicarbonate are effective but require energy-intensive processes, such as RO or thermal treatment, for water recovery, limiting the system's overall energy efficiency. Magnetic nanoparticles have emerged as a promising alternative draw solute due to their unique properties, including ease of separation and reusability. Hydrophilic coatings on magnetic nanoparticles enhance their solubility and osmotic pressure, significantly improving their performance in FO applications (Salamanca et al., 2023; Wang & Liu, 2021). This demonstrates the potential of functionalized magnetic nanoparticles to deliver high water flux while simplifying the recovery process.

In addition to serving as draw solutes, magnetic nanoparticles have been employed to recuperate other draw solutes. A study confirmed the use of $Fe_3O_4@SiO_2$ nanoparticles to recover aluminum sulfate ($Al_2(SO_4)_3$) through a flocculation process, providing an efficient method for draw solute reclamation. This innovative approach highlights the versatility of magnetic nanoparticles in enhancing FO systems (Esmaeilpour et al., 2023). Integrating magnetic nanoparticles and advanced recovery strategies significantly improves FO processes' energy efficiency, cost-effectiveness, and sustainability. These advancements position forward osmosis as a viable and environmentally friendly solution for addressing modern water treatment challenges.

2.3 Photocatalysis

Photocatalytic oxidation removes microbiological pathogens and traces pollutants from water. This technology is particularly effective as a pretreatment step, transforming hazardous and nonbiodegradable contaminants into more biodegradable forms, thereby enhancing the efficiency of subsequent biological treatments. Additionally, it is frequently used as a polishing phase to eliminate stubborn organic compounds that remain after the first treatment (Yemmireddy & Hung, 2017). However, the larger use of photocatalytic oxidation confronts difficulties due to delayed reaction kinetics caused by low light fluence and the poor activity of present photocatalysts. Recent research tackles these constraints by increasing photocatalytic reaction rates and expanding the photoactivity spectrum to encompass visible light. Strategies such as doping, surface modifications, and nano-structuring are being explored to enhance the efficiency and versatility of photocatalysts (Al-Ghafri et al., 2018; Roy et al., 2021). These advancements aim to improve photocatalytic oxidation's practicality, scalability, and cost-effectiveness, laying the groundwork for its broad use in wastewater and drinking water treatment applications.

2.3.1 Nano-photocatalyst Performance Optimization

Titanium dioxide (TiO_2) is the preferred semiconductor photocatalyst for wastewater and drinking water treatment because of its low toxicity, chemical stability, cost-effectiveness, and widespread availability. Under ultraviolet (UV) light, TiO_2 generates electron–hole (e^-/h^+) pairs, which can transfer to the surface to produce reactive oxygen species (ROS) for contaminant degradation or recombine within the material, reducing its efficiency. Research efforts have focused on enhancing TiO_2's photocatalytic performance through particle size optimization, doping, increasing reactive crystal facets, and surface modifications to improve the adsorption of contaminants. The size of TiO_2 particles significantly influences their phase stability, charge carrier dynamics, and sorption properties. Anatase, the most ROS-efficient phase, is stable for particle sizes less than 11 nm, while rutile dominates for sizes above 35 nm. Reducing the size of particles decreases the volume recombination of e^-/h^+ pairs and improves charge transfer to the surface. However, surface recombination becomes dominant at sub-nanometer sizes, lowering photocatalytic activity. This tradeoff results in an optimal particle size range for maximum efficiency (Kusworo et al., 2020; Mohadesi et al., 2024). TiO_2 nanotubes have demonstrated enhanced photocatalytic activity compared to nanoparticles, owing to shorter charge diffusion paths and faster mass transfer of reactants. Additionally, doping TiO_2 with noble metals, such as platinum, reduces e^-/h^+ recombination by trapping electrons, further improving its photocatalytic performance.

Enhancing TiO_2's reactive crystal facets is another promising approach. High-energy {001} facets are significantly more reactive than the commonly dominant {101} facets but are thermodynamically less stable. Using capping agents like fluoride can stabilize and increase the proportion of {001} facets from smaller than 10.00% to as much as 89.00%, considerably improving hydroxyl radical production and organic pollutant degradation (Mohammadi et al., 2016). This enhancement is attributed to stronger reactant adsorption and better spatial separation of charge carriers on the high-energy facets. Expanding TiO_2's light absorption into the visible spectrum has been a major area of research. Strategies such as doping with metals or anions (e.g., nitrogen), using dye sensitizers, or incorporating narrow-band-gap semiconductors have been explored (Cervantes-Avilés et al., 2018). These methods create intermediate energy levels within the band gap or inject electrons into TiO_2 under visible light excitation, initiating catalytic reactions. Nitrogen doping, in particular, has shown promise due to its cost-effectiveness and scalability, though its long-term stability under photocatalytic conditions remains a challenge.

In addition to TiO_2, alternative photocatalysts such as tungsten trioxide (WO_3) and fullerene derivatives are being developed. WO_3, with its narrower band gap, can be activated with the help of visible light (less than 450 nm). Platinum doping increased its performance by improving charge separation and enabling multi-electron O_2 reduction. Fullerene derivatives, such as amino fullerenes, generate singlet oxygen ($^1O_2$) under visible light (<550 nm), making them effective for degrading pharmaceuticals and inactivating viruses, while $^1O_2$ is less reactive than hydroxyl radicals. It is kinder to selection and less vulnerable to quenching from organic material in the background. Amino-fullerenes also offer advantages in disinfection applications due to their positive charge and ease of immobilization. However, fullerenes' high cost and limited availability remain significant barriers to widespread adoption (Cervantes-Avilés et al., 2018). Advancements in photocatalyst design, including optimization of particle size, structural modifications, and broadening light absorption capabilities, are crucial for improving photocatalytic water treatment's efficiency, scalability, and applicability. While TiO_2 continues to set the benchmark, alternative photocatalysts like WO_3 and fullerenes present exciting opportunities for specialized applications. Addressing cost, durability, and large-scale implementation challenges will be essential for integrating these advanced materials into next-generation water treatment systems.

2.3.2 Potential Utility in Water Decontamination

The design and operation of photoreactors play a critical role in determining the efficiency of photocatalytic water treatment systems. Common configurations for these systems include slurry reactors and reactors with immobilized TiO_2 catalysts. Enhancing their performance often focuses on improving catalyst dispersion, recovery, and immobilization techniques. Research at both laboratory and pilot scales has thoroughly examined factors influencing system efficiency, such as the amount of TiO_2 used, pH levels, temperature, dissolved oxygen, contaminants' characteristics, and the light source's properties, including wavelength and intensity. These factors have been reviewed in detail, offering guidance on optimizing operational parameters. A prominent commercial advancement in this field is the Purifics Photo-Cat® system, capable of treating up to 2 million gallons of water daily within a compact footprint of 678 square feet. Mayer et al. (2014) have demonstrated its effectiveness in eliminating organic contaminants without producing waste byproducts, and its energy efficiency requires approximately 4 kWh/m^3. Moreover, solar disinfection (SODIS) utilizing nano-TiO_2 has developed as a practical and scalable solution for safe drinking water (Vivar et al., 2021). This approach is mostly beneficial in resource-limited or remote areas. SODIS systems are versatile and can be adapted for small-scale personal use or larger-scale applications, such as those employing solar compound parabolic collectors.

Photocatalysis is recognized for its cost-effectiveness, sustainability, and eco-friendly nature, making it a promising technology for water treatment (Tripathy et al., 2022). Despite its potential, several technical challenges hinder its widespread implementation. These challenges include improving photocatalyst performance to increase quantum yield, expanding light absorption into the visible spectrum, designing efficient photoreactors with advanced catalyst recovery and immobilization techniques, and enhancing reaction selectivity to address specific contaminants. In addition to photocatalysis, heterogeneous catalytic ozonation has emerged as a viable method for degrading organic pollutants. Metal oxide nanomaterials, such as titanium dioxide (TiO_2) and cerium dioxide (CeO_2), alongside carbon nanotubes, have emerged as highly efficient catalysts for rapidly degrading pollutants through catalytic ozonation. These systems operate via both radical-mediated and non-radical-mediated pathways, with the adsorption of ozone and pollutants onto the catalyst surface being a critical step for effective pollutant breakdown. The exceptional catalytic performance of these nanomaterials is attributed to their high surface area and abundant active sites, which facilitate enhanced reactivity. Additionally, certain nanomaterials promote the decomposition of ozone into hydroxyl radicals, further driving pollutant degradation through radical-mediated mechanisms (Vivar et al., 2021). However, to enable industrial-scale applications, a comprehensive understanding of the mechanisms governing nanomaterial-assisted catalytic ozonation is crucial. This knowledge will be instrumental in optimizing performance, scalability, and cost-efficiency for practical deployment in water and wastewater treatment systems.

2.4 Contamination Control and Disinfection

Addressing the challenge of effective water disinfection while minimizing the production of harmful disinfection byproducts (DBPs) is a persistent issue in the water treatment sector. Traditional disinfectants, like ozone and chlorine, are highly effective against pathogens but are known to produce harmful DBPs, including halogenated compounds, carcinogenic nitrosamines, and bromates. While ultraviolet (UV) disinfection offers an alternative with minimal DBP formation, it requires high doses to effectively inactivate particular viruses, like adenoviruses. These limitations highlight the need for innovative disinfection technologies that provide robust microbial control without contributing to DBP formation. Nanotechnology presents a promising solution to this dilemma. As detailed in a previous review (Ismail et al., 2019), a range of nanomaterials, including n-Ag, n-ZnO, n-TiO_2, n-CeO_2, carbon nanotubes, and fullerenes, show potent antimicrobial properties without relying heavily on oxidation. This characteristic significantly reduces the potential for DBP formation compared to conventional methods. The review extensively examined the antimicrobial mechanisms, advantages, limitations, and applicability of these nanomaterials for water treatment while identifying critical areas for future research. This update focuses on recent advancements in nano-Ag and carbon-based nanomaterials for water disinfection. These materials demonstrate unique advantages, including high antimicrobial efficacy, reduced chemical dependence, and minimal byproduct generation, positioning them as promising candidates for developing next-generation disinfection technologies (Sahu et al., 2024).

2.4.1 Antimicrobial Mechanisms of Nanomaterials

Nano-Ag is among the most commonly used nanomaterials in antimicrobial applications due to its strong antimicrobial activity, broad-spectrum effectiveness, low human toxicity, and ease of integration into water treatment arrangement systems. Its antimicrobial mechanism properties primarily involve the release of from releasing silver ions, which lead to thiol groups in vital proteins, disrupting enzymatic functions. Furthermore, silver ions interfere with DNA replication and damage the microbial cell envelope, ultimately leading to cell death. The efficiency of nano-Ag is significantly influenced by the release rate and bioavailability of silver ions, which depend on factors such as the material's size, shape, surface modifications, and crystallographic characteristics (Fatima & Mushtaq, 2023). However, the presence of ligands in water can bind to silver ions, reducing their availability and antimicrobial efficacy. Notably, sublethal concentrations of nano-Ag or silver ions have been observed to stimulate bacterial growth, such as in *Escherichia coli*, signifying a hormetic response and limiting its disinfection efficiency in certain scenarios. Carbon nanotubes (CNTs) exhibit antimicrobial activity through various mechanisms, including physical distraction of microbial cell membranes, interference with vital cellular processes, and induction of oxidative stress via direct contact with bacterial surfaces. Similar antimicrobial effects have been observed in graphene and graphite-based materials. The antimicrobial potency of CNTs is powerfully influenced by their physicochemical characteristics, with shorter, metallic, and more dispersed CNTs with smaller diameters demonstrating higher cytotoxicity (Liu et al., 2023).

2.4.2 Applications of Antimicrobial Nanomaterials

Antimicrobial nanomaterials offer cutting-edge solutions to major water and wastewater treatment challenges, including disinfection, biofouling prevention in membranes, and biofilm control on water-contact surfaces. Nano-Ag is particularly promising for point-of-use (POU) water treatment systems, providing enhanced water quality for specialized uses and additional protection against waterborne pathogens, especially for vulnerable populations. Commercially, nano-Ag is utilized in systems such as MARATHON® and Aquapure® (Villaseñor & Ríos, 2018). The ceramic microfilters serve as reliable pathogen barriers in remote and resource-limited regions of developing countries. In parallel, carbon nanotubes (CNTs) are emerging as a powerful tool for advanced filtration systems because of their unique fibrous assembly, high conductivity, and antimicrobial properties. These filters remove bacteria through size exclusion and viruses via depth filtration, inactivating trapped microorganisms over time. Applying a low voltage (2–3 V) to multi-walled CNTs (MWNTs) significantly enhances their disinfection efficiency, rapidly oxidizing bacteria and viruses in seconds (S. Lu & Zhang, 2022). This applied voltage improves viral transport to the CNT surface, increasing filtration effectiveness. With minimal power requirements and high performance, CNT filters are particularly well-suited for POU water disinfection applications.

Antimicrobial nanomaterials provide innovative solutions to critical water and wastewater treatment challenges, such as disinfection, biofouling prevention in membranes, and control of biofilms on water-contacting surfaces. Nano-silver (nano-Ag) is a highly effective option for point-of-use (POU) water treatment systems, enhancing water quality for specialized needs and offering additional protection against waterborne pathogens, particularly for vulnerable groups. Commercially, nano-Ag is incorporated into systems like MARATHON® and Aquapure® and has been successfully embedded in ceramic microfilters, which act as effective pathogen barriers in remote and resource-limited areas of developing nations (Villaseñor & Ríos, 2018). Similarly, carbon nanotubes (CNTs) are gaining attention as advanced filtration materials due to their fibrous structure, excellent conductivity, and antimicrobial properties. These filters effectively remove bacteria via size exclusion and viruses through depth filtration, with trapped microorganisms becoming inactivated over time. Applying a low voltage (2–3 V) to multi-walled CNTs (MWNTs) further enhances their disinfection capability by rapidly oxidizing bacteria and viruses within seconds (S. Lu & Zhang, 2022). This voltage application also improves viral transport to the CNT surface, boosting filtration efficiency. With their low energy consumption and high efficiency, CNT-based filters are suitable for POU water disinfection applications (Fatima & Mushtaq, 2023).

Beyond their role in disinfection, antimicrobial nanomaterials are highly effective in reducing biofouling on membranes and other water-contact surfaces, such as storage tanks and distribution pipelines. These applications help to prevent pathogen contamination, biofilm development, and microbially induced corrosion (Verma et al., 2020; Villaseñor & Ríos, 2018). To maximize their potential, cost-efficient coating technologies that minimize nanomaterial usage while enabling regeneration are critically needed. Another innovative strategy involves creating biofouling-resistant nanoscale surface structures inspired by nature, such as the textured surfaces of shark and dolphin skin or the self-cleaning abilities of lotus leaves. Despite their numerous benefits, one limitation of nanomaterial-based disinfection systems is the absence of residual disinfectant to control microbial growth during storage and distribution. However, these systems significantly minimize the formation of disinfection byproducts (DBPs), as chemical disinfectants like chlorine are required only as secondary treatments. Long-term effectiveness remains a challenge, especially for nanomaterials that rely on the gradual release of biocidal ions, which can become depleted over time. Controlled-release systems and strategies for replenishing active agents are essential to overcoming this limitation. Encapsulating antimicrobial agents within matrices that activate in response to specific triggers, such as microorganisms or biofilms, provides a promising solution. These "on-demand" release systems can be refined with mechanisms to recognize pathogens, enabling precise and targeted applications, as shown in Fig. 2. Fouling poses a significant challenge for nanomaterials requiring direct microbial contact (Sinha et al., 2023), as it can reduce or eliminate antimicrobial activity. Developing effective antifouling strategies will ensure the long-term viability and effectiveness of nanomaterial-enabled water disinfection technologies long-term viability and effectiveness.

2.5 *Multifunctional Devices*

Advances in functional nanomaterials, combined with their integration into traditional technologies, have revolutionized the design of multifunctional water treatment systems. These nanotechnology-enabled devices can perform multiple tasks within a single platform, offering enhanced efficiency, reduced redundancy, and a significantly smaller footprint. This innovative approach is particularly suited for decentralized and small-scale water treatment applications where adaptability and compactness are essential. The integration of functional nanomaterials can be

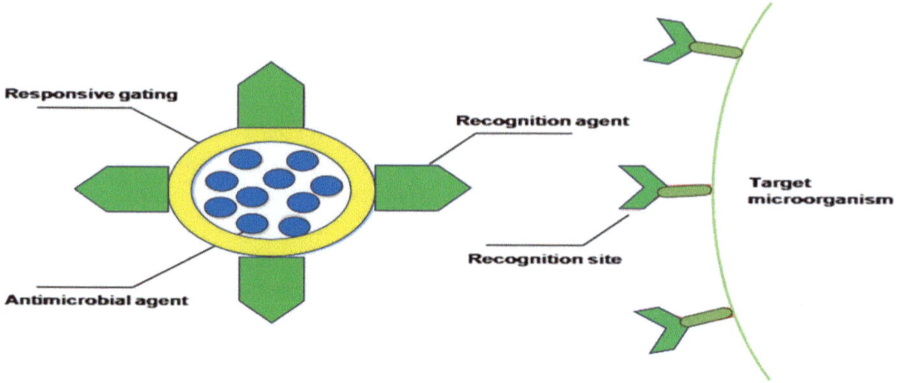

Fig. 2 On-demand microbial control

tailored to address specific treatment requirements. Magnetic nanoparticles have demonstrated wide applicability, while membranes have emerged as a highly versatile platform for multifunctional system design. Among these, electrospun nanofibers (Cui et al., 2020) are exceptional nanomaterial carriers due to their high surface area, adaptability, and efficient mass transfer capabilities. Nanotechnology-enabled water treatment systems are designed with modularity and scalability in mind (Al Harby et al., 2022). These systems incorporate functionalities as layered cartridges or serially connected modules, enabling independent optimization, maintenance, and regeneration of each functional component. The modular design also allows for easy customization, where capacity and functionality can be adjusted by adding or removing specific modules. This plug-and-play flexibility makes these systems highly versatile, allowing them to address a wide range of water treatment challenges efficiently. By combining high performance, small footprints, and modular configurations, these nanotechnology-enabled multifunctional systems represent a transformative approach to water treatment, particularly in resource-limited and decentralized environments. Their adaptability and scalability ensure they can meet diverse treatment needs while promoting sustainable and efficient water management practices.

3 Retention and Reuse of Nanomaterials

This is essential for ensuring the cost-efficiency and safety of nanotechnology-enabled water treatment systems. These strategies not only address public health concerns but also improve the overall sustainability of such systems. Retention can be achieved through advanced separation techniques or immobilization of nanomaterials within the treatment setup. Darlington Eze Ekechukwu et al. (2024) show that separation methods and membrane filtration have emerged as a promising solution due to their capability for continuous operation, minimal chemical requirements, and compact system footprint. Ceramic membranes, in particular, offer superior performance over polymeric membranes in photocatalytic and catalytic ozonation processes thanks to their enhanced resistance to UV radiation and chemical oxidants. However, the presence of suspended particles in untreated water can clog membranes, reduce reaction efficiency, and impair the performance of reactor-membrane hybrid systems. To address this, raw water pretreatment is often necessary to lower turbidity and optimize system functionality.

Immobilizing nanomaterials on platforms such as resins and membranes provides an alternative approach to retention, eliminating the need for downstream recovery processes. Despite its advantages, current immobilization techniques often compromise the efficiency of nanomaterials, highlighting the need for research into cost-effective methods that preserve performance. Magnetic nanoparticles and nanocomposites offer a unique retention advantage through low-field magnetic separation, an energy-efficient and scalable approach with significant potential for optimization. Nanomaterial release from treatment systems is another critical concern. The extent and rate of release heavily influence the immobilization technique and separation method (Bouabidi et al., 2019). Systems without downstream separation are more likely to experience rapid and complete release of nanomaterials coated on treatment surfaces, whereas nanomaterials embedded in solid matrices exhibit minimal release until disposal. For nanomaterials that release metal ions, controlling dissolution rates through coatings or optimizing size and shape is crucial to maintaining safety and efficacy. A major hurdle in evaluating nanomaterial release is the development of effective detection techniques, particularly for complex aqueous environments. Existing methods are often expensive, technically demanding, and limited in sensitivity. The creation of fast, selective, and sensitive analytical tools is imperative for monitoring nanomaterial release and enabling accurate risk assessments. While this paper does not address detection techniques in detail, readers are encouraged to consult recent reviews on the topic. Addressing these challenges will be crucial for advancing the safe, efficient, and sustainable application of nanotechnology in water treatment (Bouabidi et al., 2019).

4 Limitations and Further Research Requirements

Nanotechnology-based drinking water and waste-water treatment technologies have demonstrated significant potential in laboratory studies, but their transition to large-scale commercial implementation has been uneven. While certain solutions are ready for market deployment, others still require extensive research and development to reach maturity. Major challenges include overcoming technical limitations, reducing costs, and mitigating potential environmental and human health risks. Two critical research areas must be prioritized to enable large-scale adoption. First, rigorous testing of nanotechnologies under real-world conditions using natural and wastewater samples is essential. Such studies will provide valuable insights into their practicality, efficiency, and reliability while validating nanomaterial-based sensing methods. Also, understanding the long-term efficacy of these systems is crucial, as most of the studies have focused on short-term performance. Research into their durability and sustained functionality over extended periods is urgently needed to address this knowledge gap and facilitate comparisons with conventional treatment methods.

Cost-effectiveness and risk management are pivotal to the widespread adoption of nanotechnology in water treatment. High production costs remain a significant obstacle, with some exceptions like nano-TiO_2, nanoscale iron oxides, and polymeric nanofibers. Two strategies offer potential solutions: utilizing lower-purity nanomaterials to reduce production costs, as much of the expense stems from purification processes (Chirayil et al., 2014), and enhancing the retention and reuse of nanomaterials to maximize their economic value. Managing risks associated with nanomaterials presents unique challenges due to their particulate nature, which differs from conventional molecules or ions with established risk assessment protocols. A deeper understanding of the potential hazards posed by nanomaterials in water treatment, alongside effective mitigation strategies, will be vital for building public trust and fostering widespread acceptance of these technologies. Integration with existing water treatment infrastructure is another critical consideration. In developed regions, treatment industries and delivery systems are predictable to remain operational for recent years, making it essential for nanotechnology-enabled solutions to integrate seamlessly with minimal infrastructure modifications. In contrast, regions without established infrastructure can benefit from deploying nanotechnology-enabled systems in point-of-use (POU) devices or decentralized setups, offering immediate and scalable solutions for water quality improvement in remote or underserved areas. Addressing these challenges like real-world performance validation, long-term efficacy, cost reduction, risk management, and infrastructure compatibility will be key to unlocking the transformative potential of nanotechnology in wastewater and drinking water treatment. By overcoming these barriers, nanotechnology can deliver innovative, efficient, and sustainable solutions, driving advancements in developing regions and securing a safer, more reliable water future.

5 Conclusion

Nanotechnology is poised to revolutionize water and wastewater treatment by offering innovative solutions to pressing global challenges. When integrated into existing treatment technologies, nanomaterials' unique properties unlock opportunities for transformative advancements. While many nanotechnologies are still under development, some have already progressed to pilot testing and commercialization. Among the most promising nanotechnologies for large-scale implementation are nano-adsorbents, nanotechnology-enabled membranes, and nano-photocatalysts. These technologies stand out due to their advanced research and development stages, commercial availability, cost-effectiveness, and compatibility with existing treatment infrastructure. While these solutions have yet to see widespread application in large-scale systems, commercial products are already available. Additionally, several nanotechnology-based innovations have proven effective in point-of-use (POU) systems, addressing decentralized water treatment needs efficiently and reliably. Despite their potential, nanotechnology-enabled water treatment systems face several challenges, including technical limitations, high costs, and concerns about environmental and human health risks. However, these challenges will likely be overcome as research and innovation progress. Collaborative efforts involving research institutions, industry, government, and other stakeholders will be essential in addressing these barriers and driving the successful deployment of nanotechnology in water treatment. The responsible development and implementation of nanotechnology must prioritize safety, sustainability, and accessibility. By carefully managing potential risks and ensuring ethical innovation, nanotechnology can deliver both incremental improvements and revolutionary water and wastewater treatment advancements. With sustained investment, collaboration, and innovation, nanotechnology promises to provide cleaner, safer, and more sustainable water solutions for communities worldwide. The following conclusive bullets can be drawn from this study:

1. Nanotechnology offers transformative potential for water and wastewater treatment.
2. Nano-adsorbents, membranes, and photocatalysts are key candidates for large-scale use.
3. Addressing technical, cost, and risk challenges is essential for progress.
4. Responsible innovation can ensure sustainable and accessible water solutions globally.

References

Adhikari, C., Ghosh, S., & Das, P. K. (2023). Graphene's role in the clean-up act: Progress in industrial wastewater treatment. *Asian Journal of Chemistry*. https://doi.org/10.14233/AJCHEM.2024.30639

Ahmed, S. N., & Haider, W. (2018). Heterogeneous photocatalysis and its potential applications in water and wastewater treatment: A review. *Nanotechnology, 342001*(29), 1–30.

Al Harby, N. F., El-Batouti, M., & Elewa, M. M. (2022). Prospects of polymeric nanocomposite membranes for water purification and scalability and their health and environmental impacts: A review. *Nanomaterials*. https://doi.org/10.3390/nano12203637

Alayan, H. M., Aljumaily, M. M., Alsaadi, A., Mjalli, F. S., & Hashim, M. A. (2021). A review exploring the adsorptive removal of organic micropollutants on tailored hierarchical carbon nanotubes Haiyam. *Toxicological & Environmental Chemistry, May*. https://doi.org/10.1080/02772248.2021.1932888

Al-Ghafri, B., Bora, T., Sathe, P., Dobrestov, S., & Al-Abri, M. (2018). Photocatalytic microbial removal and degradation of organic contaminants of water using PES fibers. *Applied Catalysis B: Environmental*. https://doi.org/10.1016/j.apcatb.2018.03.095

Ali, I., Basheer, A. A., Mbianda, X. Y., Burakov, A., Galunin, E., Burakova, I., Mkrtchyan, E., Tkachev, A., & Grachev, V. (2019). Graphene based adsorbents for remediation of noxious pollutants from wastewater. *Environment International*. https://doi.org/10.1016/j.envint.2019.03.029

Bera, S. P., Godhaniya, M., & Kothari, C. (2022). Emerging and advanced membrane technology for wastewater treatment: A review. *Journal of Basic Microbiology*. https://doi.org/10.1002/jobm.202100259

Bodzek, M., Konieczny, K., & Kwiecińska-mydlak, A. (2020). The application of nanomaterial adsorbents for the removal of impurities from water and wastewaters: A review. *Desalination and Water Treatment, 185*, 1–26. https://doi.org/10.5004/dwt.2020.25454

Bouabidi, Z. B., El-Naas, M. H., & Zhang, Z. (2019). Immobilization of microbial cells for the biotreatment of wastewater: A review. *Environmental Chemistry Letters*. https://doi.org/10.1007/s10311-018-0795-7

Cervantes-Avilés, P., Ida, J., Toda, T., & Cuevas-Rodríguez, G. (2018). Effects and fate of TiO2 nanoparticles in the anaerobic treatment of wastewater and waste sludge. *Journal of Environmental Management*. https://doi.org/10.1016/j.jenvman.2018.05.074

Chirayil, C. J., Mathew, L., & Thomas, S. (2014). Review of recent research in nano cellulose preparation from different lignocellulosic fibers. *Reviews on Advanced Materials Science, 37*(1–2), 20–28.

Cui, J., Li, F., Wang, Y., Zhang, Q., Ma, W., & Huang, C. (2020). Electrospun nanofiber membranes for wastewater treatment applications. *Separation and Purification Technology*. https://doi.org/10.1016/j.seppur.2020.117116

Ekechukwu, D. E., Daramola, G. O., & Olanrewaju, O. I. K. (2024). Advancements in catalysts for zero-carbon synthetic fuel production: A comprehensive review. *GSC Advanced Research and Reviews, 19*(3), 215–226. https://doi.org/10.30574/gscarr.2024.19.3.0212

Das, A., Dey, S., Bala, K., Gnanasekaran, R., & Das, A. P. (2024a). Microbial bioremediation technology for sustainable treatment and management of synthetic microfiber waste. *Sustainable Microbial Technology for Synthetic and Cellulosic Microfiber Bioremediation, 221–238*. https://doi.org/10.1007/978-3-031-62600-5_12

Das, A., Dey, S., & Das, A. P. (2024b). Ecotoxicological impacts of synthetic microfiber pollutants and development of sustainable mitigation strategies. *Environmental Chemistry and Ecotoxicology*. https://doi.org/10.1016/J.ENCECO.2024.12.003

Das, L., Das, P., Bhowal, A., & Bhattachariee, C. (2020). Synthesis of hybrid hydrogel nanopolymer composite using graphene oxide, chitosan and PVA and its application in waste water treatment. *Environmental Technology and Innovation*. https://doi.org/10.1016/j.eti.2020.100664

De Villoria, R. G., Yamamoto, N., Miravete, A., & Wardle, B. L. (2011). Multi-physics damage sensing in nano-engineered structural composites. *Nanotechnology*. https://doi.org/10.1088/0957-4484/22/18/185502

Dey, S., Bissoyi, A., Koteswara Reddy, G., & Das, A. P. (2024a). Synthetic microfibers: An urban emerging pollutant demanding for sustainable environmental solutions. *Sustainable Microbial Technology for Synthetic and Cellulosic Microfiber Bioremediation, 107–121*. https://doi.org/10.1007/978-3-031-62600-5_7

Dey, S., Dash, A., & Das, A. P. (2024b). Bioremediation as a sustainable and circular solution for mining waste management and bioenergy generation. *Sustainable Management of Mining Waste and Tailings: A Circular Economy Approach, 107–119*. https://doi.org/10.1201/9781003442455-6/BIOREMEDIATION-SUSTAINABLE-CIRCULAR-SOLUTION-MINING-WASTE-MANAGEMENT-BIOENERGY-GENERATION-SUDESHNA-DEY-AKANSHYA-DASH-ALOK-PRASAD-DAS

Esmaeilpour, M., Larimi, A., Ghahramanafshar, M., & Faghihi, M. (2023). Ethylenediaminetetraacetic acid coated $Fe_3O_4@SiO_2$ nanocomposite: An effective adsorbent for the removal of copper ions from aqueous system. *Applied Chemistry., 17*(65), 45–54.

Ezugbe, E. O., & Rathilal, S. (2020). Membrane technologies in wastewater treatment: A review. *Membranes*. https://doi.org/10.3390/membranes10050089

Fatima, T., & Mushtaq, A. (2023). Efficacy and challenges of carbon-based nanomaterials in water treatment: A review. *International Journal of Chemical and Biochemical Sciences., 23*, 232–248.

Hristovski, K. D., & Markovski, J. (2017). Engineering metal (hydr)oxide sorbents for removal of arsenate and similar weak-acid oxyanion contaminants: A critical review with emphasis on factors governing sorption processes. *Science of the Total Environment*. https://doi.org/10.1016/j.scitotenv.2017.04.108

Ismail, M., Akhtar, K., Khan, M. I., Kamal, T., Khan, M. A., Asiri, M., & A., Seo, J., & Khan, S. B. (2019). *Pollution*. Current Pharmaceutical Design. https://doi.org/10.2174/1381612825666191021142026

Jacukowicz-Sobala, I., Ociński, D., & Kociołek-Balawejder, E. (2015). Iron and aluminium oxides containing industrial wastes as adsorbents of heavy metals: Application possibilities and limitations. *Waste Management and Research*. https://doi.org/10.1177/0734242X15584841

John, A., & Das, A. P. (2024). Government strategies for achieving circular economy for plastic waste management in Southeast Asia. *Sustainable Management of Urban Plastic Waste Through Circular Economic Approaches, 72–82*. https://doi.org/10.1201/9781003465447-6/GOVERNMENT-STRATEGIES-ACHIEVING-CIRCULAR-ECONOMY-PLASTIC-WASTE-MANAGEMENT-SOUTHEAST-ASIA-AKBAR-JOHN-ALOK-PRASAD-DAS

Khaled, B., Wided, B., Béchir, H., Elimame, E., Mouna, L., & Zied, T. (2019). Investigation of electrocoagulation reactor design parameters effect on the removal of cadmium from synthetic and phosphate industrial wastewater. *Arabian Journal of Chemistry*. https://doi.org/10.1016/j.arabjc.2014.12.012

Kong, Q., Zhang, H., Lan, Y., Shi, X., & Fang, Z. (2023). Functional graphene oxide for organic pollutants removal from wastewater: A mini review. *Environmental Technology, 44*(21), 3183–3195. https://doi.org/10.1080/09593330.2022.2053754

Kusworo, T. D., Ariyanti, N., & Utomo, D. P. (2020). Effect of nano-TiO2 loading in polysulfone membranes on the removal of pollutant following natural-rubber wastewater treatment. *Journal of Water Process Engineering*. https://doi.org/10.1016/j.jwpe.2020.101190

Lazar, M. M., Ghiorghita, C. A., Dragan, E. S., Humelnicu, D., & Dinu, M. V. (2023). Ion-imprinted polymeric materials for selective adsorption of heavy metal ions from aqueous solution. *Molecules*. https://doi.org/10.3390/molecules28062798

Liu, J. R., Wang, X. Y., Saberi, A., & Heydari, Z. (2023). The effect of Co-encapsulated GNPs-CNTs nanofillers on mechanical properties, degradation and antibacterial behavior of Mg-based composite. *Journal of the Mechanical Behavior of Biomedical Materials*. https://doi.org/10.1016/j.jmbbm.2022.105601

Lu, C., Chiu, H., & Bai, H. (2007). Comparisons of adsorbent cost for the removal of zinc (II) from aqueous solution by carbon nanotubes and activated carbon. *Journal of Nanoscience and Nanotechnology*. https://doi.org/10.1166/jnn.2007.349

Lu, C., Chiu, H., & Liu, C. (2006). Removal of zinc(II) from aqueous solution by purified carbon nanotubes: Kinetics and equilibrium studies. *Industrial and Engineering Chemistry Research*. https://doi.org/10.1021/ie051206h

Lu, S., & Zhang, G. (2022). Recent advances on inactivation of waterborne pathogenic microorganisms by (photo) electrochemical oxidation processes: Design and application strategies. *Journal of Hazardous Materials*. https://doi.org/10.1016/j.jhazmat.2022.128619

Mashkoor, F., & Nasar, A. (2020). Carbon nanotube—Based adsorbents for the removal of dyes from waters: A review. *Environmental Chemistry Letters, 18*, 605–629.

Mayer, B. K., Daugherty, E., & Abbaszadegan, M. (2014). Disinfection byproduct formation resulting from settled, filtered, and finished water treated by titanium dioxide photocatalysis. *Chemosphere*. https://doi.org/10.1016/j.chemosphere.2014.05.073

Meena, R. R., Kumar, S., & Soni, P. (2023). *Electrocoagulation of fluoride from water with Fe-based ion electrode*. Springer. https://doi.org/10.1007/978-3-031-38845-3_9

Mishra, S., & Das, A. P. (2021). Current treatment technologies for removal of microplastic and microfiber pollutants from wastewater. *Wastewater Treatment: Cutting-Edge Molecular*

Tools, Techniques and Applied Aspects, 237–251. https://doi.org/10.1016/B978-0-12-821881-5.00011-8

Mishra, S., & Das, A. P. (2023). Microbial remediation of synthetic microfiber contaminated wastewater. In M. P. Shah (Ed.), *Microbial technologies in industrial wastewater treatment* (pp. 337–350). Springer. https://doi.org/10.1007/978-981-99-2435-6_17

Mladin, G., Ciopec, M., Negrea, A., Duteanu, N., Negrea, P., Ianasi, P., & Ianaşi, C. (2022). Silica-Iron oxide nanocomposite enhanced with Porogen agent used for arsenic removal. *Materials*. https://doi.org/10.3390/ma15155366

Mohadesi, M., Sanavi Fard, M., & Shokri, A. (2024). The application of modified nano-TiO2 photocatalyst for wastewater treatment: A review. *International Journal of Environmental Analytical Chemistry*. https://doi.org/10.1080/03067319.2022.2064751

Mohammadi, L., Bazrafshan, E., Noroozifar, M., & Ansari-Moghaddam, A. (2016). Application of heterogeneous catalytic ozonation process with magnesium oxide nanoparticles for toluene degradation in aqueous environments. *Health Scope*. https://doi.org/10.17795/jhealthscope-40439

Mohammedali, D. R., Salman, H. I., Bahjat, M. N., & Abood, E. S. (2022). Synthesis of new corrosion inhibitor from nano-polymer and study its adsorption on carbon steel at different temperatures. *Egyptian Journal of Chemistry*. https://doi.org/10.21608/EJCHEM.2022.139171.6113

Mohanty, J., Dey, S., Bala, K., & Das, A. P. (2024a). Recent advances in microplastic pollution for its sustainable management. *Sustainable Management of Urban Plastic Waste Through Circular Economic Approaches, 27–48*. https://doi.org/10.1201/9781003465447-3

Mohanty, M., Mohanty, J., Dey, S., Dutta, K., Shah, M. P., & Das, A. P. (2024b). The face mask: A tale from protection to pollution and demanding sustainable solution. *Emerging Contaminants, 10*(2), 100298. https://doi.org/10.1016/J.EMCON.2023.100298

Mohanty, M., Tripathy, B., Koteswara Reddy, G., & Das, A. P. (2024c). Greener technologies for establishing restraint over microplastic pollution. *Sustainable Microbial Technology for Synthetic and Cellulosic Microfiber Bioremediation, 239–258*. https://doi.org/10.1007/978-3-031-62600-5_13

Nabi, D., Aslam, I., & Qazi, I. A. (2009). Evaluation of the adsorption potential of titanium dioxide nanoparticles for arsenic removal. *Journal of Environmental Sciences*. https://doi.org/10.1016/S1001-0742(08)62283-4

Natarajan, B., Kannan, P., Rather, J. A., & Sheikh, R. A. (2023). Recent developments in metal nanoparticles functionalized nanocomposite adsorbents for heavy metals removal from wastewaters. *Journal of the Taiwan Institute of Chemical Engineers*. https://doi.org/10.1016/j.jtice.2023.104942

Paul, A., Priyadarsini, S., Karakuş, S., Behera, J. K., Gnanasekaran, R., & Das, A. P. (2024). Fate and accumulation of synthetic microfibers in the environment. *Sustainable Microbial Technology for Synthetic and Cellulosic Microfiber Bioremediation, 61–76*. https://doi.org/10.1007/978-3-031-62600-5_4

Prakashkumar, N., Asik, R. M., Kavitha, T., Archunan, G., & Suganthy, N. (2022). Unveiling the anticancer and antibiofilm potential of catechin overlaid reduced graphene oxide/zinc oxide nanocomposites. *Journal of Cluster Science*. https://doi.org/10.1007/s10876-021-02194-2

Ra, M., Tahir, M. B., & Sadaf, I. (2019). Nanotechnology: An innovative way for wastewater treatment and purification. In *Advanced research in nanosciences for water technology, nanotechnology in the life sciences* (pp. 95–131). Springer.

Ray, S. K., Dey, S., Das, A. P., & John, A. (2024). Microplastic waste management through greener technologies: A review. *Sustainable Management of Urban Plastic Waste Through Circular Economic Approaches, 237–261*. https://doi.org/10.1201/9781003465447-14

Roy, N., Alex, S. A., Chandrasekaran, N., Mukherjee, A., & Kannabiran, K. (2021). A comprehensive update on antibiotics as an emerging water pollutant and their removal using nano-structured photocatalysts. *Journal of Environmental Chemical Engineering*. https://doi.org/10.1016/j.jece.2020.104796

Runnerstrom, E. L., Bergerud, A., Agrawal, A., Johns, R. W., Dahlman, C. J., Singh, A., Selbach, S. M., & Milliron, D. J. (2016). Defect engineering in plasmonic metal oxide nanocrystals. *Nano Letters, 16*(5), 3390–3398. https://doi.org/10.1021/acs.nanolett.6b01171

Saadi, S., Abdulhakim, M., & Zurina, W. (2019). Review on heavy metal adsorption processes by carbon nanotubes. *Journal of Cleaner Production, 230*, 783–793.

Sable, H., Kumar, V., Singh, V., Rustagi, S., Chahal, S., & Chaudhary, V. (2024). Strategically engineering advanced nanomaterials for heavy-metal remediation from wastewater. *Coordination Chemistry Reviews, 518*(July). https://doi.org/10.1016/j.ccr.2024.216079

Sahu, A., Dey, S., Mallick, K., Priyadarsini, S., Behera, I. D., & Das, A. P. (2024). A review on advanced technology for sustainable management of synthetic microplastic waste. *Sustainable Management of Urban Plastic Waste Through Circular Economic Approaches, 262–279*. https://doi.org/10.1201/9781003465447-15

Salamanca, M., Peña, M., Hernandez, A., Prádanos, P., & Palacio, L. (2023). Forward osmosis application for the removal of emerging contaminants from municipal wastewater: A review. *Membranes*. https://doi.org/10.3390/membranes13070655

Salehi, B., & Wang, L. (2022). Critical review on nanomaterials for enhancing bioconversion and bioremediation of agricultural wastes and wastewater. *Energies*. https://doi.org/10.3390/en15155387

Shi, S., Si, Y., Han, Y., Wu, T., Iqbal, M. I., Fei, B., Li, R. K. Y., Hu, J., & Qu, J. (2022). Recent progress in protective membranes fabricated via electrospinning: Advanced materials, biomimetic structures, and functional applications. *Advanced Materials*. https://doi.org/10.1002/adma.202107938

Singh, S., Garg, R., Jana, A., Bathula, C., Naik, S., & Mittal, M. (2023). Current developments in nanostructurally engineered metal oxide for removal of contaminants in water. *Ceramics International*. https://doi.org/10.1016/j.ceramint.2022.10.183

Sinha, S., Kumar, R., Anand, J., Gupta, R., Gupta, A., Pant, K., Dohare, S., Tiwari, P., Kesari, K. K., Krishnan, S., & Gupta, P. K. (2023). Nanotechnology-based solutions for antibiofouling applications: An overview. *ACS Applied Nano Materials*. https://doi.org/10.1021/acsanm.3c01539

Tiwari, A., Michalska-Domanska, M., & Dhoble, S. J. (2022). Nanocomposite-based functional materials: Synthesis, properties, and applications. *Functional Materials from Carbon, Inorganic, and Organic Sources: Methods and Advances*. https://doi.org/10.1016/B978-0-323-85788-8.00016-1

Tripathy, B., Dash, A., & Das, A. P. (2022). Detection of environmental microfiber pollutants through vibrational spectroscopic techniques: Recent advances of environmental monitoring and future prospects. *Critical Reviews of Analytical Chemistry, 1–11*. https://doi.org/10.1080/10408347.2022.2144994

Verma, N., Vaidh, S., Vishwakarma, G. S., & Pandya, A. (2020). Antimicrobial nanomaterials for water disinfection. *Nanotoxicity: Prevention and Antibacterial Applications of Nanomaterials*. https://doi.org/10.1016/B978-0-12-819943-5.00018-X

Villaseñor, M. J., & Ríos, Á. (2018). Nanomaterials for water cleaning and desalination, energy production, disinfection, agriculture and green chemistry. *Environmental Chemistry Letters*. https://doi.org/10.1007/s10311-017-0656-9

Vivar, M., Fuentes, M., Torres, J., & Rodrigo, M. J. (2021). Solar disinfection as a direct tertiary treatment of a wastewater plant using a photochemical-photovoltaic hybrid system. *Journal of Water Process Engineering*. https://doi.org/10.1016/j.jwpe.2021.102196

Wang, J., & Liu, X. (2021). Forward osmosis technology for water treatment: Recent advances and future perspectives. *Journal of Cleaner Production*. https://doi.org/10.1016/j.jclepro.2020.124354

Wang, X. N., Liu, Y., Pan, X. H., Han, J. X., & Hao, J. (2017). Parameters for seawater reverse osmosis product water: A review. *Exposure and Health*. https://doi.org/10.1007/s12403-016-0232-8

World Health Organization, World Health Organization, & World Health Organization. (2012). Regional Office for Europe (WHO, 2012). *Environmental health inequalities in Europe*. Copenhagen.

Wychowaniec, J. K., & Brougham, D. F. (2022). Emerging magnetic fabrication technologies provide controllable hierarchically-structured biomaterials and stimulus response for biomedical applications. *Advanced Science*. https://doi.org/10.1002/advs.202202278

Yalcinkaya, F. (2019). A review on advanced nanofiber technology for membrane distillation. *Journal of Engineered Fibers and Fabrics*. https://doi.org/10.1177/1558925018824901

Yemmireddy, V. K., & Hung, Y. C. (2017). Using photocatalyst metal oxides as antimicrobial surface coatings to ensure food safety—Opportunities and challenges. *Comprehensive Reviews in Food Science and Food Safety*. https://doi.org/10.1111/1541-4337.12267

Zhang, P., Tong, M., Yuan, S., & Liao, P. (2014). Transformation and removal of arsenic in groundwater by sequential anodic oxidation and electrocoagulation. *Journal of Contaminant Hydrology*. https://doi.org/10.1016/j.jconhyd.2014.06.009

Zhang, Q., Appel, E., Hu, S., Pennington, R. S., Meyer, J., Neumann, U., Burchard, M., Allstädt, F., Wang, L., & Koutsodendris, A. (2020). Nano-magnetite aggregates in red soil on low magnetic bedrock, their changes during source-sink transfer, and implications for paleoclimate studies. *Journal of Geophysical Research: Solid Earth*. https://doi.org/10.1029/2020JB020588

Zhao, X., Zhang, B., Liu, H., & Qu, J. (2010). Removal of arsenite by simultaneous electro-oxidation and electro-coagulation process. *Journal of Hazardous Materials*. https://doi.org/10.1016/j.jhazmat.2010.08.058

Innovative Skills for Agricultural Waste Management: Strategies and Future Direction of Cellulose-Based Polymers in Pharmaceutical Sciences

Manoj Kumar Sharma, Sumit Tewatia, Dipanjan Koley, Tanya Gupta, and Anupama Diwan

Abstract The crux of the matter in today's environmental discourse is the need for effective and innovative waste management strategies for increasing the meet of exponentially rising population. As pollution caused due to ignorance to correctly handle waste by us has led to massive climate changes. This negligence has led to severe environmental issues globally, including floods in India and its neighbouring countries, extreme rainfall in Southeast Asia, droughts in sub-Saharan Africa, forest fires in Australia and the Amazon, air pollution in urban centres like Beijing and Delhi, and rising global temperatures. Additionally, these environmental challenges have exacerbated contamination-related problems and diseases, highlighting the urgent need for sustainable waste management solutions. All these drastic events occurring around the globe is heavily affecting the global economy and loss of unimaginable downfall of finances and families. With an average annual growth rate of 4%, India presently produces 62 million tonnes of waste annually, comprising both recyclable and non-recyclable materials, according to the Ministry of Environment, Forests, and Climate Change. According to projections made by the United Nations Environment Programme (UNEP), the quantity of municipal solid trash produced worldwide is expected to rise from 2.1 billion tonnes in 2023 to 3.8 billion tonnes by 2050. These facts strongly advocate that conventional and informal methods are falling behind and are not able to keep up with the demands of effective waste regulation with the exponential increase in population. This chapter explores various conventional and informal approaches, their short comings, new strategies, and innovation technologies implemented to improve the management of agricultural waste like rice straw for the development of cellulose-based polymers. These novel strategies not only mitigate environmental impact but also promote

M. K. Sharma (✉) · S. Tewatia · D. Koley · A. Diwan
School of Pharmaceutical Sciences, Apeejay Stya University, Gurugram, Haryana, India

T. Gupta
School of Pharmaceutical Sciences, Apeejay Stya University, Gurugram, Haryana, India

RV Institute of Pharmacy, Bijnor, Uttar Pradesh, India

S. Rajpoot, A. P. Das (eds.), *Innovative Technologies for Waste Management*, Sustainable Environmental Waste Management Strategies,
https://doi.org/10.1007/978-3-031-97184-6_8

economic and social benefits, paving the way for a cleaner and more sustainable future.

1 Introduction

Agricultural waste, encompassing organic and inorganic byproducts generated from various farming activities, represents a significant global challenge in modern agriculture. These residues include crop remnants, stalks, straw, leaves, husks, roots, and animal waste produced during agronomic processing. The generation of agricultural waste is inherent to farming processes, occurring at various stages from harvesting to postharvest handling, storage, processing, and packaging of agricultural products (Koul et al., 2022). The classification of agricultural waste primarily falls into two categories: crop residues and animal waste, with crop residues further subdivided into field residues (remaining in fields after harvest) and process residues (generated during crop processing). The magnitude of agricultural waste generation has become increasingly concerning, particularly in emerging countries where management of the waste infrastructure is often inadequate. In India, the challenge is especially acute, with approximately 500 million tons of agricultural waste produced annually, leading to severe environmental and health consequences (Kumar & Agarwal, 2020).

The indecorous regulation of agricultural waste, particularly the prevalent practice of crop residue burning, has emerged as a critical environmental concern (Bala et al., 2024b; Dey et al., 2024a, b). This practice significantly contributes to climate change through the release of greenhouse gases and particulate matter into the atmosphere. The situation is particularly evident in Northern India, where seasonal burning of seasonal crops like straws from the rice crops creates hazardous smog conditions, severely affecting air quality in major cities including Delhi NCR, and contributing to respiratory health issues among millions of residents (Yadav et al., 2021a, b). However, agricultural waste presents numerous opportunities for sustainable utilization, particularly in the pharmaceutical industry as excipients (Yadev et al., 2022). These waste materials, rich in cellulose, hemicellulose, lignin, etc. serves viable options for transmuting into various pharmaceutical-grade ingredients through proper processing and modification. For instance, cellulose from crop residues can be modified into microcrystalline cellulose, a widely used pharmaceutical excipient in tablet formulation (Czekała & Pulka, 2024).

The transition towards sustainable agricultural waste management requires a multi-faceted approach, with farmer education and awareness at its core. Farmers need to be introduced to innovative waste management techniques and the principles of circular economy, where agricultural waste is viewed as a valuable resource rather than a disposal problem (Mishra et al., 2019; Das et al., 2011). This paradigm shift can be achieved through comprehensive training programs, demonstration projects, and financial incentives that encourage the adoption of sustainable practices. The successful implementation of agricultural waste management strategies depends on creating a sustainable ecosystem where farmers, industry stakeholders, and government agencies work collaboratively. By transforming various environmental waste into valuable products, we can address both environmental concerns and create economic opportunities for farming communities (Priyadarshini et al., 2024; Tripathy et al., 2024) This approach not only mitigates environmental degradation but also contributes to the broader goals of sustainable agriculture and circular economy, making it an imperative area for research and development in contemporary agricultural practices (Ajwani-Ramchandani 2021).

2 Types of Agricultural Waste

- **Crop Residues:** Processing of the cereals crops like rice, maize, and wheat generates a significant amount of plant leftovers such as straw, stalks, leaves, husks, and shells called as crop residues.
- **Animal Waste:** Manure, bedding material, and other byproducts from livestock farming include manure from cattle, poultry, pigs, and other farm animals.
- **Agro-processing Waste:** Byproducts from the processing of agricultural products such as sugarcane bagasse, fruit peels, vegetable waste, and rice husks. These are generated during processing of crops or agricultural products, including bagasse from sugarcane processing, and pulp and paper industry residues. These residues are rich sources of cellulose, a polysaccharide that can be utilized in various industrial applications such as biofuels, paper production, and biodegradable plastics. Agricultural residues rich in cellulose. These materials play a crucial role in sustainable biofuel production, biodegradable products, and more.
- **Food-Processing Waste:** Leftover parts of fruits, vegetables, grains, and other agricultural products after processing (e.g., canning, juicing, milling).
- **Chemical Waste:** Waste from fertilizers, pesticides, and herbicides used in farming.
- **Waste from Agro-industrial Products:** Byproducts from industries that use agricultural products to create goods, such as textile, paper, and bioenergy industries.

2.1 General Crop Residues

- **Corn Stover**: The leftover stalks, leaves, and husks after corn harvest contain cellulose. These residues are abundant and can be used for various purposes.
- **Wheat Straw**: The straw left behind after wheat harvesting is another valuable source of cellulose. It can be processed to extract this renewable polymer.
- **Rice Straw**: Rice straw, a byproduct of rice cultivation, is rich in cellulose. It can contribute significantly to sustainable materials.

- **Perennial Energy Grasses:** These grasses, such as switchgrass and miscanthus, grow year after year. They accumulate cellulose and can be harvested for biofuel production.
- **Forest Residues**: While not strictly agricultural, forest residues (such as branches, bark, and leaves) also contain cellulose. Woody biomass from forests can be utilized for various applications.

2.2 Other Agricultural Sources

- **Sugarcane Bagasse**: After extracting sugar from sugarcane, the fibrous bagasse remains. It contains cellulose and can be used for bioenergy.
- **Oil Palm**: Oil palm residues, including empty fruit bunches and fronds, contain cellulose.
- **Bamboo**: Bamboo is a fast-growing plant with abundant cellulose in its stems.
- **Coconut Coir:** The fibrous material from coconut husks contains cellulose.
- **Corn Cobs and Husks**: Leftover parts from corn processing are rich in cellulose.
- **Pineapple Crown**: The crown of pineapple plants also contains cellulose.
- **Groundnut Shells**: Residues from groundnut (peanut) processing can be a source of cellulose.

3 Barriers and Breakthroughs in Utilizing Agriculture Waste

The management and utilization of agricultural waste face numerous complex challenges that span across multiple dimensions, creating a web of interconnected problems that hinder effective solutions. At the fundamental level, infrastructure limitations pose a significant barrier, with insufficient storing services, compromised transport networks, and insufficient processing centres preventing the efficient collection and handling of agricultural waste. The traditional methods of waste management, particularly open burning and uncontrolled dumping, have become obsolete and counterproductive in the face of growing waste volumes and environmental concerns, yet they persist due to their perceived convenience and low immediate costs (Bala et al., 2024a; Das et al., 2024d; Mohanty et al., 2024a). Technical challenges present another significant hurdle, as many existing waste processing technologies require substantial initial investments, specialized knowledge, and continuous maintenance, making them inaccessible to small and medium-scale farmers. The heterogeneous nature of agricultural waste, varying in moisture content, composition, and physical properties, further complicates the development of standardized processing methods. Additionally, the seasonal nature of agricultural waste generation creates logistical challenges in maintaining consistent supply chains for waste-based products. Economic barriers represent a critical impediment, as farmers often lack the financial resources to invest in modern waste management technologies or alternative disposal methods. The absence of well-developed markets for waste-derived products, combined with high processing costs and uncertain returns on investment, discourages private sector participation. The current economic models rarely account for the environmental costs of improper waste disposal, creating a disconnect between sustainable practices and financial viability. Institutional and policy-related challenges further compound these issues, with fragmented regulatory frameworks, insufficient enforcement mechanisms, and limited coordination among various stakeholders (Das et al., 2024c; John & Das, 2024; Mohanty et al., 2024b). The lack of comprehensive policies supporting waste-to-value initiatives, coupled with inadequate financial incentives and technical support programs, hampers the adoption of innovative solutions (Dey et al., 2024a; Ray et al., 2024; Sahu et al., 2024). Knowledge gaps about potential value-addition opportunities and modern waste management techniques persist, while the absence of standardized protocols for waste collection, segregation, and processing leads to inefficient resource utilization (Gontard et al., 2018).

Environmental and social challenges add another layer of complexity, as improper waste management practices contribute to soil degradation, water pollution, and air quality deterioration (Dey et al., 2024c). The burning of agricultural waste, particularly prevalent in developing regions, releases harmful pollutants and greenhouse gases, exacerbating climate change and public health concerns. Social resistance to new waste management practices, often stemming from traditional mindsets, poses a significant barrier to implementing sustainable solutions. Logistical challenges faced during the streamlining of agricultural waste, including collection, transportation, and storage, create bottlenecks in establishing efficient waste-to-value systems. The geographical dispersion of farming activities, combined with seasonal variations in waste generation, increases the maintenance of consistent supply chains for industrial applications. Inadequate storage services for waste materials and preventing degradation during storage and transportation further complicate the utilization of agricultural waste in value-added applications (Arora et al., 2023).

4 Excipient from Waste vs Live Vegetation

While the conversion of agricultural waste into pharmaceutical excipients presents a promising solution, it's crucial to distinguish this approach from the direct harvesting of vegetation or plants solely for excipient production. The latter practice, if not carefully managed, can lead to significant environmental concerns and should not be considered a green or sustainable solution.

Direct extraction of excipients from primary plant sources, rather than from their waste products, can potentially harm local ecosystems, disrupt biodiversity, and contribute to deforestation. This approach may lead to the overexploitation of plant species, particularly if demand for these excipients increases rapidly. For instance, the harvesting of *Acacia senegal* and *Acacia seyal* trees for gum Arabic, a widely used pharmaceutical excipient, has led to concerns about desertification and loss of habitat in the Sahel region of Africa (Hammad, & Mohammed, 2018). In Sudan, which produces about 80% of the world's gum Arabic, unsustainable harvesting practices have contributed to soil degradation and loss of biodiversity (Ballal et al., 2005).

Another example is the cultivation of *Eucalyptus* trees for cellulose production. While *Eucalyptus* grows quickly and produces high yields, large-scale monoculture plantations can lead to soil depletion, reduced water availability for other plant species, and loss of native habitats (Bayle, 2019).

These examples highlight the potential negative impacts of sourcing excipients directly from vegetation rather than from agricultural waste. Such practices can lead to:

- Loss of biodiversity due to monoculture cultivation or overharvesting of specific plant species.
- Soil degradation and erosion from intensive farming practices.
- Increased water consumption, potentially leading to water scarcity in affected regions.
- Disruption of local ecosystems and loss of habitat for native flora and fauna.

Table 1 Different excipient obtained from the different sources

Agriculture waste	Excipient obtained	Use	Insights
Rice straw	Microcrystalline cellulose	Binder, disintegrant	High yield and cost-effective; reduces environmental impact of rice farming
Cocoa pod husks	Microcrystalline cellulose	Binder	Utilizes waste from cocoa production, adding value to the chocolate industry
Sugarcane bagasse	Nanocellulose	Drug carrier, stabilizer	Excellent mechanical properties; enhances drug delivery systems
Banana peels	Pectin	Binding agent, disintegrant l	Rich in pectin; promotes sustainable use of fruit waste
Mango seed kernels	Starch	Tablet binder, disintegrant	High starch content; supports waste valorization in tropical fruit industries
Apple pomace	Pectin	Gelling agent, stabilizer	Utilizes apple processing waste; enhances texture and stability in formulations
Orange peels	Pectin	Gelling agent, stabilizer	High pectin yield; supports citrus industry waste management
Grape pomace	Polyphenols	Antioxidant	Rich in antioxidants; adds value to wine industry byproducts
Wheat bran	Arabinoxylans	Dietary fibre, prebiotic	Promotes gut health; supports sustainable use of cereal byproducts
Soybean hulls	Soy polysaccharides	Tablet binder, disintegrant	High polysaccharide content; enhances tablet formulation properties

Therefore, while plant-based excipients offer many advantages, it's crucial to prioritize sourcing these materials from agricultural waste streams rather than from dedicated crops or wild harvesting. This approach ensures that the production of pharmaceutical excipients contributes to waste reduction and resource efficiency without adding additional pressure on natural ecosystems. By focusing on waste valorization, the pharmaceutical industry can truly align with principles of sustainability and circular economy, minimizing its environmental footprint while meeting the growing demand for effective and safe drug formulations. Table 1 displays the various excipient obtained from the agriculture waste.

5 Importance of Sustainable Solutions: Introducing Cellulose

5.1 Cellulose: Nature's Sustainable Solution for a Greener Tomorrow

In the face of escalating environmental challenges, including climate change and resource depletion, the pursuit of sustainable solutions has become increasingly critical. Cellulose emerges as a promising candidate in this context, offering a renewable and environmentally responsible alternative to conventional materials. As the most abundant natural polymer on Earth, primarily found in plant cell walls, cellulose represents a readily available and sustainable resource that can significantly reduce our dependence on finite materials while promoting responsible resource management practices.

The inherent biodegradability of cellulose stands as one of its most valuable environmental attributes, as it naturally decomposes into harmless substances through biological processes. This characteristic is particularly significant in addressing the growing concerns of waste accumulation and environmental pollution, especially in industries traditionally reliant on non-biodegradable materials (Das et al., 2024d; Mohanty et al., 2024b; Priyadarsini & Das, 2024; Sahu et al., 2024). The versatility of cellulose further enhances its appeal, as it can be adapted for diverse applications across multiple sectors, including packaging, textiles, biomedical applications, and pharmaceutical industries, offering sustainable alternatives to conventional materials with higher environmental footprints (Shokri & Adibkia, 2013).

The integration of cellulose-based materials into various industrial processes represents a significant step towards environmental sustainability. By replacing nonrenewable and environmentally harmful materials with cellulose-derived alternatives, industries can substantially reduce their ecological impact while contributing to the circular economy. This transition is supported by ongoing research and innovation in cellulose-based materials, which continues to uncover new applications and improve production efficiencies. In the advancement of cellulose technology not only drives progress towards environmental sustainability but also creates opportunities for developing novel solutions to existing environmental challenges (Sharma et al., 2025), through continued investment and advancement of research in the potential of cellulose in contributing environmentally conscious future, (Sharma et al., 2026), offering promising pathways for addressing global environmental concerns while meeting the growing demand for sustainable materials (Kamel et al., 2008).

5.2 *Cellulose in Modern Pharmaceuticals: Bridging Sustainability with Therapeutic Innovation*

The growing emphasis on sustainable materials has brought cellulose, a naturally abundant polymer found in plants, to the forefront of industrial innovation. As a renewable resource, cellulose presents a compelling alternative to conventional fossil fuel-derived materials, offering a pathway to reduce environmental impact while maintaining industrial efficiency. Its remarkable combination of properties, including biodegradability, structural strength, and versatility, positions cellulose as a valuable material across diverse sectors, from packaging, waste treatment and textiles to advanced biomedical applications (Norfarhana et al., 2022). This natural polymer aligns perfectly with circular economy principles, providing opportunities to decrease dependence on nonrenewable resources while promoting sustainable industrial practices.

The pharmaceutical industry, in particular, has witnessed significant advancements in cellulose-based applications, especially in drug delivery systems. The versatility of cellulose and its derivatives has revolutionized drug delivery mechanisms, offering sophisticated solutions for controlled release and targeted delivery to specific organs or tissues. Microcrystalline cellulose and powdered cellulose serve as essential excipients in pharmaceutical formulations, functioning as effective adsorbents, diluents, and stabilizing agents. The emergence of cellulose nanofibres has further expanded the possibilities in biomedicine, introducing new approaches to drug delivery systems that capitalize on their unique properties of flexibility, minimal toxicity, and reactive surface characteristics (Ajmal et al., 2022).

The application spectrum of cellulose-based materials in pharmaceuticals continues to expand, with innovations such as cellulose acetate nanofibre mats finding specific applications in topical and transdermal drug delivery systems (Orasugh et al., 2018). The FDA-approved hydroxypropyl methyl cellulose (HPMC) has demonstrated particular success in controlled-release formulations, showing promising results in sustained antibiotic delivery applications. These developments not only showcase the versatility of cellulose-based materials but also highlight their potential to revolutionize pharmaceutical delivery systems while maintaining environmental sustainability (Das et al., 2024a, b, c, d). The integration of cellulose-based materials in pharmaceutical applications continues to evolve, offering innovative solutions that balance therapeutic efficacy with environmental responsibility, ultimately contributing to a more sustainable future in healthcare and pharmaceutical industries.

5.3 Cellulose in Ophthalmic Formulations: Bridging Sustainability with Therapeutic Excellence

The remarkable versatility of cellulose extends prominently into pharmaceutical applications, particularly in ophthalmic formulations, where its derivatives have become indispensable components in modern drug delivery systems. Since its discovery in 1838, this high-molecular-weight homopolysaccharide, characterized by its strong hydrogen bonds and unique structural properties, has evolved from a simple natural polymer to a sophisticated pharmaceutical excipient. The inherent insolubility of cellulose in water and organic solvents, attributed to its molecular architecture, serves as a foundation for its modification into various functional derivatives, each tailored to specific pharmaceutical needs (Patchan et al., 2016).

In the realm of ophthalmic formulations, cellulosic polymers such as methylcellulose, hydroxyethyl cellulose, and carboxymethylcellulose have emerged as crucial components, primarily due to their exceptional viscosity-enhancing properties. These derivatives, obtained through chemical modification of the hydroxyl groups present in D-glucose units, demonstrate remarkable versatility in controlling drug release mechanisms and enhancing the overall efficacy of ocular drug delivery systems. The ability to modify cellulose through various chemical reactions provides pharmaceutical scientists with a powerful tool to develop formulations that meet specific therapeutic requirements while maintaining optimal biocompatibility and stability (Tundisi et al., 2021).

The adoption of cellulose in ophthalmic preparations is further supported by its abundance and cost-effectiveness, making it particularly attractive for large-scale pharmaceutical manufacturing. The mechanical properties inherited from its natural role in plant structures contribute to the robustness and stability of pharmaceutical formulations. This combination of sustainability, modifiability, and functional properties positions cellulose as a premier choice in modern pharmaceutical development, exemplifying how naturally derived materials can effectively meet the demanding requirements of specialized medical applications while maintaining environmental consciousness.

5.4 Cellulose in Modern Drug Delivery: Compatibility with the Current and Novel Drug Targeting

The integration of cellulose-based materials in contemporary drug delivery systems represents a significant advancement in pharmaceutical science, where sustainability meets therapeutic efficacy (Priyadarsini & Das, 2025). These materials have emerged as powerful tools in drug delivery applications, primarily due to their inherent biodegradability and biocompatibility, characteristics that are increasingly valued in modern pharmaceutical development. The versatility of cellulose becomes particularly evident in its surface modification capabilities, where cellulose-based polymers can enhance the hydrophilic properties of nanoparticles, thereby optimizing their drug delivery potential and improving therapeutic outcomes (Abeer et al., 2014).

The application spectrum of cellulose derivatives, particularly carboxymethylcellulose (CMC), extends across various pharmaceutical formulations, from nanoparticle development to innovative drug delivery systems. These derivatives play crucial roles in pharmaceutical coatings and film formation, effectively modulating drug release profiles from solid dosage forms. The advancement in sophisticated drug delivery systems, exemplified by innovations like Cellax-CBZ, demonstrates the potential of cellulose-based polymers in addressing challenging therapeutic areas, including drug-resistant cancers, where enhanced drug delivery and improved safety profiles are paramount (Dai & Si 2019).

The evolution of cellulose-based pharmaceutical excipients continues to push boundaries in drug delivery innovation, particularly in the development of polysaccharide hydrogel capsules (Auriemma et al., 2020). These advanced systems, utilizing cellulose derivatives, offer precise control over drug release at specific pH levels, making them particularly valuable for oral drug delivery applications. The exceptional binding and disintegration properties of cellulose excipients, combined with their ability to enhance drug stability and release profiles, make them indispensable in modern pharmaceutical formulations. This alignment with green and sustainable practices in healthcare not only addresses environmental concerns but also ensures the development of more effective and environmentally responsible therapeutic solutions, marking a significant step forward in sustainable pharmaceutical development.

6 Future Prospectives

The future of cellulose extraction from agricultural waste presents a promising pathway that addresses dual challenges: waste management and sustainable pharmaceutical development. In regions like Northern India, where agricultural waste burning contributes significantly to air pollution and seasonal smog, the transformation of crop residues into pharmaceutical-grade cellulose offers a sustainable solution. This approach not only provides an economic incentive for farmers to avoid crop burning but also establishes a circular economy. The progress of efficient, cost-effective extraction methods and the establishment of decentralized processing units could revolutionize agricultural waste management while creating new revenue streams for farming communities. In the pharmaceutical domain, the future of agriculture-derived cellulose holds immense potential, particularly in advanced drug delivery systems and novel formulation development. Research is increasingly focused on optimizing extraction processes to obtain high-purity cellulose that meets pharmaceutical standards. Future developments may include innovative modification techniques to enhance the functionality of agricultural waste-derived cellulose, making it more suitable for specialized pharmaceutical applications. The emergence of nanotechnology and advanced processing methods could overcome current limitations in particle size control and uniformity, leading to more efficient drug delivery systems (Ansari et al., 2024).

However, several challenges need to be addressed through focused research efforts. The variability in cellulose quality from different agricultural sources, seasonal availability, and the presence of impurities remain significant concerns. Future research directions should focus on developing standardized extraction protocols, implementing robust quality control measures, and establishing consistent supply chains. Advanced characterization techniques and automated processing systems could help maintain uniform quality standards. Additionally, research into green extraction methods and environmentally friendly modification processes could further enhance the sustainability aspect of cellulose production from agricultural waste. The integration of machine learning and AI in process optimization, coupled with advances in biotechnology, could revolutionize the extraction and modification of cellulose from agricultural waste. These technological advancements may lead to more efficient processes, reduced production costs, and improved product quality. Furthermore, the development of novel applications in areas such as targeted drug delivery, tissue engineering, and personalized medicine could expand the scope of agricultural waste-derived cellulose in pharmaceutical applications (Khalife et al., 2024). As research progresses and technology evolves, the transformation of agricultural waste into high-value pharmaceutical excipients represents not just an environmental solution but a significant step towards sustainable healthcare innovation, promising a future where waste management and pharmaceutical advancement go hand in hand, creating a more sustainable and healthier world.

Acknowledgement We are highly acknowledged to Haryana State Council for Science, Innovation and Technology (HSCSIT), Government of Haryana, India, for providing research grant with file number: HSCSIT/R&D/2022/2976.

References

Abeer, M. M., Mohd Amin, M. C., & Martin, C. (2014). A review of bacterial cellulose-based drug delivery systems: Their biochemistry, current approaches and future prospects. *Journal of Pharmacy and Pharmacology, 66*(8), 1047–1061. https://doi.org/10.1111/jphp.12234

Ajmal, G., Yadav, N., Kumawat, M. K., Sharma, M. K., & Iqbal, M. R. (2022). Application of electrospun nanofiber in wound healing: Trends and recent patents analysis (2023). *International Journal of Life Science and Pharma Research, 13*(1), L37–L47. https://doi.org/10.22376/ijlpr.2023.13.1.SP1.L37-47

Ajwani-Ramchandani, R., Figueira, S., de Oliveira, R. T., Jha, S., Ramchandani, A., & Schuricht, L. (2021). Towards a circular economy for packaging waste by using new technologies: The case of large multinationals in emerging economies. *Journal of Cleaner Production, 281*, 125139. https://doi.org/10.1016/j.jclepro.2020.125139

Ansari, M. M., Heo, Y., Do, K., Ghosh, M., & Son, Y. O. (2024). Nanocellulose derived from agricultural biowaste by-products–Sustainable synthesis, biocompatibility, biomedical applications, and future perspectives: A review. *Carbohydrate Polymer Technologies and Applications, 8*, 100529. https://doi.org/10.1016/j.carpta.2024.100529

Arora, J., Ramawat, K. G., & Mérillon, J. M. (2023). Disposal of agricultural waste and its effects on the environment, production of useful metabolites and energy: Potential and challenges. In *Agricultural waste: Environmental impact, useful metabolites and energy production* (pp. 3–20). Springer. https://doi.org/10.1007/978-981-19-8774-8_1

Auriemma, G., Russo, P., Del Gaudio, P., García-González, C. A., Landín, M., & Aquino, R. P. (2020). Technologies and formulation design of polysaccharide-based hydrogels for drug delivery. *Molecules, 25*(14), 3156. https://doi.org/10.3390/molecules25143156

Bala, K., Anshul, V. S., Dogra, A., Das, A. P., Kumar, R., & Nandini. (2024a). Sustainable management for urban plastic waste generation. *Sustainable Management of Urban Plastic Waste Through Circular Economic Approaches, 211–236.* https://doi.org/10.1201/9781003465447-13

Bala, K., Dogra, A., Lata, P., & Das, A. P. (2024b). Discussing sources and biological-chemical remediation approaches for microplastic pollution. *Sustainable Microbial Technology for Synthetic and Cellulosic Microfiber Bioremediation, 37–59.* https://doi.org/10.1007/978-3-031-62600-5_3

Ballal, M. E., El Siddig, E. A., Elfadl, M. A., & Luukkanen, O. (2005). Relationship between environmental factors, tapping dates, tapping intensity and gum arabic yield of an Acacia senegal plantation in western Sudan. *Journal of Arid Environments, 63*(2), 379–389. https://doi.org/10.1016/j.jaridenv.2005.01.024

Bayle, G. (2019). Ecological and social impacts of eucalyptus tree plantation on the environment. *Journal of Biodiversity Conservation and Bioresource Management, 5*(1), 93–104. https://doi.org/10.3329/jbcbm.v5i1.42189

Czekała, W., & Pulka, J. (2024). Water, wastewater, waste management in agriculture and agri-food industry. *Water, 16*(13), 1817. https://doi.org/10.3390/w16131817

Dai, L., & Si, C. (2019). Recent advances on cellulose-based nano-drug delivery systems: Design of prodrugs and nanoparticles. *Current Medicinal Chemistry, 26*(14), 2410–2429. https://doi.org/10.2174/0929867324666170711131353

Das, A. P., Sukla, L. B., Pradhan, N., & Nayak, S. (2011). Manganese biomining: A review. *Bioresource Technology, 102*(16), 7381–7387. https://doi.org/10.1016/J.BIORTECH.2011.05.018

Das, M., Lalsangi, S., Santra, S., & Banerjee, R. (2024a). Nanocellulose as a carrier for improved drug delivery: Progresses and innovation. *Journal of Drug Delivery Science and Technology, 9*, 105743. https://doi.org/10.1016/j.jddst.2024.105743

Das, A. P., Behera, I. D., & Das, N. P. (2024b). *Renewable energy generation and value addition from environmental microfiber pollution through advanced greener solution* (p. 278). Springer.

Das, A., Dey, S., & Das, A. P. (2024c). Ecotoxicological impacts of synthetic microfiber pollutants and development of sustainable mitigation strategies. *Environmental Chemistry and Ecotoxicology*. https://doi.org/10.1016/J.ENCECO.2024.12.003

Das, A., Dey, S., Bala, K., Gnanasekaran, R., & Das, A. P. (2024d). Microbial bioremediation technology for sustainable treatment and management of synthetic microfiber waste. *Sustainable Microbial Technology for Synthetic and Cellulosic Microfiber Bioremediation, 221–238.* https://doi.org/10.1007/978-3-031-62600-5_12

Dey, S., Bissoyi, A., Koteswara Reddy, G., & Das, A. P. (2024a). Synthetic microfibers: An urban emerging pollutant demanding for sustainable environmental solutions. *Sustainable Microbial Technology for Synthetic and Cellulosic Microfiber Bioremediation, 107–121.* https://doi.org/10.1007/978-3-031-62600-5_7

Dey, S., Dash, N., Ray, S. K., Priyadarsini, S., Chhotaray, C., & Das, A. P. (2024b). Occurrence and quantification of tire wear plastics pollutants from highway city dust of Bhubaneswar City of Odisha. *Sustainable Microbial Technology for Synthetic and Cellulosic Microfiber Bioremediation, 77–89.* https://doi.org/10.1007/978-3-031-62600-5_5

Dey, S., Dash, A., & Das, A. P. (2024c). Bioremediation as a sustainable and circular solution for mining waste management and bioenergy generation. *Sustainable Management of Mining Waste and Tailings: A Circular Economy Approach, 107–119.* https://doi.org/10.1201/9781003442455-6/BIOREMEDIATION-SUSTAINABLE-CIRCULAR-

SOLUTION-MINING-WASTE-MANAGEMENT-BIOENERGY-GENERATION-SUDESHNA-DEY-AKANSHYA-DASH-ALOK-PRASAD-DAS

Gontard, N., Sonesson, U., Birkved, M., Majone, M., Bolzonella, D., Celli, A., Angellier-Coussy, H., Jang, G. W., Verniquet, A., Broeze, J., & Schaer, B. (2018). A research challenge vision regarding management of agricultural waste in a circular bio-based economy. *Critical Reviews in Environmental Science and Technology, 48*(6), 614–654. https://doi.org/10.1080/10643389.2018.1471957

Hammad, Z. M., & Mohammed, M. H. (2018). Management practices of gum arabic–producing trees. In *Gum Arabic* (pp. 13–21). Academic Press. https://doi.org/10.1016/B978-0-12-812002-6.00002-6

John, A., & Das, A. P. (2024). Government strategies for achieving circular economy for plastic waste management in Southeast Asia. *Sustainable Management of Urban Plastic Waste Through Circular Economic Approaches, 72–82*. https://doi.org/10.1201/9781003465447-6/GOVERNMENT-STRATEGIES-ACHIEVING-CIRCULAR-ECONOMY-PLASTIC-WASTE-MANAGEMENT-SOUTHEAST-ASIA-AKBAR-JOHN-ALOK-PRASAD-DAS

Kamel, S., Ali, N., Jahangir, K., Shah, S. M., & El-Gendy, A. A. (2008). Pharmaceutical significance of cellulose: A review. *Express Polymer Letters, 2*(11), 758–778. https://doi.org/10.3144/expresspolymlett.2008.90

Khalife, E., Sabouri, M., Kaveh, M., & Szymanek, M. (2024). Recent advances in the application of agricultural waste in construction. *Applied Sciences, 14*(6), 2355. https://doi.org/10.3390/app14062355

Koul, B., Yakoob, M., & Shah, M. P. (2022). Agricultural waste management strategies for environmental sustainability. *Environmental Research, 206*, 112285. https://doi.org/10.1016/j.envres.2021.112285

Kumar, A., & Agrawal, A. (2020). Recent trends in solid waste management status, challenges, and potential for the future Indian cities–A review. *Current Research in Environmental Sustainability, 2*, 100011.

Mishra, S., Rath, C., & charan, & Das, A. P. (2019). Marine microfiber pollution: A review on present status and future challenges. *Marine Pollution Bulletin, 140*, 188–197. https://doi.org/10.1016/J.MARPOLBUL.2019.01.039

Mohanty, J., Dey, S., Bala, K., & Das, A. P. (2024a). Recent advances in microplastic pollution for its sustainable management. *Sustainable Management of Urban Plastic Waste Through Circular Economic Approaches, 27–48*. https://doi.org/10.1201/9781003465447-3/RECENT-ADVANCES-MICROPLASTIC-POLLUTION-SUSTAINABLE-MANAGEMENT-JYOTIRMAYEE-MOHANTY-SUDESHNA-DEY-KIRAN-BALA-ALOK-PRASAD-DAS

Mohanty, M., Tripathy, B., Koteswara Reddy, G., & Das, A. P. (2024b). Greener technologies for establishing restraint over microplastic pollution. *Sustainable Microbial Technology for Synthetic and Cellulosic Microfiber Bioremediation, 239–258*. https://doi.org/10.1007/978-3-031-62600-5_13

Norfarhana, A. S., Ilyas, R. A., & Ngadi, N. J. (2022). A review of nanocellulose adsorptive membrane as multifunctional wastewater treatment. *Carbohydrate Polymers, 291*, 119563. https://doi.org/10.1016/j.carbpol.2022.119563

Orasugh, J. T., Saha, N. R., Rana, D., Sarkar, G., Mollick, M. M., Chattoapadhyay, A., Mitra, B. C., Mondal, D., Ghosh, S. K., & Chattopadhyay, D. (2018). Jute cellulose nano-fibrils/hydroxypropylmethylcellulose nanocomposite: A novel material with potential for application in packaging and transdermal drug delivery system. *Industrial Crops and Products, 112*, 633–643. https://doi.org/10.1016/j.indcrop.2017.12.069

Patchan, M. W., Chae, J. J., Lee, J. D., Calderon-Colon, X., Maranchi, J. P., McCally, R. L., Schein, O. D., Elisseeff, J. H., & Trexler, M. M. (2016). Evaluation of the biocompatibility of regenerated cellulose hydrogels with high strength and transparency for ocular applications. *Journal of Biomaterials Applications, 30*(7), 1049–1059. https://doi.org/10.1177/0885328215616273

Priyadarshini, B., Santhosh Kumar, M., & Das, A. P. (2024). Microbial degradation of synthetic plastics released from COVID-19 generated single use disposable face masks. *Environmental Science and Engineering, Part, F2827*, 1–16. https://doi.org/10.1007/978-3-031-50840-0_1

Priyadarsini, S., & Das, A. P. (2024). Analytical and structural characterization of waste lithium-ion batteries for their effective recycling strategy. *Environmental Chemistry and Ecotoxicology*. https://doi.org/10.1016/j.enceco.2024.12.004

Priyadarsini, S., & Das, A. P. (2025). Lithium bioleaching: A review on microbial-assisted sustainable technology for lithium bio-circularity. *Journal of Water Process Engineering, 69*, 106744. https://doi.org/10.1016/j.jwpe.2024.106744

Rajpoot, S. (2024). Eco-friendly solutions for sustainable waste management: An approach towards a cleaner and greener future. In *Waste management and life cycle assessment for sustainable business practice* (pp. 43–64). IGI Global.

Ray, S. K., Dey, S., Das, A. P., & John, A. (2024). Microplastic waste management through greener technologies: A review. *Sustainable Management of Urban Plastic Waste Through Circular Economic Approaches, 237–261*. https://doi.org/10.1201/9781003465447-14

Sahu, A., Dey, S., Mallick, K., Priyadarsini, S., Behera, I. D., & Das, A. P. (2024). A review on advanced technology for sustainable management of synthetic microplastic waste. *Sustainable Management of Urban Plastic Waste Through Circular Economic Approaches, 262–279*. https://doi.org/10.1201/9781003465447-15

Shokri, J., & Adibkia, K. (2013). Application of cellulose and cellulose derivatives in pharmaceutical industries. In *Cellulose-medical, pharmaceutical and electronic applications*. IntechOpen. https://doi.org/10.5772/55178

Sharma, M.K., Diwan, A., Sardana, S., Yadav, N., Gupta, T., & Kumawat, M.K. (2025). Isolation and characterization of cellulose derived from prominent agricultural waste (Sugarcane bagasse) and its utilization in various biomedical field. *J. Appl. Pharm.* Res. 13, 136–146. https://doi.org/10.69857/joapr.v13i1.942

Sharma MK, Diwan A, Sardana S, Kumar K. Derivation of hazardous wastes: Pharmaceutical and food applications. Foods and Raw Materials. 2026;14(1):104–116. https://doi.org/10.21603/2308-4057-2026-1-661

Tripathy, B., Dey, S., Dubey, N. K., Lo, W.-C., & Das, A. P. (2024). *An omics-based approach for managing the risk of COVID-19-generated wastes* (pp. 51–66). https://doi.org/10.1007/978-3-031-50840-0_4

Tundisi, L. L., Mostaço, G. B., Carricondo, P. C., & Petri, D. F. (2021). Hydroxypropyl methylcellulose: Physicochemical properties and ocular drug delivery formulations. *European Journal of Pharmaceutical Sciences, 159*, 105736. https://doi.org/10.1016/j.ejps.2021.105736

Yadav, N., Diwan, A., Sharma, M. K., Ajmal, L. G., & Kumawat, M. K. (2021a). Biological activity of rice straw-derived materials: An overview. *Uttar Pradesh Journal of Zoology*, 42(24), 1256–1264. https://doi.org/10.56557/upjoz/2021/v42i243243

Yadav, N., Rani, S., Sharma, M., Kumawat, M. K., & Diwan, A. (2021b). Pharmaceutical applications of RS: An overview. *Uttar Pradesh Journal of Zoology*, 42(24), 1335–1343. https://doi.org/10.56557/upjoz/2021/v42i243258

Yadav, N., Sharma, M. K., & Diwan, A., inventors; Apeejay Stya University, assignee. (2022). Rice straw-based biomaterials for tablet coating and a method thereof. Indian patent 202211031062 A.

Mechanistic Approach and Ecotoxicological Aspects for Use of Nanomaterials in Sustainable Water Remediation

Ansar Anjum, Reshu Chaudhary, Rajni Garg, Rishav Garg, and Nnabuk Okon Eddy

Abstract Increasing surface and groundwater pollution is a major problem due to ever-increasing population growth, industrialization, proliferation of microorganisms, and intensive agricultural practices causing a shortage of potable water. Due to this problem, more attention has to be paid to wastewater recycling. Traditional wastewater treatment technologies can be expensive and sometimes have low efficiency. Hence, there is the utmost requirement to continuously evaluate innovative technologies that complement and improve these traditional wastewater treatment processes to solve these problems. Methods based on exploiting nanomaterials are among the major focuses of improving wastewater management. The chapter emphasizes the potential of various nanomaterials in treating wastewater. The underlying mechanisms have been explored in detail with the help of specific examples. The toxicity and ecological impact of nanomaterials have been discussed along with their future prospective for sustainable and safe water management.

A. Anjum · R. Chaudhary
Department of Applied Chemistry, Delhi Technological University, Delhi, India

R. Garg (✉)
Department of Applied Science and Humanities, Galgotias College of Engineering and Technology, Greater Noida, Uttar Pradesh, India
e-mail: rajnigarg@science.org.in

R. Garg
Department of Civil Engineering, Galgotias College of Engineering and Technology, Greater Noida, Uttar Pradesh, India

N. O. Eddy
Department of Pure and Industrial Chemistry, University of Nigeria, Nsukka, Nigeria

S. Rajpoot, A. P. Das (eds.), *Innovative Technologies for Waste Management*, Sustainable Environmental Waste Management Strategies,
https://doi.org/10.1007/978-3-031-97184-6_9

1 Introduction

Water pollution is a foremost concern all over the world. The mounting human population enhanced anthropogenic activities, and climate change has raised a serious concern about supplying clean drinking water to the growing demand (Das & Mishra, 2009; Meidl et al., 2024). The quality of water in oceans, rivers, and other water bodies is highly damaged due to human activities (Mishra & Das, 2023). Toxic chemicals such as polyaromatic hydrocarbons, heavy metals, pathogens, pesticides, organic and inorganic solvents, and other pollutants are flushed into water bodies (Das & Mishra, 2010; Mishra, Dash, Al-Tawaha, & Das, 2022a; Anjum et al., 2024). To meet this challenge, highly advanced nanotechnology promises new opportunities for wastewater treatment processes (Mishra & Das, 2021; Mishra, Singh, Rout, & Das, 2022c). Nanotechnology can be used to decontaminate water by treating surface water, groundwater, and wastewater. Nanotechnology-based solutions can be more cost-effective, sustainable, energy-saving, scalable, and competent than existing methods and can help to treat contaminants that were previously untreatable (Nazir et al., 2022).

The physicochemical properties and behaviors of nanoparticles (NPs) change as their size decreases. The NPs may exhibit different mechanical, optical, electrical, and chemical characteristics as compared to their bulk equivalents. These changes in properties may be due to improved surface area-to-volume ratios and quantum effects. Nanotechnology provides techniques such as nanofabrication and band manipulation of NPs to engineer and control particles at a very small scale (Del Prado-Audelo et al., 2021). NPs offer broad applicability in the adsorption of dyes, adsorption of heavy metals, dye degradation, polymer removal, antibacterial effects, and more, therefore leading to the remediation and purification of water resources. The conventional methods for water treatment are disinfection, sedimentation, filtration, flocculation, etc. (Hafeez et al., 2022).

Conventional methods for water treatment are not always very effective for eliminating contaminants. The conventional methods for water treatment have high implementation costs, and low removal efficiency, which leads to the discharge of harmful compounds into the surroundings (Mishra, Singh, Rout, & Das, 2022c). Another problem is the formation of byproducts which can detriment human health. The use of NPs is favorable as they can be cost-effective, easy to prepare, nonpolluting, and recyclable (Caixeta, 2023). Several NPs are used for the treatment of wastewater by the mechanism of adsorption, photodegradation or photocatalysis, membrane filtration, etc. (Saleem et al., 2020). Most of the NPs also have antimicrobial activities. Silver (Ag) and gold (Au) metals were used in ancient times for their medical values. So, the NPs are proven to be suitable for water treatment in several ways, as listed in Table 1.

This chapter explores applications of several types of NPs like metal and metal oxide NPs, carbon nanotubes (CNTs), graphenes, magnetic NPs, and different composite NPs that can be used for the treatment of wastewater. The underlying mechanisms of removal of contaminants by various nanomaterials (NMs) have also been discussed in detail. The environmental impact of NPs with their large-scale use has also been signified.

Table 1 Impact of nanotechnology on water treatment

Aspect	Nanotechnology application	Impact	References
Contaminant removal	Nano-adsorbents (e.g., graphene oxide, iron oxide)	Efficient sequestration of heavy metals (e.g., inorganic, organic pollutants, and toxins	Meidl et al. (2024)
Pathogen removal	Ag NPs, titanium dioxide (TiO_2), and zinc oxide (ZnO)	Kills bacteria, viruses, and other microorganisms, ensuring microbiological safety	Nazir et al. (2022)
Desalination	Nanofiltration membranes (e.g., graphene-based, aquaporin)	Enhanced salt rejection, fouling resistance, and energy efficiency in desalination processes	Anjum et al. (2024)
Advanced oxidation	Photocatalysts (e.g., TiO_2, ZnO)	Breaks down pesticides, dyes, and pharmaceuticals into harmless substances under sunlight/UV light	Del Prado-Audelo et al. (2021)
Nutrient removal	Metal oxide NPs for phosphate capture	Prevents eutrophication by reducing phosphates and nitrates in water bodies	Hafeez et al. (2022)
Heavy metal removal	Nano-zero-valent iron (NZVI), graphene oxide, CNTs	Removes toxic heavy metals like arsenic, mercury, and chromium from water	Caixeta (2023)
Oil and grease removal	Nanosponges, graphene, superhydrophobic nanocoatings	Cleans industrial wastewater and oil spills, improving water clarity and quality	Saleem et al., (2020)
Disinfection	Antimicrobial nanomaterials (e.g., ag NPs)	Effective in killing pathogens and preventing waterborne diseases	Haijiao Lu et al. (2016)
Energy efficiency	Nanostructured membranes for low-pressure water filtration	Reduces energy consumption in filtration and desalination processes	Su et al. (2024)
Cost-effectiveness	Reusable nanomaterials and long-lasting membranes	Lowers operational costs for water treatment systems	Zamani et al. (2019)
Scalability	Portable nano-enabled filters for household and industrial use	Provides water purification solutions for both small-scale and large-scale applications	Cohen et al. (2019)
Sustainability	Solar-powered nanocatalysts (e.g., TiO_2)	Promotes green solutions by utilizing renewable energy sources for pollutant degradation	Anucha et al. (2022)
Challenges	Potential toxicity, high production costs, and lack of regulations	Requires further research to address safety, affordability, and regulatory standards	Mageswari et al. (2016)

2 NPs for Water Purification

All domestic and agricultural runoff/wastewater contains metals, biodegradable organic matter, dissolved inorganic solids, suspended solids, nutrients, pesticides, and detergents (Mishra & Das, 2022). The complex structure of wastewater treatment plants can make the removal of contaminants very difficult as they often do not take water from a single source Su et al. (2024). Traditional water purification processes include physicochemical techniques, viz., coagulation, precipitation, adsorption, membrane filtration, and chemical oxidation or reduction (Kumar et al., 2023). The development of nanotechnology applications can improve or replace these conventional technologies (Zamani et al., 2019). Nanomaterials are nano-scaled materials with at least one dimension less than 100 nm. The nanoscale size of these materials and, most importantly, their precise surface area to volume ratio are what give them their utility. Due to their small process footprint and large specific surface area, nanomaterials hold great potential for liquid-solid interactions (Cohen et al., 2019). A major advantage of nanomaterials is their adaptability in terms of fabrication and design to meet the specific properties required for effective water treatment. Additionally, these materials can be used to control various facets of water treatment and purification impacting the water quality (Anucha et al. 2022).

NPs are used in water purification owing to their small size, enhanced surface area, and reactivity. The NPs that are used in water purification are grouped under three categories (Mageswari et al., 2016), viz., metal-based NPs (inorganic NPs), carbon NPs (organic NPs including CNTs, graphene, carbon dots, and fullerenes), and polymeric NPs (Table 2).

2.1 Inorganic NPs

Metal-based NPs mainly include metal and metal oxide NPs like Zn, Ag, Fe, MnO_2, Al_2O_3, MgO, TiO_2, Fe_3O_4/Fe_2O_3, etc. which are used for the purification of water. Metal and metal oxide NPs are used to sequestrate various pollutants, suspended substances or bed sediments, and pathogens causing water contamination (Fig. 1) from water by several mechanisms like adsorption, photocatalysis, selective capture, molecular level interaction, targeted removal (Zamani et al., 2019).

Ag NPs have antimicrobial properties as they are highly toxic to various microorganisms like viruses (Karthik et al., 2022), bacteria (Zaheer, 2018), and fungi (Soltys et al., 2021). In ancient times Ag articles were used for antibacterial protection and disinfection of water (Shittu & Ihebunna, 2017). Although the exact action of antimicrobial properties of Ag NPs is not understood according to one of the theories it binds with the cell membrane and penetrates the cell damaging the structure and making microorganisms inactive (Garg et al., 2021). However, Ag NPs have the limitation that they aggregate in water solution as a result of which their efficiency gets reduced over a long time. To overcome this Ag NPs can be

Table 2 Type of nanomaterials (NMs) used in water remediation

Type of nanomaterial	Composition/ structure	Mechanism of action	Advantages	References
Inorganic NMs				
Metallic NPs	Pure metals	Adsorption, redox reactions, disinfection	High reactivity, and antimicrobial properties	Mageswari et al. (2016)
Metal oxide NPs	Oxides of metals	Photocatalytic degradation, adsorption, magnetic separation	High stability, and photocatalytic efficiency	Zamani et al. (2019)
Magnetic NPs	Magnetic metal oxides	Adsorption followed by magnetic separation	Easy recovery, reusable	Garg et al. (2024)
Quantum dots (QDs)	Semiconductor nanocrystals	Photocatalytic degradation under UV/visible light	High efficiency in light-driven reactions	Eddy et al. (2024)
NZVI	Iron in zero oxidation state	Reduction and adsorption of contaminants	Cost-effective, high reactivity	Mondal and De (2022)
Metal-organic frameworks (MOFs)	Porous hybrid structures of metals and organic ligands	Adsorption, catalysis, and separation	High porosity, tunable chemistry	Singh and Madras (2013)
Organic NMs				
Activated carbon	Porous carbon-based material	Adsorption of pollutants via porous structure and high surface area	High adsorption capacity, cost-effective	Chen et al. (2017)
CNTs	Cylindrical carbon allotropes	Adsorption, degradation, and membrane filtration	Large surface area, high conductivity	Arora and Attri (2020)
Graphene	A single layer of carbon atoms	Adsorption, catalytic oxidation, and reduction of contaminants	Exceptional surface area and high mechanical strength	Ijaz et al. (2020)
Graphene oxide	Oxidized form of graphene	Adsorption through functional groups, photocatalytic degradation	High chemical reactivity, and tunable surface properties	Yang et al. (2022)
Organic polymers	Synthetic or natural polymers	Adsorption, ion exchange, or encapsulation of pollutants	High selectivity, tunable properties, and reusability	Anucha et al. (2022)

combined/deposited on some material/membranes like ceramic membranes, clays, sawdust, etc. and can be used effectively (Cabello-Ribota et al., 2021).

Many zero-valent metal NPs have drawn interest in water pollution remediation over successive years (Garg et al., 2024). The great reducing capacity of nano-zero-valent Al, oxides, and hydroxides causes them to develop on the metal surface in the

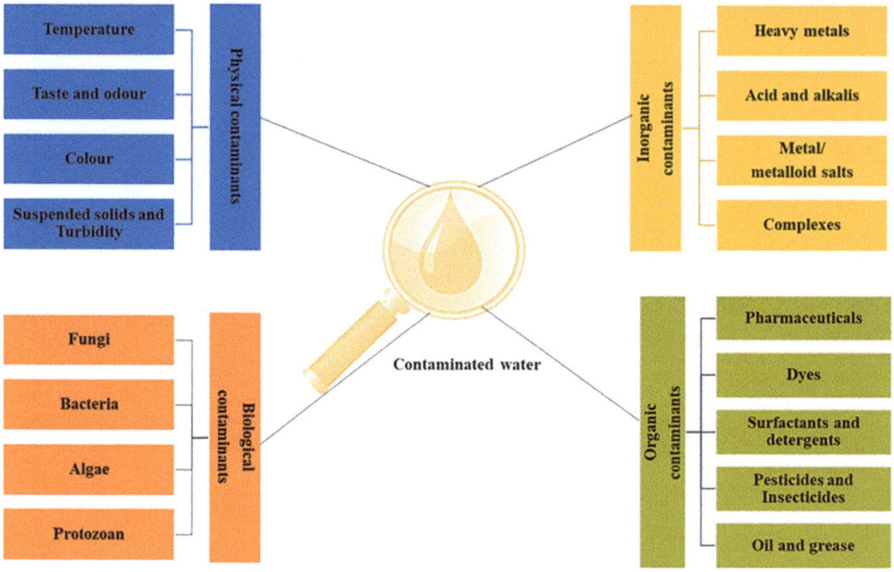

Fig. 1 Pollutants present in water

presence of water, so thermodynamically unstable and impeding the electron migration from the metal surface to the impurities (Rivero-Huguet and Marshall 2009). Ni has limited reducing ability as its value of standard reduction potential is low (Bratsch 1989). Appropriate reducing agents for the elimination of pollutants include nano-zero-valent Fe and Zn. Fe offers numerous benefits over Zn for water treatment despite its lower reduction potential: great adsorption qualities, precipitation, oxidation, and inexpensive cost (Cohen et al., 2019). Many pollutants including aromatic chemicals, dyes, halogenated organic compounds, phenols, radioactive compounds, heavy metals, and inorganic ions (Zou et al., 2016) are therefore removed using NPs of zero-valence metals.

Regardless of having several advantages, zero-valent metal NPs also have disadvantages of oxidation, aggregation, and difficulty of separation from the system. To overcome the above problems, several modifications are proposed to intensify the performance of zero-valent metal NPs in wastewater treatment (Lv et al., 2019). Common modifications like tampering with other metals are combining with supports, surface coating, encapsulation, and emulsification. The significant reactivity of zero-valent metal NPs is improved when tampered with other metals (Garg et al., 2024). Coating on the surface, capping, and conjugation with supports prohibit accumulation and increase the dispersibility of zero-valent metal NPs. Combining with eluting agents and encapsulation in the matrix facilitates the removal of zero-valent metal NPs from the degraded system (Stefaniuk et al., 2016). Zero-valent Zn NPs are also used in wastewater treatment. Zn has more standard reduction potential than Fe and hence can degrade the contaminants faster than Fe. However, studies have shown that nano-zero-valent zinc is mainly used in dehalogenation reactions, i.e., it can break down CCl_4 more quickly and completely as compared to other NPs (Tratnyek et al., 2010). So, it is mainly used for the breaking down of halogenated organic compounds, CCl_4.

The metal oxide NPs are stable chemically and are used for different processes such as adsorption, photocatalytic activities, and antibacterial as well as antifungal activities. TiO_2 NPs are designated as good photocatalysts. It has high photocatalytic activity, reasonable price, and photochemical as well as biological stability (Singla et al., 2024). Photocatalytic degradation capability has been used in the last few years for the decomposition of pollutants in wastewater. The contaminants are degraded and oxidized in the presence of light and catalyst, into intermediate products having low molecular weight and in the end converted into CO_2, H_2O, and anions like NO_3^-, PO_4^{3-}, and Cl^-. TiO_2 NPs can be used for the breakdown of all types of contaminants such as heavy metals, dyes, pesticides, phenols, chlorinated organic compounds, polycyclic aromatic hydrocarbons, as well as microorganisms (Bashambu et al., 2021). Besides these, there are certain disadvantages of TiO_2 NPs as they need more energy (in the UV region) for excitation due to a large bandgap which reduces their activity in visible light. Moreover, the recovery of TiO_2 NPs from treated wastewater is difficult, particularly when they are castoff in suspension. These problems can be encountered by doping with some metals (Cerrato et al., 2022) and coupling with membranes (Asli & Neumann, 2009). ZnO NPs are good photocatalysts for water treatment owing to their high chemical stability, affordability, nontoxicity, abundance, and excellent photocatalytic activity (Sharma et al., 2020). However, similar to TiO_2 NPs, ZnO NPs also have a large bandgap which restricts its activity. Therefore, ZnO NPs can also be doped with metal ions (Modi et al., 2023) or coupled with semiconductors for increased efficiency (Odler & Abdul-Maula, 1983).

Recently, iron oxide NPs have become very popular for sequestrating heavy metals from wastewater as these are easy to prepare and applicable to diverse contaminants. These NPs can be nonmagnetic (Fe_2O_3) or magnetic iron oxides (Fe_3O_4, Fe_2O_4). The nonmagnetic ones are very difficult to remove from the remedial systems due to their small size. However, magnetic NPs can be easily separated with the help of a magnet, making them efficaciously useable in sequestrating contaminants from wastewater (Sharma et al., 2023). Their adsorption properties can be tailored suitably by functionalization with different ligands (Mansur et al., 2023).

2.2 *Carbon NPs*

Carbon-based NPs, such as activated carbon, CNTs, graphene, and other materials related to graphene, are among the finest adsorbent materials for eliminating both organic and inorganic pollutants from wastewater. They possess a wide surface area, excellent porosity, great mechanical, chemical, and thermal stability in aqueous media, irrespective of solution pH (Ijaz et al., 2020). In CNTs, carbon atoms are grouped in hexagonal sheets that are coiled up like cylindrical tubes

(Chen et al., 2017). Both, single-walled and multi-walled CNTs are suitable for treating water. Combining CNTs with other metals or supports that increase their surface area and dispersibility can improve their adsorption and other qualities. When compared to other carbon-based NPs such as activated carbon, one of the main disadvantages of CNTs as adsorbents is their expensive cost. However, this may be addressed by functionalizing CNTs, which lowers the cost of water treatment (Luo et al., 2013).

Oxidation in the presence of air or treatment with specific solutions can accomplish functionalization by conjugating functional groups such as hydroxy, carbonyl, and acid onto the surfaces of CNTs. CNT catalyzes in addition to being an adsorbent. When combined with ZnO and TiO_2, CNT also functions as a photocatalyst, increasing its thermal conductivity, total surface area, and photocatalytic activity (Arora & Attri, 2020).

Polymer functionalized nanocomposites (PFNCs), which are polymer-conjugated nanocomposites, have the great qualities of both polymers and NPs. Some of the characteristics that make them special are their high dynamic strength, high surface area-to-volume ratio, and strong interfacial reaction. The polymer mixture can also be used again and again. PFNCs are mostly used to remove metals and dyes from wastewater through adsorption. They are also used because they kill germs due to their antibacterial action. It is possible to distribute NPs evenly in a polymer matrix to make nanomembranes. These membranes then form strong links with the matrix and can separate heavy metals, dyes, pesticides, and other toxins from wastewater (Anucha et al. 2022).

3 Mechanisms for Water Treatment and Purification by Nanomaterials

Nanotechnology has given us the most advanced and highly effective ways to get rid of contaminants in water, which is also one of the biggest humanitarian issues in the world (Mishra et al., 2022b, 2024). Heavy metals like arsenic, chromium, and lead can be easily and cheaply removed from the environment using NPs, which include zero-valent Ag, iron, and functionalized metal oxide NPs. A different approach is to use semiconductor nanomaterials like TiO_2 and ZnO NPs to break down organic pollutants using photocatalysis. These materials harness UV or visible light to break down persistent organic compounds, such as pesticides and dyes, into harmless molecules like water and carbon dioxide. NPs have also been used in combination or consecutively with some microbes wherein they behave as nanocarriers and have proved to give significant results. The first-stage mechanism of nanobioremediation involves the phenomenon of adsorption, dissolution, absorption, and chemical catalysis of photocatalytic reactions. Different nanomaterial-based technologies have been reported for remediating various pollutants from water (Table 3).

Table 3 Applications of nanotechnology in wastewater remediation

Category	Adsorption	Photocatalysis	Membrane filtration	References
Nanomaterials used	Graphene oxide (GO), and CNTs. Metal oxides (Fe_2O_3, TiO_2, ZnO). Nanocomposites (GO metal hybrids) functionalized biobased nanomaterials (chitosan NPs)	TiO_2, ZnO, WO_3, and Fe_2O_3 NPs. Hybrid photocatalysts (graphene supported TiO_2, $AgTiO_2$ nanocomposites) quantum dots (QDs) for visible light-driven catalysis. Plasmonic NPs (ag, au)	Graphene oxide (GO) membranes. Carbon nanotube (CNT)-enhanced membranes. Metalorganic frameworks (MOFs). Polymeric membranes with nanomaterial coatings	Anjum et al. (2024), Eddy et al. (2024)
Pollutants targeted	Heavy metals: Lead (Pb), chromium (Cr), arsenic (as). Organic pollutants: Dyes (methylene blue, azo dyes), pesticides. Emerging contaminants: Pharmaceuticals, hormones. Industrial effluents: Oils, surfactants. Inorganic ions: Fluoride, nitrate, phosphate	Organic pollutants: Dyes, antibiotics, phenols. Toxic metals: Cr^{6+}, Hg^{2+}. Emerging pollutants: Hormones, endocrine disruptors, microplastics. Industrial wastes: Textile dyes, chemical solvents. Pathogens: Bacteria, viruses	Heavy metals: Cadmium (cd), lead (Pb), mercury (hg). Desalination: Removal of salts (brine). Pathogens: Bacteria, viruses. Emerging contaminants: Pharmaceuticals, microplastics, personal care products. Organic pollutants: Pesticides, hydrocarbons, phenols	Ahmadpour et al. (2015), Chen et al. (2017)
Mechanism	Physical adsorption: Pollutants adhere to high surface area nanomaterials via Van der Waals forces. Ion exchange: Functional groups exchange ions with pollutants. Chelation: Formation of stable complexes with metal ions electrostatic interactions: Oppositely charged ions are attracted to the surface of nanomaterials surface modification: Functional groups enhance selectivity and binding capacity for specific pollutants	ROS generation: UV or visible light excites photocatalysts, generating ROS like •OH radicals that oxidize pollutants. Electron-hole pair mechanism: Electrons in the catalyst are excited to conduction bands, initiating redox reactions. Plasmonic effects: Metal NPs enhance visible light absorption and catalytic activity. Hybrid photocatalysis: Combines adsorption and photocatalytic degradation for enhanced efficiency	Size exclusion: Nanoscale pores selectively block larger contaminants while allowing water to pass. **Charge-based** separation: Surface charges repel specific pollutants. Hybrid systems: Integrated adsorption or photocatalysis for multifunctionality. Hybrid filtration: Combines nanoscale filtration with adsorption or photocatalysis for improved removal efficiency. Thermal/ mechanical stability: Nanomaterials improve membrane durability in harsh environments	Kabay et al. (2022), Vishnu et al. (2021)

(continued)

Table 3 (continued)

Category	Adsorption	Photocatalysis	Membrane filtration	References
Advantages	High adsorption efficiency due to nanoscale surface area. Selective pollutant removal through functionalization. Reusability and cost-effectiveness for large-scale applications. Environmentally friendly (e.g., biobased nanomaterials). Capable of handling a wide range of pollutants	Highly effective for breaking down persistent organic pollutants. Converts harmful substances into harmless byproducts (e.g., CO_2, H_2O). Solar photocatalysis enables energy efficiency. Combines pollutant removal and disinfection in a single step. Highly versatile for various types of water contaminants	High contaminant rejection rates. Low energy consumption compared to conventional methods. Long-term durability and fouling resistance. Effective for desalination and industrial wastewater treatment. Can handle complex contaminant mixtures	Yang et al. (2022), Wadhwa et al. (2020)
Limitations	Regeneration of nano-adsorbents may involve secondary pollution. Performance can degrade with high contaminant loads or complex water matrices. Limited adsorption capacity for some nanomaterials. Effectiveness depends on pH, temperature, and water composition	Requires UV or visible light, limiting application in lowlight conditions. Possible nanoparticle leaching, causing secondary pollution. Catalyst recovery and reuse can be challenging. Efficiency may decrease with the presence of scavenging ions or complex mixtures	Fouling remains a concern for certain membranes. High production costs for advanced membranes (e.g., GO, MOFs). Nanomaterial integration can increase costs and complexity. Some nanomaterials may leach, raising environmental concerns	Azad and Mohsennia (2020)

3.1 Adsorption

The most widely used technique that has proved to be economical and environment-friendly and can remove pollutants with a size less than 300 nm is known to be adsorption. It is a crucial method for eliminating heavy metals, dyes, and antibiotics from aquatic systems. During this process, the contaminant adheres to the adsorbent's surface, effectively immobilizing it while offering an extensive surface area and necessary functionalities. However, lately, the synthesis of adsorbents is based on the achievement of selective adsorption, regeneration capability, scalability, and multifunctionality of the adsorbent due to which functionalized NPs are being prepared with customized surface properties (Anjum et al., 2024). For this purpose, the metamorphosed materials are used including carbon-based materials, zero-valent metals, clays, zeolite, magnetic materials, polymer-based materials, and metal oxides, as well as their nanocomposites that have shown improved efficiency as an adsorbent for heavy metal remediation (Anjum et al., 2014).

Dyes are commonly used organic compounds in textile industries and are released in substantial quantities in dye-laden wastewater that significantly pollutes aquatic systems. To mitigate this, adsorption has been materialized as one of the oldest, simplest, and most steadfast methods due to its characteristic feature of being a surface-based phenomenon and cost-effective; however, its efficiency is often limited by the surface properties of adsorbent. Recent advancements have introduced nano manganese dioxide with a highly amorphous structure, reduced surface charge, and large specific surface area to enhance its adsorption efficiency and accelerated reaction kinetics, making them promising candidates for effective dye remediation. Moreover, the fast oxidation kinetics of the manganese-incorporated composites have shown better dye removal efficiency than those of manganese oxide itself (Islam et al., 2019). Qin et al. (2017) investigated, varying the sorbent dose, the adsorption of methylene blue dye on manganese dioxide (MnO_2). With a maximal adsorption capacity of MnO_2 as 765 mg/g, the dye removal efficiency increased up to 95%. The availability of active surface sites with the increased specific surface area has been attributed to changes in the behavior of sorbent by promoting the adsorption process.

The choice of the adsorbent is mostly defined by their intrinsic adsorption capacity and their inherent affinity for the particular ion or compound of interest; these adsorbents help in the separation of chemical species from the solution phase and their subsequent concentration onto the surface of adsorbent material (Garg & Anjum, 2024). The mechanism involves three primary steps: initially, the adsorbed species interact with the adsorbent at the interface between a liquid phase (wastewater) and a solid phase (adsorbent), each possessing distinct compositions, under controlled conditions for a defined period, which is followed by separation of the adsorbent material from the liquid layer. There are specific examples of NPs for water treatment by adsorption. Different types of dyes are adsorbed on AgNPs-coated activated carbon. Heavy metal ions like Cu and as ions have been removed using iron NPs. Additionally, useful for the elimination of Cu and As ions are ZnO NPs. Different heavy metals and dyes are also removed using CNTs (Mandal et al., 2023).

Followed by the uptake of pollutant/s onto the surface of the adsorbent, the residual amount of pollutant in water is ascertained taking into consideration the supernatant liquid and any pressurized fluid phase that are measured to assess the process. According to the research, there are four primary phenomena responsible for the adsorption process, namely, ion exchange, electrostatic adsorption mainly steered by an attractive force between oppositely charged species, surface physisorption, and complexation/chelation in which a stable complex is formed with functional groups extant on the surface of the adsorbent. Physical adsorption of adsorbate depends on weak *van der Waals* forces that establish temporary bonds between the sites of the adsorbent and metal ions. Conversely, chemical adsorption involves the formation of stronger covalent or ionic bonds, resulting in a more stable attachment of metal ions. Ion exchange is another important process wherein metal ions replace the current cations on the surface of the adsorbent thereby enabling the selective binding of target metals (Eddy et al., 2024).

The separation of dyes and heavy metals by using inorganic nanomaterials mainly depends on the adsorption of contaminants, a process that takes advantage of the large surface area of nanomaterials paralleled to their volume and the existence of many active binding sites. This allows them to capture pollutants effectively through various interactions (Fig. 2). Electrostatic interactions between the metal ions and the surfaces of the NPs are crucial (Eddy et al., 2023). Another kind of interactions, like hydrogen bonding, *van der Waals* forces, and covalent bonding, also help. Additionally, heavy metal ions can bind with functional groups like hydroxyl or oxygen ions on the surface of NPs, which improves their ability to bind to these metals. Several factors affect the adsorption process, including temperature, pH, size, and surface structure. The concentration and structure of the adsorbate also play a role. Different adsorbents have unique properties based on their active surface, pore size, pore distribution, and surface functional groups. Most adsorbents, especially NPs, can be recycled, and the pollutants can easily be removed from their surfaces (Garg et al., 2024).

Removal of phenol and 3-chlorophenol from water has been exploited using activated carbon made from plum kernels. The surface of this activated carbon had various functional groups, such as hydroxyl, carboxylic, and carbonyl groups, suggesting that different types of interactions could occur between the adsorbent and the pollutants. Given the pKa values of phenol (9.92) and 3-chlorophenol (9.43), these compounds primarily exist in their undissociated forms in water under conditions

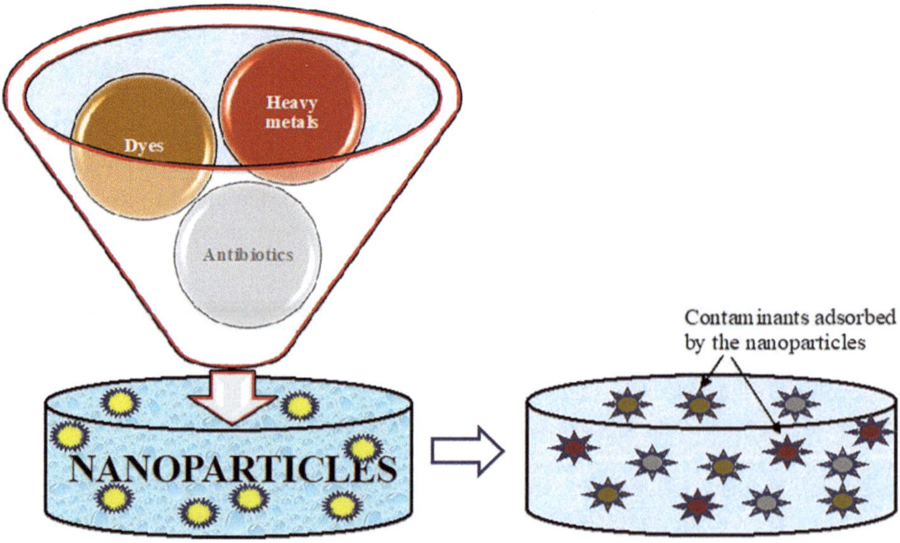

Fig. 2 Schematic representation of adsorption of contaminants in water

where the pH is less than 8.19, both before and after adsorption. Hydrogen bonding was proposed as a key mechanism, taking place in the hydroxyl groups of the phenol and functional groups, like carboxyl groups, on the surface of activated carbon. In desorption studies, the making of stable chemical bonds between the adsorbent surface and MB molecules was revealed, preventing most dye molecules from being eluted. Only a small fraction of MB molecules (~0.6%) was eluted, indicating the presence of weaker hydrogen bonding in MB molecules and the adsorbent surface. However, the majority of the dye adsorbed was strongly attached due to the formation of ionic bond. These findings highlight the interplay of weak and strong bonding mechanisms in the adsorption and retention processes for different solutes and adsorbents.

But NPs have very small sizes with ultrafine particles; therefore they can easily leach into the water during purification, hence making separation difficult. This difficulty can be minimized by conjugating the NPs on the appropriate surface (support) to form a nanocomposite. NPs whether metal and metal oxide, carbon-based, or composite all work by the process of adsorption for wastewater treatment (Fig. 3). The considerable influence of operational constraints including pH, temperature, and concentration of metal ions on adsorption is further observed; therefore optimizing these parameters becomes a necessary step to improve the metal removal

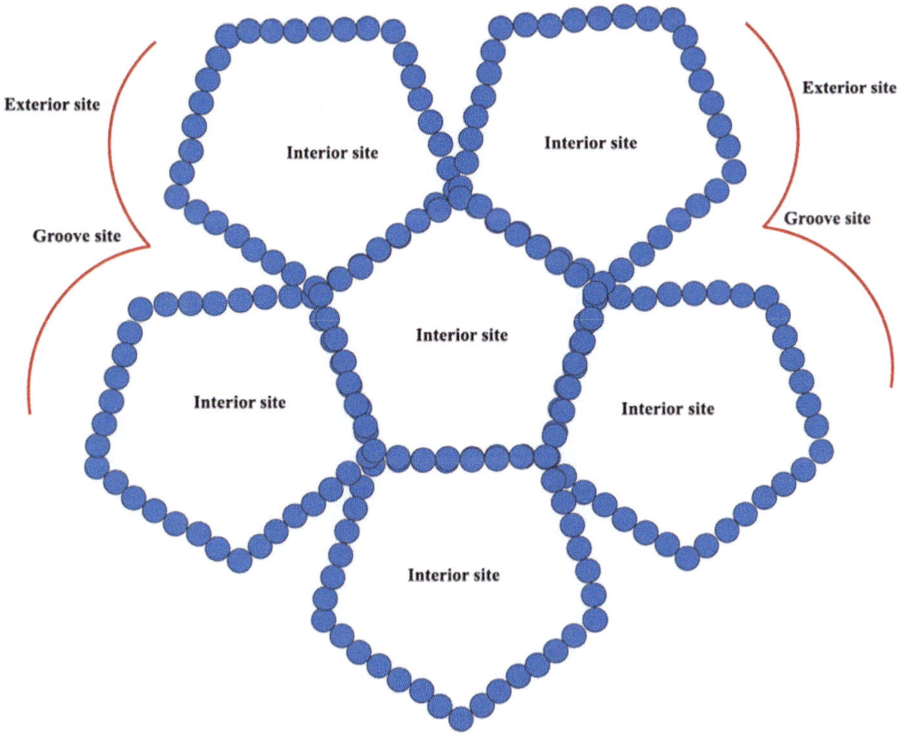

Fig. 3 Sites of adsorption present in CNT

efficiency of the adsorbent (Das et al., 2012). The two main considerations that control the adsorption rate of the adsorbate are adsorption isotherms and kinetics. These elements not only define the efficiency of the adsorbent but also give the necessary criteria for describing the adsorption process (Eddy et al., 2023). Cai et al. produced Fe_3O_4 NPs for the adsorptive removal of antibiotic rifampicin (RIF). Unlike conventional Fe_3O_4 with an adsorption effectiveness of just 7%, the synthesized Fe_3O_4 attained an adsorption efficiency of 98.4% at an ideal pH of 5.5. Fe_3O_4 NPs showed magnetic features that helped them to be easily separated and allowed their reuse; the greatest adsorption capacity for rifampicin was 84.9 mg/g (Cai et al., 2019).

Whereas CNTs are tubular nanostructures generated from sp^2-bonded carbon atoms, graphene is famous for its layered structure. The former is a very promising alternative to activated carbon due to their well-ordered pore size dispersal and high ratio of active sites to volume, carrying active sites for the binding of pollutants. CNTs are excellent adsorbents for the separation of various water contaminants. There are four major adsorption sites in CNTs including interstitial sites, inner holes, grooves, and outer surfaces of the CNT (Fig. 3) (Arora & Attri, 2020). This facilitates the adsorption of large organic molecules, enabling selective interactions through π–π stacking, hydrogen bonding, and electrostatic interactions that enhance its efficiency compared to traditional powdered or granular activated carbons (Pyrzynska, 2008). The improved interaction of the pollutant with the surface of the adsorbent due to its integrated blending provides valuable insights for advancing the tailoring of nano-adsorbents, thereby enabling more efficient removal of pollutants from aquatic resources (Garg et al., 2022).

Graphene oxide (GO), which is made from graphene, is hydrophilic because it has substantially enhanced surface area and oxygen-containing functional groups that help it absorb pollutants. Their amazing ability to get rid of different pollutants in water shows how useful they could be for treating water (Yang et al., 2022). Jayanthi et al. (2016) created a three-dimensional graphene oxide (3D GO) material using a lyophilic freeze-drying method to increase the surface area of GO foam. This made it better at absorbing substances and killing bacteria, especially when used to break down dyes. The 3D GO foam was effective in killing bacteria in both water and salty water. Adsorption studies done at different pH levels showed that the ability to bind was different depending on how the GO and dye molecules interacted with each other. The special 3D porous structure of GO foam makes it more useful for purifying water, especially in salty bodies of water where salt ions are common. In saline environments, where graphene oxide typically loses activity due to interactions with extant ions, the structural organization of the GO layers effectively reduced particle aggregation, maintaining its functional properties.

The removal of methylene blue dye from aqueous solutions using manganese ferrite-graphene oxide composites was examined in another study. The adsorption efficiency increased to 91–95% when the GO content was increased from 10 to 50 weight percent, according to the Langmuir isotherm. The process of adsorption was mainly mediated by electrostatic interactions between the negatively charged OH^- groups and positively charged dye molecules on the adsorbent's surface. Another study synthesized GO-functionalized Fe_3O_4@SiO_2@CS-TETA (triethylenetriamine) NPs for the removal of methylene blue (dye) and Cu^{2+} ions (heavy metal), with adsorption kinetics and isotherm data fitting with pseudo-second-order model, achieving a maximum adsorption capacity of 529.1 mg/g within 20 minutes under optimal conditions. This superior performance is attributed to the functional groups contributed by GO, TETA, and CS, which enhance active site availability through amidation reactions between carboxyl groups of GO and amino groups of TETA. The adsorbent showed excellent regeneration and maintained performance across five to six cycles. The adsorption efficiency, however, decreased at low pH due to competition between hydrogen ions and contaminants for active sites but improved as pH increased, thereby making the active sites free for adsorption (Wang et al., 2017).

3.2 Photodegradation

Because of their huge surface-to-volume ratio, which improves their contact with reactants, photocatalysts are useful at accelerating chemical processes. Because of their large surface area in relation to volume, nanomaterials are employed as catalysts, particularly the nanoparticle-based photocatalysts that are gaining prominence in water purification technologies for breaking down and mineralizing harmful organic pollutants, converting them into benign byproducts like carbon dioxide and water. Also, it reduces the toxic metal ions to their nontoxic counterparts (Wolff et al., 2018). Photocatalytic NPs, mainly metal NPs (Ag, Au), TiO_2 NPs, ZnO NPs, and graphene oxide are used for degradation of organic pollutants and disinfect wastewater under UV irradiation. Photocatalysis involves semiconductor material (often metal oxide NPs) that, when exposed to light (usually ultraviolet or visible), absorbs energy and generates charge carriers as electron-hole pairs. Electrons get transferred to the conduction band from the valence band, leaving holes (h^+) in the valence band. These species can then react with water and oxygen molecules to produce reactive oxygen species (ROS), like hydroxyl radicals (•OH), and superoxide anions (O_2•−), which are highly efficient in degrading organic pollutants, bacteria, and other harmful substances in water (Sharma et al., 2023).

Due to the proper position of valance and conduction bands, TiO_2 is a semiconductor and is mainly used for photocatalytic decontamination of wastewater. The potent oxidizing agents such as hydroxyl radicals (OH·) are produced in aqueous environment by using ultraviolet (UV) radiation in combination with TiO_2-based catalysts (Fig. 4). However, the TiO_2 is also used in membranes wherein the process is dependent on the use of UV light for excitation of electrons that subsequently produce hydroxyl radicals and the performance of membrane is governed by on the amount of TiO_2 used, the source of radiation, and the duration of exposure (Ahmadpour et al., 2015). The support materials used for such membranes include ceramics, polymeric substances, and other inorganic materials. Solar radiation is

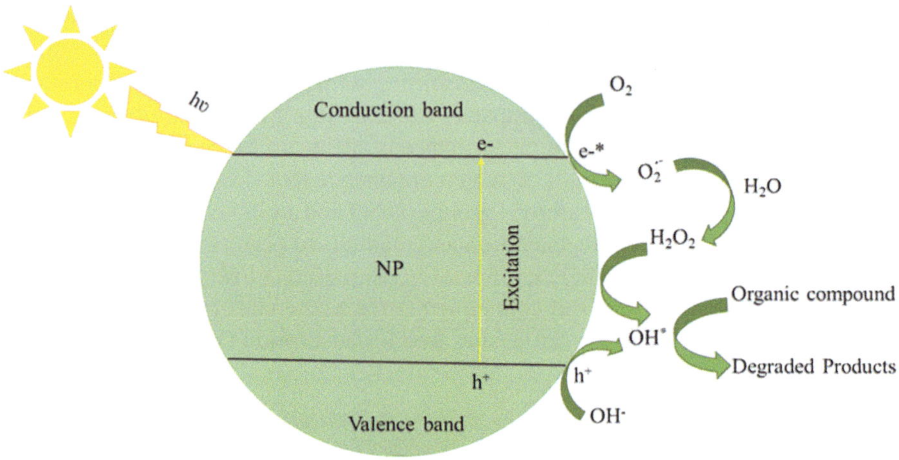

Fig. 4 Illustrative mechanism of photocatalytic degradation of organic pollutants by NPs

used in place of UV due to the instability of polymers under UV irradiation. Tungsten oxide (WO_3) is also used as a catalyst due to its lower point of charge and high surface affinity for dye adsorption which is responsible for improved degradation rates. A study by S. A. Singh and colleagues demonstrated that modifying WO_3 onto commercial TiO_2 enhances the stability and efficacy of photocatalytic degradation, resulting in superior dye degradation performance as compared to unmodified commercial TiO_2 (Singh & Madras, 2013).

Another study investigates the use of titania-zeolite Y composite, which effectively integrates the high porosity of zeolites with the strong photocatalytic capabilities of titania NPs. Research shows that a composition containing approximately 10% of this nanocomposite is capable of enhancing the degradation of methylene blue dye by 50% compared to standard methods. In 60 minutes of treatment, titania P25 NPs demonstrated great efficacy by totally removing the dye. It was found that the ideal concentration for maximal dye degradation was around 10% weight of zeolite Y in the composite. When the concentration of the composite was raised to 20%, it displayed substantial antibacterial properties, therefore preventing *Escherichia coli* and *Staphylococcus aureus* from growing. These results underline the two synergistic functions of the composite: it is a useful photocatalyst and a material with great antibacterial properties. Studies confirm that using 10–20% weight of the composite provided excellent photocatalytic and antimicrobial performance, so enabling the development of cost-effective materials suitable for environmental and healthcare applications, by combining exceptional dye degradation efficiency with significant antibacterial activity (Wadhwa et al., 2020).

Iron oxide (FeO_3) is a fundamental component for photocatalytic degradation based on its stability, small bandgap, efficient use of the visible light spectrum, and recyclability. Recent developments in photocatalytic degradation underscore the relevance of Fe_2O_3 and its preparation techniques which affect its morphological properties and hence improve its photocatalytic performance. Improved versions of this catalyst have been created that sufficiently increase photocatalytic activity by combining Fe_2O_3 with other photocatalysts with similar bandgap properties or by employing support materials. Changing the bandgap and lowering electron-hole pair recombination can help to improve both conventional and Z-scheme heterojunctions. Using reduced graphene oxide (rGO)-modified Ag@ZnO nanocomposite, the photodegradation of antibiotics (roxarsone, nonylphenol, rhodamine B, and ciprofloxacin) and a dye (methylene blue) was investigated. By using reduced graphene oxide and intrinsic plasmonic characteristics of Ag NPs, the nanocomposite was found sensitive to visible light, thereby increasing the absorption capacity over the visible spectrum when 20 ppm polluted solution was applied. Underlining the transformational capacity of nanostructured materials in pollution degradation, the effective fabrication and utilization of the Ag@ZnO@rGO nanocomposite reinforce the effects. This nanocomposite shows that current nanotechnology may overcome the constraints of old technologies including poor efficiency and huge operating expenses using its high photocatalytic efficacy, vast surface area, and durability over many cycles. Ag@ZnO@rGO NPs had a maximum of 85% degree of effectiveness greatly facilitated the breakdown of methylene blue. This implies that by including rGO into Ag@ZnO NPs, their photocatalytic performance has improved greatly due of the linked graphene sheets and nanocomposite that provides an increased surface area for enhanced organic dye adsorption and facilitates more effective separation of electron-hole pairs (Yaou Balarabe et al. 2023). Moreover, the higher composite efficiency underlined its resilience and likely uses during five consecutive cycles. The investigations on the photodegradation of methylene blue dye underscored the critical function O_2• perform in the process.

Moreover, the understanding of the function of reactive oxygen species in the degradation of pollutants offers a basis for the future creation of more focused and efficient nanomaterials. Such developments open the path for scalable, affordable, and environmentally benign remedial solutions, therefore highlighting the essential role of nanotechnology in reaching world environmental sustainability Pan et al. 2022).

Also, the photocatalytic rate is improved through metal defects and oxygen vacancies in Fe_2O_3 that increase active sites for reaction and improve the degradability toward dyes and other pollutants (Hitam & Jalil, 2020).

3.3 Membrane Filtration

Engineered nanomaterials have lately been produced using membranes, which are small sieves that specifically filter metal ions from water. This water treatment approach has great separation efficiency and continuous operation capacity (Kabay et al., 2022) and uses less energy. Compared to traditional water treatment systems, the newest developments in membrane technology are their endurance and selectivity. Employing metal-organic frameworks and ceramic membranes, the efficiency and selectivity of such membranes have been attained, thereby fulfilling certain functional needs and overcoming obstacles in separation processes (Othman et al., 2022). Furthermore, highly permeable nanomaterials allow effective solute and water transport across the membrane. Their efficiency in fouling control helps to limit the build-up of pollutants on membrane surfaces, therefore lowering the demand for regular maintenance and cleaning. Another major benefit is improved selectivity, which lets particular pollutants be selectively eliminated from complicated wastewater matrices while reducing interference from other molecules (Eddy et al., 2024).

Low-pressure membrane systems, namely, microfiltration (MF) and ultrafiltration (UF), which run within a pressure range of 10–30 psi, and high-pressure systems, including nanofiltration (NF) and reverse osmosis (RO), which run comparatively higher pressures in the range of 75–250 psi, are categorized dependent on the pore size and operational pressure. One significant advantage of including nanomaterials into membranes is their natural hydrophilicity which improves water flow and hence increases water recovery and lowers energy consumption. Moreover, the remarkable thermal stability of NPs makes them suitable for many wastewater treatments as it helps them function effectively under demanding conditions. By improving membrane adsorption capacity, the higher surface roughness of NPs helps to effectively remove organic contaminants like heavy metals and organic pollutants. Moreover, the NPs show hydraulic stability, therefore ensuring the long-term survival of membrane systems (Mondal and De 2022).

Wang et al. (2017) examined the effectiveness of Fe-doped TiO_2/polys sulfone (PSF) composite UF membranes using the phase inversion technique. Composite membranes made of bisphenol A (BPA) included a light-responsive photocatalyst made using the hydrothermal process. The impacts of hydrothermal reaction duration and other reaction restrictions were carefully examined in situ in this work. According to the findings, the hybrid composite degraded BPA more quickly than TiO_2. The composite membranes serve two purposes: visible-light photocatalysis and ultrafiltration. This hybrid functioning also produced self-cleaning capabilities, highlighting its potential for sophisticated wastewater treatment applications. Flexible, large-area membranes were made possible by Zhang and Liu (2021) by fabricating hydrogen titanate nanobelts using a modified hydrothermal method. The photocatalytic activity of undoped TiO_2 membranes in visible light was attributed to the self-sensitization of methylene blue molecules. The photocatalytic activity was enhanced by the production of photoactivated holes and electrons under UV light. The effectiveness of the photocatalytic activity was affected by copper doping. Improved photocatalytic materials may result from an understanding of how doping and processing parameters impact the photocatalytic performance of TiO_2-based membranes. Utilizing filler TiO_2 NPs, a physicochemical characterization of the membrane was performed utilizing the phase inversion approach (Mobarakabad et al., 2015).

4 Ecotoxicological Assessment and Environmental Impact of Nanomaterials

Nanomaterials possess distinct physicochemical properties, including high surface reactivity, and nanoscale dimensions, and are frequently modified or undergo chemical alterations, significantly affecting their toxicological profiles related to their bulk counterparts (Crane et al., 2008). Hence, the toxicity data derived from bulk materials cannot be directly applied to nanosized materials. Such materials are capable of accumulating in water bodies, posing risks to aquatic making the understanding of their ecotoxicological impact and assessment important (Table 4). Also, the contamination of aquatic ecosystems with increased presence of nanomaterials highlights the need to evaluate their potential impact on aquatic organisms, as the behavior of nanomaterials varies under different environmental conditions. Therefore, standardized testing and predictive models are essential for accurate risk assessment (Schwab et al., 2016).

The production and application of engineered nanomaterials have rapidly expanded in recent years, including consumer and industrial goods. Throughout the lifecycle of nano-enabled goods, these nanomaterials are discharged into the environment through different pathways- product weathering; intentional releases, as in environmental remediation efforts; and accidental releases, such as spills during manufacturing, transportation, or disposal (Eddy et al., 2024). Research on the inhalation of released ultrafine particles in the range of 14 nm to 250 nm shows adverse effects at concentrations from a few mg/m^3 to as high as 250 mg/m^3. These findings highlight the integration of ecotoxicity assessments into a "safe-by-design" framework, aiming to evaluate and mitigate risks throughout their creation, production, use, and disposal of engineered nanomaterials. Another study shows that chemical stability and strong adsorption properties make the wide application of ZnO nanomaterials in coatings, paints, and sunscreens. However, such extensive use of ZnO poses a substantial risk to the environment due to its high potential to contaminate water bodies, often reaching elevated concentrations in surface waters causing dual toxicity at varying concentrations. At lower concentrations, these NPs exert toxic effects primarily through the release of Zn^{2+}, compromising the membrane integrity and decreasing cell viability in algae. On the contrary, at higher concentrations, the least sensitive algal species (*Phaeodactylum tricornutum*) shows inhibited growth due to increased interaction between algal cells and NPs, which intensifies the toxicological impact (Kumari et al., 2011). The environmental toxicity assessments that are used generally are of two types- acute (short-term) and chronic (long-term) assessments. Acute toxicity analyses include an initial step of assessment of the survival rates of exposed organisms and provide foundational data on ecotoxicological impacts. Chronic assays, on the other hand, are employed when acute test results, combined with conservative safety factors indicate potential environmental risks. These tests focus on evaluating the sublethal effects, including the impact of nanomaterials on development and

Table 4 Ecotoxicological assessment and environmental impact of nanomaterials

Aspect	Details	Environmental impact	References
Bioavailability	Nanomaterials have high surface area and reactivity, making them easily bioavailable	May accumulate in organisms, leading to potential toxicity in aquatic and terrestrial ecosystems	Crane et al. (2008)
Toxicity to aquatic life	Nanomaterials like ag NPs and CNTs can be toxic to fish, algae, and microorganisms	Disrupts aquatic food chains, reduces biodiversity, and impacts ecosystem stability	Schwab et al. (2016)
Uptake in plants	NPs can be absorbed by plant roots and transported to other parts of the plant	May alter plant growth, nutrient absorption, and introduce toxicity into the food chain	Eddy et al. (2024)
Sedimentation and aggregation	NPs can settle in sediments or aggregate in water bodies, altering their reactivity	Impacts benthic organisms and reduces the availability of NPs for degradation processes	Kumari et al. (2011)
Bioaccumulation	Persistent nanomaterials can accumulate in organisms over time (e.g., fish, invertebrates)	Leads to long-term ecological risks, including biomagnification through the food chain	Kyriakides et al. (2021)
Interaction with microorganisms	Nanomaterials may inhibit or disrupt microbial communities in soil and water	Affects nutrient cycling, organic matter decomposition, and ecosystem balance	Romero-Franco et al. (2017)
Generation of ROS	Certain nanomaterials generate ROS, causing oxidative stress in organisms	Damages cellular components like DNA, proteins, and lipids, leading to toxicity and cell death	Kyriakides et al. (2021)
Transformation and persistence	Nanomaterials may undergo transformations (e.g., oxidation, dissolution, agglomeration) in the environment	Alters their toxicity and environmental behavior; persistent forms may have prolonged impacts	van de Meent et al. (2023)
Soil contamination	NPs can accumulate in soils through agricultural use or industrial discharge	Alters soil properties, reduces fertility, and impacts soil-dwelling organisms (e.g., earthworms)	(Rajput et al. 2020)
Airborne NPs	NPs released into the air can settle on land and water or be inhaled by organisms	May cause respiratory issues in humans and animals and contaminate nearby ecosystems	Suter and Norton (2019)
Impact on water quality	Nanomaterials can alter pH, turbidity, and chemical composition of water	Reduces water quality, affects aquatic organisms, and creates challenges for water treatment processes	Wang et al. (2023)

(continued)

Table 4 (continued)

Aspect	Details	Environmental impact	References
Persistence in ecosystems	Some nanomaterials degrade very slowly, remaining in the environment for long periods	Leads to cumulative ecological risks and difficulty in remediation	Gamoń et al. (2023)
Regulatory and knowledge gaps	Limited understanding of long-term effects and environmental fate of nanomaterials	Delays in creating effective regulations and standards for safe usage and disposal	Schwirn et al. (2020)
Biodegradable nanomaterials	Some emerging nanomaterials are designed to degrade into nontoxic byproducts	Reduces long-term environmental impact and improves sustainability	Eddy (2020)

reproduction of organisms (Romero-Franco et al., 2017). Such an impact of nanomaterials in water bodies necessitates robust guidelines to ensure their safe application in potable and wastewater systems.

Since the current frameworks for conventional chemicals need to be significantly improved, applying Ecological Risk Assessment (ERA) approaches to engineered nanomaterials presents significant hurdles (Suter & Norton, 2019). The main difficult factors in predicting their persistence and degradation patterns, modeling, and transport in order to create toxicological benchmarks include dissolution, aggregation, agglomeration, and adsorption (Gamoń et al., 2023). Although there are some established frameworks for water quality and safety, there are currently few guidelines specific to nanomaterials because they primarily report general water contaminants and their precautionary principles of use (Meyer et al., 2020). However, the Environmental Protection Agency (EPA) acknowledges the unique characteristics of nanomaterials while emphasizing the need for more study on their toxicity, bioaccumulation, and environmental persistence (Boros and Ostafe 2020). Therefore, in order to overcome these obstacles, a multidisciplinary evaluation strategy that incorporates developments in environmental chemistry, toxicology, and nanoscience is needed to improve risk assessment frameworks and promote the long-term development of nanotechnology (Kyriakides et al., 2021). The evaluation of ecotoxicity needs to be both transferable and predictive. The results of predictive tests are more broadly applicable as they guarantee that the effects of nanomaterials are extrapolated outside of the evaluated concentration range. Conversely, transferable tests improve the dependability of test results for a variety of environmental conditions by guaranteeing that the effects of nanomaterials detected are consistent or comparable among organisms that have not been studied (Kyriakides et al., 2021).

The contributions have also been made for testing protocols by European Union and ISO (International Organization for Standardization) toward developing preliminary standards for nanomaterial safety. But these standards are filled with considerable gaps when subjected to real-world exposure and long-term impacts. It is a complex mechanism to calculate the exposure concentrations for nanomaterials due to the heterogeneous characteristics of different environmental compartments and are known to significantly influence key endpoints such as dissolution, dispersion stability, and transformation (Wang et al., 2023). The conditions in marine environments are different than those in freshwater. Furthermore, the organic matter content in these environments complicates the analysis due to which the data derived from one medium or condition cannot effectively represent the diverse scenarios found in natural ecosystems. There is also one regional model named as SimpleBox, for Predicted environmental concentration assessment (van de Meent et al., 2023). However, the adaptation of such models for nanomaterials is deterred due to lack of methodologies to generate nanospecific input parameters. Unlike the conventional qualitative structure-activity relationships (QSARs) approach, a potential pathway to address this gap is required (Eddy, 2020). The approach introduces additional complexities, as raw data may require significant processing and management. Advancing these methodologies is critical for improving the risk assessment and ensuring their safe and sustainable use in compliance with regulatory frameworks such as REACH (Schwirn et al., 2020).

5 Future Scenario of Integrated Technologies for Water Treatment

Nanomaterials enhance chemical reactions and help resolve the tough challenges in treatment of wastewater. They are able to monitor contaminants in real-time with high sensitivity when used in sensors, or can break down harmful pollutants more effectively as nanocatalysts, leading to improved treatment results. As the science of nanomaterials advances, their application in wastewater treatment is poised to drive transformative improvements, ensuring the development of smarter, more resilient systems for water resource management (Madhogaria et al., 2024). The integrated use of predictive modeling through machine learning and artificial intelligence into the horizons of wastewater treatment using nanomaterials is an emerging technology with enhanced efficiency, suitability, and innovation that help to fabricate ecological and scalable solutions to create multifunctional and sustainable nanotechnology solutions. This amalgamated technology with prevailing water treatment methods is capable of providing safe design for detection and removal of emerging pollutants such as microplastics and pharmaceuticals. In this direction, an improvement of such models is required to fully understand the behavior and properties of nanomaterials (Nehra et al., 2020). At the same time, the economical aspect of monitoring systems with easy implementation specifically in areas where limited resources are present is important that can be achieved with the help of new technologies like the Internet of Things (IoT) and big data wherein nanosensors when connected to IoT allow concurrent data collection and processing to produce advanced data analysis sufficient to detect and predict changes in contaminant levels and the effectiveness of treatments (Sharma et al., 2021).

6 Conclusion

More research is required for improvement in the predictive models so as to achieve sustainable development in wastewater treatment. To achieve such goals, nanotechnology plays an important role to support global sustainability and create smarter, more resilient water treatment systems using nanomaterials. These materials have unique properties that help monitor systems, predict maintenance needs, and remove all types of pollutants effectively to make the treatment processes better. However, there are challenges including the scaling up of system, understanding environmental impacts, and safety. Also, the developments in risk assessment methods and life cycle analysis are important for developing future regulations. The preparation, usage, and disposal of such materials are equally significant to ensure safe waste management system due to which research should be focused to create eco-friendly, biodegradable nanomaterials and set clear regulations to protect public health and the environment. Clear guidelines for nanomaterials are vital to balancing their ability to improve water treatment with the need to reduce ecological and health risks.

References

Ahmadpour, A., Zare, M., Behjoomanesh, M., & Avazpour, M. (2015). Photocatalytic decolorization of methyl orange dye using nano-photocatalysts. *Advances in Environmental Technology, 1*, 121–127. https://doi.org/10.22104/aet.2015.371

Anjum, A., Kumar Seth, C., & Datta, M. (2014). Characterization of montmorillonite biocomposite and its application for trace level removal of Sb3+: Equilibrium and kinetic studies. *International Journal of Chemistry, 7*. https://doi.org/10.5539/ijc.v7n1p28

Anjum, A., Gupta, D., Singh, B., Garg, R., Pani, B., Kashif, M., & Jain, S. (2024). Clay–polymer nanocomposites for effective water treatment: Opportunities, challenges, and future prospects. *Environmental Monitoring and Assessment, 196*, 666. https://doi.org/10.1007/s10661-024-12823-8

Anucha, C. B., Altin, I., Bacaksiz, E., & Stathopoulos, V. N. (2022). Titanium dioxide (TiO_2)-based photocatalyst materials activity enhancement for contaminants of emerging concern (CECs) degradation: In the light of modification strategies. *Chemical Engineering Journal Advances, 10*(February), 100262. https://doi.org/10.1016/j.ceja.2022.100262

Arora, B., & Attri, P. (2020). Carbon Nanotubes (CNTs): A potential nanomaterial for water purification. *Journal Of Composites Science, 4*, 135. https://doi.org/10.3390/jcs4030135

Asli, S., & Neumann, P. M. (2009). Colloidal suspensions of clay or titanium dioxide nanoparticles can inhibit leaf growth and transpiration via physical effects on root water transport. *Plant, Cell and Environment, 32*, 577–584. https://doi.org/10.1111/j.1365-3040.2009.01952.x

Azad, H., & Mohsennia, M. (2020). A novel free-standing polyvinyl butyral-polyacrylonitrile/ZnAl-layered double hydroxide nanocomposite membrane for enhanced heavy metal removal from wastewater. *Journal of Membrane Science, 615*, 118487. https://doi.org/10.1016/j.memsci.2020.118487

Bashambu, L., Singh, R., & Verma, J. (2021). Metal/metal oxide nanocomposite membranes for water purification. *Materials Today: Proceedings, 44*, 538–545. https://doi.org/10.1016/j.matpr.2020.10.213

Boros, B. V., & Ostafe, V. (2020). Evaluation of ecotoxicology assessment methods of nanomaterials and their effects. *Nanomaterials, 10*(4). https://doi.org/10.3390/nano10040610

Bratsch, S. G. (1989). Standard electrode potentials and temperature coefficients in water at 298.15 K. *Journal of Physical and Chemical Reference Data, 18*(1), 1–21. https://doi.org/10.1063/1.555839

Cabello-Ribota, B. S., Farías, R., & Reyes-López, S. Y. (2021). Surface enhanced infrared absorption studies of sio2 –tio2 –ag nanofibers: Effect of silver electrodeposition time on the amplification of signals. *Crystals, 11*. https://doi.org/10.3390/cryst11050563

Caixeta, M. P. (2023). Contaminants of emerging concern: A review of risk assessment and treatment strategies. *UPorto Journal of Engineering, 9*, 191–228. https://doi.org/10.24840/2183-6493_009-001_001282

Cai, W., Weng, X., & Chen, Z. (2019). Highly efficient removal of antibiotic rifampicin from aqueous solution using green synthesis of recyclable nano-Fe_3O_4. *Environmental Pollution, 247*, 839–846. https://doi.org/10.1016/j.envpol.2019.01.108

Cerrato, E., Gaggero, E., Calza, P., & Paganini, M. C. (2022). The role of Cerium, Europium and Erbium doped TiO2 photocatalysts in water treatment: A mini-review. *Chemical Engineering Journal Advances, 10*, 100268. https://doi.org/10.1016/j.ceja.2022.100268

Chen, S., Shan, B., Yang, Y., Yuan, G., Huang, S., Lu, X., Zhang, Y., Fu, Y., Ye, L., & Liu, J. (2017). An overview of carbon nanotubes based interconnects for microelectronic packaging. In *2017 IMAPS Nordic Conference on Microelectronics Packaging, NordPac 2017, January 2022* (pp. 113–119). https://doi.org/10.1109/NORDPAC.2017.7993175

Cohen, M., Yakirevich, A., & Weisbrod, N. (2019). The role of stabilizer concentration in the mobility of carbon-supported nanozerovalent iron (nZVI) in fractured media. *Water Resources Research, 55*(10), 8164–8177. https://doi.org/10.1029/2019WR025553

Crane, M., Handy, R. D., Garrod, J., & Owen, R. (2008). Ecotoxicity test methods and environmental hazard assessment for engineered nanoparticles. *Ecotoxicology, 17*, 421–437. https://doi.org/10.1007/s10646-008-0215-z

Das, A. P., & Mishra, S. (2009). Hexavalent chromium [Cr (VI)]: Yellow water pollution and its remediation. *Sarovar Saurabh ENVIS Newsletter Wetland Ecosystem, 5*.

Das, A., & Mishra, S. (2010). Biodegradation of the metallic carcinogen hexavalent chromium Cr(VI) by an indigenously isolated bacterial strain. *Journal of Carcinogenesis, 9*, 6. https://doi.org/10.4103/1477-3163.63584

Das, A. P., Sukla, L. B., & Pradhan, N. (2012). Microbial recovery of manganese using *Staphylococcus Epidermidis*. *International Journal of Nonferrous Metallurgy, 2012*, 9–12. https://doi.org/10.4236/IJNM.2012.12002

Del Prado-Audelo, M. L., García Kerdan, I., Escutia-Guadarrama, L., Reyna-González, J. M., Magaña, J. J., & Leyva-Gómez, G. (2021). Nanoremediation: Nanomaterials and nanotechnologies for environmental cleanup. *Frontiers in Environmental Science, 9*.

Eddy, N. O. (2020). Theoretical chemistry study on the toxicity of some polychlorobiphenyl (PCB) compounds using molecular descriptors. *Scientific African, 10*, e00587. https://doi.org/10.1016/j.sciaf.2020.e00587

Eddy, N. O., Ukpe, R. A., Garg, R. R., Garg, R. R., Odiongenyi, A., Ameh, P., Akpet, I. N., & Udo, S. E. (2023). *Review of in-depth knowledge on the application of oxides nanoparticles and nanocomposites of Al, Si and Ca as photocatalyst and antimicrobial agents in the treatment of contaminants in water*. Springer.

Eddy, N. O., Garg, R., Ukpe, R. A., Akpanudo, N. W., Abugu, H., Garg, R., Anjum, A., & Anand, B. (2024). Ethical and environmental implications associated with the application of nanotechnology. In *Smart and sustainable applications of nanocomposites* (pp. 274–299). IGI Global.

Gamoń, F., Ziembińska-Buczyńska, A., Łukowiec, D., & Tomaszewski, M. (2023). Ecotoxicity of selected carbon-based nanomaterials. *International Journal of Environmental Science and Technology, 20*(9), 10153–10162. https://doi.org/10.1007/s13762-022-04692-w

Garg, R., & Anjum, A. (2024). *Smart and sustainable applications of nanocomposites*. IGI Global.

Garg, R. R., Rani, P., Garg, R. R., & Eddy, N. O. (2021). Study on potential applications and toxicity analysis of green synthesized nanoparticles. *Turkish Journal of Chemistry, 45*, 1690–1706. https://doi.org/10.3906/kim-2106-59

Garg, R. R., Garg, R. R., Okon Eddy, N., Ibrahim Almohana, A., Fahad Almojil, S., Amir Khan, M., & Ho Hong, S. (2022). Biosynthesized silica-based zinc oxide nanocomposites for the sequestration of heavy metal ions from aqueous solutions. *Journal of King Saud University - Science, 34*, 101996. https://doi.org/10.1016/j.jksus.2022.101996

Garg, R., Mittal, M., Tripathi, S., & Eddy, N. O. (2024). Core to concept: Synthesis, structure, and reactivity of nanoscale zero-valent iron (NZVI) for wastewater remediation. *Environmental Science and Pollution Research, 31*, 67496. https://doi.org/10.1007/s11356-024-33197-x

Hafeez, M., Afyaz, S., Khalid, A., Ahmad, P., Khandaker, M. U., Sahibzada, M. U. K., Ahmad, I., Khan, J., Alhumaydhi, F. A., Bin, E. T., & Idris, A. M. (2022). Synthesis of cobalt and sulphur doped titanium dioxide photocatalysts for environmental applications. *Journal of King Saud University—Science, 34*, 102028. https://doi.org/10.1016/j.jksus.2022.102028

Hitam, C. N. C., & Jalil, A. A. (2020). A review on exploration of Fe2O3 photocatalyst towards degradation of dyes and organic contaminants. *Journal of Environmental Management, 258*, 110050. https://doi.org/10.1016/j.jenvman.2019.110050

Ijaz, I., Gilani, E., Nazir, A., & Bukhari, A. (2020). Detail review on chemical, physical and green synthesis, classification, characterizations and applications of nanoparticles. *Green Chemistry Letters and Reviews, 13*(3), 59–81. https://doi.org/10.1080/17518253.2020.1802517

Islam, M. A., Ali, I., Karim, S. M. A., Hossain Firoz, M. S., Chowdhury, A.-N., Morton, D. W., & Angove, M. J. (2019). Removal of dye from polluted water using novel nano manganese oxide-based materials. *Journal of Water Process Engineering, 32*, 100911. https://doi.org/10.1016/j.jwpe.2019.100911

Jayanthi, S., Krishnarao Eswar, N., Singh, S. A., Chatterjee, K., Madras, G., & Sood, A. K. (2016). Macroporous three-dimensional graphene oxide foams for dye adsorption and antibacterial applications. *RSC Advances, 6*(2), 1231–1242. https://doi.org/10.1039/c5ra19925e

Kabay, N., Shirazi, M. M. A., Güler, E., & Bryjak, M. (2022). Grand challenges in membrane modules and processes. *Frontiers in Membrane Science and Technology, 1*. https://doi.org/10.3389/frmst.2022.913597

Karthik, C., Punnaivalavan, K. A., Prabha, S. P., & Caroline, D. G. (2022). *Multifarious global flora fabricated phytosynthesis of silver nanoparticles: A green nanoweapon for antiviral approach including SARS-CoV-2*. Springer.

Kumar, M., Pandit, S., Patel, V., Khanna, N., Nag, M., Lahiri, D., Ray, R. R., Das, A., & Das, D. (2023). Maximization of energy recovery from starch processing wastewater by thermophilic dark fermentation coupled with microbial fuel cell technology. *Geomicrobiology Journal, 1–12*, 392. https://doi.org/10.1080/01490451.2023.2209555

Kumari, M., Khan, S. S., Pakrashi, S., Mukherjee, A., & Chandrasekaran, N. (2011). Cytogenetic and genotoxic effects of zinc oxide nanoparticles on root cells of Allium cepa. *Journal of Hazardous Materials, 190*, 613–621. https://doi.org/10.1016/j.jhazmat.2011.03.095

Kyriakides, T. R., Raj, A., Tseng, T. H., Xiao, H., Nguyen, R., Mohammed, F. S., Halder, S., Xu, M., Wu, M. J., Bao, S., & Sheu, W. C. (2021). Biocompatibility of nanomaterials and their immunological properties. *Biomedical Materials, 16*, 042005. https://doi.org/10.1088/1748-605X/abe5fa

Lu, H., Wang, J., Stoller, M., Wang, T., Bao, Y., & Hao, H. (2016). An overview of nanomaterials for water and wastewater treatment. *Advances in Materials Science and Engineering, 2016*, 4964828, 10 pages. https://doi.org/10.1155/2016/4964828

Luo, C., Tian, Z., Yang, B., Zhang, L., & Yan, S. (2013). Manganese dioxide/iron oxide/acid oxidized multi-walled carbon nanotube magnetic nanocomposite for enhanced hexavalent chromium removal. *Chemical Engineering Journal, 234*, 266–275. https://doi.org/10.1016/j.cej.2013.08.084

Lv, D., Zhou, J., Cao, Z., Xu, J., Liu, Y., Li, Y., Yang, K., Lou, Z., Lou, L., & Xu, X. (2019). Mechanism and influence factors of chromium(VI) removal by sulfide-modified nanoscale zerovalent iron. *Chemosphere, 224*, 306–315. https://doi.org/10.1016/j.chemosphere.2019.02.109

Madhogaria, B., Banerjee, S., Kundu, A., & Dhak, P. (2024). Efficacy of new generation biosorbents for the sustainable treatment of antibiotic residues and antibiotic resistance genes from polluted waste effluent. *Infectious Medicine, 3*, 100092. https://doi.org/10.1016/j.imj.2024.100092

Mageswari, A., Srinivasan, R., Subramanian, P., Ramesh, N., & Gothandam, K. M. (2016). Nanomaterials: Classification, biological synthesis and characterization. In S. Ranjan, N. Dasgupta, & E. Lichtfouse (Eds.), *Nanoscience in food and agriculture* (pp. 31–71). Springer.

Mandal, S., Hwang, S., Marpu, S. B., Omary, M. A., Prybutok, V., & Shi, S. Q. (2023). Bioinspired synthesis of silver nanoparticles for the remediation of toxic pollutants and enhanced antibacterial activity. *Biomolecules, 13*, 1054. https://doi.org/10.3390/biom13071054

Mansur, A. A. P., Custódio, D. A. C., Dorneles, E. M. S., Coura, F. M., Carvalho, I. C., Lage, A. P., & Mansur, H. S. (2023). Nanoplexes of ZnS quantum dot-poly-L-lysine/iron oxide nanoparticle-carboxymethylcellulose for photocatalytic degradation of dyes and antibacterial activity in wastewater treatment. *International Journal of Biological Macromolecules, 231*, 123363. https://doi.org/10.1016/j.ijbiomac.2023.123363

Meidl, P., Lehmann, A., Bi, M., Breitenreiter, C., Benkrama, J., Li, E., Riedo, J., & Rillig, M. C. (2024). Combined application of up to ten pesticides decreases key soil processes. *Environmental Science and Pollution Research, 31*, 11995–12004. https://doi.org/10.1007/s11356-024-31836-x

Meyer, J. S., Lyons-Darden, T., Garman, E. R., Middleton, E. T., & Schlekat, C. E. (2020). Toxicity of nanoparticulate nickel to aquatic organisms: Review and recommendations for improvement of toxicity tests. *Environmental Toxicology and Chemistry, 39*(10), 1861–1883. https://doi.org/10.1002/etc.4812

Mishra, S., & Das, A. P. (2021). Current treatment technologies for removal of microplastic and microfiber pollutants from wastewater. *Wastewater Treatment: Cutting-Edge Molecular Tools, Techniques and Applied Aspects, 237–251*. https://doi.org/10.1016/B978-0-12-821881-5.00011-8

Mishra, S., & Das, A. P. (2022). Treatment of the wastewater polluted with synthetic microfiber released from washing machine. In B. B. Das, H. Hettiarachchi, P. K. Sahu, & S. Nanda (Eds.), *Lecture notes in civil engineering* (pp. 109–117). Springer.

Mishra, S., & Das, A. P. (2023). Microbial remediation of synthetic microfiber contaminated wastewater. In M. P. Shah (Ed.), *Microbial technologies in industrial wastewater treatment* (pp. 337–350). Springer.

Mishra, S., Dash, D., Al-Tawaha, A. R. M. S., & Das, A. P. (2022a). A review on heavy metal ion adsorption on synthetic microfiber surface in aquatic environments. *Applied Biochemistry and Biotechnology, 194*, 4639–4654. https://doi.org/10.1007/S12010-022-04029-W/METRICS

Mishra, S., Dash, D., & Das, A. P. (2022b). Detection, characterization and possible biofragmentation of synthetic microfibers released from domestic laundering wastewater as an emerging source of marine pollution. *Marine Pollution Bulletin, 185*, 114254. https://doi.org/10.1016/J.MARPOLBUL.2022.114254

Mishra, S., Singh, R. P., Rout, P. K., & Das, A. P. (2022c). Membrane bioreactor (MBR) as an advanced wastewater treatment technology for removal of synthetic microplastics. *Development in Wastewater Treatment Research and Processes, 45–60*. https://doi.org/10.1016/B978-0-323-85583-9.00022-3

Mishra, S., Dash, D., & Das, A. P. (2024). Isolation and molecular characterization of polyester-degrading bacterial strains from municipal wastewater and their application in microplastic degradation in contaminated wastewater. *Geomicrobiology Journal, 41*, 732. https://doi.org/10.1080/01490451.2024.2376101

Mobarakabad, P., Moghadassi, A. R., & Hosseini, S. M. (2015). Fabrication and characterization of poly(phenylene ether-ether sulfone) based nanofiltration membranes modified by titanium dioxide nanoparticles for water desalination. *Desalination, 365*, 227–233. https://doi.org/10.1016/j.desal.2015.03.002

Mondal, R., & De, S. (2022). Removal of copper(II) from aqueous solution using zinc oxide nanoparticle impregnated mixed matrix hollow fiber membrane. *Environmental Technology and Innovation, 26*, 102300. https://doi.org/10.1016/j.eti.2022.102300

Modi, S., Yadav, V. K., Amari, A., Alyami, A. Y., Gacem, A., Harharah, H. N., & Fulekar, M. H. (2023). Photocatalytic degradation of methylene blue dye from wastewater by using doped zinc oxide nanoparticles. *Water (Switzerland), 15*, 2275. https://doi.org/10.3390/w15122275

Nazir, M. M., Li, Q., Noman, M., Ulhassan, Z., Ali, S., Ahmed, T., Zeng, F., & Zhang, G. (2022). Calcium oxide nanoparticles have the role of alleviating arsenic toxicity of barley. *Frontiers in Plant Science, 13*. https://doi.org/10.3389/fpls.2022.843795

Nehra, S., Raghav, S., & Kumar, D. (2020). Biomaterial functionalized cerium nanocomposite for removal of fluoride using central composite design optimization study. *Environmental Pollution, 258*, 113773. https://doi.org/10.1016/j.envpol.2019.113773

Odler, I., & Abdul-Maula, S. (1983). Polymorphism and hydration of tricalcium silicate doped with ZnO. *Journal of the American Ceramic Society, 66*, 1–04. https://doi.org/10.1111/j.1151-2916.1983.tb09956.x

Othman, N. H., Alias, N. H., Fuzil, N. S., Marpani, F., Shahruddin, M. Z., Chew, C. M., Ng, K. M. D., Lau, W. J., & Ismail, A. F. (2022). A review on the use of membrane technology systems in developing countries. *Membranes, 12*(1), 30. https://doi.org/10.3390/membranes12010030

Pan, Y., Peng, Z., Liu, Z., Shao, B., Liang, Q., He, Q., Wu, T., Zhang, X., Zhao, C., Liu, Y., Ge, L., & He, M. (2022). Activation of peroxydisulfate by bimetal modified peanut hull-derived porous biochar for the degradation of tetracycline in aqueous solution. *Journal of Environmental Chemical Engineering, 10*(2), 107366. https://doi.org/10.1016/j.jece.2022.107366

Pyrzynska, K. (2008). Carbon nanotubes as a new solid-phase extraction material for removal and enrichment of organic pollutants in water. *Separation & Purification Reviews, 37*, 372–389. https://doi.org/10.1080/15422110802178843

Qin, Q., Sun, T., Yin, W., & Xu, Y. (2017). Rapid and efficient removal of methylene blue by freshly prepared manganese dioxide. *Cogent Engineering, 4*(1), 1345289. https://doi.org/10.1080/23311916.2017.1345289

Rajput, V., Minkina, T., Mazarji, M., Shende, S., Sushkova, S., Mandzhieva, S., Burachevskaya, M., Chaplygin, V., Singh, A., & Jatav, H. (2020). Accumulation of nanoparticles in the soil-plant systems and their effects on human health. *Annals of Agricultural Sciences, 65*(2), 137–143. https://doi.org/10.1016/j.aoas.2020.08.001

Rivero-Huguet, M., & Marshall, W. D. (2009). Reduction of hexavalent chromium mediated by micron- and nano-scale zero-valent metallic particles. *Journal of Environmental Monitoring, 11*(5), 1072–1079. https://doi.org/10.1039/b819279k

Romero-Franco, M., Godwin, H. A., Bilal, M., & Cohen, Y. (2017). Needs and challenges for assessing the environmental impacts of engineered nanomaterials (ENMs). *Beilstein Journal of Nanotechnology, 8*, 989–1014. https://doi.org/10.3762/bjnano.8.101

Saleem, H., Trabzon, L., Kilic, A., & Zaidi, S. J. (2020). Recent advances in nanofibrous membranes: Production and applications in water treatment and desalination. *Desalination, 478*, 114178. https://doi.org/10.1016/j.desal.2019.114178

Schwab, F., Zhai, G., Kern, M., Turner, A., Schnoor, J. L., & Wiesner, M. R. (2016). Barriers, pathways and processes for uptake, translocation and accumulation of nanomaterials in plants—Critical review. *Nanotoxicology, 10*, 257–278. https://doi.org/10.3109/17435390.2015.1048326

Schwirn, K., Voelker, D., Galert, W., Quik, J., & Tietjen, L. (2020). Environmental risk assessment of nanomaterials in the light of new obligations under the REACH regulation: Which challenges remain and how to approach them? *Integrated Environmental Assessment and Management, 16*, 706–717. https://doi.org/10.1002/ieam.4267

Sharma, R., Garg, R., & Kumari, A. (2020). A review on biogenic synthesis, applications and toxicity aspects of zinc oxide nanoparticles. *EXCLI Journal, 19*, 1325–1340. https://doi.org/10.17179/excli2020-2842

Sharma, M. D., Rayalu, S. S., Kolev, S. D., & Krupadam, R. J. (2021). Graphene/fluorescein dye-based sensor for detecting As(III) in drinking water. *Scientific Reports, 11*, 1–11. https://doi.org/10.1038/s41598-021-96968-3

Sharma, R., Garg, R., Bali, M., & Eddy, N. O. (2023). Potential applications of green-synthesized iron oxide NPs for environmental remediation. *Environmental Monitoring and Assessment, 195*. https://doi.org/10.1007/s10661-023-12035-6

Shittu, K. O., & Ihebunna, O. (2017). Purification of simulated waste water using green synthesized silver nanoparticles of Piliostigma thonningii aqueous leave extract. *Advances in Natural Sciences: Nanoscience and Nanotechnology, 8*. https://doi.org/10.1088/2043-6254/aa8536

Singh, S. A., & Madras, G. (2013). Photocatalytic degradation with combustion synthesized WO3 and WO3TiO2 mixed oxides under UV and visible light. *Separation and Purification Technology, 105*, 79–89. https://doi.org/10.1016/j.seppur.2012.12.010

Singla, S., Garg, R., & Garg, R. (2024). Removal of priority pollutants using photocatalysis and wastewater remediation. In *Sustainable management of land, water and pollution of built-up area* (pp. 179–188). Springer.

Soltys, L., Olkhovyy, O., Tatarchuk, T., & Naushad, M. (2021). Green synthesis of metal and metal oxide nanoparticles: Principles of green chemistry and raw materials. *Magnetochemistry, 7*. https://doi.org/10.3390/magnetochemistry7110145

Stefaniuk, M., Oleszczuk, P., & Ok, Y. S. (2016). Review on nano zerovalent iron (nZVI): From synthesis to environmental applications. *Chemical Engineering Journal, 287*, 618–632. https://doi.org/10.1016/j.cej.2015.11.046

Su, Z., Chen, L., & Wen, D. (2024). Impact of wastewater treatment plant effluent discharge on the antibiotic resistome in downstream aquatic environments: A mini review. *Frontiers of Environmental Science and Engineering, 18*(3), 1–11. https://doi.org/10.1007/s11783-024-1796-3

Suter, G. W., & Norton, S. B. (2019). Ecological risk assessment. In *Encyclopedia of ecology* (pp. 402–406). Elsevier.

Tratnyek, P. G., Salter, A. J., Nurmi, J. T., & Sarathy, V. (2010). *Environmental applications of Zerovalent metals: Iron vs. Zinc* (pp. 165–178). American Chemical Society.

van de Meent, D., de Zwart, D., Struijs, J., Hermens, J. L. M., van Straalen, N. M., den Haan, K. H., Parsons, J. R., de Voogt, P., & van Leeuwen, K. (2023). Expected risk as basis for assessment of safe use of chemicals. *Environmental Sciences Europe, 35*, 16. https://doi.org/10.1186/s12302-023-00713-8

Vishnu, D., Dhandapani, B., Kannappan Panchamoorthy, G., Vo, D. V. N., & Ramakrishnan, S. R. (2021). Comparison of surface-engineered superparamagnetic nanosorbents with low-cost adsorbents of cellulose, zeolites and biochar for the removal of organic and inorganic pollutants: A review. *Environmental Chemistry Letters, 19*(4), 3181–3208. https://doi.org/10.1007/s10311-021-01201-2

Wadhwa, S., Mathur, A., Pendurthi, R., Singhal, U., Khanuja, M., & Roy, S. S. (2020). Titaniabased porous nanocomposites for potential environmental applications. *Bulletin of Materials Science, 43*(1), 47. https://doi.org/10.1007/s12034-019-2009-8

Wang, F., Zhang, L., Wang, Y., Liu, X., Rohani, S., & Lu, J. (2017). Fe 3 O 4 @SiO 2 @CS-TETA functionalized graphene oxide for the adsorption of methylene blue (MB) and Cu(II). *Applied Surface Science, 420*, 970–981. https://doi.org/10.1016/j.apsusc.2017.05.179

Wang, Q., Yan, S., Chang, C., Qu, C., Tian, Y., Song, J., & Guo, J. (2023). Occurrence, potential risk assessment, and source apportionment of polychlorinated biphenyls in water from Beiluo River. *Water (Switzerland), 15*, 459. https://doi.org/10.3390/w15030459

Wolff, C. M., Frischmann, P. D., Schulze, M., Bohn, B. J., Wein, R., Livadas, P., Carlson, M. T., Jäckel, F., Feldmann, J., Würthner, F., & Stolarczyk, J. K. (2018). All-in-one visible-light-driven water splitting by combining nanoparticulate and molecular co-catalysts on CdS nanorods. *Nature Energy, 3*, 862–869. https://doi.org/10.1038/s41560-018-0229-6

Yang, J., Shojaei, S., & Shojaei, S. (2022). Removal of drug and dye from aqueous solutions by graphene oxide: Adsorption studies and chemometrics methods. *Npj Clean Water, 5*(1), 1–10. https://doi.org/10.1038/s41545-022-00148-3

Yaou Balarabe, B., Irédon, A., Hassimi, M., Illiassou Oumarou, M. N., Masiyambiri, V., & Gunda, T. J. (2023). Effective removal of emerging organic pollutants using hybrid Ag@ZnO supported reduced-graphene oxide nanocomposite under visible light. *Hybrid Advances, 4*, 100114. https://doi.org/10.1016/j.hybadv.2023.100114

Zaheer, Z. (2018). Biogenic synthesis, optical, catalytic, and in vitro antimicrobial potential of Ag-nanoparticles prepared using palm date fruit extract. *Journal of Photochemistry and Photobiology B: Biology, 178*, 584–592. https://doi.org/10.1016/j.jphotobiol.2017.12.002

Zamani, A., Marjani, A. P., & Mousavi, Z. (2019). Agricultural waste biomass-assisted nanostructures: Synthesis and application. *Green Processing and Synthesis, 8*, 421–429. https://doi.org/10.1515/gps-2019-0010

Zhang, H., & Liu, B. (2021). Preparation, characterization, and photocatalytic properties of self-standing pure and Cu-doped TiO_2 nanobelt membranes. *ACS Omega, 6*(7), 4534–4541. https://doi.org/10.1021/acsomega.0c03873

Zou, Y., Wang, X., Khan, A., Wang, P., Liu, Y., Alsaedi, A., Hayat, T., & Wang, X. (2016). Environmental remediation and application of nanoscale zero-valent iron and its composites for the removal of heavy metal ions: A review. *Environmental Science and Technology, 50*(14), 7290–7304. https://doi.org/10.1021/acs.est.6b01897

Promoting the Global Economies by Sustainable Use and Management of Agricultural Waste

Ajit Singh, Monica, and Subhadra Rajpoot

Abstract The global population has dramatically increased from 3.7 billion in 1970 to 7.9 billion in 2021, and it's expected to reach 9 billion by 2050 and 11 billion by 2100. This rapid growth poses a significant challenge for food security. To meet the rising demand for food, there has been a substantial increase in both crop and livestock production, leading to a corresponding increase in agricultural waste generation. Countries like China, India, and Africa have experienced rapid population growth, economic development, and a rise in agricultural waste production. India, for example, produces a massive amount of solid waste annually, with agricultural waste being the primary component, estimated at around 350–990 million tonnes per year. As the world's second-largest producer of agricultural waste after China, India generates over 130 million tonnes of paddy straw, half of which is used as fodder and the other half is discarded. The practice of burning rice residue (parali) in the northwestern region of India contributes significantly to air pollution and public health concerns. Agriculture plays a vital role in the global economy and is the sole provider of food and nutrition. However, it also produces a significant amount of waste, including waste from food processing, crop waste, animal waste, and hazardous waste from farming practices. The growing global concern over environmental degradation and resource depletion has prompted the adoption of sustainable agricultural practices to reduce agricultural waste. Agricultural waste, a byproduct of agricultural production, poses a significant challenge to global sustainability. The unsustainable management of agricultural waste not only results in the loss of valuable resources but also contributes to environmental degradation, cli-

A. Singh (✉)
Mahatma Jyotiba Phule Rohilkhand University, Bareilly, India

Monica
Department of Economics, M.M.H. College, Ghaziabad, India

S. Rajpoot
Department of Applied Sciences, Amity University, Greater Noida, India

S. Rajpoot, A. P. Das (eds.), *Innovative Technologies for Waste Management*, Sustainable Environmental Waste Management Strategies,
https://doi.org/10.1007/978-3-031-97184-6_10

mate change, and public health issues. Conversely, its sustainable management can offer numerous benefits, including improved soil health, reduced greenhouse gas emissions, and the creation of valuable products.

The present chapter highlights the importance of promoting sustainable agricultural waste management on a global scale, and prominence of its economic, environmental, and social implications. This chapter explores the classification of agricultural waste from various sources, strategies for utilizing and valorizing these wastes, and composting as a sustainable approach to managing agricultural waste. The study also examines the current state of agricultural waste management, discuss the benefits of sustainable practices, and propose policy recommendations to support the transition toward a more circular and sustainable agricultural economy.

Agricultural waste can be a double-edged sword, posing significant environmental and health risks if mishandled, but also offering valuable benefits when managed responsibly. It can be concluded that effective management of agricultural waste is crucial for sustainable agriculture. This involves reducing waste generation, utilizing valuable resources, and recycling materials. By minimizing pollution and conserving resources, these practices not only benefit the environment but also boost agricultural productivity and long-term sustainability. As we strive for a more sustainable future, it is essential to integrate the principles of reduce, reuse, and recycle into agricultural practices to protect the planet and ensure the well-being of future generations. To achieve this, key strategies such as composting, biogas production, mulching, biomass conversion, and recycling packaging materials can be employed. Sustainable agricultural waste management is a global imperative that offers significant economic, environmental, and social benefits. Effective waste management systems can help farmers reduce their environmental impact, improve soil health, generate renewable energy, create valuable products, and enhance food security. By promoting the adoption of sustainable practices, we can contribute to a more resilient, equitable, and sustainable future. Furthermore, technological advancements and innovations are making it possible to manage agricultural waste in a more efficient and sustainable manner, ultimately contributing to a more sustainable and resilient future for the agricultural sector.

Keywords Agricultural waste · Sustainable use · Management · Environmental · Economic and global economies

1 Introduction

The escalating global population, coupled with rapid urbanization and economic development, has placed unprecedented pressure on agricultural systems. This surge in demand for food has led to a significant increase in agricultural production, resulting in a corresponding rise in waste generation. One of the most pressing concerns for human health and the well-being of civilization as a whole is waste (Banerjee et al., 2019). The alarming consequences of waste, particularly man-made waste, have sparked serious concerns about the protection of the current generation and the planet at large. The rapid growth of the population and the level of education among individuals have emerged as significant factors contributing to waste generation (Himabindu et al., 2015). Agricultural waste, a byproduct of agricultural activities, poses a significant challenge to global sustainability. Its unsustainable management can contribute to environmental degradation, climate change, and public health issues. However, agricultural waste is primarily linked to soil and water pollution, which can degrade soil health and lead to infertility by exerting pressure on the soil (Sims & Wolf, 1994). Conversely, its sustainable management can offer numerous benefits, including improved soil health, reduced greenhouse gas emissions, and the creation of valuable products.

The energy crisis and environmental degradation are the two most pressing global concerns hindering sustainable development (Khosla et al., 2022). The majority of energy consumption (over 80%) comes from fossil fuels, which are widely acknowledged to contribute to climate change, global warming, and the rapid depletion of natural energy supplies (Ni et al., 2006). As a result, nations are seeking alternative, sustainable energy sources. In recent decades, scientists have focused on generating green fuels like bioethanol and biodiesel to address this issue (Batra et al., 2022; Sonu & Chaudhary, 2022). These first-generation biofuels, produced from sources like sugar cane, palm oil, and corn, offer an economic alternative to reduce dependence on fossil fuels (Angelidaki et al., 2007). However, it is crucial to develop second-generation biofuels by converting entire plants, including agricultural waste, into biofuels to achieve sustainable energy. The development of alternative, sustainable, and clean energy sources has become a top priority due to the severe threats posed by resource depletion and climate change. Efforts are being made to provide abundant, sustainable, and eco-friendly energy sources to power the growth of advanced technologies like smart gadgets, wearable electronics, and wireless sensor networks (Chaudhary, 2021a, b, 2022; Zhang et al., 2014; Chaudhary et al., 2022, 2023). Future portable and smart devices will utilize soft, biodegradable, and elastomeric materials to ensure seamless integration with the natural world and the human body (Zou et al., 2018).

Agricultural waste management involves the coordination, handling, and control of waste generated from agricultural activities, with the primary objective of preventing soil and water pollution, greenhouse gas emissions, and health risks for humans and animals. An effective agricultural waste management system employs one or more of the following strategies: waste reduction, recycling, and reusing. These methods convert waste into valuable resources, such as organic fertilizers or biogas, resulting in a mutually beneficial outcome for the environment, agricultural organizations, and the communities they serve.

This chapter delves into the critical importance of promoting sustainable agricultural waste management on a global scale. It explores the classification of agricultural waste, strategies for its utilization and valorization, and composting as a sustainable approach. Additionally, the chapter examines the current state of agricultural waste management, discusses the benefits of sustainable practices, and proposes policy recommendations to support the transition toward a more circular and sustainable agricultural economy.

2 The Growing Challenge of Agricultural Waste

The global population has dramatically increased from 3.7 billion in 1970 to 7.9 billion in 2021, and it's expected to reach 9 billion by 2050 and 11 billion by 2100. This rapid growth has necessitated a substantial increase in agricultural production to meet the rising demand for food. Consequently, agricultural waste generation has also surged. Countries like China, India, and Africa, experiencing rapid population growth and economic development, have witnessed a significant increase in agricultural waste production.

India, for instance, generates a massive amount of solid waste annually, with agricultural waste being the primary component, estimated at around 350–990 million tonnes per year. As the world's second-largest producer of agricultural waste after China, India generates over 130 million tonnes of paddy straw, half of which is used as fodder and the other half is discarded. The practice of burning rice residue (parali) in the northwestern region of India contributes significantly to air pollution and public health concerns.

According to Bernal (2017), modern intensive agricultural systems generate enormous quantities of manure in concentrated areas, necessitating both on-farm and off-farm management strategies. The traditional linear economic model of "take-make-consume-dispose" is transitioning to a circular system that connects resource use with waste management, aligning with Europe's Circular Economy and Bioeconomy Strategy initiatives.

The scientific community faces significant pressure to address questions stemming from these emerging concepts, such as the advancement of treatment technologies, the improvement of environmental conservation and sustainability, and the assurance of food safety. The major challenges in managing waste within agroecosystems include:

- Guaranteeing the availability of nutrients (N and P) from organic waste: optimizing fertilizer efficiency and soil nutrient cycles to effectively replace mineral fertilizers (Burton and Turner, 2003; Sommer et al., 2013).
- Creating treatment technologies for waste reuse: recovering nutrients and valuable compounds (biorefinery), utilizing anaerobic digestion, and composting to produce high-quality products (Alburquerque et al., 2012; Bernal et al., 2017).
- Tackling the issue of pollutants or hazardous substances in organic waste: assessing bioavailability, toxicity, soil risk, and biological transformation during waste treatment processes to ensure food safety.
- Reducing the environmental impact of gaseous emissions—such as ammonia (NH3), greenhouse gases (GHGs), bioaerosols, particulate matter, odors, etc.

- Comprehending the effects of organic waste on soil carbon conservation, microbial diversity, and function.

Agriculture's reliance on synthetic fertilizers, specifically nitrogen and phosphorus, leads to considerable environmental consequences on local, regional, and global levels. Consequently, there is a pressing need to replace nutrients from synthetic mineral fertilizers with those derived from waste products. However, assessing the fertilizer efficiency of organic materials poses a challenge due to the complexities of mineralization-immobilization cycles, dissolution and absorption processes, and the pathways through which nutrients are lost from soil systems (Burton & Turner, 2003; Sommer et al., 2013).

Innovative technologies are being developed to recover nutrients and valuable compounds from organic waste. For instance, phosphorus can be recovered through precipitation or the formation of struvite (Martí et al., 2010; Szogi et al., 2015), while ammonia can be extracted from waste using stripping or gas-permeable membranes (Serna-Maza et al., 2015; Vanotti et al., 2017). These methods have the potential to create marketable fertilizers from sources like animal manure, slurry, and sewage sludge. Nonetheless, it is crucial to ensure the quality of these products for safe commercialization.

Incorporating organic waste into agricultural soils can enhance crop yields, produce quality food, and support sustainable ecosystems while mitigating pollution (Burton & Turner, 2003). Nonetheless, managing organic waste within soil–plant systems presents notable environmental challenges, such as the risk of soil and water contamination from nutrients, organic carbon, heavy metals, pathogens, and organic pollutants. Additionally, wastewater, sewage sludge, and animal manures may introduce pharmaceutical substances and veterinary products into agricultural ecosystems (Hu et al., 2010). Biosolids, which are high-quality stabilized sewage sludges, are frequently utilized to boost soil fertility, yet they may also contain pharmaceutical and personal care products, endocrine-disrupting chemicals, and other emerging pollutants (Sabourin et al., 2012; Mohapatra et al., 2016).

These contaminants can accumulate in crops when wastewater is used for irrigation (Calderón-Preciado et al., 2011). Recent research has concentrated on evaluating the effectiveness of various waste treatment methods, such as composting and anaerobic digestion, in breaking down veterinary antibiotics and pharmaceuticals (Carballa et al., 2007; Ho et al., 2013; Mitchell et al., 2015; Wang et al., 2015; Butkovskyi et al., 2016). However, additional studies are necessary to confirm the quality and safety of digestates and composts for commercial applications (Bernal et al., 2017).

Persistent organic pollutants (POPs), including polybrominated diphenyl ethers and polycyclic aromatic hydrocarbons (PAHs), present considerable environmental and health hazards due to their low water solubility, persistence, and tendency to bioaccumulate in food chains (Katsoyiannis & Samara, 2005; Haritash & Kaushik, 2009). The breakdown of PAHs is influenced by several factors, such as pH, temperature, oxygen levels, microbial communities, and nutrient availability (Haritash & Kaushik, 2009). Nonetheless, the process of PAH degradation in soil remains intricate and not completely understood (Serrano et al., 2009). The application of municipal biosolids to land can result in the bioaccumulation and amplification of both emerging and persistent contaminants, disrupting the balance of biological systems. To tackle this issue, researchers need to create monitoring protocols and innovative treatment methods to ensure the safe use of biosolids and manures as fertilizers. The management of organic waste significantly contributes to human-induced emissions, which lead to regional and global environmental problems like eutrophication, acidification, and climate change (Chadwick et al., 2011).

The microbial breakdown of organic carbon compounds in waste generates and releases greenhouse gases (GHGs) such as nitrous oxide (N2O) and methane (CH4) (VanderZaag et al., 2011). To effectively reduce gaseous emissions from solid waste, actions are needed at all stages, from animal feed and manure management to storage, treatment systems, and soil application (Chadwick et al., 2011; Gerber et al., 2013; Pardo et al., 2015; Bernal et al., 2017). Feeding strategy is regarded as a crucial approach for reducing GHGs (Chadwick et al., 2011). However, the variability in results due to differences in manure characteristics, climate, soil types, and farming practices necessitates the use of modeling to estimate emissions under various conditions and climates (Petersen et al., 2013).

Greenhouse gas emissions from wastewater and sewage sludge treatment facilities are influenced by various processes, such as thickening, anaerobic digestion, dewatering, incineration, and melting (Soda et al., 2010). Emissions from sewage sludge can be reduced through stabilization methods like liming, anaerobic digestion, and composting, as well as by using additives (Yoshida et al., 2015; Wang et al., 2017). To ensure that soil waste management practices are sustainable, it is crucial to minimize their environmental impact related to climate change and ecosystem acidification (Pardo et al., 2015). The effects of organic waste on soil ecosystems remain uncertain. Although organic waste can enhance soil humus content and microbial activity, it may also negatively impact soil microorganisms due to contaminants (Mossa et al., 2017). Long-term biosolid disposal can influence soil microbial diversity and function, potentially altering ecosystem operations (Coolon et al., 2013). Grasping the ongoing environmental changes and the effects of human activities on the environment and climate is essential for future agricultural practices (Santoyo et al., 2017).

In agroecosystems, waste management faces five critical challenges according to Bernal (2017): improving how nutrients become available and cycle through soil, creating new technologies that allow for effective nutrient recovery and reuse, addressing contamination issues to ensure food remains safe for consumption, reducing harmful environmental emissions, and enhancing overall soil health and biological functioning. These interconnected challenges require interdisciplinary approaches as agricultural systems transition toward more circular and sustainable models that transform waste materials into valuable resources (Fig. 1).

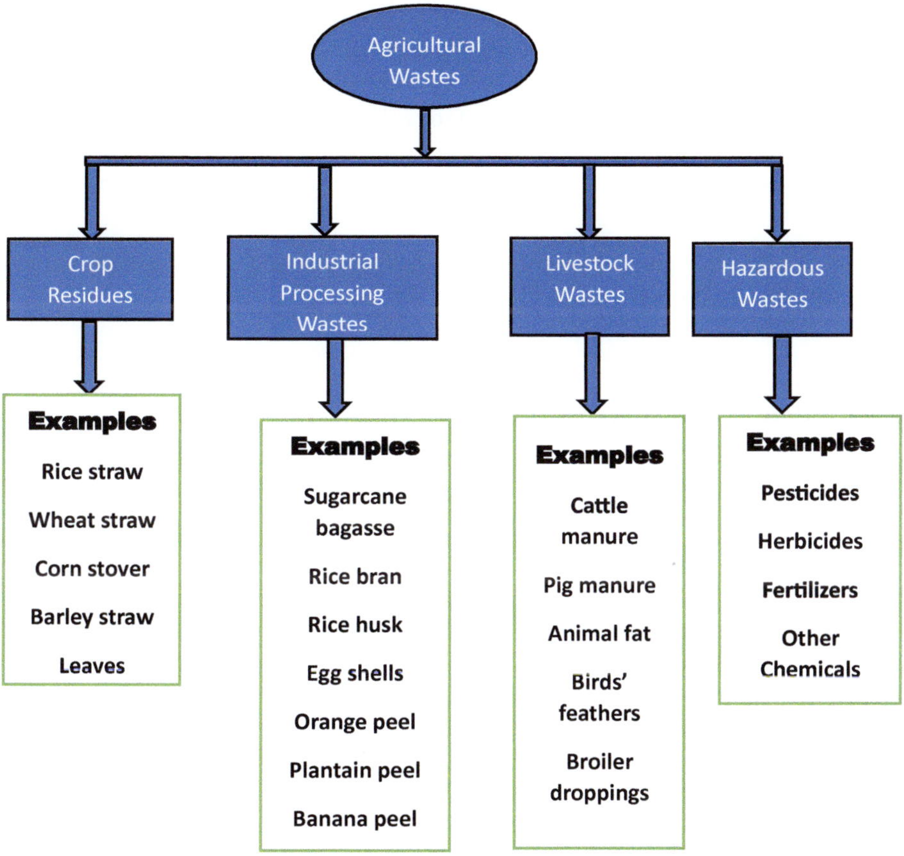

Fig. 1 Agricultural waste classification. (Source Awogbemi and Daramy (2022))

3 Classification of Agricultural Waste

Agricultural waste is a significant environmental concern, and its proper classification is crucial for developing effective strategies for its management and reduction. Agricultural waste can be broadly categorized into four main types based on its source and composition. These categories are as follows:

3.1 Crop Waste

Crop waste is generated from the residues of crops, including stalks, leaves, and chaff. This type of waste is typically generated during the harvesting and processing of crops. Examples of crop waste include:

- **Paddy straw**: The dry stalks and leaves of rice plants left over after harvesting
- **Wheat straw**: The dry stalks and leaves of wheat plants left over after harvesting
- **Corn stover**: The leaves, stalks, and cobs of corn plants left over after harvesting
- **Sugarcane bagasse**: The fibrous residue left over after sugarcane is crushed to extract its juice

Crop waste is a significant source of biomass energy and can be used as animal feed, compost, or biofuel.

3.2 Food Processing Waste

Food processing waste is generated during the processing of agricultural products, such as fruits, vegetables, dairy products, and meat. This type of waste is typically generated in food processing factories and can be rich in nutrients. Examples of food processing waste include:

- **Fruit and vegetable peels**: The outer layers of fruits and vegetables removed during processing
- **Husks**: The outer coverings of grains, such as rice, wheat, and corn
- **Byproducts from dairy processing**: Whey, lactose, and other byproducts generated during the processing of milk and dairy products
- **Byproducts from meat processing**: Bones, fat, and other byproducts generated during the processing of meat and poultry

Food processing waste can be used as animal feed, compost, or biofuel.

3.3 Animal Waste

Animal waste consists of manure, bedding materials, and carcasses from livestock and poultry. This type of waste is typically generated on farms and can be used as fertilizer or biofuel. Examples of animal waste include:

- **Manure**: The feces and urine of livestock and poultry
- **Bedding materials**: The materials used to line animal enclosures, such as straw, sawdust, and sand
- **Carcasses**: The bodies of dead animals

Animal waste is a significant source of nutrients and can be used as fertilizer or biofuel.

3.4 Hazardous Waste

Hazardous waste includes pesticides, herbicides, fertilizers, and other chemicals used in agriculture. This type of waste is typically generated during the use of chemicals in farming and can be harmful to the environment and human health if not disposed of properly. Examples of hazardous waste include:

- **Pesticides**: Chemicals used to control pests and diseases in crops
- **Herbicides**: Chemicals used to control weeds in crops
- **Fertilizers**: Chemicals used to provide nutrients to crops
- **Other chemicals**: Other chemicals used in agriculture, such as fungicides and insecticides

Hazardous waste requires special handling and disposal to prevent environmental pollution and health risks.

Understanding these categories is important for developing effective strategies for managing and reducing agricultural waste. By categorizing agricultural waste, we can identify opportunities for waste reduction, recycling, and reuse and develop sustainable practices for waste management.

4 Strategies for Utilizing and Valorizing Agricultural Waste

Agricultural waste poses a considerable environmental challenge, with millions of tons produced annually from various farming activities, including crop residues, livestock manure, and food processing byproducts. This waste not only contributes to pollution and greenhouse gas emissions but also represents a lost opportunity for resource recovery. However, if managed effectively, agricultural waste can be transformed into a valuable resource.

Valorization is the process of converting waste materials into useful products, which can include energy, chemicals, fertilizers, and even bioplastics. By implementing effective valorization strategies, we can mitigate the negative environmental impacts of agricultural waste while simultaneously creating economic opportunities.

In this chapter, we will explore a range of innovative approaches for utilizing and valorizing agricultural waste. These strategies may include composting, anaerobic digestion, bioenergy production, and the extraction of high-value compounds.

4.1 Anaerobic Digestion

Anaerobic digestion is a biological process that decomposes and stabilizes organic material without oxygen, transforming it into a renewable energy source called biogas. This process involves a consortium of microorganisms that break down the organic material, which can include a variety of agricultural waste such as crop residues, livestock manure, and food processing byproducts.

Biogas Production The primary output of anaerobic digestion is biogas, a renewable energy source primarily composed of methane (CH4) and carbon dioxide (CO2). This biogas can be harnessed for various applications, including cooking, heating, and electricity generation. It serves as an environmentally friendly alternative to fossil fuels, helping to reduce reliance on non-renewable energy sources and lower greenhouse gas emissions.

Conversion of Agricultural Waste The anaerobic digestion process effectively transforms agricultural waste into valuable resources. For instance, crop residues, which might otherwise contribute to soil degradation or be burned, can be processed to extract energy. Similarly, livestock manure, often a source of pollution, can be converted into biogas, mitigating its environmental impact while providing energy.

Digestate as Fertilizer In addition to biogas, the anaerobic digestion process yields digestate, which is rich in nutrients such as nitrogen, phosphorus, and potassium. This nutrient-rich byproduct can be used as a natural fertilizer, enhancing soil fertility and promoting sustainable agricultural practices. By returning these nutrients to the soil, digestate not only supports crop growth but also helps to close the nutrient loop in agricultural systems.

In summary, anaerobic digestion presents a sustainable approach to managing agricultural waste by converting it into biogas for energy production and digestate for fertilization. This dual benefit supports energy transition efforts while promoting environmental sustainability and resource efficiency in agriculture.

4.2 Composting

Composting is a natural, aerobic process that facilitates the decomposition of organic matter, transforming it into a nutrient-rich soil amendment known as compost. This process harnesses the activity of microorganisms, such as bacteria and fungi, along with microorganisms like earthworms and insects, to break down organic materials into a stable, humus-like product.

Nutrient-Rich Fertilizer
One of the primary benefits of composting is its ability to convert various forms of organic waste—such as crop residues, food scraps, yard waste, and livestock manure—into a high-quality, nutrient-rich fertilizer. This compost is rich in essential nutrients, including nitrogen, phosphorus, and potassium, as well as micronutrients that are vital for plant growth. By using compost as a fertilizer, farmers and gardeners can enhance soil fertility and health, significantly reducing their reliance on synthetic fertilizers. This shift not only lowers input costs but also minimizes the

environmental impact associated with chemical fertilizers, such as soil degradation and water pollution.

Improved Soil Health
The application of compost goes beyond simply providing nutrients; it also significantly improves overall soil health. Compost enhances soil structure, leading to better aeration and root penetration. It increases the soil's ability to retain moisture, reducing the need for frequent irrigation and helping to mitigate drought conditions. Additionally, compost fosters a thriving microbial community, which is essential for nutrient cycling and disease suppression. The diverse array of beneficial microorganisms found in compost helps to outcompete harmful pathogens, promoting a healthier growing environment for crops.

Composting Agricultural Waste
Agricultural waste, including crop residues (such as stalks, leaves, and husks) and food processing waste (like peels, seeds, and pulp), can be effectively composted to produce valuable fertilizer. By composting these materials, farmers can reduce the volume of waste generated on their farms, thereby minimizing disposal costs and environmental impacts. The composting process also eliminates pathogens and weed seeds, resulting in a safe, stable product that can be confidently applied to agricultural fields.

Environmental Benefits
Beyond its agricultural advantages, composting contributes to environmental sustainability. By diverting organic waste from landfills, composting reduces methane emissions—a potent greenhouse gas generated during the anaerobic decomposition of organic materials in landfills. Additionally, composting enriches soil biodiversity, enhances carbon sequestration, and promotes a circular economy by returning nutrients to the soil.

In summary, composting is a powerful tool for transforming organic waste into a nutrient-rich fertilizer that enhances soil fertility and health. By composting agricultural waste, farmers not only improve their soil's physical and biological properties but also contribute to a more sustainable agricultural system. This practice supports environmental health, reduces dependency on synthetic fertilizers, and fosters resilience in food production systems.

4.3 Biochemical Conversion

Biochemical conversion is a sophisticated process that leverages the metabolic capabilities of microorganisms to transform agricultural waste into a variety of valuable products, including biofuels, chemicals, and bioproducts. This method not only provides an efficient means of waste management but also contributes to the development of renewable energy sources and sustainable materials.

4.3.1 Microbial Mechanisms in Biochemical Conversion

The process of biochemical conversion relies on specific microorganisms—such as bacteria, yeasts, and fungi—that can break down complex organic materials found in agricultural waste. These microorganisms utilize enzymes to decompose polysaccharides, proteins, and lipids present in biomass into simpler compounds. The two primary pathways for biochemical conversion are fermentation and anaerobic digestion.

1. **Fermentation:** In this process, microorganisms convert sugars derived from biomass into alcohols and organic acids. For example, yeast can ferment sugars from corn or sugarcane to produce ethanol, while certain bacteria can convert lignocellulosic biomass (like straw or corn stover) into various organic acids and solvents.
2. **Anaerobic Digestion:** This process involves the breakdown of organic matter by microorganisms in an oxygen-free environment, resulting in the production of biogas (primarily methane) and digestate. While biogas can be used for energy, the digestate can be further processed to extract valuable chemicals or nutrients.

4.3.2 Biofuels from Agricultural Waste

One of the most significant applications of biochemical conversion is the production of biofuels from agricultural waste. Biomass, which includes leftover crop materials, plant residues, and food processing byproducts, can be converted into various biofuels, most notably ethanol and biodiesel.

- **Ethanol Production:** Ethanol is commonly produced through the fermentation of sugars extracted from crops like corn, sugarcane, and wheat. However, second-generation biofuels focus on using lignocellulosic biomass, which is more sustainable and does not compete with food production. This includes agricultural residues like straw, corn stover, and sugarcane bagasse. The conversion process typically involves pretreatment to break down the tough lignin structure, followed by enzymatic hydrolysis to release fermentable sugars, which are then fermented by yeast into ethanol. The resulting ethanol can be blended with gasoline to reduce greenhouse gas emissions and dependence on fossil fuels.
- **Biodiesel Production:** Biodiesel is produced through the transesterification of triglycerides found in vegetable oils or animal fats. Agricultural waste, such as used cooking oil or oilseed press cake, can also be utilized as feedstock for biodiesel production. The process involves reacting the oil with an alcohol (usually methanol) in the presence of a catalyst to produce fatty acid methyl esters (FAME), which constitute biodiesel. This renewable fuel can be used in diesel engines, providing a cleaner alternative to conventional diesel fuel.

4.3.3 Chemicals and Bioproducts

In addition to biofuels, biochemical conversion can yield a range of valuable chemicals and bioproducts. These include:

- **Organic Acids:** Compounds such as acetic acid, lactic acid, and succinic acid can be produced through microbial fermentation. These organic acids have various applications in the food, pharmaceutical, and chemical industries.
- **Bioplastics:** Certain microorganisms can produce biopolymers like polylactic acid (PLA) from agricultural waste, which can be used to create biodegradable plastics. This offers a sustainable alternative to petroleum-based plastics, reducing environmental pollution.
- **Enzymes and Proteins:** The fermentation process can also lead to the production of enzymes and proteins that have applications in food processing, animal feed, and bioremediation.

4.3.4 Environmental and Economic Benefits

The biochemical conversion of agricultural waste into biofuels and other valuable products offers significant environmental and economic advantages. By utilizing waste materials, this process helps to reduce the volume of organic waste sent to landfills, thereby decreasing methane emissions. Additionally, the production of biofuels contributes to energy security and reduces greenhouse gas emissions compared to fossil fuels. Economically, the development of a bioproducts industry can create jobs and stimulate rural economies while providing farmers with an additional revenue stream through the sale of agricultural waste.

In summary, biochemical conversion is a transformative process that harnesses the power of microorganisms to convert agricultural waste into biofuels, chemicals, and bioproducts. This innovative approach not only addresses waste management challenges but also promotes renewable energy production, reduces dependence on fossil fuels, and contributes to a more sustainable and circular economy. As research and technology continue to advance, the potential for biochemical conversion to play a pivotal role in achieving environmental and energy goals will only increase.

4.4 Thermochemical Conversion

Thermochemical conversion is a process that employs high temperatures to transform agricultural waste into a range of valuable products, including biofuels, chemicals, and bioproducts. This method capitalizes on the thermal decomposition of organic materials, breaking them down into simpler compounds through various thermochemical processes such as pyrolysis, gasification, and combustion.

4.4.1 Thermochemical Processes

1. **Pyrolysis:** Pyrolysis is the thermal decomposition of organic material in the absence of oxygen. During this process, agricultural waste—such as crop residues, wood chips, and other biomass—is heated to temperatures typically between 300 °C and 700 °C. This results in the production of bio-oil, char, and syngas (synthetic gas). Bio-oil is a liquid fuel that can be utilized as a substitute for fossil fuels in various applications, including transportation and heating. Additionally, the char produced can be used as a soil amendment (biochar) to enhance soil fertility and sequester carbon.
2. **Gasification:** Gasification involves converting organic materials into syngas through partial oxidation at high temperatures (around 700 °C to 1,000 °C). In this process, agricultural waste is subjected to a limited supply of oxygen, resulting in the production of a mixture of carbon monoxide, hydrogen, and other gases. Syngas can be further processed to produce biofuels like methanol or even converted into hydrogen for fuel cells. This process is advantageous because it can utilize a variety of feedstocks and produce energy-dense fuels.
3. **Combustion:** Combustion is the complete oxidation of organic materials in the presence of oxygen, generating heat and producing carbon dioxide and water as byproducts. While this process is less focused on producing liquid fuels, it is commonly used for generating heat and electricity from biomass. Agricultural waste can be burned directly in biomass power plants to produce steam that drives turbines, contributing to renewable energy generation.

4.4.2 Applications of Bio-oil

Bio-oil, produced through pyrolysis, is a versatile product that can be used in several ways:

- **Substitute for Fossil Fuels:** Bio-oil can be directly used as a liquid fuel in boilers and engines, providing a renewable alternative to traditional fossil fuels. Its properties can be modified to enhance its stability and compatibility with existing fuel infrastructure.
- **Feedstock for Chemical Production:** Bio-oil contains a complex mixture of organic compounds that can be further refined into valuable chemicals, such as phenols, acids, and alcohols. These chemicals have applications in the pharmaceutical, cosmetic, and agricultural industries.
- **Energy Storage:** Due to its liquid form, bio-oil can serve as a means of energy storage, allowing for the transport and utilization of energy derived from biomass in a more manageable way compared to solid biomass.

4.4.3 Environmental and Economic Benefits

The thermochemical conversion of agricultural waste offers significant environmental and economic advantages:

- **Waste Management:** By converting agricultural residues into valuable products, thermochemical conversion helps reduce the volume of waste that would otherwise contribute to landfill accumulation and greenhouse gas emissions.
- **Renewable Energy Production:** The process contributes to the development of renewable energy sources, decreasing reliance on fossil fuels and helping to mitigate climate change.
- **Economic Opportunities:** The establishment of thermochemical conversion facilities can create jobs in rural areas, providing farmers with an additional revenue stream through the sale of agricultural waste.

In summary, thermochemical conversion is a powerful technology that utilizes heat to transform agricultural waste into valuable products such as biofuels, chemicals, and bioproducts. Through processes like pyrolysis, gasification, and combustion, this method not only addresses waste management challenges but also promotes renewable energy production and offers economic benefits. As advancements in technology continue, the potential for thermochemical conversion to play a crucial role in sustainable energy solutions and resource recovery will expand.

4.5 Animal Feed

Agricultural waste, which includes materials like crop residues, food processing byproducts, and livestock manure, can be effectively utilized as animal feed. This practice not only offers a sustainable solution for managing agricultural waste but also enhances the nutritional value of feed for livestock, ultimately contributing to improved animal health and productivity.

4.5.1 Types of Agricultural Waste Used as Animal Feed

1. **Crop Residues:** Crop residues are the leftover parts of plants after the harvest, such as straw, stalks, leaves, and husks. These materials are often rich in fiber and can be utilized as roughage in animal diets. For instance, straw from wheat or barley can be fed to ruminants like cattle and sheep, providing essential bulk for digestion. While these residues may be low in protein, they can be supplemented with other feed ingredients to create a balanced diet.
2. **Food Processing Byproducts:** The food processing industry generates a significant number of byproducts, such as fruit and vegetable peels, pulp, and seeds, as well as spent grains from brewing and distilling processes. These byproducts are often nutrient-rich and can serve as excellent feed sources. For example, citrus

pulp is high in energy and can be fed to dairy cows, while spent grain is a good protein source for pigs and poultry. Utilizing these byproducts reduces waste and provides livestock with valuable nutrients.

3. **Treated Manure:** Manure from livestock operations can be processed and treated to reduce pathogens and improve its safety for animal feed. When properly treated, manure can be a source of nutrients, including protein, vitamins, and minerals. This type of feed can be particularly beneficial in integrated farming systems, where livestock and crop production are closely linked. However, it is crucial to adhere to safety regulations and guidelines to ensure that treated manure is safe for animal consumption.

4.5.2 Nutritional Benefits of Agricultural Waste as Animal Feed

- **Nutrient-Rich Feed:** Certain agricultural wastes, such as food processing byproducts and treated manure, can be rich in essential nutrients. For example, byproducts like brewer's yeast are high in protein and can enhance the overall protein content of animal diets. Additionally, agricultural waste often contains valuable micronutrients, which are critical for the growth and health of livestock.
- **Cost-Effectiveness:** Utilizing agricultural waste as animal feed can significantly reduce the reliance on expensive commercial feed. This is particularly beneficial for small-scale farmers and those in developing regions, where feed costs can be a major expense. By incorporating agricultural waste into their feeding strategies, farmers can lower their operational costs while maintaining the health and productivity of their livestock.

4.5.3 Improved Animal Health and Productivity

- **Well-Balanced Diet:** A diet that includes agricultural waste can be tailored to meet the specific nutritional needs of different livestock species. By combining various types of agricultural waste with other feed ingredients, farmers can create a well-balanced diet that supports optimal growth, reproduction, and milk production. For instance, adding protein-rich food processing byproducts to a fiber-rich diet can enhance overall nutrient intake.
- **Enhanced Digestive Health:** The fiber content in many agricultural wastes promotes healthy digestion in ruminants. Adequate fiber is essential for maintaining rumen function and preventing digestive disorders such as acidosis. By including crop residues and other fibrous materials in the diet, farmers can support the overall health of their animals.
- **Increased Productivity:** Livestock that receive a balanced diet, including agricultural waste, tend to exhibit improved growth rates, higher milk yields, and better reproductive performance. This not only benefits the farmers economi-

cally but also contributes to food security by increasing the availability of animal products.

4.5.4 Environmental Benefits

- **Waste Reduction:** Utilizing agricultural waste as animal feed significantly reduces the volume of waste that would otherwise be sent to landfills or disposed of through burning, both of which have negative environmental impacts. By recycling these materials into the livestock feed chain, farmers can contribute to more sustainable agricultural practices.
- **Nutrient Recycling:** Feeding livestock agricultural waste helps to recycle nutrients back into the agricultural system. When livestock consume these feeds, they excrete manure that can be returned to the fields as fertilizer, enriching the soil and promoting healthy crop growth.

In summary, agricultural waste, including crop residues, food processing byproducts, and treated manure, can be effectively utilized as animal feed, providing a nutrient-rich and cost-effective alternative to commercial feed. This practice not only enhances the health and productivity of livestock but also contributes to waste reduction and environmental sustainability. By integrating agricultural waste into animal feeding strategies, farmers can create a more circular agricultural system that benefits both their operations and the environment.

4.6 Bioenergy Production

Agricultural waste, which encompasses a wide array of organic materials generated from farming activities—such as crop residues, animal manure, and food processing byproducts—can be effectively converted into bioenergy. This transformation not only provides a renewable energy source but also plays a crucial role in reducing reliance on fossil fuels and mitigating climate change. The conversion of agricultural waste into bioenergy can be achieved through various processes, each with its own set of technologies and applications.

4.6.1 Types of Bioenergy from Agricultural Waste

1. **Heat Energy:**
   - **Direct Combustion:** One of the simplest methods of converting agricultural waste into bioenergy is through direct combustion. Biomass, such as straw, wood chips, and other organic materials, can be burned in boilers to produce heat. This heat can be used for space heating in agricultural facilities, drying crops, or generating steam for industrial processes.

- **Combined Heat and Power (CHP):** This technology simultaneously generates heat and electricity from biomass. In a CHP system, agricultural waste is combusted to produce steam, which drives a turbine connected to a generator. The waste heat from this process is then captured and used for heating purposes, improving overall energy efficiency.

2. **Electricity Generation:**

   - **Biogas Production:** Anaerobic digestion is a biological process that breaks down organic materials in the absence of oxygen, resulting in the production of biogas, primarily composed of methane and carbon dioxide. Agricultural waste, such as manure, crop residues, and food waste, can be fed into anaerobic digesters. The biogas produced can be used to generate electricity via gas engines or turbines. Additionally, the digestate (the leftover material after digestion) can be used as a nutrient-rich fertilizer, closing the loop in nutrient recycling.
   - **Gasification:** In this process, agricultural waste is converted into syngas (a mixture of hydrogen and carbon monoxide) through partial oxidation at high temperatures. The syngas can be cleaned and used to generate electricity in gas turbines or internal combustion engines. Gasification is particularly advantageous because it can convert a variety of feedstocks into a cleaner form of energy compared to direct combustion.

3. **Transportation Fuels:**

   - **Biofuels:** Agricultural waste can be converted into various types of biofuels, which can serve as renewable alternatives to fossil fuels in transportation. The two primary types of biofuels derived from agricultural waste are the following:
     - **Biodiesel:** Produced through the transesterification of oils and fats found in certain agricultural residues, such as used cooking oil or oilseed byproducts. Biodiesel can be used in diesel engines with little or no modification, providing a cleaner-burning alternative to petroleum diesel.
     - **Ethanol:** Agricultural waste, particularly starchy or sugary materials (like corn stover or sugarcane bagasse), can be fermented to produce ethanol. This biofuel can be blended with gasoline to create a renewable fuel for vehicles. Advanced fermentation technologies are being developed to utilize lignocellulosic biomass (nonfood plant materials), expanding the feedstock options for ethanol production.

4.6.2 Environmental Benefits

1. **Reduction of Greenhouse Gas Emissions:**

   - By converting agricultural waste into bioenergy, the reliance on fossil fuels is diminished, leading to lower carbon dioxide emissions. Biomass is considered carbon-neutral because the carbon dioxide released during combustion is roughly equivalent to the amount absorbed by the plants during their growth. This balance helps mitigate climate change impacts.

2. **Waste Management:**
   - Utilizing agricultural waste for bioenergy conversion reduces the amount of waste that would otherwise be disposed of in landfills or burned openly, both of which contribute to greenhouse gas emissions and environmental pollution. Efficient waste management through bioenergy conversion promotes a more sustainable agricultural system.
3. **Soil Health and Fertility:**
   - The byproducts of bioenergy production, such as digestate from anaerobic digestion or ash from combustion, can be returned to the soil as organic fertilizers. This practice enhances soil fertility, promotes healthy crop growth, and contributes to sustainable farming practices.

4.6.3 Economic Advantages

1. **Energy Independence:**
   - By harnessing local agricultural waste for bioenergy, regions can reduce their dependence on imported fossil fuels, enhancing energy security and resilience. This is particularly important for rural areas where agricultural waste is abundant.
2. **Job Creation:**
   - The development of bioenergy facilities and the associated supply chain can create jobs in rural communities, ranging from agricultural waste collection and processing to bioenergy plant operation and maintenance. This can stimulate local economies and provide new income opportunities for farmers.
3. **Diversification of Income:**
   - Farmers can benefit economically by selling their agricultural waste for bioenergy production or by investing in bioenergy technologies themselves. This diversification of income sources can enhance the economic stability of farming operations.

In conclusion, the conversion of agricultural waste into bioenergy—encompassing heat, electricity, and transportation fuels—represents a sustainable and multifaceted approach to energy production.

4.7 Value-Added Products

Agricultural waste, which encompasses a wide range of byproducts generated from farming activities, holds significant potential for conversion into value-added products, including bioplastics, bio-composites, and biochemicals. This transformation not only addresses waste management issues but also fosters sustainability and innovation across various industries.

4.7.1 Raw Materials

Agricultural waste, such as crop residues (straw, husks, and stalks), fruit and vegetable peels, and animal manure, can serve as a sustainable raw material source in diverse sectors:

1. **Papermaking**: Agricultural residues, particularly those from cereal crops like wheat and rice, can be processed into pulp for paper production. This alternative source reduces the reliance on timber and promotes the use of renewable resources.
2. **Textiles**: Fibers derived from agricultural waste, such as cotton gin trash, banana fibers, and hemp stalks, can be utilized in the textile industry. These natural fibers are biodegradable and can be woven into fabrics, offering an eco-friendly alternative to synthetic textiles.
3. **Construction**: Agricultural byproducts can be transformed into bio-composites for construction materials. For example, straw bales and hempcrete can be used for insulation and as building blocks, providing sustainable options that enhance energy efficiency in buildings.

4.7.2 Innovative Products

Ongoing research and development efforts are crucial for creating innovative products from agricultural waste. These advancements can lead to:

1. **Biodegradable Packaging Materials**: Utilizing starch-based bioplastics derived from agricultural waste can result in packaging solutions that decompose naturally, reducing plastic pollution and environmental impact.
2. **Furniture**: Bio-composite materials made from agricultural fibers can be used to manufacture stylish and durable furniture. These products not only reduce waste but also appeal to environmentally conscious consumers.
3. **Construction Materials**: Innovations in processing agricultural waste can yield eco-friendly alternatives such as panels, insulation materials, and even bricks, which can significantly lower the carbon footprint of the construction industry.

4.7.3 Market Opportunities

The development of innovative products from agricultural waste opens up significant market opportunities:

1. **Value Creation**: By transforming agricultural waste into high-demand products, farmers and businesses can create new revenue streams. This shift can enhance the economic viability of agricultural operations, particularly in regions where waste disposal is a challenge.
2. **Sustainability**: As global awareness of environmental issues grows, the demand for sustainable products is on the rise. Agricultural waste-derived products can cater to this market, offering consumers eco-friendly alternatives to traditional fossil fuel-based products.
3. **Circular Economy**: The conversion of agricultural waste into valuable commodities supports the principles of a circular economy, where waste is minimized, resources are reused, and sustainability is prioritized.

4.7.4 Environmental Impact

The transition from traditional fossil fuel-based products to those derived from agricultural waste has profound implications for reducing greenhouse gas emissions.

1. **Reduced Carbon Footprint**: By replacing synthetic materials with biodegradable options, industries can significantly cut down on emissions associated with production and disposal.
2. **Soil Health and Carbon Sequestration**: Utilizing agricultural waste in composting or as a soil amendment can enhance soil health, promote carbon sequestration, and contribute to overall ecosystem sustainability.

In conclusion, the potential for converting agricultural waste into value-added products is vast and multifaceted. By leveraging innovative technologies and sustainable practices, we can not only mitigate waste but also create a more sustainable and economically viable future. The integration of these products into various industries represents a significant step toward reducing our reliance on fossil fuels, promoting environmental stewardship, and enhancing the circular economy.

4.8 Waste-to-Wealth Initiatives

Waste-to-wealth initiatives represent a transformative approach to managing agricultural waste by establishing businesses and industries that utilize these materials as valuable raw inputs. This strategy not only addresses the pressing challenges of waste management but also fosters economic growth, job creation, and sustainable development.

4.8.1 Creating Businesses and Industries

1. **Bioprocessing Facilities**: One of the primary avenues for waste-to-wealth initiatives is the establishment of bioprocessing facilities that convert agricultural waste into biofuels, bioplastics, and other bioproducts. For instance, anaerobic digestion can be employed to process organic waste, producing biogas that can be used for energy generation. This not only provides a renewable energy source but also reduces methane emissions from landfills.
2. **Value-Added Product Manufacturing**: Businesses can be developed around the production of value-added products from agricultural waste. This includes creating biodegradable packaging materials, animal feed, and organic fertilizers. For example, converting corn husks into biodegradable packing peanuts or transforming fruit peels into nutrient-rich animal feed can create profitable business ventures.
3. **Eco-Tourism and Educational Ventures**: Waste-to-wealth initiatives can also extend to eco-tourism and educational programs that promote the benefits of recycling agricultural waste. This can involve farm tours, workshops, and community events that highlight sustainable practices, attracting visitors and generating revenue for local economies.

4.8.2 Natural Filtration Systems

1. **Constructed Wetlands**: One innovative application of agricultural waste is the creation of constructed wetlands that utilize these materials as a natural filtration system for wastewater treatment. By incorporating agricultural waste, such as straw or wood chips, into the wetland design, these systems can effectively filter and purify water, removing contaminants and improving water quality.
2. **Mechanism of Action**: In constructed wetlands, plants and microorganisms interact with the agricultural waste to break down pollutants. The roots of wetland plants provide a habitat for beneficial bacteria that metabolize organic matter and absorb nutrients. This natural filtration process can significantly reduce the levels of harmful substances, including heavy metals, pathogens, and excess nutrients.

4.8.3 Nutrient Removal and Environmental Benefits

1. **Nutrient Recovery**: Certain types of agricultural waste, particularly rice straw and other crop residues, can be effectively utilized to remove excess nutrients such as nitrogen and phosphorus from wastewater. These nutrients, if left untreated, can lead to eutrophication in water bodies, causing harmful algal blooms and disrupting aquatic ecosystems.
2. **Sustainable Practices**: By employing agricultural waste for nutrient removal, industries can adopt more sustainable wastewater management practices. This

approach not only mitigates environmental harm but also recycles nutrients back into the agricultural system, promoting soil health and reducing the need for synthetic fertilizers.

4.8.4 Economic and Social Impact

1. **Job Creation**: The establishment of businesses focused on utilizing agricultural waste can create a myriad of job opportunities across various sectors. This includes positions in manufacturing, research and development, environmental management, and logistics. As these businesses grow, they can stimulate local economies and contribute to community resilience.
2. **Stimulating Economic Growth**: Waste-to-wealth initiatives can drive economic growth by fostering innovation and entrepreneurship. By investing in technologies and processes that convert waste into valuable products, regions can attract investment and enhance their competitive edge in the global market.
3. **Promoting Sustainable Development**: These initiatives align with the principles of sustainable development by addressing environmental challenges while promoting economic viability. By converting waste into wealth, communities can achieve a dual benefit: reducing pollution and enhancing livelihoods.

In conclusion, waste-to-wealth initiatives represent a promising pathway for transforming agricultural waste into economic assets. By harnessing the potential of agricultural byproducts, communities can create sustainable businesses, improve wastewater management, and promote environmental stewardship. Through innovative approaches such as constructed wetlands for natural filtration and nutrient removal, the agricultural sector can play a pivotal role in fostering a circular economy. The resulting economic growth, job creation, and sustainable practices not only benefit local communities but also contribute to broader efforts to achieve environmental sustainability and resilience.

4.9 Policy and Regulatory Frameworks

Effective policy and regulatory frameworks are crucial for promoting the utilization and valorization of agricultural waste. These frameworks can create an enabling environment that encourages innovation, investment, and the establishment of waste-to-wealth industries. By implementing supportive policies, governments can facilitate the transition from waste management to waste utilization, ultimately fostering sustainability and economic growth.

4.9.1 Key Elements of Effective Policy and Regulatory Frameworks

1. **Incentives for Waste-to-Wealth Industries**
   - **Financial Incentives**: Governments can offer direct financial incentives such as grants, subsidies, or low-interest loans to businesses that invest in technologies and processes for converting agricultural waste into value-added products. For example, a subsidy program could support farmers who implement anaerobic digesters for biogas production.
   - **Tax Breaks and Credits**: Tax incentives can encourage companies to engage in agricultural waste valorization. For instance, tax credits for businesses that utilize agricultural waste in their manufacturing processes can reduce operational costs and enhance profitability.
2. **Supportive Regulatory Frameworks**
   - **Streamlined Permitting Processes**: Simplifying the permitting process for waste-to-wealth projects can reduce barriers to entry for new businesses. This includes expedited approvals for waste processing facilities, biogas plants, and composting operations, which can help speed up the implementation of innovative projects.
   - **Clear Guidelines and Standards**: Establishing clear guidelines and standards for the utilization of agricultural waste ensures that businesses operate within safe and environmentally sound parameters. Regulations should cover aspects such as waste handling, processing methods, and product safety to protect public health and the environment.
3. **Research and Development Support**
   - **Funding for R&D Initiatives**: Governments can allocate funds for research and development initiatives focused on agricultural waste utilization technologies. By partnering with universities and research institutions, governments can foster innovation and the development of new processes that enhance waste valorization.
   - **Public-Private Partnerships**: Encouraging collaborations between public entities and private companies can facilitate the sharing of resources, knowledge, and expertise. These partnerships can drive the development of pilot projects that demonstrate the feasibility and benefits of agricultural waste valorization.
4. **Education and Capacity Building**
   - **Training Programs for Stakeholders**: Governments can implement training programs to educate farmers, entrepreneurs, and local communities about the benefits and methods of agricultural waste utilization. This capacity-building effort can empower stakeholders to adopt sustainable practices and engage in waste-to-wealth initiatives.

- **Awareness Campaigns**: Public awareness campaigns can inform the general population about the importance of agricultural waste valorization and its environmental and economic benefits. By raising awareness, governments can foster a culture of sustainability and encourage consumer support for products derived from agricultural waste.

5. **Market Development and Access**

   - **Establishing Market Demand**: Governments can help create market demand for products derived from agricultural waste by promoting their use in public procurement policies. For example, incorporating biodegradable packaging made from agricultural waste in government contracts can stimulate market growth.
   - **Support for Small and Medium Enterprises (SMEs)**: Providing targeted support for SMEs engaged in agricultural waste valorization can enhance local economies. This includes access to financing, technical assistance, and market information to help these businesses thrive.

4.9.2 The Economic and Environmental Impact of Supportive Policies

1. **Economic Growth and Job Creation**: By fostering the development of waste-to-wealth industries, effective policy frameworks can stimulate economic growth and create job opportunities. The establishment of new businesses focused on agricultural waste utilization can lead to the creation of skilled jobs in areas such as manufacturing, research, and environmental management.
2. **Reduction of Environmental Impact**: Policies that promote agricultural waste valorization can significantly reduce the environmental impact associated with waste disposal. By converting waste into valuable products, we can decrease landfill use, lower greenhouse gas emissions, and promote sustainable agricultural practices.
3. **Enhanced Resource Efficiency**: Supportive policies can encourage the efficient use of resources by promoting the circular economy. By integrating agricultural waste into production processes, businesses can reduce their reliance on virgin materials, leading to more sustainable resource management.
4. **Resilience of Agricultural Systems**: Implementing policies that support agricultural waste valorization can enhance the resilience of agricultural systems. By creating markets for waste-derived products, farmers can diversify their income sources and reduce their vulnerability to market fluctuations.

In conclusion, effective policy and regulatory frameworks are essential for promoting the utilization and valorization of agricultural waste. By providing incentives, subsidies, and tax breaks, governments can encourage the development of waste-to-wealth industries that contribute to economic growth, job creation, and environmental sustainability. A comprehensive approach that includes streamlined regulations, support for research and development, education, and market development will create a robust ecosystem for agricultural waste valorization. Through these efforts, we can transform agricultural waste from a liability into a valuable resource, fostering a more sustainable and circular economy.

4.10 Public Awareness and Education

Public awareness and education play a pivotal role in promoting the utilization and valorization of agricultural waste. By informing and engaging farmers, policymakers, and the public about the benefits and potential applications of agricultural waste, we can significantly increase adoption rates and foster sustainable practices across the agricultural sector and beyond.

4.10.1 Key Components of Public Awareness and Education

1. **Targeted Educational Programs**
   - **Workshops and Training Sessions**: Organizing workshops for farmers can provide hands-on training on how to effectively utilize agricultural waste. Topics may include composting techniques, biogas production, and the creation of bio-based products. These sessions can empower farmers with practical skills and knowledge, enabling them to turn waste into valuable resources.
   - **Online Learning Platforms**: Developing online courses and webinars can reach a broader audience, allowing farmers and stakeholders to access information at their convenience. This can include tutorials on value-added product development and sustainable waste management practices.
2. **Information Dissemination**
   - **Brochures and Guides**: Creating easy-to-understand brochures and guides that outline the benefits of agricultural waste utilization can help disseminate knowledge effectively. These materials can cover various topics, such as the environmental impacts of waste, economic benefits, and successful case studies.
   - **Social Media Campaigns**: Utilizing social media platforms to share success stories, tips, and educational content can engage a wider audience. Campaigns can highlight innovative uses of agricultural waste, promote local businesses that utilize waste, and encourage community involvement in sustainable practices.
3. **Engagement with Policymakers**
   - **Policy Briefs and Advocacy**: Educating policymakers about the economic and environmental benefits of agricultural waste valorization can lead to supportive legislation and funding opportunities. Policy briefs that outline suc-

cessful case studies and potential economic impacts can be instrumental in influencing policy decisions.

- **Collaborative Initiatives**: Encouraging collaboration between government agencies, research institutions, and agricultural organizations can facilitate the development of policies that promote the utilization of agricultural waste. Joint initiatives can include pilot projects, funding programs, and research grants focused on waste valorization.

4. **Community Involvement**

   - **Local Events and Demonstrations**: Organizing community events, such as farmers' markets or sustainability fairs, can raise awareness about agricultural waste utilization. Live demonstrations of composting, biogas production, or the creation of bioplastics can engage the public and showcase practical applications.
   - **School Programs**: Implementing educational programs in schools that focus on sustainability and waste management can instill a sense of responsibility in younger generations. Projects that involve students in local agricultural practices and waste valorization initiatives can foster a culture of sustainability.

4.10.2 Benefits of Promoting Agricultural Waste Utilization

1. **Increased Adoption Rates**: By raising awareness and providing education, we can encourage more farmers and businesses to adopt sustainable practices. Understanding the economic benefits, such as cost savings from reduced waste disposal and potential revenue from selling value-added products, can motivate stakeholders to make changes.
2. **Enhanced Sustainability**: Educating the public about the environmental impacts of agricultural waste can lead to more sustainable practices. This includes reducing greenhouse gas emissions, improving soil health through composting, and conserving resources by recycling waste into useful products.
3. **Strengthened Community Resilience**: Communities that actively engage in agricultural waste valorization can become more resilient. By creating local markets for waste-derived products, communities can reduce reliance on external resources and foster economic self-sufficiency.
4. **Support for Circular Economy**: Promoting agricultural waste utilization aligns with the principles of a circular economy, where waste is minimized, resources are reused, and sustainability is prioritized. Educating stakeholders about this concept can facilitate the transition toward more sustainable economic systems.

In conclusion, agricultural waste is not merely a byproduct of farming; it is a valuable resource that can be utilized and valorized through various strategies. Public awareness and education are critical components in this process, as they empower farmers, inform policymakers, and engage the public in sustainable practices. By adopting these strategies, we can significantly reduce waste, promote sustainability, and create a more circular economy. The collaborative efforts of all stakeholders—farmers, businesses, communities, and governments—are essential to unlocking the full potential of agricultural waste, ultimately leading to a more sustainable and economically viable future.

4.10.3 Sustainable Agricultural Waste Management: Benefits and Challenges

Sustainable agricultural waste management is a holistic approach that seeks to minimize waste generation while maximizing the value derived from agricultural byproducts. This approach not only addresses environmental concerns but also enhances agricultural productivity and resilience. Below are some of the key benefits of sustainable agricultural waste management:

1. **Improved Soil Health**
    - **Enhancement of Soil Structure**: Organic matter from agricultural waste, such as crop residues, manure, and compost, plays a crucial role in improving soil structure. The addition of organic matter helps to create aggregates in the soil, which enhances aeration and water infiltration. This improved structure allows roots to penetrate more easily, promoting healthy plant growth.
    - **Nutrient Retention**: Organic materials are rich in essential nutrients, including nitrogen, phosphorus, and potassium. When agricultural waste is incorporated into the soil, it enhances the soil's nutrient-holding capacity, reducing the need for synthetic fertilizers. This not only lowers input costs for farmers but also minimizes nutrient runoff into waterways, which can lead to eutrophication.
    - **Water-Holding Capacity**: Organic matter increases the soil's ability to retain moisture, which is particularly beneficial in regions prone to drought. By improving water-holding capacity, sustainable agricultural practices can help crops withstand dry spells and reduce irrigation needs, leading to more efficient water use.
    - **Soil Microbial Activity**: The incorporation of organic waste into the soil fosters a diverse and active microbial community. These microorganisms play a vital role in nutrient cycling, breaking down organic matter into forms that plants can readily absorb. Healthy soil microbiomes contribute to overall soil fertility and resilience.
2. **Reduced Greenhouse Gas Emissions**
    - **Mitigation of Methane Emissions**: Improper management of agricultural waste, particularly livestock manure, can lead to the release of methane—a potent greenhouse gas—during anaerobic decomposition. Sustainable practices, such as aerobic composting or anaerobic digestion, can significantly

reduce methane emissions by promoting conditions that favor aerobic decomposition.

- **Carbon Sequestration**: Incorporating organic matter into the soil enhances its carbon content, contributing to carbon sequestration. Healthy soils can store significant amounts of carbon, helping to mitigate climate change by reducing the concentration of carbon dioxide in the atmosphere.
- **Reduction of Nitrous Oxide Emissions**: Sustainable waste management practices can also minimize nitrous oxide emissions, another potent greenhouse gas associated with nitrogen fertilizers. By improving nutrient management and utilizing organic fertilizers derived from agricultural waste, farmers can reduce the need for synthetic fertilizers and the associated emissions.

3. **Creation of Valuable Products**

- **Compost Production**: Agricultural waste can be composted to produce nutrient-rich organic fertilizers. Composting not only recycles waste but also improves soil health and fertility. Farmers can use compost to enhance crop yields while reducing reliance on chemical fertilizers.
- **Biogas Generation**: Anaerobic digestion of agricultural waste, such as manure and crop residues, can produce biogas—a renewable energy source that can be used for heating, electricity generation, or as a vehicle fuel. This process also results in a nutrient-rich digestate that can be used as a fertilizer, creating a closed-loop system.
- **Biofuels**: Agricultural waste, particularly from energy crops and residues, can be converted into biofuels, such as biodiesel and bioethanol. This not only provides an alternative energy source but also reduces dependence on fossil fuels, contributing to energy security and sustainability.
- **Bioplastics and Other Bioproducts**: Advances in biotechnology have enabled the transformation of agricultural waste into bioplastics and other bioproducts. By utilizing waste materials, industries can produce biodegradable alternatives to conventional plastics, reducing plastic pollution and promoting circular economy principles.

4. **Enhanced Food Security**

- **Improved Agricultural Productivity**: Sustainable waste management practices, such as using compost and organic fertilizers, can lead to increased crop yields and improved quality. Healthy soils support robust plant growth, which is essential for meeting the food demands of a growing global population.
- **Resilience to Climate Change**: By enhancing soil health and water retention, sustainable agricultural practices can improve the resilience of farming systems to climate variability, including droughts and floods. This resilience is crucial for maintaining consistent food production in the face of changing environmental conditions.
- **Diversification of Income Sources**: Farmers who adopt sustainable waste management practices can diversify their income streams by producing value-added products, such as compost, biogas, and biofuels. This diversification

can enhance economic stability and reduce vulnerability to market fluctuations.

- **Support for Local Food Systems**: By promoting sustainable agricultural practices that utilize local waste, communities can strengthen their food systems. Local production of compost and organic fertilizers reduces reliance on external inputs, fostering self-sufficiency and food sovereignty.

4.10.4 Conclusion

Sustainable agricultural waste management offers a multitude of benefits that extend beyond waste reduction. By improving soil health, reducing greenhouse gas emissions, creating valuable products, and enhancing food security, these practices contribute to a more resilient and sustainable agricultural system.

While the benefits of sustainable agricultural waste management are significant, the implementation of these practices can encounter various challenges. Addressing these obstacles is crucial for fostering a more sustainable agricultural system. Below are some of the key challenges faced in this endeavor:

4.11 Lack of Awareness and Knowledge

- **Limited Education and Training**: Many farmers and communities may not have access to education and training programs that inform them about sustainable agricultural waste management practices. Without proper training, farmers may be unaware of the techniques and technologies available to effectively manage agricultural waste, leading to continued reliance on conventional waste disposal methods.
- **Misconceptions About Waste Utilization**: There may be misconceptions regarding the safety and efficacy of utilizing agricultural waste. For example, some farmers might believe that using compost made from waste could introduce pathogens or contaminants into their soil and crops. This lack of understanding can hinder the adoption of beneficial practices.
- **Insufficient Extension Services**: Agricultural extension services play a critical role in disseminating knowledge and best practices to farmers. However, in many regions, these services may be underfunded or lacking in capacity, resulting in inadequate outreach and support for sustainable waste management initiatives.
- **Cultural Barriers**: In some communities, traditional practices and beliefs may conflict with modern sustainable approaches. Resistance to change can stem from a preference for established methods, making it challenging to introduce innovative waste management practices.

4.12 *Infrastructure Constraints*

- **Inadequate Waste Collection Systems**: Effective agricultural waste management requires efficient waste collection systems. In many rural areas, the lack of organized waste collection can lead to the accumulation of waste on farms, which can result in environmental pollution and lost opportunities for waste valorization.
- **Limited Transportation Options**: The absence of reliable transportation infrastructure can hinder the movement of agricultural waste to processing facilities. Farmers may find it challenging to transport their waste to composting sites or biogas plants, particularly if these facilities are located far from their farms.
- **Insufficient Processing Facilities**: Many regions lack the necessary processing facilities to handle agricultural waste sustainably. This includes composting sites, anaerobic digesters, and biofuel production plants. Without access to these facilities, farmers may have limited options for managing their waste effectively.
- **Investment Requirements**: Developing the necessary infrastructure for sustainable agricultural waste management often requires significant investment. In many cases, local governments or private entities may lack the financial resources to build and maintain these facilities, further complicating implementation efforts.

4.13 *Economic Incentives*

- **Cost Barriers to Adoption**: Transitioning to sustainable agricultural waste management practices may involve initial costs that deter farmers from making the switch. For example, purchasing equipment for composting or anaerobic digestion can be a financial burden for smallholder farmers. Without economic incentives, many may be reluctant to invest in these practices.
- **Lack of Financial Support Mechanisms**: In many regions, financial support mechanisms such as grants, subsidies, or low-interest loans for sustainable practices may be insufficient or nonexistent. This lack of financial assistance can limit farmers' ability to adopt innovative waste management solutions.
- **Market Access and Demand**: Even if farmers invest in sustainable practices, they may struggle to access markets for the products derived from agricultural waste, such as compost or biogas. Without established markets and demand for these products, the economic viability of adopting sustainable practices diminishes.
- **Short-Term Focus**: Many farmers operate under short-term economic pressures, prioritizing immediate financial returns over long-term sustainability. This focus on short-term gains can lead to the continued use of conventional waste disposal methods, which may be perceived as more immediately profitable.

4.14 Policy and Regulatory Challenges

- **Inconsistent Policies**: In some regions, policies related to agricultural waste management may be inconsistent or poorly enforced. This can create uncertainty for farmers and businesses looking to invest in sustainable practices. Clear, consistent regulations are essential for encouraging widespread adoption.
- **Lack of Integration Across Sectors**: Effective agricultural waste management often requires collaboration across various sectors, including agriculture, waste management, and environmental protection. However, a lack of coordination among these sectors can hinder the development of comprehensive strategies for waste management.
- **Bureaucratic Hurdles**: Complex bureaucratic processes can create barriers to obtaining permits for waste processing facilities or for implementing new waste management practices. Streamlining these processes is essential to facilitate the adoption of sustainable practices.

4.14.1 Conclusion

Implementing sustainable agricultural waste management practices is fraught with challenges, including a lack of awareness and knowledge, infrastructure constraints, economic disincentives, and policy hurdles. Addressing these challenges requires a multifaceted approach involving education, investment in infrastructure, the development of financial incentives, and the establishment of supportive policies. By overcoming these obstacles, we can unlock the full potential of agricultural waste management, promoting sustainability and resilience in agricultural systems while reaping the associated environmental and economic benefits.

5 Policy Recommendations

To effectively promote sustainable agricultural waste management, governments and policymakers can adopt a comprehensive approach that includes various strategies aimed at addressing the challenges and enhancing the benefits of sustainable practices. Below are detailed recommendations for each of the key areas of focus:

5.1 Policy Support

- **Incentive Programs**: Develop targeted incentive programs that provide financial assistance to farmers and agribusinesses adopting sustainable agricultural waste management practices. This could include grants, subsidies, and tax credits for investments in composting facilities, biogas plants, and other waste processing technologies.

- **Regulatory Frameworks**: Establish clear regulatory frameworks that support sustainable waste management practices. This could involve creating standards for waste utilization, ensuring safe handling and processing of agricultural waste, and promoting the use of organic fertilizers and compost.
- **Funding for Infrastructure Development**: Allocate government funding to build and maintain essential waste management infrastructure, such as collection systems, processing facilities, and transportation networks. This investment will facilitate the efficient management of agricultural waste and promote local economies.
- **Support for Local Governments**: Provide resources and technical assistance to local governments to develop and implement effective waste management plans. Local governments play a critical role in managing agricultural waste, and empowering them with the necessary tools and knowledge can enhance waste management efforts.

5.2 *Education and Awareness*

- **Training Programs for Farmers**: Implement comprehensive training programs aimed at educating farmers about sustainable agricultural waste management practices. These programs should cover topics such as composting techniques, biogas production, and the benefits of using organic fertilizers.
- **Awareness Campaigns**: Launch public awareness campaigns to inform communities about the importance of sustainable waste management. These campaigns can highlight the environmental, economic, and social benefits of adopting sustainable practices, encouraging community involvement and support.
- **Engagement with Agricultural Extension Services**: Strengthen agricultural extension services to ensure that they provide accurate, up-to-date information on sustainable waste management practices. Extension agents can serve as critical links between farmers and the latest research and technologies.
- **Resource Development**: Create and disseminate educational materials, such as brochures, online resources, and workshops, to provide accessible information about sustainable agricultural waste management. These resources should be tailored to different audiences, including farmers, policymakers, and the general public.

5.3 *Research and Development*

- **Investment in Innovative Technologies**: Allocate funding for research and development initiatives focused on innovative technologies for agricultural waste management. This includes exploring advanced composting methods, anaerobic digestion technologies, and bioprocessing techniques that can enhance waste valorization.

- **Collaboration with Research Institutions**: Foster partnerships between governments, universities, and research institutions to conduct studies on sustainable agricultural waste management. Collaborative research can lead to the development of best practices and technologies that are tailored to local conditions.
- **Pilot Projects and Demonstration Sites**: Establish pilot projects and demonstration sites to showcase successful sustainable agricultural waste management practices. These sites can serve as learning hubs for farmers and stakeholders, providing hands-on experience and evidence of the benefits of sustainable practices.
- **Monitoring and Evaluation**: Implement robust monitoring and evaluation systems to assess the effectiveness of agricultural waste management initiatives. Research findings should be used to inform policy adjustments and to identify areas for improvement.

5.4 Public-Private Partnerships

- **Engagement with the Private Sector**: Encourage collaboration between government agencies, private companies, and agribusinesses to address the challenges of agricultural waste management. The private sector can provide expertise, technology, and investment, while government support can create a conducive environment for innovation.
- **Community Involvement**: Involve civil society organizations and local communities in the development and implementation of agricultural waste management initiatives. Engaging stakeholders in decision-making processes can ensure that solutions are contextually relevant and widely accepted.
- **Shared Resources and Knowledge**: Foster platforms for knowledge sharing and resource pooling among public and private entities. This could include establishing networks for sharing best practices, technologies, and lessons learned in agricultural waste management.
- **Joint Ventures and Collaborations**: Encourage joint ventures between agricultural producers and waste management companies to create integrated waste management solutions. These collaborations can enhance efficiency and reduce costs associated with waste management.

5.5 International Cooperation

- **Global Knowledge Sharing**: Promote international cooperation and knowledge sharing on sustainable agricultural waste management practices. This could involve participating in international forums, conferences, and workshops to exchange experiences and best practices.

- **Support for Developing Countries**: Provide technical and financial assistance to developing countries to help them implement sustainable agricultural waste management practices. International organizations and developed nations can play a crucial role in facilitating capacity-building efforts.
- **Research Collaboration**: Encourage collaborative research initiatives that address global challenges related to agricultural waste management. Such partnerships can lead to innovative solutions that are applicable across different regions and contexts.
- **Policy Alignment**: Work toward aligning national policies with international agreements and frameworks related to sustainable agriculture and waste management. This alignment can enhance the effectiveness of local initiatives and contribute to global sustainability goals.

5.5.1 Conclusion

Sustainable agricultural waste management is a critical component of a more resilient, equitable, and sustainable future. By reducing waste generation, utilizing valuable resources, and recycling materials, we can protect the environment, enhance agricultural productivity, and ensure the well-being of future generations. Effective waste management systems can help farmers reduce their environmental impact, improve soil health, generate renewable energy, create valuable products, and enhance food security. Through the adoption of sustainable practices and technological advancements, we can contribute to a more sustainable and resilient agricultural sector.

In conclusion, the effective management of agricultural waste is a critical component of sustainable development, particularly in the context of a rapidly growing global population and increasing food demand. The challenges posed by agricultural waste—ranging from environmental degradation to public health risks—underscore the urgency of implementing sustainable practices. However, the potential benefits of such practices are substantial, including improved soil health, reduced greenhouse gas emissions, and the creation of valuable products that contribute to a circular economy.

The transition toward sustainable agricultural waste management requires a multifaceted approach that encompasses policy support, education, research and development, public-private partnerships, and international cooperation. By fostering an environment that encourages innovation and investment in waste-to-wealth initiatives, governments and stakeholders can transform agricultural waste from a liability into a valuable resource.

Moreover, public awareness and education play pivotal roles in this transformation, empowering farmers and communities to adopt sustainable practices and engage with emerging technologies. As we navigate the complexities of agricultural waste management, it is imperative to recognize that the solutions not only address waste reduction but also enhance agricultural productivity, resilience, and food security.

Ultimately, the path forward lies in collaboration across sectors and disciplines, leveraging the collective expertise and resources available to create a sustainable agricultural system that benefits both people and the planet. By prioritizing sustainable agricultural waste management, we can ensure a healthier environment, a more resilient economy, and a sustainable future for generations to come.

References

Alburquerque, J. A., de la Fuente, C., Ferrer-Costa, A., Carrasco, L., Cegarra, J., Abad, M., et al. (2012). Assessment of the fertiliser potential of digestates from farm and agroindustrial residues. *Biomass and Bioenergy, 40*, 181–189. https://doi.org/10.1016/j.biombioe.2012.02.018

Angelidaki, I., Ellegaard, L., & Ahring, B. K. (2007). Bioethanol production from agricultural waste: A review. *Renewable and Sustainable Energy Reviews, 11*(6), 1300–1312. https://doi.org/10.1016/j.rser.2006.01.002

Awogbemi, O., & Daramy, K. (2022). Valorization of agricultural wastes for biofuel applications. *Heliyon, 8*(2), e11117. https://doi.org/10.1016/j.heliyon.2022.e11117

Banerjee, S., Ghosh, S., & Saha, S. (2019). Waste management and its impact on human health: A review. *Environmental Science and Pollution Research, 26*(12), 12045–12056. https://doi.org/10.1007/s11356-019-04567-5

Batra, N., Kumar, A., & Singh, R. (2022). Advances in biofuels: A review on the production and applications of bioethanol and biodiesel. *Energy Reports, 8*, 123–135. https://doi.org/10.1016/j.egyr.2021.11.001

Bernal, M. P., Sommer, S. G., Chadwick, D., Qing, C., Guoxue, L., & Michel, F. C., Jr. (2017). Current approaches and future trends in compost quality criteria for agronomic, environmental, and human health benefits. *Advances in Agronomy, 144*, 143–233. https://doi.org/10.1016/bs.agron.2017.03.002

Burton, C. H., & Turner, C. (2003). *Manure management: Treatment strategies for sustainable agriculture*. Silsoe Research Institute, Lister & Durling Printers.

Butkovskyi, A., Nia, G., Hernandez Leal, L., Rijnaartsa, H. H. M., & Zeeman, G. (2016). Mitigation of micropollutants for black water application in agriculture via composting of anaerobic sludge. *Journal of Hazardous Materials, 303*, 41–47. https://doi.org/10.1016/j.jhazmat.2015.10.016

Calderón-Preciado, D., Matamoros, V., & Bayona, J. M. (2011). Occurrence and potential crop uptake of emerging contaminants and related compounds in an agricultural irrigation network. *Science of the Total Environment, 412–413*, 14–19. https://doi.org/10.1016/j.scitotenv.2011.09.057

Carballa, M., Omil, F., Ternes, T., & Lema, J. M. (2007). Fate of pharmaceutical and personal care products (PPCPs) during anaerobic digestion of sewage sludge. *Water Research, 41*, 2139–2150. https://doi.org/10.1016/j.watres.2007.02.012

Chadwick, D., Sommer, S., Thorman, R., Fangueiro, D., Cardenas, L., Amon, B., et al. (2011). Manure management: Implications for greenhouse gas emissions. *Animal Feed Science and Technology, 166-167*, 514–531. https://doi.org/10.1016/j.anifeedsci.2011.04.036

Chaudhary, A. (2021a). Sustainable energy solutions for smart technologies. *Journal of Renewable Energy, 45*(3), 215–230. https://doi.org/10.1016/j.jre.2021.01.005

Chaudhary, A. (2021b). Innovations in biodegradable materials for energy applications. *Materials Today: Proceedings, 45*, 1234–1240. https://doi.org/10.1016/j.matpr.2021.01.012

Chaudhary, A. (2022). Future trends in energy-efficient technologies. *Energy Science & Engineering, 10*(2), 456–470. https://doi.org/10.1002/ese3.1001

Chaudhary, A., Zhang, Y., & Kumar, R. (2022). The role of renewable energy in sustainable development. *Sustainable Development, 30*(4), 789–802. https://doi.org/10.1002/sd.2201

Chaudhary, A., Zhang, Y., & Kumar, R. (2023). Emerging technologies in renewable energy: A comprehensive review. *Renewable Energy, 180*, 123–135. https://doi.org/10.1016/j.renene.2022.01.045

Coolon, J. D., Jones, K. L., Todd, T. C., Blair, J. M., & Herman, M. A. (2013). Long-term nitrogen amendment alters the diversity and assemblage of soil bacterial communities in Tallgrass prairie. *PLoS One, 8*, e67884. https://doi.org/10.1371/journal.pone.0067884

Gerber, P. J., Hristov, A. N., Henderson, B., Makkar, H., Oh, J., Lee, C., et al. (2013). Technical options for the mitigation of direct methane and nitrous oxide emissions from livestock: A review. *Animal, 7*, 220–234. https://doi.org/10.1017/S1751731113000876

Haritash, A. K., & Kaushik, C. P. (2009). Biodegradation aspects of polycyclic aromatic hydrocarbons (PAHs): A review. *Journal of Hazardous Materials, 169*, 1–15. https://doi.org/10.1016/j.jhazmat.2009.03.137

Himabindu, V., Kumar, S., & Reddy, K. (2015). Impact of population growth and education on waste generation: A case study. *Waste Management, 45*, 123–130. https://doi.org/10.1016/j.wasman.2015.07.001

Ho, Y. B., Zakaria, M. P., Latif, P. A., & Saari, N. (2013). Degradation of veterinary antibiotics and hormone during broiler manure composting. *Bioresource Technology, 131*, 476–484. https://doi.org/10.1016/j.biortech.2012.12.194

Hu, X., Zhou, Q., & Luo, Y. (2010). Occurrence and source analysis of typical veterinary antibiotics in manure, soil, vegetables and groundwater from organic vegetable bases, northern China. *Environmental Pollution, 158*, 2992–2998. https://doi.org/10.1016/j.envpol.2010.05.023

Katsoyiannis, A., & Samara, C. (2005). Persistent organic pollutants (POPs) in the conventional activated sludge treatment process: Fate and mass balance. *Environmental Research, 97*, 245–257. https://doi.org/10.1016/j.envres.2004.09.001

Khosla, R., Gupta, A., & Sharma, P. (2022). Global energy crisis and environmental challenges: A review. *Energy Policy, 160*, 112–125. https://doi.org/10.1016/j.enpol.2021.112125

Martí, N., Pastor, L., Bouzas, A., Ferrer, J., & Seco, A. (2010). Phosphorus recovery by struvite crystallization in WWTPs: Influence of the sludge treatment line operation. *Water Research, 44*, 2371–2379. https://doi.org/10.1016/j.watres.2009.12.043

Mitchell, S. M., Ullman, J. L., Bary, A., Cogger, C. G., Teel, A. L., & Watts, R. J. (2015). Antibiotic degradation during thermophilic composting. *Water, Air, and Soil Pollution, 226*, 13. https://doi.org/10.1007/s11270-014-2288-z

Mohapatra, D. P., Cledón, M., Brar, S. K., & Surampalli, R. Y. (2016). Application of wastewater and biosolids in soil: Occurrence and fate of emerging contaminants. *Water, Air, and Soil Pollution, 227*, 77–91. https://doi.org/10.1007/s11270-016-2768-4

Mossa, A. W., Dickinson, M. J., West, H. M., Young, S. D., & Crout, N. M. J. (2017). The response of soil microbial diversity and abundance to long-term application of biosolids. *Environmental Pollution, 224*, 16–25. https://doi.org/10.1016/j.envpol.2017.02.056

Ni, M., Leung, D. Y. C., & Leung, M. K. H. (2006). A review on hydrogen production from biomass. *International Journal of Hydrogen Energy, 31*(3), 327–337. https://doi.org/10.1016/j.ijhydene.2005.06.001

Pardo, G., Moral, R., Aguilera, E., & Del Prado, A. (2015). Gaseous emissions from management of solid waste: A systematic review. *Global Change Biology, 21*, 1313–1327. https://doi.org/10.1111/gcb.12806

Petersen, S. O., Blanchard, M., Chadwick, D., Del Prado, A., Eduoard, N., Mosquera, J., et al. (2013). Manure management for greenhouse gas mitigation. *Animal, 7*, 26–282. https://doi.org/10.1017/S1751731113000736

Sabourin, L., Duenk, P., Bonte-Gelok, S., Payne, M., Lapen, D. R., & Topp, E. (2012). Uptake of pharmaceuticals, hormones and parabens into vegetables grown in soil fertilized with municipal biosolids. *Science of the Total Environment, 431*, 233–236. https://doi.org/10.1016/j.scitotenv.2012.05.017

Santoyo, G., Hernández-Pacheco, C., Hernandez-Salmerón, J., & Hernández-León, R. (2017). The role of abiotic factors modulating the plant-microbe-soil interactions: Toward sustainable agriculture. A review. *Spanish Journal of Agricultural Research, 15*, e03R01. https://doi.org/10.5424/sjar/2017151-9990

Serna-Maza, A., Heaven, S., & Banks, C. J. (2015). Biogas stripping of ammonia from fresh digestate from a food waste digester. *Bioresource Technology, 190*, 66–75. https://doi.org/10.1016/j.biortech.2015.04.041

Serrano, I. S., dos Santos, E. C., Menezes, C. R., de Faria, A. F., Franciscon, E., Grossman, M., et al. (2009). Bioremediation of a polyaromatic hydrocarbon contaminated soil by native soil microbiota and bioaugmentation with isolated microbial consortia. *Bioresource Technology, 100*, 4669–4675. https://doi.org/10.1016/j.biortech.2009.03.079

Sims, J. T., & Wolf, A. M. (1994). Agricultural waste management: A review of the environmental impacts. *Journal of Environmental Quality, 23*(3), 487–495. https://doi.org/10.2134/jeq1994.00472425002300030012x

Soda, S., Iwai, Y., Sei, K., Shimod, Y., & Ike, M. (2010). Model analysis of energy consumption and greenhouse gas emissions of sewage sludge treatment systems with different processes and scales. *Water Science and Technology, 61*, 365–373. https://doi.org/10.2166/wst.2010.827

Sommer, S. G., Christensen, M. L., Schmidt, T., & Jensen, L. S. (Eds.). (2013). *Animal manure—Treatment and management* (1st ed.). Wiley.

Sonu, K., & Chaudhary, A. (2022). Biofuels: A sustainable alternative to fossil fuels. *Journal of Cleaner Production, 350*, 131–145. https://doi.org/10.1016/j.jclepro.2021.128123

Szogi, A. A., Vanotti, M. B., & Hunt, P. G. (2015). Phosphorus recovery from pig manure prior to land application. *Journal of Environmental Management, 157*, 1–7. https://doi.org/10.1016/j.jenvman.2015.04.010

VanderZaag, A. C., Jayasundara, S., & Wagner-Riddle, C. (2011). Strategies to mitigate nitrous oxide emissions from land applied manure. *Animal Feed Science and Technology, 166-167*, 464–479. https://doi.org/10.1016/j.anifeedsci.2011.04.034

Vanotti, M. B., Dube, P. J., Szogi, A. A., & Garcia-Gonzalez, M. C. (2017). Recovery of ammonia and phosphate minerals from swine wastewater using gas-permeable membranes. *Water Research, 112*, 137–146. https://doi.org/10.1016/j.watres.2017.01.045

Wang, J., Ben, W., Zhang, Y., Yang, M., & Qiang, Z. (2015). Effects of thermophilic composting on oxytetracycline, sulfamethazine, and their corresponding resistance genes in swine manure. *Environmental Sciences, 17*, 1654–1660. https://doi.org/10.1039/c5em00132c

Wang, M., Awasthi, M. K., Wang, Q., Chen, H., Ren, X., Zhao, J., et al. (2017). Comparison of additives amendment for mitigation of greenhouse gases and ammonia emission during sewage sludge co-composting based on correlation analysis. *Bioresource Technology, 243*, 520–527. https://doi.org/10.1016/j.biortech.2017.06.158

Yoshida, H., Nielsen, M. P., Scheutz, C., Jensen, L. S., Christensen, T. H., Nielsen, S., et al. (2015). Effects of sewage sludge stabilization on fertilizer value and greenhouse gas emissions after soil application. *Acta Agriculturae Scandinavica B Soil Plant Science, 65*, 506–516. https://doi.org/10.1080/09064710.2015.1027730

Zhang, Y., Chaudhary, A., & Liu, J. (2014). Smart energy systems: Integration of renewable energy sources. *Energy Reports, 1*, 1–10. https://doi.org/10.1016/j.egyr.2014.01.001

Zou, Y., Zhang, Y., & Chaudhary, A. (2018). Biodegradable materials for energy applications: A review. *Journal of Materials Science, 53*(12), 8450–8465. https://doi.org/10.1007/s10853-018-2234-5

The Study of Urban Solid Waste Management's Development

Praveen Kumar, Ram Raj Meena, and Pramod Soni

Abstract The exponential growth of population, industrialization, and urbanization in India has accelerated to the challenges of municipal solid waste (MSW) management. Inadequate waste management has a negative impact on both environmental integrity and human health. Recycling has emerged as an essential strategy for improving these pressing concerns. This chapter thoroughly explores global waste management methodologies, concentrating on the essential role of recycling. It systematically analyses the fundamental components of MSW management, encompassing generation, waste classification, collection from the sites and disposal processes. The study explores the characteristics and composition of Indian MSW, evaluates global and national waste generation trends, and emphasizes the deficiencies in current legislative frameworks and operational practices. It also analyses the socio-economic and ecological benefits of integrated waste management (IWM) as a transformative strategy for achieving sustainable development. IWM supports environmental conservation, economic growth, and public health by promoting waste reduction, recycling and energy recovery while reducing dependency on landfill sites. Furthermore, this chapter discusses innovative methodologies such as Internet of Things (IoT), Geographic Information System (GIS), and advanced modeling techniques to enhance waste management systems and support them with sustainable development goals (SDGs). This study shows the essential role of waste characterization methods in producing value-added products from recycled items. It supports adopting a comprehensive waste management framework to support socio-economic and environmental progress. The conclusions delineated in this chapter aim to assist municipal authorities and scholars in formulating more successful approaches for effective MSW management.

P. Kumar · R. R. Meena
Department of Civil Engineering, Motilal Nehru National Institute of Technology Allahabad, Prayagraj, India

P. Soni (✉)
Department of Civil Engineering, Indian Institute of Technology (BHU) Varanasi, Varanasi, India

S. Rajpoot, A. P. Das (eds.), *Innovative Technologies for Waste Management*, Sustainable Environmental Waste Management Strategies,
https://doi.org/10.1007/978-3-031-97184-6_11

Keywords Integrated waste management · Recycling · Municipal solid waste · Sustainability

1 Introduction

The problems of establishing urban solid waste management (USWM) in India have increased in the present decade due to rising urbanization, population expansion, and changing living standards. Municipalities are experiencing failure in the collection, segregation, remediation, and overall management of solid waste, underscoring the need for a comprehensive and consistent approach. Critical difficulties such as the lack of source waste segregation, social judgment, public indifference, inadequate evaluation procedures, and disjointed initiatives are recognized as enhancing factors. Furthermore, the informal sector is essentially unstructured, and insufficient financial planning coupled with ineffective execution of government regulations is decreasing progress.

Addressing these issues requires an adjustment in pattern of how urban solid waste is managed in India. This study highlights the pressing need for integrated treatment and recycling approaches specifically designed to address the various characteristics of urban solid waste (Bala et al., 2024b; Das et al., 2024b; Dey et al., 2024). A dual approach by adding centralized and decentralized waste management systems using scientific methods develops a promising pathway to sustainability. For this to succeed, the active involvement of municipalities, informal sectors, and private agencies is essential. By developing collaboration among these stakeholders, India can overcome the barriers to effective waste management and lay the foundation for a sustainable and strong USWM system.

The production, collection, processing, and subsequent release of waste into the environment adversely impact ecological systems and public health (Morrissey & Browne, 2004). Municipal solid waste (MSW) mainly comprises of nondegradable, degradable, and partially degradable materials (Jha et al., 2011). Therefore, for managing this a sustainable and effective solid waste management (SWM) approach is necessary. Compliance with waste management regulations is inconsistent and frequently inadequate (Gupta and Arora 2016). Urban flooding is contributed to by the spread of plastic waste, and microplastics are introduced into water bodies deteriorating water quality (Anderson et al., 2016). Microplastics disturb aquatic ecosystems and contribute to global warming, leading to indigenous species extinction (Barlow et al., 2018). Due to the improper management of hazardous chemical waste create severe health risks, including respiratory issues and an increased risk of early death (Giusti, 2009; Suk et al., 2016). While India has established itself as a leading recycler, its practices usually fall short of regulatory standards, addressing significant gaps in compliance (Bhattacharya et al., 2018). Improper handling of MSW further increases the health risk of waste collectors and frontline workers (Rémi, Bercegol & Nguyen., 2017) while also contributing to the spread of infectious diseases (Kumar et al., 2017). Effective MSW management is crucial for sustainable development and remains a priority area of research in developed nations (Dijkgraaf & Gradus, 2004; Shekdar, 2009). Population growth, urbanization, industrialization, and changing waste compositions increase the demand for improved MSWM systems (Idris, et al., 2004; Troschinetz and Mihelcic 2009). Addressing these challenges requires adopting socially, economically, and environmentally responsible waste management practices (Misra & Pandey, 2005; Uiterkamp et al., 2011; Batista et al., 2021). In India, the lack of waste segregation and the unsystematic disposal of construction and demolition debris, plastics, and e-waste reduce recycling efforts (Gomez et al., 2008; CPCB, 2013). Municipalities usually focus on waste collection over treatment, requiring the urgent need for system modernization (Mani & Singh, 2016). Incorporating the informal sector into the waste management system presents a viable strategy for enhancing waste recovery. However, this requires financial support, greater public awareness, and efforts to address social responsibility associated with waste-related work. This study examines MSWM practices in Indian cities, exploring the barriers and opportunities for effective recycling and management. A sustainable waste management framework is employed, optimizing advanced data analysis techniques to evaluate the prospects and challenges of solid waste recycling (Bala et al., 2024a; Mohanty et al., 2024a). Techniques such as the fuzzy Delphi model, complexity analysis, and uncertainty analysis have been used to assess the potential of reduce environmental impact of the product, waste-to-energy (WtE) initiatives, and material flow optimization (Bui et al., 2022). These strategies support the United Nations Sustainable Development Goals (SDGs) for 2050, focusing on reduced environmental impacts through green processes, sustainable products, and eco-friendly community policies (Zorpas, 2020). Additionally, a multi-objective inventory model incorporating green investments has been proposed to facilitate effective decision-making within the manufacturing sector, thereby improving environmental protection (Ahmadini et al., 2021). Together, these measures focus on the importance of innovation, policy development, and community involvement in achieving sustainable waste management.

MSWM encounters significant challenges particularly the generation of hazardous chemical waste from industries, hospitals, and urban sectors, which contributes to respiratory diseases and premature mortality (Joshi & Ahmed, 2016; Mohan; 2019) (Mishra et al., 2019). Hence, India has emerged as a major player in the global recycling market and its recycling practices generally fail to fulfill with regulatory standards, focusing on critical gaps in waste management systems (Bhattacharya et al., 2018). Furthermore, inadequately managed MSW at landfill sites often attracts wildlife, including vultures, rodents, scavengers, and mosquitoes, causing serious human health risks and increasing the likelihood of injuries and fatalities among frontline workers and waste collectors (Bercegol et al., 2017).

Despite these challenges, many cities, such as Bobbili, Alleppey, Surat, Pune, and Panaji, have implemented preventive actions in selecting appropriate solid waste management systems, as detailed in the present study. Municipalities mainly focus on waste collection, while advanced treatment methods remain inadequate. Improving these systems is essential to address the MSWM challenge (Annepu,

2012). Integrating the informal sector into the official municipal solid waste management system may minimize the significant challenge of effective materials recovery (John & Das, 2024; Mohanty et al., 2024b). Achieving this requires governmental support, public awareness, active participation, and the reduction of societal stress associated with waste management. These variables can substantially improve the creation of clean, sustainable urban environments. This study provides an in-depth analysis of historical and potential MSWM solutions focusing on the challenges and opportunities pertinent to urban regions in India.

2 Scientific Relevance of the Study

Numerous efforts have been undertaken to address the issues related to solid waste management; however, a comprehensive approach to handling these challenges remains difficult to determine. Common problems, such as landfill disposal, insufficient trash incineration, and extensive littering, continue in many developing countries where centralized or decentralized waste treatment facilities are frequently absent (Singh, 2020; Vij, 2012). Different kinds of waste are still collected in several Indian cities with limited efficient centralized or decentralized treatment systems established (Ahluwalia & Patel, 2018, Singh, 2020). Ongoing challenges related to MSWM in India have yet to be effectively resolved regarding waste treatment. This chapter also addresses the gaps and irregularities in the present scenario. The circumstances have increased for many municipal authorities thorough evaluation of the SWM system has not been conducted before the conduct of strategies (Zhu et al., 2008). As a result, unscientific dumpsites are created from the disposal of sustainable waste without suitable treatment.

3 Current Scenario of MSW Management in India

The present state of MSW management in India is complex and has huge challenges. Increasing production of MSW secondary to rapid urbanization and increase of population is beyond measure hence place existing waste management system under enormous pressure. Most of the cities and towns in India have inadequate infrastructure, little or no funds, and poor implementation of waste management policies. The problem is lack of segregation at the source, unscientific methods of disposal, and inadequate focus on recycling and resource recovery. Besides, the informal sector is very important in waste management, but it is poor, unorganized, and unused. To practically handle the waste management, it demands an urgent shift toward the integrated and sustainable waste management practices based on both technological advancement as well as in active stakeholder participation.

3.1 Evaluation of MSW Quantity and Generation Rates

India produces around 143,449 metric tonnes (MT) of MSW daily, with over 111,000 metric tonnes collected and over 35,602 metric tonnes processed (S. Kumar et al., 2017). The per capita garbage creation among cities shows significant diversity, with an exponential increase from 0.24 to 0.85 kg per capita/day between 2001 and 2018, according to the Central Pollution Control Board (CPCB) annual report 2018. This behavior is likely to continue and increase in the near future (Kumar et al., 2017; CPCB, 2018a). Table 1 summarizes the standing of MSW management throughout the 29 states of India.

Cities differ greatly regarding waste characteristics and composition due to various factors, including geography, climate, and socio-economic status. Urban areas with high population density typically produce more significant volumes of municipal trash. The variability in waste content presents further issues for municipalities, indicating that a standardized waste management plan is inapplicable across all metropolitan regions. As a result, most municipalities find it challenging to manage the situation successfully, with just a limited number demonstrating superior performance. Around 70% to 80% of the daily refuse comprises heterogeneous solid waste in region with higher population (MNRE, 2018).

According to the Ministry of New and Renewable Energy (MNRE) 2018 study, the national waste load is caused by the significant garbage production of highly populated states such as Tamil Nadu, Karnataka, Maharashtra, Gujarat, Uttar Pradesh, West Bengal, and the National Capital Region (NCR).

Table 1 MSWM and its present status

Sr. no.	Parameter	Status
1	Population density	382 persons/sqkm.
2	Door-to-door collection	18 states out of 29
3	Waste segregation at source	05 states out of 29
4	Unsanitary landfill sites (number)	1,285
5	Compost/vermicomposting facilities (urban local body)	95
6	Compost/vermicomposting facilities (under construction)	173
7	Pipe composting facilities (operational)	7,000
8	RDF facilities (operating)	12
9	Biogas plants (operating)	645
10	Energy generation plants (operational/total)	6 operational/11
11	Solid waste (SW) generation	143,449 MT/day
12	SW collection	111,000 MT/day (77.6% of total generated SW)
13	Solid waste processing/treatment	35,602 MT/day (24.8% of total generated SW)

Source: data from Census 2011 India, CPCB 2016 (Sharma and Jain (2018))

3.2 Comparative Analysis of Global SW Generation Trends

Global waste generation is projected to reach 27 BT yearly by 2050. Asia represents approximately 1/3 of total waste, with significant contributions from big countries like India and China. China produces waste in the range of 0 to 0.49 kg per capita daily, while India generates waste between 0.50 to 0.90 kg per capita daily (Kaza et al., 2018). If current trends continue, waste production in India is expected to range from 334 to 661 million tonnes/day by the year 2050. During this period (2016–2050), waste generation in China and the East Asia-Pacific region is anticipated to reach figures between 468 and 714 million tonnes per day. Recent findings show that the daily average waste generation per person varies between 150 gm and 300 gm. Globally, it is projected that the entire amount of municipal waste, which includes industrial, construction, and demolition (C&D) waste, will rise to 7–10 billion tonnes each year. The World Bank has analyzed the current situation and projections for future waste generation across various regions. In 2016, the daily per capita waste production varied between 0.09–0.60 kg in lower-income regions, 0.16–0.78 kg in lower-middle-income regions, and 0.10–1.20 kg in upper-middle income regions (Waste Generation-World Bank Org., 2015). A report from the World Bank Economist in 2018 indicates that these regional waste generation trends are expected to double by 2050.

3.3 Quantification of MSW in India

The quantity of municipal solid waste production depends on numerous factors, like the standard of living, the scale and nature of commercial activities, food habits, geographical features, and climate conditions. The arrival of migrants pursuing economic opportunities intensifies the situation. The latest reports of MNRE reveal that India generates around 145 million tonnes of waste annually, with forecasts suggesting that this amount could rise to 260 and 300 million tonnes per day by 2047. The CPCB indicated that in 2016, approximately 117,644 MT of waste was collected per day, yet only 49,401 MT were treated (Ahluwalia & Patel, 2018).

By the year 2051, it is projected that waste disposal sites will need an area of around 1,400 square kilometers. Figure 1 illustrates that waste production, the necessary land for its disposal, and the increasing population growth rate have all played a crucial role in creating a notable deficit of available landfill sites.

The data in Table 2 illustrates the percentage variation in waste generation from 1999–2000 to 2015–16. Over time, there has been a notable increment in the waste generation rate, especially within metropolitan areas. For example, in Delhi and Bangalore, the amount of waste generated rise by 2,175% and 1,850%, respectively, from 1999–2000 to 2015–2016.

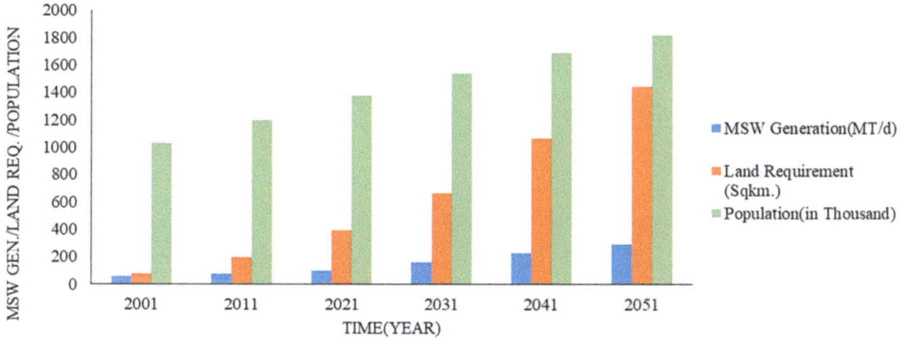

Fig. 1 Projected trends in MSW generation, land utilization, and population growth in India (2001–2051). (Joshi & Ahmed., 2016)

4 Characteristics and Composition of MSW in India

In India, MSW is characterized by a high organic content (40–60%), primarily from biodegradable materials like food and agricultural residues. Recyclables such as plastics, paper, metals, and glass constitute 20–30%, while inert materials form the rest (Sahu et al., 2024). High moisture content and low calorific value present challenges for effective waste management. Understanding these characteristics is vital for developing sustainable strategies focused on recycling, resource recovery, and energy generation. Municipal solid waste is generally categorized into several major following types:

- **Biodegradable waste** is characterized by organic materials, including waste generated from the kitchen, food scraps, green waste (such as flowers, vegetables, fruits, and leaves), and paper products, naturally decomposed through biological processes.
- **Recyclable materials** are defined as those that can be reprocessed and reused, such as paper, glass, bottles, cans, metals, and certain types of plastics.
- **Inert waste** material is identified as dirt, construction and demolition debris, and nonbiodegradable substances that remain largely unchanged in their chemical, biological, or physical properties.
- **Domestic hazardous waste**, also known as household hazardous waste, is classified as toxic materials that must be carefully handled and disposed of due to the potential risks they pose to human health and the environment.
- **Composite Waste** Comprises various materials, including discarded clothing, Tetra Pak containers, and plastics that cannot be recycled.

This includes fluorescent tubes, chemicals, expired medicines, electronic waste (e-waste), paints, spray cans, batteries, containers of fertilizers, pesticides, and products such as shoe polish. The composition of Indian MSW generally includes 10–30% recyclable materials, around 40–60% compostable organic waste, and 30–50% inert waste. According to the National Environmental Engineering

Table 2 Solid waste generation trends and changes in percentage over time

Sr. no.	City name	Population (Census, 2011)	Waste generation (tonnes/day)				% Change (wrt 1999–2000)
			(1999–2000)	(2000–2005)	(2010–2011)	(2015–2016)	
1	Mumbai	12,442,373	5355	5320	6500	11,000	205.42
2	Hyderabad	6,731,790	1566	2187	4200	4000	255.43
3	Delhi	11,034	400	5922	6800	8700	2175
4	Kolkata	4,496,694	3692	2653	3670	4000	108.34
5	Bangalore	8,443,675	200	1669	3700	3700	1850
6	Pune	3,124,458	700	1175	1300	1600	288.57
7	Chennai	7,088,000	3124	3036	4500	5000	160.05
8	Lucknow	2,817,105	1010	475	1200	1200	188.81
9	Nagpur	2,405,665	443	504	650	1000	225.73
10	Ahmadabad	5,577,940	1683	1302	2300	2500	148.54
11	Indore	1,960,631	350	557	720	850	242.86
12	Patna	1,683,200	330	511	220	450	136.36
13	Surat	4,467,797	900	1000	1200	1680	186.67
14	Ludhiana	1,613,878	400	735	850	850	212.50
15	Madurai	1,561,129	370	275	450	450	121.62
16	Jaipur	3,046,163	580	904	310	1000	172.41
17	Varanasi	1,201,815	412	425	450	500	121.36
18	Coimbatore	1,601,438	350	530	700	850	242.86
19	Kanpur	2,765,348	1200	1100	1600	1500	125
20	Vadodara	1,666,703	400	357	600	700	175
21	Bhopal	1,798,218	546	574	350	700	128.21
22	Visakhapatnam	2,035,992	300	584	334	350	116.67

Source: CPCB India (2018b)

Research Institute, the chemical composition of Indian waste is reported to have an average nutrient content of phosphorus (0.67 ± 0.15%), a (C/N) ratio of (26 ± 5%), potassium (0.68 ± 0.15%), and nitrogen (0.64 ± 0.8%).

5 The Policy Contexts

India's waste management framework is guided by a series of national missions and regulatory rules aimed at addressing the growing challenges of waste generation, segregation, and disposal. These policies seek to promote sustainability, environmental protection, and the efficient management of resources.

5.1 The National Policy

The Indian government has recently turned municipal solid waste management into a distinct policy priority. For the first time, the central government gave municipal authorities comprehensive waste management regulations when the Solid Waste Management Rules were passed in 2000. As a result, numerous national programs have been implemented to address the comprehensive municipal challenges in India and innovative mechanisms have been developed to promote sustainable waste management practices.

5.2 Solid Waste Management Rules, 2016

The Solid Waste Management Rules of 2016 supersede the 2000 version and establish a comprehensive framework for waste management. However, adherence to these rules remains markedly insufficient. Despite the regulations focusing on safe and reliable disposal procedures, only about 10% of the collected garbage is treated, with certain urban areas reporting that less than half of the waste is collected (Mani et al., 2016). The regulations are significant for recognizing the part of the informal sector in waste management. Still, limited guidance is provided on how garbage pickers should be integrated into the waste management system.

5.3 The Swachh Bharat Mission (SBM)

The Swachh Bharat Mission, also called the "Clean India" Mission, started in October 2014. It is a government-financed, statewide program to improve sanitation and waste management throughout India, with a projected completion year 2019. As a premier initiative of the Honorable Prime Minister, SBM has attracted considerable attention owing to its prominent political support. A significant accomplishment of the program has been its ability to enhance public awareness via a comprehensive marketing and media initiative. Critics contend that SBM has predominantly framed waste management as an aesthetic issue, failing to sufficiently tackle fundamental municipal and infrastructure shortcomings or substantial socio-economic issues (Luthra, 2018).

The mission guidelines specifically address the informal sector, stating: In their efforts to enhance and formalize SWM systems, municipalities shall prioritize the improvement of working conditions for informal sector workers (rag pickers) and integrate them into the formal solid waste management system within urban areas. In 2018, the Ministry of Housing and Urban Affairs issued supplementary guidelines to implement this order, merging the Swachh Bharat Mission (SBM) with the National Urban Livelihoods Mission (NULM), a program aimed at reducing poverty and enhancing employment opportunities for the urban poor. The extent of municipalities' implementation of these recommendations remains uncertain.

5.4 The Smart Cities Mission

The Indian government started Smart Cities Mission in 2015 to boost decaying metropolitan places with a sustainable, peer-inclusive, and hi-tech infrastructure creation of 100 cities. The aim of the initiative is to enhance urban living condition, economic growth, and environmental sustainability (Das et al., 2024a, b; Dey et al., 2023; Jagatee et al., 2024; Mohanty et al., 2024a). The effort is intended to support such development of solutions as waste-to-energy systems and source segregation technologies but does not take account of the critical role played by informal garbage collectors in sorting waste in India (CAEP report, 2011). The local version of the program is implemented through special-purpose vehicles created specifically for this purpose by the central government. The privatization of this type of approach is deemed to be a way to reduce public responsibility and transparency.

6 Key Processes and Operational Challenges in MSWM

In effective MSWM in India, waste segregation, collection, transportation, storage, and treatment are linked processes. But resource constraints, infrastructural limitations, and inefficient implementation make these processes quite challenging to operate. These processes and underlying challenges are outlined in this section.

6.1 Collection and Segregation

Waste segregation at sources serves to process and serve the waste and prevent environmental hazards, which are among the fundamental components of MSWM. However, it is critical for source level segregation, but it has little practice available in India and it performs poorly during the waste treatment when available. Municipalities collect unsorted waste from around the cities every day in minimal staffing and poor resource (Ghosh, 2016). The lack of segregation in these finances goes a long way in making it difficult financially for this treatment process and giving rise to environmental degradation. Public awareness and participation are crucial to promoting source segregation, ensuring effective door-to-door collection, and utilizing designated collection bins appropriately. For example, in 2011–2012, India generated around 127,486 tonnes of MSW every day, with only 70% (89,334 tonnes) collected. MSW generation is expected to peak at 6.1 million metric tonnes per day by 2025 (Balasubramanian, 2018). While some states improved waste collection efficiency between 2010 and 2015, others, such as Meghalaya, Maharashtra, and Assam, lagged behind, highlighting regional disparities in MSWM performance. These issues continue among rising waste generation and inadequate treatment infrastructure.

6.2 Transportation and Storage

Generally, both individual and community bins are used to store solid waste, frequently in unhygienic and disorganized ways. Single-bin systems dominate residential areas, leading to improper waste storage, particularly during the rainy season when leachate generation becomes a significant issue. Poorly designed waste management services create gaps between residential zones and efficient waste collection systems, further increasing the problem.

Waste recycling and treatment processes perform better with the availability of needed resources and infrastructural availability (Ray et al., 2024). However, there exist few suitable sites for waste disposal, and landfills are generally located far away from urban center thereby adding transportation costs and operating problems (A. Kumar & Agrawal, 2020). Residential developments that are close to landfill sites are a serious health and environmental risk to developing countries. It is imperative to implement effective systematic improvement in waste storage, transportation, and disposal in India for increasing efficiency of Municipal Solid Waste Management (MSWM).

6.3 Treatment and Disposal

In 2024, India has 29 waste-to-energy (WtE) plants in operation of RDF and biomethanation type. The day-to-day 5,250 tonnes of waste are processed through the three operational WtE plants of Ghazipur, Narela-Bawana, and Okhla which generate 52 megawatts power. For example, these facilities that are adopted to energy recovery of wastes are used for energy recovery waste and will help in waste management energy and its production to the city as a whole. However, states such as Himachal Pradesh are not pacing up with this which points to the fact that there are no uniform adoption of these waste-to-energy (WtE) methods all over the country. The mixed behavior of MSW in India leads to inefficiency of MSW treatment and its associated energy recovery technology. Yet open waste dumping remains extensive, and there is grave environmental degradation and land scarcity, particularly in the states of Tamil Nadu, Madhya Pradesh, and Maharashtra where dumpsites are being reported in large numbers. Reclaiming and turning these dump sites into land that can be used is important to solve land scarcity problem and have less impact to environment. India has seven megacities which contribute more than one half of India's waste disposal site methane emissions and that too waste not disposed of properly or properly managed is a serious environmental concern as per the CPCB report (2015). Hazardous waste category includes e waste, batteries, and sanitary items, and it has to be processed specially to reduce contamination and health hazard (Lv et al. 2018; Rarotra et al. 2020). However, several states like Karnataka, Madhya Pradesh, and Maharashtra have more waste treatment facilities, and they are not very effective. All of these sites have inadequate leachate management, poor ambient air quality, weak groundwater protection, and inefficient compost processing, all of which constitute serious environmental sustainability and public health concerns (Bala et al., 2024a). To obtain sustainable waste treatment and final disposal, India must strongly adopt advanced waste-to-energy technologies, enhance treatment infrastructure, and reclaim existing dumpsites. These challenges will be addressed and a sustainable waste management system promoted through operational improvement in waste processing facilities as well as integrating into modern waste management strategies.

6.4 Challenges and Technological Gaps in SWM Planning and Investment

India's management and investment in SWM is lacking in critical planning and technologic coordination. Set up of technologies is done without assessing their suitability or assuming that these technologies fit in with comprehensive waste management structure. A lot of its resources go toward waste collection and transportation. At the same time, minimal focus is placed on critical processes such as segregation of waste and the development of advanced WtE technologies (Rana et al., 2015). The use of outdated equipment and unsuitable technologies results in a failure in operational as well service coverage. All these factors reduce the efficiency of the waste management systems in developing nations, where high maintenance cost, improper technology selection, and frequent equipment break down due to untrained personnel are generally the case. In order to address these challenges, it is necessary to develop techno-economically sustainable solutions that integrating advanced approaches including scientific disposal systems, Internet of Things integration, and GIS/remote sensing-based approaches for real time monitoring and optimization (Mundhe et.al, 2014). Furthermore, designing a comprehensive economic strategy that gives priority to waste segregation and collection at the source is essential for enhancing solid waste management infrastructure's operational efficiency and sustainability.

6.5 Land Scarcity and Operational Challenges in SWM

Municipalities in India face difficulties in managing manage solid waste due to an inexperienced and dispersed waste management sector. Inefficient technologies and outdated equipment contribute to limited service coverage and operational inefficiencies, weakening efforts to achieve sustainable waste management (Wilson et al., 2006). The rapid population increase and urbanization have further stretched resources, resulting in a severe shortage of land for housing, infrastructure, and garbage disposal. By 2047 projections, untreated urban trash will require around 1,400 km^2 of additional land for disposal; growing demand, coupled with rising land prices, presents an impressive challenge to optimizing land use for waste management while addressing the competing needs of urban development.

7 Integrated Waste Management: Advancing Sustainable Solutions for MSWM

Effective management of MSW requires a comprehensive and systematic method as no single methodology can address the complex difficulties involved. IWM emerges as a crucial strategy for the assessment, analysis, and integration of all the critical components of waste management to improve the efficiency and sustainability. Researchers suggest that IWM is important for incorporating advanced strategies, state-of-the-art methodologies, and robust programs to streamline municipal solid waste management. IWM is investigating the most critical pathways to resource recovery to minimize dependency on raw materials and reduce their disposal price, in order to extract the maximum amount of potential recovery (Hajar et al., 2020). It also avoids waste from reaching the waste stream at source and improves energy efficiency.

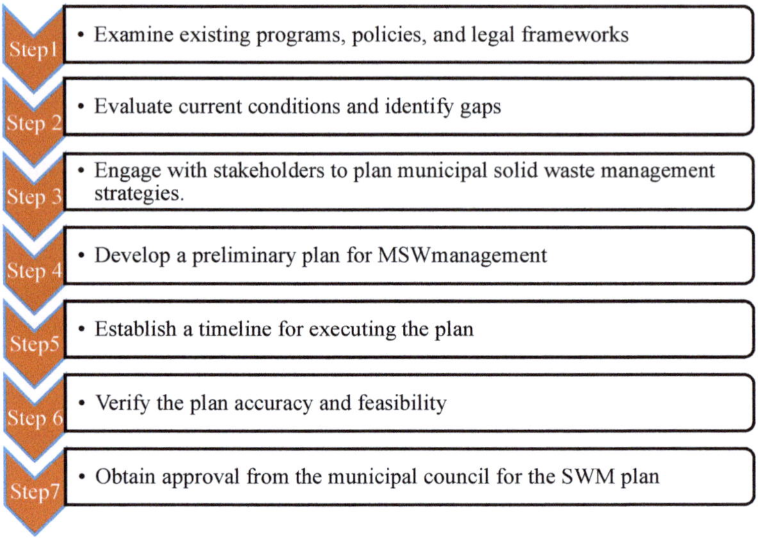

Fig. 2 Seven-stage framework for effective SWM

The effectiveness of IWM is constrained by the inconsistencies between waste composition and the technical requirements for available processes and low efficiency (Singh et al., 2014). As shown in Fig. 2, a seven-step IWM framework is quoted to overcome these challenges. A thorough review and analysis of existing policies and schemes are followed by a comprehensive study of the current situation and definitive gaps in the system detected. The following consists of drafting targeted methods, beginning implementation schedules, and validating these plans through repetition of testing and refinement. This methodical approach provides confirmation that waste management policies and operational frameworks are practical and scalable to a significant extent. By harmonizing waste properties with process capabilities and using advanced methodologies of planning, IWM can deliver a technologically and economically feasible pathway for sustainable MSW management.

Integrated waste management (IWM) incorporates five core components: waste reduction, reuse, recycling, incineration, and landfilling. Among these, recycling is particularly significant as it repurposes a substantial portion of waste into raw materials for the production of new goods, thereby contributing to socio-economic progress and environmental conservation. IWM aligns seamlessly with sustainable development goals, offering an eco-efficient and economically viable framework for managing municipal solid waste. Figure 3 illustrates the drawbacks and advantages of integrated waste management.

The hierarchical structure of IWM, illustrated in Fig. 4, prioritizes waste reduction as the most desirable and effective strategy. This involves minimizing waste generation at the source and encouraging material reuse. Recycling is a critical step for resource recovery, while incineration provides an energy-efficient solution for

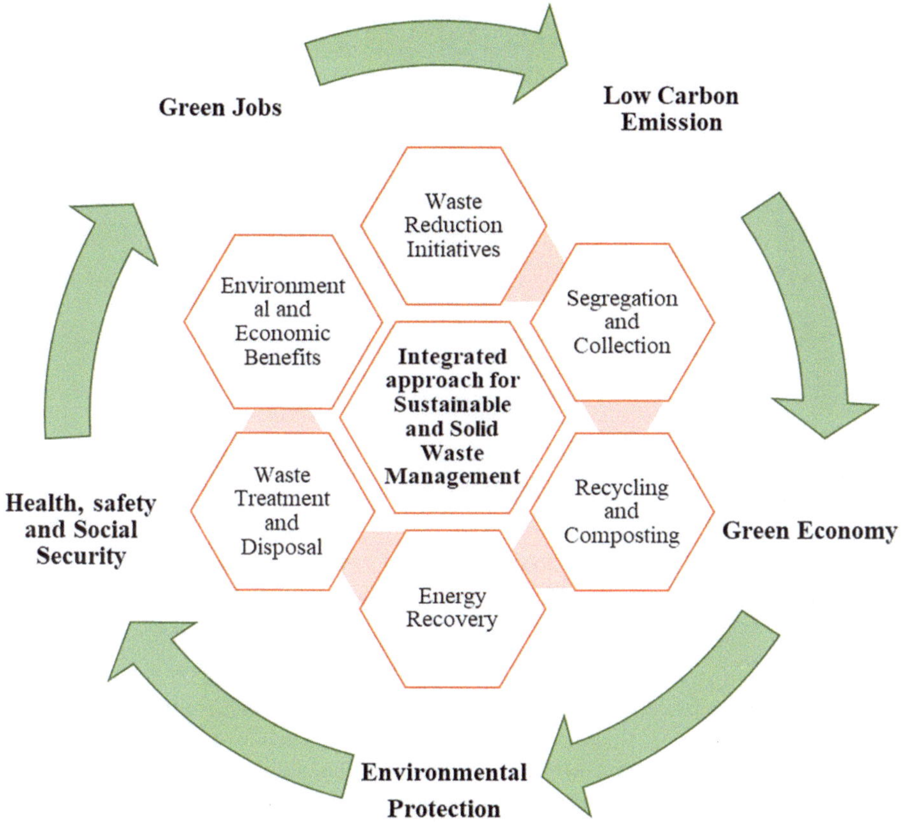

Fig. 3 Integrated framework for sustainable urban solid waste management in India

nonrecyclable waste. Landfilling, the least favorable option, is employed only for residual waste that other methods cannot process.

This hierarchy highlights the importance of prioritizing resource conservation and decreasing environmental impact through waste management observations. Implementation of the IWM framework is important for reducing environmental difficulties and achieving a sustainable society. Its accomplishment depends on the coordinated approach of technological features, waste composition, and environmental objectives. Municipalities are key in implementing IWM principles into local waste management systems, resource recovery, confirming proper handling, and adherence to environmental standards.

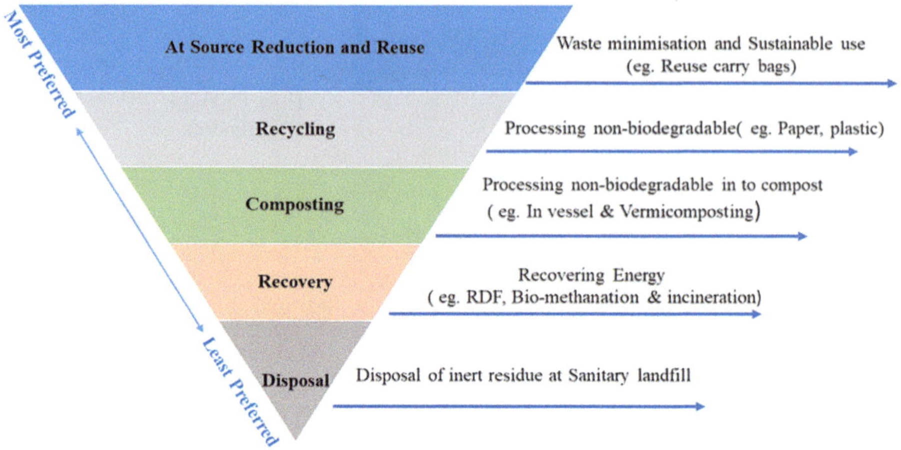

Fig. 4 Hierarchy of integrated waste management

8 Socio-economic and Ecological Advantages of a Comprehensive Waste Management Approach

Sustainable solid waste management can be easily accomplished by applying solutions that significantly decrease pollution, conserve energy and resources, and benefit society. Recycling is most crucial for this process since it considerably reduces the quantity of waste transported to landfills and plays a significant role in socio-economic growth and environmental sustainability (Das et al., 2024b; Paul et al., 2024; Satapathy et al., 2024). Valuable resources can be recovered and the environmental impact of waste is reduced by recycling and reusing them properly. Further employing waste avoidance measures at the consumer level enhances sustainable consumption and reduces the production of waste at the source. This approach to waste management highlights the combination of social, environmental, and economic systems, confirming that sustainability efforts result in balanced benefits across all sectors. Adopting such advanced practices is vital for developing a sustainable and efficient waste management system supporting with wider socio-economic goals and environmental advancement.

The IWM approach suggests a multifaceted framework with significant socio-economic and environmental benefits. By producing employment opportunities IWM contributes to poverty mitigation and improves the economic resilience of communities fostering national economic growth. This economic gain is further supported by lower expenses associated with inefficient waste disposal systems. From an environmental perspective IWM alleviates pollution by reducing reliance on landfilling and incineration thus limiting greenhouse gas emissions and reducing ecological degradation. Improved environmental conditions directly affect human health by reducing exposure to hazardous pollutants dropping the existence of waste-related diseases and improving overall life quality. This highlights the fundamental link between economic prosperity, social well-being, and environmental integrity. As a holistic concept, sustainability encourages environmental conservation and promotes advancing social equity and economic stability. Adopting IWM observations aligns with the principles of a green economy which endorses sustainable economic growth, job creation, and improved human health. Integrating social, economic, and environmental dimensions, IWM catalyzes transformative change empowering the change toward a sustainable future that balances economic development with ecological preservation and social equity.

9 Conclusions

MSW generation has become a serious national issue as a result of the improvement of living standards, rapid population growth, urbanization, and economic development. The growing amount of waste and the variety of waste constitute challenges but also great opportunities in order to develop efficient waste management systems to deal with the same. Major challenges to efficient MSW management are shown through this study, particularly with public participation deficiency, low awareness and behavior gap, low adoption of complicated WtE technology, and limited involvement in recyclable waste management strategy generally. MSW flexibility shows its need for efficient waste segregation so that the framework for solid waste management can be optimized. Enhancing civic awareness and encouraging sustainable waste management strategies are essential to overcome these complications. Collective efforts that involve the citizens, government agencies, and industry leaders are important to address unorganized environmental and human health risks due to solid waste. Additionally, this chapter examines the current status and characterization of the legislative framework, waste production, and institutional governance structures for MSW management. It emphasizes improved data acquisition and identifies research gaps, operational limitations, and critical performance indicators to improve flexibility and effectiveness of recycling. This wide-ranging analysis highlights the importance of integrating technology advancement, civic participation, and policy-driven methods to establish sustainable and effective waste management solutions. Some conclusions derived from the study are as follows:

1. MSW presents significant potential for developing sustainable, value-added products, offering a pathway to alleviate environmental burdens associated with waste accumulation.
2. Sustainable solid waste management involves a holistic approach that considers technical, social, economic, legal, institutional, and environmental factors, as highlighted by global data.
3. Local authorities must ensure the effective enforcement and implementation of the guidelines and rules set forth by the governing ministry responsible for environmental protection and waste management.

4. Future research should focus on advancing solid waste management through enhanced data collection, life cycle assessments of recycled products, optimization of recycling processes, and evaluation of treatment techniques. Key areas include leveraging IoT for waste characterization, promoting innovative technologies, developing predictive models, and advancing waste-to-energy solutions to drive sustainable waste management practices.

Acknowledgment We would like to acknowledge that this study was conducted independently without financial support or funding from any external agencies or organizations. We are grateful for the institutional resources and infrastructure that facilitated this study.

References

Ahluwalia, I. J., & Patel, U. (2018). *Solid waste management in India: An assessment of resource recovery and environmental impact* (No. 356). Working paper. https://icrier.org/pdf/Working_Paper_356.pdf.

Ahmadini, A. A. H., Modibbo, U. M., Shaikh, A. A., & Ali, I. (2021). Multi-objective optimization modelling of sustainable green supply chain in inventory and production management. *Alexandria Engineering Journal, 60*(6), 5129–5146.

Anderson, J. C., Park, B. J., & Palace, V. P. (2016). Microplastics in aquatic environments: Implications for Canadian ecosystems. *Environmental Pollution, 218*, 269–280. https://doi.org/10.1016/j.envpol.2016.06.074

Annepu, R. K. (2012). Sustainable solid waste management in India. *MS Dissertation, 1–189*. https://doi.org/10.1007/978-981-4451-73-4

Troschinetz, A. M., & Mihelcic, J. R. (2009). Sustainable recycling of municipal solid waste in developing countries. Waste Management, 29(2), 915–923. https://doi.org/10.1016/j.wasman.2008.04.016

Bala, K., Anshul, V. S., Dogra, A., Das, A. P., Kumar, R., & Nandini. (2024a). Sustainable management for urban plastic waste generation. *Sustainable Management of Urban Plastic Waste Through Circular Economic Approaches, 211–236*. https://doi.org/10.1201/9781003465447-13

Bala, K., Dogra, A., Lata, P., & Das, A. P. (2024b). Discussing sources and biological-chemical remediation approaches for microplastic pollution. *Sustainable Microbial Technology for Synthetic and Cellulosic Microfiber Bioremediation, 37–59*. https://doi.org/10.1007/978-3-031-62600-5_3

Balasubramanian, M. (2018). Municipal solid waste management in India: Status, problems and challenges. *International Journal of Environment and Waste Management, 21*(4), 253–268.

Barlow, J., França, F., Gardner, T. A., Hicks, C. C., Lennox, G. D., Berenguer, E., Castello, L., Economo, E. P., Ferreira, J., Guénard, B., Leal, C. G., Isaac, V., Lees, A. C., Parr, C. L., Wilson, S. K., Young, P. J., & Graham, N. A. J. (2018). The future of hyperdiverse tropical ecosystems. Nature, 559(7715), 517–526. https://doi.org/10.1038/s41586-018-0301-1

Barlow, J., França, F., Toby, A. Gardner Christina C., Hicks Gareth D., Lennox Erika, Berenguer Leandro, Castello Evan P., Economo Joice, Ferreira Benoit, Guénard Cecília, Gontijo Leal Victoria, Isaac Alexander C., Lees Catherine L., Parr Shaun K., Wilson Paul J., Young Nicholas A. J., Graham (2018) The future of hyperdiverse tropical ecosystems Nature 559(7715) 517-526 https://doi.org/10.1038/s41586-018-0301-1

Batista, M., Caiado, R. G. G., Quelhas, O. L. G., Lima, G. B. A., Leal Filho, W., & Yparraguirre, I. T. R. (2021). A framework for sustainable and integrated municipal solid waste management: Barriers and critical factors to developing countries. *Journal of Cleaner Production, 312*, 127516.

Bhattacharya, R. R. N. S., Chandrasekhar, K., Deepthi, M. V., Roy, P., & Khan, A. (2018). Challenges and opportunities: Plastic waste management in India. *The Energy and Resources Institute., 24*(2) https://www.teriin.org/sites/default/files/2018-06/plasticwaste-management_0.pdf

Bui, T.-D., Tseng, J.-W., Tseng, M.-L., & Lim, M. K. (2022). Opportunities and challenges for solid waste reuse and recycling in emerging economies: A hybrid analysis. *Resources, Conservation and Recycling, 177*, 105968.

Census_of_India_2011_GOI (2011). Census of India 2011 provisional population totals. https://censusindia.gov.in/nada/index.php/catalog/42611/download/46274.

Central Pollution Control Board. (2016). Consolidated Annual Review report on Implementation of Municipal Solid Wastes (Management and Handling) Rules, 2000. *8*(2). https://cpcb.nic.in/status-of-implementation-of-solid-waste-rules/

Chinese Academy for Environmental Planning, and the Energy and Resources Institute. (2011). *Environment and development: China and India*. TERI Press.

CPCB. (2013). Status report on municipal solid waste management: CPCB .

CPCB India, 2018a. MSW- generation & composition CPCB 2018. https://cpcb.nic.in/uploads/MSW/Waste_generation_Composition.pdf.

CPCB India, 2018b. MSW trend in 46 cities in India list 2018-CPCB report 2018. https://cpcb.nic.in/uploads/MSW/trend_46_cities_list.pdf.

Das, A., Ray, S. K., Mohanty, M., Mohanty, J., Dey, S., & Das, A. P. (2024a). Ecotoxicity of microplastic wastes and their sustainable management: A review. *Environmental Chemistry and Ecotoxicology, 6*, 144–152. https://doi.org/10.1016/J.ENCECO.2024.05.003

Das, A., Dey, S., Bala, K., Gnanasekaran, R., & Das, A. P. (2024b). Microbial bioremediation technology for sustainable treatment and management of synthetic microfiber waste. *Sustainable Microbial Technology for Synthetic and Cellulosic Microfiber Bioremediation, 221–238.* https://doi.org/10.1007/978-3-031-62600-5_12

De Bercegol, R., Cavé, J., & Nguyen Thai Huyen, A. (2017). Waste municipal service and informal recycling sector in fastgrowing Asian cities: Co-existence, opposition or integration? Resources, 6(4), 70. https://doi.org/10.3390/resources6040070

Rémi, De Bercegol Jérémie, Cavé Arch, Nguyen Thai Huyen (2017) Waste Municipal Service and Informal Recycling Sector in Fast-Growing Asian Cities: Co-Existence Opposition or Integration? Resources 6(4) 70–10.3390/resources6040070Dey, S., Das, A., Mallick, K., Sahu, A., & Das, A. P. (2023). *Environmental petroleum waste: Pollution, toxicity, sustainable remediation* (pp. 159–175). Springer. https://doi.org/10.1007/978-3-031-48220-5_7

Dey, S., Bissoyi, A., Koteswara Reddy, G., & Das, A. P. (2024). Synthetic microfibers: An urban emerging pollutant demanding for sustainable environmental solutions. *Sustainable Microbial Technology for Synthetic and Cellulosic Microfiber Bioremediation, 107–121*. https://doi.org/10.1007/978-3-031-62600-5_7

Dijkgraaf, E., & Gradus, R. H. (2004). Cost savings in unit-based pricing of household waste: The case of The Netherlands. *Resource and Energy Economics, 26*(4), 353–371.

Ghosh, S. K. (2016). Swachhaa Bharat Mission (SBM)–A paradigm shift in waste management and cleanliness in India. *Procedia Environmental Sciences, 35*, 15–27.

Giusti, L. (2009). A review of waste management practices and their impact on human health. *Waste Management, 29*(8), 2227–2239.

Gomez, G., Meneses, M., Ballinas, L., & Castells, F. (2008). Characterization of urban solid waste in Chihuahua Mexico. *Waste Management, 28*(12), 2465–2471.

Gupta, B., & Arora, S. K. (2016). Municipal solid waste management in Delhi—The capital of India. International Journal of Innovative Research in Science, Engineering and Technology, 5(4), 5130–5138.

Hajar, H. A. A., Tweissi, A., Hajar, Y. A. A., Al-Weshah, R., Shatanawi, K. M., Imam, R., & Hajer, M. A. A. (2020). Assessment of the municipal solid waste management sector development in Jor dan towards green growth by sustainability window analysis. *Journal of Cleaner Production, 258*, 120539.

Idris, A., Inanc, B., & Hassan, M. N. (2004). Overview of waste disposal and landfills/dumps in Asian countries. *Journal of Material Cycles and Waste Management, 6*(2), 104–110.

Jagatee, S., Priyadarshini, S., Rath, C. C., & Das, A. P. (2024). Synthetic microfiber: An enduring environmental problem linked to sustainable development. *Environmental Science and Engineering, Part F2333*, 93–112. https://doi.org/10.1007/978-3-031-51792-1_6

Jha, A. K., Singh, S. K., Singh, G. P., & Gupta, P. K. (2011). Sustainable municipal solid waste management in low income group of cities: A review. *Tropical Ecology, 52*(1), 123–131.

John, A., & Das, A. P. (2024). Government strategies for achieving circular economy for plastic waste management in Southeast Asia. *Sustainable Management of Urban Plastic Waste Through Circular Economic Approaches, 72–82.* https://doi.org/10.1201/9781003465447-6/GOVERNMENT-STRATEGIES-ACHIEVING-CIRCULAR-ECONOMY-PLASTIC-WASTE-MANAGEMENT-SOUTHEAST-ASIA-AKBAR-JOHN-ALOK-PRASAD-DAS

Joshi, R., & Ahmed, S. (2016). Status and challenges of municipal solid waste management in India: A review. *Cogent Environmental Science, 2*(1), 1139434.

Kaushal, R. K., Varghese, G. K., & Chabukdhara, M. (2012). Municipal solid waste management in India current state and future challenges: A review. *International Journal of Engineering Science and Technology, 4*(4), 1473–1489.

Kaza, S., Yao, L., Bhada-Tata, P., & Van Woerden, F. (2018). What a waste 2.0: A global snapshot of solid waste management to 2050. World Bank Publications.

Kumar, A., & Agrawal, A. (2020). Recent trends in solid waste management status, challenges, and potential for the future Indian cities–A review. *Current Research in Environmental Sustainability, 2*, 100011.

Kumar, J. S., Subbaiah, K. V., & Rao, P. V. V. P. (2014). Municipal solid waste management scenario in India. *Australian Journal of Engineering and Technology Research, 2*, 1–8.

Kumar, S., Smith, S. R., Fowler, G., Velis, C., Kumar, S. J., Arya, S., Rena, K. R., & Cheeseman, C. (2017). Challenges and opportunities associated with waste management in India. *Royal Society Open Science, 4*(3). https://doi.org/10.1098/rsos.160764

Luthra, A. (2018). "Old habits die hard": Discourses of urban filth in Swachh Bharat Mission and the ugly Indian. *Journal of Multicultural Discourses, 13*, 120–138.

Lv, W., Wang, Z., Cao, H., Sun, Y., Zhang, Y., & Sun, Z. (2018). A critical review and analysis on the recycling of spent lithium-ion batteries. *ACS Sustainable Chemistry & Engineering, 6*(2), 1504–1521.

Mani, S., & Singh, S. (2016). Sustainable municipal solid waste management in India: A policy agenda. Procedia Environmental Sciences, 35, 150–157. https://doi.org/10.1016/j.proenv.2016.07.064

Ministry of Housing and Urban Affairs. 2018. Empowering marginalised groups: Convergence between SBM and DAY-NULM, Delhi. https://nulm.gov.in/PDF/NULM_Mission/SBM_NULM_Convergence_Guideline.pdf.

Mishra, S., Rath, C., & charan, & Das, A. P. (2019). Marine microfiber pollution: A review on present status and future challenges. *Marine Pollution Bulletin, 140*, 188–197. https://doi.org/10.1016/J.MARPOLBUL.2019.01.039

Misra, V., & Pandey, S. (2005). Hazardous waste, impact on health and environment for development of better waste management strategies in future in India. *Environment International, 31*(3), 417–431.

MNRE India, 2018. Government of India Ministry of New and Renewable Energy. http://www.indiaenvironmentportal.org.in/files/file/GenerationofSolidWaste.pdf.

Mohan, V. (2019, May 10). Topic: 91% premature deaths due to air pollution in low and middle-income countries: UNEP. 1. *The Economics Times*. https://economictimes.indiatimes.com/news/environment/pollution/91-premature-deaths-due-to-air-pollution in-low-and-middle-income-countries-unep/articleshow/69264417.cms

Mohanty, M., Tripathy, B., Koteswara Reddy, G., & Das, A. P. (2024b). Greener technologies for establishing restraint over microplastic pollution. *Sustainable Microbial Technology for Synthetic and Cellulosic Microfiber Bioremediation, 239–258.* https://doi.org/10.1007/978-3-031-62600-5_13

Mohanty, J., Dey, S., Bala, K., & Das, A. P. (2024a). Recent advances in microplastic pollution for its sustainable management. *Sustainable Management of Urban Plastic Waste Through Circular Economic Approaches, 27–48*. https://doi.org/10.1201/9781003465447-3/RECENT-ADVANCES-MICROPLASTIC-POLLUTION-SUSTAINABLE-MANAGEMENT-JYOTIRMAYEE-MOHANTY-SUDESHNA-DEY-KIRAN-BALA-ALOK-PRASAD-DAS

Morrissey, A. J., & Browne, J. (2004). Waste management models and their application to sustainable waste management. *Waste Management, 24*(3), 297–308. https://doi.org/10.1016/j.wasman.2003.09.005

Mundhe, N., Jaybhaye, R., & Dorik, B. (2014). Assessment of municipal solid waste management of Pune city using geospatial tools. *International Journal of Computer Applications, 100*(10), 24–32.

Paul, A., Priyadarsini, S., Karakuş, S., Behera, J. K., Gnanasekaran, R., & Das, A. P. (2024). Fate and accumulation of synthetic microfibers in the environment. *Sustainable Microbial Technology for Synthetic and Cellulosic Microfiber Bioremediation, 61–76*. https://doi.org/10.1007/978-3-031-62600-5_4

Rajpoot, S. (2019). Current technologies for wastewater treatment in chemical industries in India. *International Journal for Modern Trends in Science and Technology, 05*(05). https://doi.org/10.2139/ssrn.3390363

Rajpoot, S. (2020). A review on ways to manage biomedical waste at different locations in Faizabad. *International Journal for Modern Trends in Science and Technology, 06*(01), 33–36.

Rajpoot, S. (2021). Impact of nanotechnology on environment—A review. *International Journal for Modern Trends in Science and Technology, 7*, 159–164. https://doi.org/10.46501/IJMTST0710026

Rajpoot, S. (2024). Eco-friendly solutions for sustainable waste management: An approach towards a cleaner and greener future. *Waste Management and Life Cycle Assessment for Sustainable Business Practice*, 43–64.

Rajpoot, S. (2025). Smart and innovative nanotechnology applications for water purification: Exploring technologies for a cleaner future. In *Emerging trends and technologies in water* (pp. 87–120). IGI Global.

Rana, R., Ganguly, R., & Gupta, A. K. (2015). An assessment of solid waste management system in Chandigarh City India. *Electronic Journal of Geotechnology Engineering, 20*(6), 1547–1572.

Rarotra, S., Sahu, S., Kumar, P., Kim, K. H., Tsang, Y. F., Kumar, V., & Lisak, G. (2020). Progress and challenges on battery waste management: A critical review. *Chemistry Select, 5*(20), 6182–6193.

Ray, S. K., Dey, S., Das, A. P., & John, A. (2024). Microplastic waste management through greener technologies: A review. *Sustainable Management of Urban Plastic Waste Through Circular Economic Approaches, 237–261*. https://doi.org/10.1201/9781003465447-14

Satapathy, S., Kar, D., Priyadarsini, S., Das, A. P., & Kuanar, A. (2024). Application of microorganisms in emerging waste management strategies and remediation of synthetic and cellulosic microfibers. *Sustainable Microbial Technology for Synthetic and Cellulosic Microfiber Bioremediation, 183–206*. https://doi.org/10.1007/978-3-031-62600-5_10

Sharma, K. D., & Jain, S. (2018). Overview of municipal solid waste generation, composition, and management in India. *Journal of Environmental Engineering, 145*(3), 1–18. https://doi.org/10.1061/(asce)ee.1943-7870.0001490.04018143

Shekdar, A. V. (2009). Sustainable solid waste management: An integrated approach for Asian countries. *Waste Management, 29*(4), 1438–1448.

Singh, S. (2020). *Decentralized solid waste management in India: A perspective on technological options* (pp. 290–304). National Institute of Urban Affairs.

Singh, R. K., Yabar, H., Mizunoya, T., Higano, Y., & Rakwal, R. (2014). Potential benefits of introducing inte grated solid waste management approach in developing countries: A case study in Kathmandu city. *Journal of Sustainable Development, 7*(6), 70.

Suk, W. A., Ahanchian, H., Asante, K. A., Carpenter, D. O., Diaz-Barriga, F., Ha, E.-H., & da Silva, E. R. (2016). Environmental pollution: An under-recognized threat to children's health, especially in low-and middle-income countries. *Environmental Health Perspectives, 124*(3), A41–A45.

Sahu, P., Upadhyay, S. C., Liya, K., Gohil, D., Srinivasan, S. V., & Kumar, A. (2024). Waste management and salt recovery from solid residue of tannery common effluent treatment plant. Indian Chemical Engineer, 66(3), 221–233. https://doi.org/10.1080/00194506.2023.2300148

Uiterkamp, B. J. S., Azadi, H., & Ho, P. (2011). Sustainable recycling model: A comparative analysis between India and Tanzania. *Resources, Conservation and Recycling, 55*(3), 344–355.

Vij, D. (2012). Urbanization and solid waste management in India: Present practices and future challenges. *Procedia-Social and Behavioral Sciences, 37*, 437–447.

Waste generation world bank Org. (2015). *Urban Development Series– Knowledge Papers Waste* (pp. 8–12). https://doi.org/10.1787/9789264227385-graph32-en

Wilson, D. C., Velis, C., & Cheeseman, C. (2006). Role of informal sector recycling in waste management in developing countries. *Habitat International, 30*(4), 797–808.

Zhu, D., Asani, P.U., Zurbrugg, C., Anapolsky, S., Mani, S., 2008. Improving municipal solid waste management in India, a sourcebook for policy makers and practitioners by The World Bank Washington, DC In The World Bank. https://doi.org/10.1596/978-08213-7361-3.

Zorpas, A. A. (2020). Strategy development in the framework of waste management. *Science of the Total Environment, 716*, 137088.

Challenges and Opportunities of Solid Waste Management in Vulnerable Communities: Experiences from a Resettlement Colony in Delhi, India

T. C. Prathna, Ashwathy Anand, Siddharth Thyagarajan, Harshita Srivastava, Mirza Firoz Beg, and Prerna Vijaykumar Mehta

Abstract Solid waste is a growing global challenge, and inadequate management poses significant risks to both the environment and human health. Sustainable solid waste management plays a crucial role in achieving several United Nations' Sustainable Development Goals (SDGs), such as clean water and sanitation (SDG6), sustainable cities and communities (SDG11), climate action (SDG13), life on land (SDG15), and responsible consumption and production (SDG12). By fostering a circular urban economy, it promotes the reduction of finite resource consumption, encourages materials reuse and recycling, and contributes to waste elimination, pollution reduction, and cost savings.

In developing countries, rapid urbanization, along with financial and institutional constraints faced by municipalities, makes solid waste management a significant challenge for city authorities. According to the Central Pollution Control Board (CPCB) annual report for 2020–2021, only 50% of the total waste generated in India is treated, while approximately 30% remains unaccounted for mostly coming from unplanned settlements. Inadequate solid waste management often leads to contamination of water resources at both a surface and underground level impacting the entire hydrology.

This chapter explores a bottom-up approach aimed at institutionalizing Joint Accountability Mechanisms (JAM) to improve solid waste management in a vulnerable community in Delhi. JAM offers significant potential for successful community engagement and the long-term effectiveness of water, sanitation, and hygiene

T. C. Prathna · A. Anand · M. F. Beg
World Resources Institute, Delhi, India
e-mail: Prathna.TC@wri.org

S. Thyagarajan · P. V. Mehta (✉)
WRI India, Delhi, India
e-mail: Prerna.mehta@wri.org

H. Srivastava
Kintsugi Regenerating Ecosystems, Unnao, Uttar Pradesh, India

S. Rajpoot, A. P. Das (eds.), *Innovative Technologies for Waste Management*, Sustainable Environmental Waste Management Strategies,
https://doi.org/10.1007/978-3-031-97184-6_12

(WASH) programmes. The chapter elaborates on the experiences of implementing the JAM in Bakkarwala resettlement colony which focused on a people-centred approach, integrating technical and governance aspects.

Focusing on local institutional coherence, user inclusivity, and the empowerment of vulnerable communities—by ensuring their representation and voice in local and city-level decision-making processes—helped bring about positive waste management outcomes.

Keywords Solid waste management · Vulnerable communities · Community engagement

1 Introduction

1.1 A Global Challenge

Solid waste management is a global challenge due to the increasing diversity of waste characteristics and lack of effective implementation of waste policies accompanied by changing lifestyles, increased production processes, and rapid urbanization (Mmereki et al., 2016; Dey et al., 2023; Mishra et al., 2019). In 2020, nearly 2.24 billion tonnes of solid waste was generated worldwide, amounting to a footprint of 0.79 kg/person/day. With rapid increase in population and urbanization, annual waste generation is expected to increase by 73% from 2020 levels to at least 3.88 billion tonnes in 2050 (World Bank, 2022).

1.2 Management in Developing Countries

Compared to the developed world, people in developing countries, especially the disadvantaged, are severely impacted by waste that remains unmanaged or is managed in unsustainable ways. A growing economy has accelerated this, migration to urban centres and the rise of living standards which ultimately increase municipal solid waste generation (Minghua et al., 2009). In lower-income countries, most of the waste generated (over 90%) is often disposed in unregulated dump sites, water ecosystem, or burnt. These practices lead to serious health, safety, and environmental consequences (Sharholy et al., 2008). In addition, inadequately managed waste serves as a breeding ground for disease causing vectors and contributes to climate change. Moreover, unregulated dumping of solid waste into the water ecosystem (Bhat et al., 2022) can result in biomagnification in plants and animals which feed on it while uncollected waste from drains can have an impact on storm water run-off and groundwater quality (Kumar & Prakash, 2020).

Municipalities, usually responsible for waste management in cities, have the challenge to provide an effective and adequate system to its residents. However, these stakeholders often face challenges which are beyond their ability to tackle (Sujauddin et al., 2008) mainly due to lack of organization and financial resources with solid waste management at times comprising 20–50% of the budget (World Bank, 2022). Large amounts of the municipal budgets are usually spent on waste collection leaving very less for waste treatment (Lohri et al., 2014).

Improvement in solid waste management practices has a multidimensional direct and indirect impact on life (Das et al., 2024; Das & Ghosh, 2022). Adequate solid waste management services at various levels can help to realize some of major Sustainable Development Goals (SDGs). For example, better management of waste can ultimately reduce the pollution of water, soil, and air significantly leading to improved health outcomes "for all" the citizens. This can ensure better health outcomes and promote well-being for all at all ages (SDG 3) and improve water quality through minimizing the release of harmful solid waste in the water ecosystem (SDG 6). In addition, it has a great role in making cities and human settlements inclusive, safe, resilient, and sustainable (SDG 11) (United Nations, 2015).

1.3 The Indian Scenario

Considering the scenario of India, data from 2021 to 2022 shows 170,339 tonnes per day (TPD) of waste is generated with a per capita waste generation per person of 123.45 g/day of which 92% is collected. However, only 54% of the total waste generated undergoes some form of treatment (Central Pollution Control Board, 2022). Estimates indicate that of the 8–12 million tonnes of plastic waste which enter the oceans and seas globally, India contributes a significant portion to the problem due to mishandling of waste (Jambeck et al., 2015). As per World Urbanization Prospects of the United Nations, India's urban population is expected to reach 800 million by 2050 (United Nations, 2018) which will further increase the load of solid waste generated. In addition to the significant load of solid waste generated, there is a lack of adequate garbage collection infrastructure, inefficient sorting of recyclable materials, limited recycling rates, and improper disposal practices (Mishra et al., 2024; Tripathy et al., 2024). This results in large quantities of waste accumulating in landfills, leading to environmental pollution and health hazards. If the problem of solid waste management is not tackled within preventive time, it may lead to many other environmental and health issues for communities making the vulnerable communities and women even more vulnerable.

2 The Need for Efficient Solid Waste Management in India

In addition to an ever-increasing load of waste generated, India also faces unique challenges marked by various geographical and socio-economic conditions which complicate the implementation of a uniform waste management strategy. Each region may have distinct waste generation and disposal patterns which necessitates distinct solutions for adequate solid waste management. For example, the per capita waste generation per person per day in Delhi is nearly five times the national average (Central Pollution Control Board, 2022). Moreover, the quantity of municipal solid waste generated also depends on living standards, the extent and type of commercial activity, eating habits, and season. The local economy impacts on waste composition, as high-income groups use more packaged products, resulting in higher volumes of inorganic waste. The challenge, therefore, extends beyond the technical aspects of waste management; it requires a nuanced understanding of the regions to design and implement strategies that resonate with the local population (Kapoor & Chakma, 2024).

In India, waste collection is the responsibility of the urban local bodies (municipal corporations), with these agencies typically spending around USD 6–10 per tonne on solid waste management (SWM) with 70% of this amount spent on collection and 20% spent on transport (Kumar et al., 2017), while less than 5% is spent on treatment and disposal. In general, primary waste collection occurs at the door-to-door level and collection from community bins. In most cases this waste is transported to a secondary storage facility which is usually a covered street bin, enclosure, or concrete bin serving as a transfer point from which it is transported for processing and the residual waste is dumped in sanitary landfills (Ministry of Urban Development, 2016). In addition, the informal sector which includes waste pickers collect household or commercial waste extracting potential value from waste bins, trucks, streets, waterways, and landfills (Kumar et al., 2017). Municipal solid waste comprises 40–60% biodegradable (organic) matter, while recyclable fraction is 10–25% (Ministry of Urban Development, 2016), while the remaining is inert. Composition of waste however varies with size of city, season, and income group.

3 Policy Regulations on Solid Waste Management in India

The Government of India has undertaken several measures to address SWM within urban areas with the 74th Constitutional Amendment Act (which came into force in 1993) providing constitutional status to urban local bodies by initiating a process of democratic decentralization which can pave the way for efficient solid waste management. This was followed by the guidelines provided under the Solid Waste Management Rules 2016 (Central Pollution Control Board, 2021). While the onus of providing SWM services in urban areas lies under the purview of the urban local bodies (ULB), the central and state governments are mandated to define the frameworks under which these ULBs can operate. The central departments that are principally involved with the management of solid waste in urban areas include the Ministry of Environment, Forests and Climate Change (MoEFCC), the Ministry of Housing and Urban Affairs (MoHUA), and the Central Pollution Control Board (CPCB). While the Ministry of Environment, Forests and Climate Change governs the overall monitoring and implementation of SWM rules across the country, the

Ministry of Housing and Urban Affairs and the Central Pollution Control Board periodically review the measures taken by states and ULBs towards implementing and managing solid waste management practices, the implementation, and their adherence to the prescribed norms. At the state and district level, SWM is led by the state departments and State Pollution Control Board which guide the municipal bodies, wherein they prepare a state policy and state level waste management strategy (Ministry of Urban Development, 2016).

Looking at the policy ecosystem, the National Environment Policy (2006) underscored the significance of establishing robust waste management rules and regulations across the country. The National Urban Sanitation Policy (2008) marked a milestone in advancing sanitation and public health initiatives, while the National Action Plan on Climate Change (2018) has highlighted the critical role of effective waste management in mitigating climate change impacts (Dave, 2022). The Plastic Waste Management Rules, 2016, and its amendments in 2018 have laid down the challenges and opportunities in terms of collection, segregation, and recycling of plastic waste. In addition, Swachh Survekshan launched in 2016, is an annual survey of cleanliness, hygiene, and sanitation in villages, cities, and towns across India, launched under the Swachh Bharat Abhiyan. The Swachh Bharat Mission 1.0 (Urban) in 2014, re-launched as Swachh Bharat Mission 2.0 (Urban) during Sept 2021, has been a country-wide campaign which has been pivotal in guiding and creating awareness towards improving solid waste management and sanitation practices within the country.

4 Solid Waste Management in Vulnerable Communities

India's growing urban population exerts immense pressure on urban waste management systems, particularly in slums and marginalized communities where housing and sanitation infrastructure are often rudimentary. Vulnerable communities in India are disproportionately impacted by solid waste management issues because of inadequate infrastructure and lack of civic awareness. The escalating volume of municipal solid waste (MSW) is posing a significant threat to public health, environmental sustainability, and the overall quality of life of citizens in these communities (Chalifour, 2013).

Vulnerable communities often lack the necessary infrastructure and resources to effectively manage the waste generated (Vinti & Vaccari, 2022). Many vulnerable neighbourhoods do not have access to regular waste collection services (Kapoor & Chakma, 2024), leading to waste being indiscriminately dumped on streets, open spaces, and water bodies. This not only pollutes the environment but also poses health risks to residents, particularly children and the elderly. Furthermore, the low awareness and education levels in these communities contribute to improper waste disposal practices. Residents may not understand the importance of segregating waste or recycling materials, leading to higher levels of waste generation and environmental pollution. These deficits severely impact affected residents, who are often forced to manage waste through makeshift dumping grounds (Ferronato & Torretta. 2019). Such practices perpetuate poverty, precarious working conditions, and hazardous exposure to disease vectors, toxins, and environmental contaminants. Local authorities need to implement more strategic and sustainable solid waste management strategies to mitigate these issues.

Addressing solid waste management challenges in vulnerable communities necessitates the simultaneous promotion of sustainable options like segregation, recycling, and communal waste management areas along with integration of digital systems and proper policy implementation. Federal and state-level government authorities should collaborate to develop policies that ensure adequate budgetary allocation to underpin SWM infrastructure improvements, raise public awareness about the seriousness of sludge and waste exposure and the right social practices towards household waste management, and reduce socio-economic vulnerabilities associated with waste management networks. Joint participation from government, stakeholders, and residents will be required to ensure effective and environmentally conscious waste management practices.

Recent initiatives such as the Australia India Water Security Initiative (AIWASI), waste management systems in Sikkim, door-to-door waste collection system in Chennai, etc. demonstrate the potential for effective functioning of such model. They aim to engage residents in waste management practices, such as reducing, reusing, and recycling waste materials. Educational awareness programmes can improve, among residents, knowledge of waste disposal, health risks, and the benefits of segregation. This chapter explores a bottom-up approach model that was used in a vulnerable community in Delhi aimed to improve solid waste management within the settlement with a Gender Equality, Disability and Social Inclusion (GEDSI) perspective which can further help in the transition of the community towards water security.

5 Case Study: Bakkarwala

5.1 About the Settlement

Bakkarwala is a locality in the western part of Delhi with the resettlement colony being a cluster of 4000 households with an urban fabric comprising of 7 blocks. The primary occupations within the settlement are predominantly construction workers, domestic labourers, factory workers, and street vendors.

Bakkarwala has a well-distributed network of open spaces which are used by the community for recreation, gatherings, and celebrating local festivals. Most of these spaces suffer from multiple challenges that include lack of vegetation, poor maintenance, constant waterlogging, and dumping of garbage, the last being one of the most predominant. Challenges of solid waste management have also spurred other issues such as localized flooding and a growing incidence of vector borne diseases because of solid waste collecting within open drains in the community. While many of these challenges can be attributed to issues within the solid waste management system such as irregularity in waste collection, lack of spaces for waste collection, and no provisions for collection of segregated waste, there are issues that also exist at the community level, such as limited awareness and willingness to segregate waste at source and dumping of wastes within the community areas due to lack of spaces for waste collection.

The Australia-India Water Security Initiative (AIWASI), a collaborative project, is a promising endeavour to help this settlement to transition towards a water secure community (AIWASI, 2022) through initiatives focused on water supply, water quality, water logging, and healthy liveable spaces. Solid waste management is a critical component of the initiative which ultimately has an impact on water quality, water logging, and healthy liveable spaces. The demonstration project was informed by overarching Gender Equality, Disability and Social Inclusion (GEDSI) principles entailing a thorough understanding of the different risks, opportunities, barriers, and impacts faced by all sections of the society, including women, persons with disabilities, and other marginalized groups. This nuanced understanding ensured that the activities and solutions benefit everyone equitably and strive for transformative changes and empowerment of vulnerable communities.

5.2 The Project Approach Towards Solid Waste Management in Bakkarwala

To enable the vision of the project which is helping transition the Bakkarwala community towards a water secure and healthier liveable settlement, one of the pathways is to help transition the community into "Zero litter Bakkarwala" by effective solid waste management practices that will contribute to a reduction in localized flooding, improved ground water quality, and achieving healthy public spaces. The project from its inception also focused on meaningful community participation with its focus on women and persons with disabilities that involved co-development of community-level vision, priorities, plans, and implementation of Water Sensitive Urban Design.

Steps towards an integrated and effective SWM can have the following benefits at a household and settlement level in Bakkarwala:

- Integrated and sustainable solid waste management raises the quality of life by reducing water logging, cleaner parks, streets, drains, and better visual appeal.
- Safe disposal of solid waste and the reduction of garbage open burning and its emitting rates of pollution are among the most important measures related to climate change.
- Reducing pollution caused by dumping plastic and hazardous wastes into or near waterways. Increases access to better quality water (ground water, surface water and storm water).

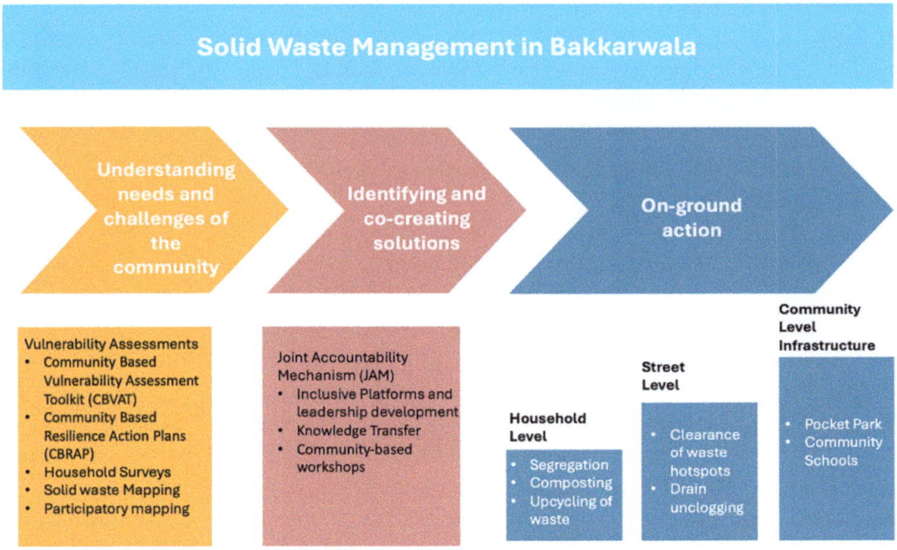

Fig. 1 Approach adapted through the project for solid waste management in Bakkarwala

- Safe handling of waste and reduced dumping in storm water drains can lead to reduced instances of localized water logging.
- Proper handling of solid waste and disposal methods supports a healthy, disease-free environment.
- Recycling of organic solid waste produces very good organic fertilizers and soil nourishers which can be reused for increasing the health of the pocket parks. In addition, the green circular economy model can also reduce the quantum of waste and related emissions incurred in the transfer to landfill.
- GEDSI-focused approach integrates technical solutions and institutional capacities, ensuring that the project addresses the specific needs of vulnerable groups, promotes inclusivity, and empowers the community to actively participate in creating sustainable and resilient urban spaces.

The approach that was followed by the project for solid waste management in the settlement is summarized in Fig. 1.

5.2.1 Vulnerability Assessment

Solid waste management in vulnerable communities demands a nuanced understanding of the interconnected social, economic, and environmental factors at play (Bala et al., 2024; Mishra & Das, 2021). In resettlement colonies like Bakkarwala, these challenges are compounded by inadequate infrastructure, irregular waste collection services, and limited community awareness. The sections below highlight the role of vulnerability mapping as a critical tool for identifying risk zones, systemic inefficiencies, and socio-spatial disparities that impact waste management outcomes.

To achieve this, a range of assessments was conducted. Household surveys established baseline conditions in the settlement, while community-based vulnerability assessments provided a deeper, spatial understanding of the challenges. Additionally, participatory mapping exercise was organized, enabling community members to share their lived experiences and highlight challenges through their perspectives. These approaches collectively offer actionable insights to drive inclusive and sustainable interventions.

5.2.1.1 Household Surveys

A baseline household survey was conducted with a 10% randomized sample from the 4000 households in the community. The survey covered over 100+ questions across multiple themes pertaining to water supply, sanitation, storm water management, solid waste management, open spaces, community participation, and gender equality disability and social inclusion. The analysis from the study on solid waste management in the community revealed that 63% of the respondents stored waste in dustbins without segregation, while 31% stored it in bags without segregating the waste. Another finding that emerged from the study was that 42% of respondents said that public spaces in the community were used as solid waste dumping areas. The baseline survey indicated a need for sensitization and capacity building focused on segregation at source and to understand the root cause of solid waste dumping in public spaces.

5.2.1.2 Community-Based Vulnerability Assessment Toolkit (CBVAT)

A more nuanced understanding of the solid waste management practice was understood through the community-based vulnerability assessments (CBVAT) which was a tool used to understand the community and their vulnerabilities through their perspective using various assessments, like histogram, Moser framework, impact matrix, adaptive capacity, and root-cause analysis. The assessment equally focused on facilitating the ideation process with the community so that they can develop and suggest resilient plans to cope with their challenges. CBVAT analysis indicated that at least 50% of the community was vulnerable to inefficient solid waste management. Moreover 40% of the respondents indicated their vulnerability to incidents of localized flooding. It is significant to note that water logging events witnessed in the settlement were exacerbated by instances of solid waste dumping in internal drains obstructing storm water run-off.

5.2.1.3 Participatory Mapping Exercise

An extensive mapping exercise was conducted in the community which significantly guided in acquiring spatial and temporal information through crowd sourcing from the communities on different amenities and resources related to aspects including open spaces and liveable conditions. This activity accompanied a better identification and understanding of vulnerable spots in the community thereby highlighting avenues for work.

A social mapping semi-structured interview schedule was used as the tool to gather information on solid waste management, health, and hygiene from the community. This exercise further confirmed that open spaces were left under-utilized or used as dump yards for solid waste as recorded from the household survey.

5.2.1.4 Solid Waste Tipper Route Mapping

With the household surveys, CBVAT, and participatory mapping exercises indicating underutilization of public spaces due to solid waste dumping and vulnerabilities existing in the community due to inefficient SWM, detailed studies were conducted to examine the waste collection system, identify its gaps, and explore improvements through community participation. The waste collecting tipper routes, waste hotspots, and the existing waste collection system were analysed and mapped.

The key findings of this mapping indicated that the community did not face challenges with the route of the waste tipper however with the halt time being less the community faced challenges with solid waste disposal in the tipper. The issue that came to light during the mapping was the lack of a fixed schedule/timetable followed by the collectors which hindered in the daily activity planning and livelihood generation for a household. Hence the need was highlighted for the frequency and timing of the collection needed to be fixed. The existing SWM route map prepared during this exercise is shown in Fig. 2.

After delving into the issues related to water, sanitation, and hygiene in their area, a microplanning exercise, called community-based resilience action plan (CBRAP) was done with the participants to cocreate relevant solutions and identify persons responsible to work on each specific solution and tentative timelines to act on those solutions.

5.2.2 Empowering the Change Makers

In the Indian context, as guided by government mandates, agencies are primarily responsible for providing and facilitating services, while communities have a more limited role, typically restricted to paying taxes and, in exceptional cases, raising issues specific to their local areas. These rare dialogues regarding area level concerns become even more scarce in vulnerable communities where there is meagre to nil knowledge or understanding about such discussion opportunities. Besides, there

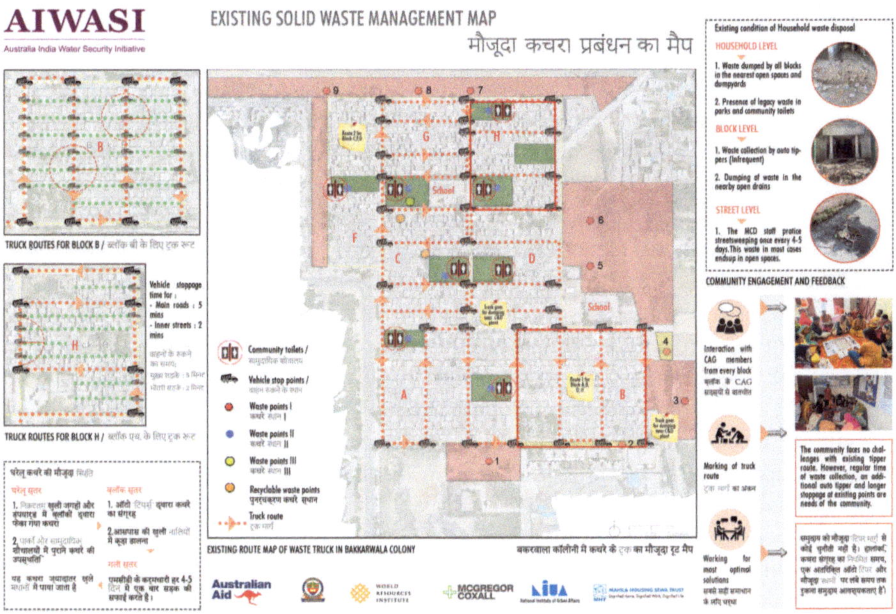

Fig. 2 Solid waste tipper route map in the resettlement

is a lack of processes that can facilitate cocreation of need-based solutions that involves multiple stakeholders including community members.

The project, by deploying consortia approach, setting up collaborative governance mechanisms, community capacity building activities, and awareness building programmes, has made an attempt to facilitate the need-based, cocreation process for solutions around healthy urban spaces. The project helped in ensuring representation and voices of marginalized communities in governance structures to bring about lasting positive changes This approach is not only bridging the gap between government and community but is also evolving as a breakthrough technique for empowering women, youth, and persons with disability (PwD). This is an important transition quite contrary to otherwise a conservative social setting where women and PwD have limited to no role in community matters. Despite all sociocultural barriers, AIWASI project has inspired a significant on ground change, where vulnerable communities including women and PwD are claiming important spaces of dialogues with stakeholders including government representatives to resolve their community challenges.

As identified in the previous section, one of the major challenges of the community was the absence of healthy liveable open spaces due to solid waste dumping. Empowering the community to work towards efficient solid waste management was powered through (1) sensitization of the existing community action groups to reach out to government authorities and work together as part of Joint Accountability Mechanism and (2) capacity building through trainings to reduce waste generation at source with opportunities to reuse together with (3) on-ground action.

5.2.2.1 Community Sensitization for Joint Accountability

One significant challenge that emerged from the community dialogues was the absence of a mechanism or platform for residents to raise concerns about solid waste management with government agencies. To address this, a Joint Accountability Mechanism (JAM) was conceived, enabling community members and government authorities to collaboratively identify the root causes of these issues and work together to resolve them (Chandra et al., 2022). The JAM is a platform that relies on trust and transparency and promotes collaboration by creating a shared space for dialogue and negotiations.

As part of this process, a Community Water Forum was established, serving as a platform for residents to regularly convene to discuss their concerns and present these to the respective government authorities. This one-of-a-kind forum in the community has convened key members of the community including community action groups, local government bodies, ward-level elected representatives, and youth (Chandra et al., 2022). This forum was designed to ensure representation from all blocks within the community with focus on involving women and persons with disability, ensuring that challenges from across different areas were brought to the table.

These gatherings not only facilitated constructive dialogue between community members and government officials but have also provided a means for the community to track progress on commitments made by the authorities and demand work that required urgent action by government departments. Simultaneously, it also offered government representatives an avenue to seek support from the community when needed, fostering a collaborative and accountable approach to addressing solid waste management challenges within the community.

Through the project the members of the community were sensitized and trained on modules such as existing governance structures, maintenance of services, health and hygiene, as well as concepts of basic climate change. Modules on map sharing helped introduce the community to maps as an effective tool of communication with government stakeholders. Trainings incorporated through these modules helped the community understand the power of collectives, demystify technical interventions, and better understand the governance structures and ways to approach government officials with applications, maps for communication, and registering complaints for identified challenges in the community.

5.2.2.1.1 Cleaning of Waste Hotspots and Drains

One of the major issues identified in the participatory mapping exercises and solid waste mapping was the presence of solid waste hotspots with at least eight hotspots identified by the community. The action group members were sensitized to reach out to government representatives and have been advocating and following up for clean-up of these hotspots with the ULBs to clean these hotspots. The action groups are primarily led by women from different blocks of the settlement and play a crucial role in identifying and addressing community concerns. With the action groups being sensitized on Joint Accountability, the community did not stop with this and have organized clean-up drives in the pocket parks of the community. The community led initiative of waste clean-up had one primary objective that "solid waste management" starts at home. Through these drives the action groups are sensitizing the community regarding the issues they face due to ineffective practices of handling solid waste and what can be the most appropriate yet simple solutions. The drives have been simple yet effective—the participants took it upon themselves to (a) segregate their household waste into dry and wet waste and (b) clean the parks of the neighbourhood with the support from the ULB officials where the plastic waste could be weighed and sent for recycling.

One of the major reasons behind localized water logging was due to clogging of internal drains within the community. This was the result of solid waste dumping in the drains. To improve sanitation services within communities through registration of complaints for uptake and action by civic authorities, there exists an app of the ULB called the MCD 311 app. The community through the project was trained to register complaints to ensure the drains cleaned of solid waste. The skilled, trained, and empowered community members have, during the project, acted on the plans and advocated for improvement in the solid waste management issues with the government agencies. The issue of solid waste management has been acknowledged at multiple levels, with the action groups actively working towards solutions to address these challenges. Parallelly the need to train the community on practices to reduce waste generation at source and waste reuse was recognized.

5.2.2.2 Capacity Building Workshops

5.2.2.2.1 Waste Segregation

Sorting and segregating waste at the household level is vital in implementing successful and sustainable waste management systems (Lunag et al., 2021). The first step in waste management is waste segregation. Waste classified as biodegradable or non-biodegradable can then help in facilitating waste management. However, very few households in India separate their waste (Sahoo et al., 2022). The situation in Bakkarwala is no different with over 90% of households not segregating waste at the baseline of the project.

Every member of the family generates waste at the household level and should therefore also act as a waste segregator (Hage et al., 2009). In line with this, youth, men, women, and physically challenged members of the community partook in these trainings with an average of 50–60 members attending each workshop.

These workshops started with clean-up drives in two pocket parks. Interesting observations were made during these clean up drives with notable differences in the waste types dumped in both the parks emphasizing the need to tailor waste management strategies and helped better guide the waste segregation training. The waste collected through these drives was thereafter segregated into certain categories such as paper waste, organic waste, soil and hard plastic waste, non-recyclable waste, glass, and mixed waste. The organic wet and paper waste collected were added to the compost bin established during the workshop to demonstrate composting.

The waste segregation workshops aimed to reinforce best practices that facilitate efficient recycling and minimize landfill contributions. This hands-on sorting and cleaning the waste process not only provided immediate environmental impact but also illustrated the volume and composition of everyday waste, making the training highly tangible. Studies show that effective waste management starts with segregation at the source; our activities echoed this by involving communities in detailed sorting tasks, creating a foundational understanding of how waste is processed, recycled, or composted. The categorization of waste helped furthermore during the following workshops on composting and upcycling of waste. Through all these workshops, the primary objective was to tap into existing knowledge of the community members and streamline the potential in a more effective output.

5.2.2.2.2 Composting

Composting is essential for converting organic waste into valuable resources, reducing landfill use, and improving soil health (Yu et al., 2019). As organic waste generation rises globally, various composting techniques have emerged, from aerobic methods like Berkeley and Indore composting to anaerobic techniques like Bokashi and biogas digesters. Each method, whether rapid aerobic processes or slower anaerobic approaches, is tailored to different waste types, climates, and goals. With composting being considered an effective way to safely manage organic waste (Ayilara et al., 2020), training workshops on composting were provided.

The training workshops were practical in nature with each participant bringing in household wet waste every day, and by the end high-quality, odour-free compost was produced. This success inspired participants who later adopted the technique at home, confidently producing their compost and creating nutrient-rich manure for household plants.

5.2.2.2.3 Upcycling of Waste

Bioenzymes

One of the most innovative aspects of the workshops were the trainings on bioenzyme production. Bioenzymes, also known as eco-enzymes or garbage enzymes, are organic solutions produced by fermenting organic waste (such as fruit peels), jaggery (or brown sugar), and water (Dhavale et al., 2020). They are an environmentally friendly alternative to many chemical-based cleaning products and can be used in a variety of ways, including as household cleaners, plant fertilizers, or even natural pesticides (Joshi & Gupta, 2023). The production of bioenzymes is a sustainable practice that not only reduces organic waste but also generates a powerful cleaning

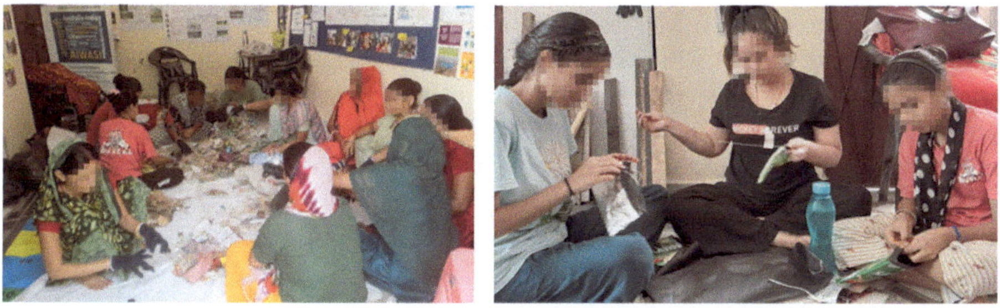

Fig. 3 Women from the community being trained on waste segregation and upcycling of waste

solution full of active enzymes, beneficial bacteria, and organic acids. These byproducts have powerful cleaning, disinfecting, and decomposing properties, which make bioenzymes versatile for household and agricultural use.

Eco-Bricks and Creating Pouches

Plastic waste is a global environmental problem. Plastic is widely used in everyday life because it has advantages such as being strong, light, and stable (Agustang et al., 2022). With majorly plastic waste finding its way into the drains and the waste hotspots of Bakkarwala, the need to upcycle this plastic waste into usable materials through creating livelihood opportunities as well was identified.

Eco-bricks is a product of upcycled one time use plastic waste made from stuffing plastic wastes to the optimum capacity of the plastic bottle. In doing so the bottle gets heavier and is called an eco-brick. Many such eco-bricks put together can, for example, make play installations for children. Similarly, the community were trained to make pouches and purses through upcycling of plastic wrappers thereby creating a livelihood opportunity as well.

Through structured training workshops as part of capacity building the community members were empowered with both practical and ecological knowledge to adopt sustainable practices (Fig. 3).

5.2.3 Pilot Demonstration Activities

5.2.3.1 Composting Bin in Pocket Park

As the organic fraction of the waste in India makes up for a significant portion of the total waste (Kalamdhad et al., 2009), composting at a household and community level can significantly reduce the waste ending in sanitary landfills. In addition to the capacity building workshops on household composting, a community composting bin was set up in one of the pocket parks to treat horticultural waste. This was done to further demonstrate composting at a community level. The system was a cold composting system which is a nutrient rich natural composting method. The system is easy to maintain and does not require much area and is therefore ideal for vulnerable communities which have the problem of organic waste in their parks and community spaces. These low-cost, low-maintenance composting approaches have paved the way for ongoing, community-driven sustainability practices, with members increasingly engaged in eco-friendly habits and eager to innovate within their own households.

5.2.3.2 Waste Segregation Bins in Schools

Reconnaissance visits to the schools in the community indicated that while they had provisions for regular solid waste collection from their premises, they lacked proper large capacity dustbins which could facilitate waste segregation and storage. To help facilitate waste segregation at source, large capacity waste segregation bins were installed in the school premises. The students are being trained on good practices of solid waste management. This demonstration activity can potentially reduce the challenges of solid waste disposal within community areas.

5.2.4 Challenges in the Journey

The journey towards making Bakkarwala resettlement colony as a zero-waste colony has just begun. While there has been a witnessable transformation towards improved solid waste management practices in the community, there are many challenges that continue to persist. The Joint Accountability Mechanism (JAM), facilitated through the community forum, has played a crucial role in urging government agencies to acknowledge and address the community's concerns. These dialogues have successfully highlighted the community's struggles to government officials, prompting actions at various levels. However, these responses have been predominantly reactive, underscoring the need for a more comprehensive and strategic approach to solid waste management from the government.

The community's persistent advocacy and action for improved waste collection services has delivered positive outcomes, but these measures are yet to be scaled and systematized across all areas of the community. Similarly, the initiative in one of the Pocket Parks with the composting bin set up has set a strong example of community-driven efforts to create cleaner and more accessible public spaces. However, this success needs to be extended to other open areas in the community, where waste dumping remains a significant issue.

Finally, while many community members have been educated on good waste management practices such as segregation, composting, and turning waste into resources, the adoption of these behaviours is still limited, indicating the need for sustained awareness and engagement efforts.

6 Conclusion and Way Forward

Efficient solid waste management and informed and empowered community play a critical role in creating water secure neighbourhoods. Waste dumping is a challenge in Indian cities, and maintaining good solid management practices is intrinsically related to secure water infrastructures and ecosystems. There can be various approaches in working towards efficient solid waste management, and this chapter provides one approach which can be used in vulnerable settlements. Solid waste management has always been the responsibility of public sector and civic authorities; however, it is more of a shared responsibility with the community. In addition to targeted capacity building of the community, the project demonstrated an example of Joint Accountability where members of the community have made use of the forums established for co-creating solutions with the government stakeholders, taking ownership of solutions as well as encouraging other community participants to practice good solid waste management at their household level such as waste segregation at source and waste recycling. The approaches have also provided upskilling which have opened livelihood opportunities. Training youth and women members in the community on better solid waste management practices have an impact on their liveability and reduce their health impacts.

As a way forward, the learnings from the project have demonstrated how an empowered community can be the ears and eyes in the settlement to see how waste is efficiently managed. A thorough understanding of the ground realities from a GEDSI lens can play a significant role in informing and shaping the ongoing programme and projects. Providing technical and financial support and addressing the intersectional inequalities are equally important for sustainability. Enhanced participation and empowerment of communities especially women and people with disability as contributors, leaders, and decision makers can contribute to and sustainability of such initiatives. Managing solid waste requires concerted efforts at the household level ensuring that there is waste segregation at the source, recycling, and at the community level ensuring the waste is not dumped in drains and open spaces. This has an impact on the water quality, also reducing instances of localized water logging as well as a step in the direction towards reclaiming liveable spaces. In addition to upskilling of vulnerable communities, a setup which can provide vocational opportunities within the community also helping in creating livelihood opportunities is an action the public sector can look at to ensure reduced waste load reaching landfills. Lastly, identifying and nurturing leadership especially from women and vulnerable sections of the community is essential for achieving desired outcomes and ensuring representation and voice of marginalized communities in governance structures can bring about lasting positive changes.

Acknowledgements This work was supported by the Australia-India Water Security Initiative. The authors acknowledge that this is a consortium led project with partners from World Resources Institute, Mc Gregor Coxall Private Limited, Mahila Housing and Trust and National Insititute of Urban Affairs.

References

Agustang, A., Oruh, S., & Agustang, S.D.M.P. (2022). Building environmental awareness through the Makassar eco-brick community social movement in plastic waste management. In *SHS Web of Conferences* (Vol. 149, p. 02005).

Australia India Water Security Initiative (AIWASI). (2022). Retrieved November 20, 2024, from https://aiwasi-cdp.wricitiesindia.org/

Ayilara, M. S., Olanrewaju, O. S., Babalola, O. O., & Odeyemi, O. (2020). Waste management through composting: Challenges and potentials. *Sustainability, 12*(11), 4456.

Bala, K., Anshul, Verma, S., Dogra, A., Das, A. P., Kumar, R., & Nandini. (2024). Sustainable management for urban plastic waste generation. In *Sustainable management of urban plastic waste through circular economic approaches* (pp. 211–236). Routledge. https://doi.org/10.1201/9781003465447-13

Bhat, R. A., Singh, D. V., Qadri, H., Dar, G. H., Dervash, M. A., Bhat, S. A., Unal, B. T., Ozturk, M., Hakeem, K. R., & Yousaf, B. (2022). Vulnerability of municipal solid waste: An emerging threat to aquatic ecosystems. *Chemosphere, 287*(3), 132223.

Central Pollution Control Board. (2021). *Solid waste management rules, 2016*. Retrieved November 20, 2024, from https://cpcb.nic.in/uploads/MSW/SWM_2016.pdf

Central Pollution Control Board. (2022). *Annual report 2021–2022 on implementation of solid waste management rules, 2016*. Retrieved November 12, 2024, from https://cpcb.nic.in/uploads/MSW/MSW_AnnualReport_2021-22.pdf

Chalifour, N. (2013). The economic costs of solid waste management: Case studies from India. *Resources, Conservation and Recycling, 74*, 86–95.

Chandra, P., Prathna, T. C., & Anand, A. (2022). Localizing joint accountability mechanism to achieve SDG 6.4. *eGov Newsletter, 18*(12), 11–13.

Das, A., & Ghosh, S. (2022). Role of microorganisms in extenuation of mining and industrial wastes. *Geomicrobiology Journal, 39*, 173–175. https://doi.org/10.1080/01490451.2022.2038953

Das, A., Ray, S. K., Mohanty, M., Mohanty, J., Dey, S., & Das, A. P. (2024). Ecotoxicity of microplastic wastes and their sustainable management: A review. *Environmental Chemistry and Ecotoxicology, 6*, 144–152. https://doi.org/10.1016/J.ENCECO.2024.05.003

Dave, D. (2022). Journey of waste management regulations in India: A historical account. *International Journal of Basic and Advanced Sciences, 11*(4), 96–101.

Dey, S., Das, A., Mallick, K., Sahu, A., & Das, A. P. (2023). *Environmental petroleum waste: Pollution, toxicity, sustainable remediation* (pp. 159–175). Springer. https://doi.org/10.1007/978-3-031-48220-5_7

Dhavale, V., Shimpi, T., Koli, V., Kamble, R., & Patil, S. B. (2020). Bioenzymes from organic waste. *Journal of Emerging Technologies and Innovative Research, 7*(8), 196–200.

Ferronato, N., & Torretta, V. (2019). Waste mismanagement in developing countries: A review of global issues. *International Journal of Environmental Research and Public Health, 16*, 1060.

Hage, O., Söderholm, P., & Berglund, C. (2009). Norms and economic motivation in household recycling: Empirical evidence from Sweden. *Resources Conservation and Recycling, 53*, 155–165.

Jambeck, J. R., Geyer, R., Wilcox, C., Siegler, T. R., Perryman, M., Andrady, A., Narayan, R., & Law, K. L. (2015). Plastic waste inputs from land into the ocean. *Science, 347*, 768–771.

Joshi, V. S., & Gupta, S. M. (2023). Production of bioenzyme from kitchen waste by anaerobic fermentation using mixture of yeast and fungal culture. *World Journal of Pharmacy and Pharmaceutical Sciences, 12*(8), 1137–1141.

Kalamdhad, A. S., Singh, Y. K., Ali, M., Khwairakpam, M., & Kazmi, A. A. (2009). Rotary drum composting of vegetable waste and tree leaves. *Bioresource Technology, 100*, 6442–6450.

Kapoor, A., & Chakma, N. (2024). Challenges of solid waste management in Urban India. In *EAC-PM working paper series* (pp. 1–25).

Kumar, M., & Prakash, V. (2020). A review on solid waste: Its impact on air and water quality. *Journal of Pollution Effects and Control, 8*(4), 1–3.

Kumar, S., Smith, S. R., Fowler, G., Velis, C., Kumar, J., Arya, S., Rena, Kumar, R., & Cheeseman, C. (2017). *Challenges and opportunities associated with waste management in India* (Vol. 4, p. 4160764). Royal Society Open Science.

Lohri, C. R., Camenzind, E. J., & Zurbrügg, C. (2014). Financial sustainability in municipal solid waste management—Costs and revenues in Bahir Dar, Ethiopia. *Waste Management, 34*, 542–552.

Lunag, M. N., Elauria, J. C., & Burguillos, J. D. (2021). Community-based bin design approach: An initial stage toward urban composting at a hill station, Philippines. *Environment, Development and Sustainability, 23*, 3832–3852.

Minghua, Z., Xiumin, F., Rovetta, A., Qichang, H., Vicentini, F., Bingkai, L., Giusti, A., & Yi, L. (2009). Municipal solid waste management in Pudong New Area, China. *Waste Management, 29*(3), 1227–1233.

Ministry of Urban Development. (2016). Swachh Bharat Mission. *Municipal solid waste management manual. Part II: The manual.* Retrieved November 14, 2024, from https://mohua.gov.in/upload/uploadfiles/files/Part2.pdf

Mishra, S., & Das, A. P. (2021). Current treatment technologies for removal of microplastic and microfiber pollutants from wastewater. In *Wastewater treatment: Cutting-edge molecular tools, techniques and applied aspects* (pp. 237–251). https://doi.org/10.1016/B978-0-12-821881-5.00011-8

Mishra, S., Rout, P. K., & Das, A. P. (2019). Solar photovoltaic panels as next generation waste: A review. *Biointerface Research in Applied Chemistry, 9*(6), 4539.

Mishra, S., Dash, D., & Das, A. P. (2024). Isolation and molecular characterization of polyester-degrading bacterial strains from municipal wastewater and their application in microplastic degradation in contaminated wastewater. *Geomicrobiology Journal, 41*, 732. https://doi.org/10.1080/01490451.2024.2376101

Mmereki, D., Baldwin, A., & Li, B. (2016). A comparative analysis of solid waste management in developed, developing and lesser developed countries. *Environmental Technology Reviews, 5*(1), 120–141.

National Environment Policy (2006). Retrieved October 28, 2024, from https://ibkp.dbtindia.gov.in/DBT_Content_Test/CMS/Guidelines/20190411103521431_National%20Environment%20Policy,%202006.pdf

National Urban Sanitation Policy (2008). Retrieved October 28, 2024, from https://www.mohua.gov.in/upload/uploadfiles/files/NUSP_0.pdf

National Action Plan for Climate Change and Human Health (2018). Retrieved October 28, 2024, from https://ncdc.mohfw.gov.in/wpcontent/uploads/2024/04/27505481411548674558.pdf

Sahoo, K. C., Soni, R., Kalyanasundaram, M., Singh, S., Parashar, V., Pathak, A., Purohit, M. R., Sabde, Y., Stålsby Lundborg, C., Sidney Annerstedt, K., Atkins, S., Rousta, K., & Diwan, V. (2022). Dynamics of household waste segregation behaviour in Urban Community in Ujjain, India: A framework analysis. *International Journal of Environmental Research and Public Health, 19*(12), 7321.

Sharholy, M., Ahmad, K., Mahmood, G., & Trivedi, R. C. (2008). Municipal solid waste management in Indian cities—A review. *Waste Management, 28*(2), 459–467.

Sujauddin, M., Huda, S. M. S., & Hoque, A. T. M. R. (2008). Household solid waste characteristics and management in Chittagong, Bangladesh. *Waste Management, 28*(9), 1688–1695.

Tripathy, B., Dey, S., Dubey, N. K., Lo, W.-C., & Das, A. P. (2024). *An omics-based approach for managing the risk of COVID-19-generated wastes* (pp. 51–66). Springer. https://doi.org/10.1007/978-3-031-50840-0_4

United Nations. (2015). *Transforming our world: The 2030 agenda for sustainable development.* Retrieved October 26, 2024, from https://sdgs.un.org/publications/transforming-our-world-2030-agenda-sustainable-development-17981

United Nations. (2018). *World urbanization prospects*. Retrieved October 26, 2024, from https://population.un.org/wup/

Vinti, G., & Vaccari, M. (2022). Solid waste management in rural communities of developing countries: An overview of challenges and opportunities. *Clean Technologies, 4*(4), 1138–1151.

World Bank. (2022). Solid waste management. Retrieved October 28, 2024, from https://www.worldbank.org/en/topic/urbandevelopment/brief/solid-waste-management

Yu, H., Xie, B., Khan, R., & Shen, G. (2019). The changes in carbon, nitrogen components and humic substances during organic-inorganic aerobic co-composting. *Bioresource Technology, 271*, 228–235.

Agricultural Waste: A New Leap Toward Encouraging Global Economy

Sayak Chakraborty, Swastika Rakshit, Ashmaul Husna, Prajita Kundu, Aditi Nayak, and Snehalata Majumdar

Abstract Agriculture, with its allied sectors, is undoubtedly the largest livelihood provider in India and contributes a large figure to the gross domestic product (GDP). Sustainability in agriculture is therefore important in terms of food security, rural employment, and environmentally sustainable technologies such as soil conservation, sustainable natural resource management, and biodiversity protection. Agricultural waste (AW) generally comprises of wastes like food processing waste (about 20% of maize is canned and 80% is waste), crop waste, animal waste, and toxic waste used in farming operations. Globally, around 998 Mt of AW is generated annually. Every year, in India, around 500 Mt of crop residue is produced; a major portion of it is used as fodder and fuel. Yet, an excess amount of 140 Mt is remained and about 92 Mt is burned annually. Paradoxically, generations failed to recycle or utilize this energy efficiently in agriculture sector. Population growth, food preferences, and economic wealth are variables that drive agricultural production, although agriculture does produce raw materials other than for food. The world population of 9 billion projected for 2050 will require food production to increase by 50–70%, with raw materials and waste increasing by the same percentages. There will be no significant changes in the amount of arable land; the increase in production will come about through agricultural intensification, the disparity between real and potential yield being known as the yield gap. Innovation in management techniques will focus on optimizing water use, both for irrigated and rain-fed crops. In addition, nutrient management will have to be intensified to fill the production gap. Sustainable management focuses on improving nutrient efficiency, but also includes waste management. Sustainable agriculture must introduce a circular economy model and reverse logistics for waste management, and in particular to the nutrient cycle.

S. Chakraborty · S. Rakshit · A. Husna · A. Nayak (✉) · S. Majumdar (✉)
Guru Nanak Institute of Pharmaceutical Science and Technology, Kolkata, India
e-mail: bm23.0106sayc@gnipst.ac.in; bm23.0074swar@gnipst.ac.in; mbt23.0487ashh@gnipst.ac.in; aditi.nayak@gnipst.ac.in; snehalata.majumdar2023@gnipst.ac.in

P. Kundu (✉)
Chemistry Department, Rajabazar Science College, University of Calcutta, Kolkata, India

S. Rajpoot, A. P. Das (eds.), *Innovative Technologies for Waste Management*, Sustainable Environmental Waste Management Strategies,
https://doi.org/10.1007/978-3-031-97184-6_13

Keywords Agriculture · Agricultural waste · Management · Gross domestic product · Global economy

1 Introduction

The global population has almost tripled in the last 50 years, which presents the need for increased food demand and productivity (Koul et al., 2022). To meet the increasing demand, crop productivity and livestock have also steadily increased, eventually leading to an elevated release of agricultural waste (Tripathi et al., 2019). Along with a steep increase in population, countries like Africa, India, and China have also seen exponential growth in agricultural waste production. The annual production rate of solid waste in India is quite high; however, agricultural wastes holds the major share accounting for approximately 1000 Mt/year. China is followed by India as the second-highest producer of agricultural wastes, of which paddy straw is one of the significant contributors, as almost half of the straw wastes is discarded after the other half is utilized as animal fodder (Wang & Tao, 2020). However, some of the traditional burning practices of the crop residues lead to a substantial amount of pollution (Shyamsundar et al., 2019). In this process, several greenhouse gases are released like methane, nitrous oxide, carbon dioxide, and other greenhouse gases (Searchinger et al., 2020).

The considerable types of agricultural wastes that have immobilized the stability of agricultural sustainability include crop residues (weeds, seed husks, straw remains, leaf residues, stalks, seed pods), livestock discards (waste fodder, dung, urine, residual milk, wash water), poultry remains (bedding material, droppings, feathers, food remains), slaughterhouse remains (flesh, blood, hair, hides, etc.), industrial-agro wastes (fruit and vegetable peels, molasses, fruit and vegetable pulps), bagasse, and oil-seed cakes (palm, coconut, soybean, mustard, etc.) (Seidavi et al., 2021).

The highest biomass-producing biological sector which holds a position of paramount importance in the global economy is agriculture. The various bioeconomic strategies facilitate agricultural waste management and can further address several concerns like underutilized livestock remains and frequent burning of crop residues to satisfy the food and safety demand. This also has immense possibilities for improving the value-added products, job prospects, and agricultural sustainability (Duque-Acevedo et al., 2020). The major goal of researchers and world leaders is to meet the exceeding food and energy demand for the increasing population that is predicted to exceed 10 billion in the next 25 years. However, the demand should be met with lowered fuel consumption, decreased pollutants, and minimal solid waste (Bracco et al., 2018). Several management bodies have already materialized this challenge and are aimed to provide sustainable growth to the European countries. The implementation further ensures the execution of the policies toward a sustainable future, with special emphasis to circular economy and bioeconomy (Duque-Acevedo et al., 2020). All the perspectives are focused on a sustainable economy concerning both economic aspects and societal sustainability.

The generated agricultural wastes frequently remain un-moved from their site of generation, many a time leading to environmental pollution. There are a wide array of pollution caused by the agricultural wastes that require urgent redressal and mitigation. The present book chapter focuses on the different types of agricultural waste and their pollution-causing effects, economic approaches toward a robust bioeconomy, and various government policies issued for the implementation of the government policies and rules.

2 Pollution Caused by Agricultural Wastes and Importance as well as the Need for the Practice

Agricultural wastes are generally termed agro-waste, generated in large quantities yearly on farms in India. A large quantity of crop waste is generated each year in India. Out of which about 50% is contributed by agricultural and commercial purposes like livestock feed, paper manufacturing, roof sheeting, and power production (Kumar Sarangi et al., 2022). Agricultural wastes or residues can cause pollution of aquatic and territorial resources causing little changes in air quality. As a result, agricultural waste can be a cause of the destruction of surface and underwater resources (Nagendran, 2011). The freshwater habitat of basins in the region of Tibet is bound with the protection of the Asian Water Tower. In the recent study, the export of nitrogen and phosphorus to freshwater systems was assessed to estimate the seasonal and annual trends of aquatic pollution. Nutrient leaching from the agricultural residues, especially nitrogen and phosphorus, changes the chemical composition of the freshwater and contributes to increased toxicity levels in aquatic predators (Li et al., 2022). Chemicals carried by runoff after rainfall which are used in pesticides and fertilizers development are the main contributors to the deteriorating quality of nearby water bodies (Nagendran, 2011). Agricultural discharge or wastes can lead to human health hazards, for example, pathogen content may significantly rise in stagnant aquatic bodies, upon receiving such runoff. When these contaminated water supplies are applied to meet thirst needs, water-related illnesses may arise, mostly in rural areas (Nagendran, 2011). Figure 1 includes the types of possible pollution caused by agricultural wastes.

2.1 Water Pollution

Agriculture waste is now considered one of the main sources of global water pollution ultimately overtaking cities and industries in many areas. Farm pollutants include crop chemicals, food components, soil enhancers (nitrates, phosphates), insecticides, biomass, and drug wastes from animal and aquafarming, degradation, saltwater seepage, plastic wastes, and pathogens from animal manure. This leads to reduced water flow in rivers and lakes, intensifying pollution. Extra nutrients can

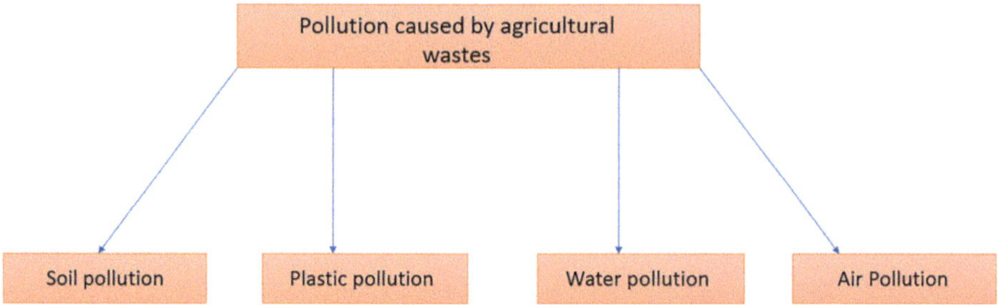

Fig. 1 Types of agricultural wastes

lead to algal blooms in aquatic bodies, damaging ecosystems and agriculture driving main water pollution in the USA or China, and igniting concerns in developing countries. Waste-driven water standard issues in agriculture are complicated and need collaborative effort and approach for a suitable understanding of them. As an overview, agricultural wastes lead to extra nutrients like nitrogen and phosphorus in aquatic bodies. This process, called eutrophication, induces the overgrowth of algae, decreases oxygen levels, and damages aquatic life (Evans et al., 2019). Water pollution can lead to various metabolic disruptions in humans and animals, for example, eutrophication can lead to the killing of fish present in aquatic bodies, or humans depending on contaminated water for drinking, irrigation, or recreation may face serious health hazards, including gastrointestinal issues, reproductive diseases, and neurological abnormalities due to exposure to toxic chemicals and pathogens. Agricultural wastes can also aggregate in the food chain, targeting animals and humans who take up contaminated fish and crops. Additionally, pollution of freshwater resources poses risks to the availability of purified and clean water for households as well as industries, escalating global water scarcity. These underline the necessity for proper agricultural practices to shield water sources and guarantee the well-being of all species. The pollution of groundwater with nitrates—an agricultural toxin—can cause serious risks, especially to infants, as it can cause conditions like methemoglobinemia (blue baby syndrome). The animal world being exposed to contaminated water resources can suffer from reproductive issues and changed behavior, possessing risks for their survival. Polluted water applied for irrigation toxifies crops, affecting food security and health. Beyond health-related issues and ecological threats, agricultural water imposes extensive economic burdens, including the price of water treatment, healthcare, and losses in aquaculture and tourism industries. Climate alterations can relate to challenges, deepening runoff, and expanding pollutant loads in aquatic bodies. Addressing these issues of water pollution requires proper practices like integrated pest development, controlled fertilizer usage, and riparian buffer areas to secure water sources and safeguard the safety of humans, animals, as well as ecosystems.

2.2 Soil Pollution

Microplastic pollution deriving from agricultural residues can reduce the overall fertility of the soil; also animal wastes containing pathogens might change the pH of the soil. Plastic wastes left in fields can be a main cause of micro- and nanoplastic which combine with the soil, distort microbial communities, and damage soil structure and moisture-holding capacity. The problem of soil contamination from plastics and micro-nano formation depends on the type of agricultural plastic used and how it is utilized. The danger is greater with single-use, customary plastics meant for direct connection with the soil or deliberately combined into it (Briassoulis, 2024). With time, chemical wastes aggregate in the soil, making it harmful to plants and animals that depend on soil for survival. Heavy metals like cadmium, arsenic, and lead, often present in pesticides, lead to a significant threat as they tend to enter the food chain, affecting both land and aquatic life. Humans being exposed to soil contaminated with agricultural residues can face severe health problems. Intake of crops grown in intoxicated soil may lead to chronic diseases, including cancer, neurological problems, and hormonal issues. Children are particularly prone to this, as exposure to soil pollutants like lead can cause developmental delays and behavioral issues. Contaminated soil also impacts the protection of water sources, as agricultural biochemicals leach into groundwater, further expanding their toxic effects. Communities relying on contaminated soil for farming often deal with decreased agricultural productivity, causing food insecurity and economic problems. Animals are similarly impacted by soil pollution. Grazing animals may ingest toxic materials from the soil, causing bioaccumulation of toxic biochemicals in their bodies. This not only harms the health and reproductive success of these animals but also threatens humans who intake their meat and dairy products. Wildlife, such as insects, birds, and mammals, deal with habitat loss and intoxication due to polluted soil, damaging ecosystems, and decreasing biodiversity.

2.3 Plastic Pollution

Mulch films and silage wraps can direct the development of microplastic upon further damage if not properly disposed of. Agricultural plastics can be useful for the utilization of agriculture, but inappropriate disposal of agricultural plastics could release greenhouse gases or damage into micro- or nanoplastics, and often contain micro-nanoplastics from purified water. Trash, surface runoff, and wind also bring nanoplates to agro-land. These plastics disintegrate further under sunlight and structural stress and are sometimes exported to remote areas and aquatic bodies, leading to marine pollution. Also, indirect nanoplastic may lead significantly even suppressing soil pollution (Koul et al., 2022). For humans, plastic intoxication in agricultural development can pollute food and water supplies. These microplastics may contain harmful chemicals, such as endocrine damage substances or carcinogens, causing health risks, including hormonal issues, infertility, and expanding susceptibility to chronic illnesses like cancer. Moreover, plastics applied in farming can leach harmful additives into the soil, further polluting crops and water supplies. Animals, including wild and domesticated, are severely impacted by agricultural plastic pollution. Grazing animals can ingest plastic fragments or small particles left in fields, causing digestive issues, malnutrition, or even death. Aquatic animals, such as fish and shellfish, are exposed to micro- and nanoplastics washed into rivers, lakes, and oceans, causing the bioaccumulation of harmful substances in their tissues. This not only risks aquatic ecosystems but also can impact humans who intake seafood polluted with plastics. Terrestrial wildlife can also suffer from entanglement in plastic debris, causing injuries, decreased mobility, and impaired reproduction. In addition to environmental and biological damage, plastic pollution from agricultural waste poses economic burdens. Farmers may deal with reduced crop production and expanded costs for soil restoration and waste disposal. The broader economic effects include losses in fisheries, tourism, and public health costs associated with treating pollution-related illnesses. Proper waste management systems and policies to restrict single-use plastics in agriculture are necessary. Education and awareness programs can encourage farmers to adopt better practices for handling and disposing of plastic residues. Additionally, research into the effects of plastic contamination and the development of innovative solutions can help to mitigate its consequences.

Although the impact of air pollution through agricultural wastes has little effect, inappropriate disposal of agricultural residues in large amounts can intoxicate the air, and burning and disintegrating agricultural waste can emit harmful greenhouse gases. Humans, being exposed to air pollution from agricultural sources, pose serious health issues. Particulate matter exposed during crop waste burning can be inserted deep into the lungs, leading to respiratory issues, cardiovascular disorders, and decreased lung function. Ammonia emissions from animal waste and fertilizer usage react with other chemical compounds in the atmosphere to form fine particulates, posing a threat to air quality issues. Ground-level ozone, formed from volcanoes and nitrogen oxides in sunlight, can disrupt the respiratory system, trigger asthma, and impair lung development, particularly in children and the elderly. Longer exposure to polluted air accelerates the risk of chronic illness, including heart attacks and cancer, and decreases overall life expectancy. Animals are also prone to air pollution from agricultural residues. Animals exposed to high amounts of ammonia and particulate matter may face respiratory issues, decreased growth rates, and lower productivity. Wildlife living near the agricultural regions are similarly impacted, experiencing habitat disruption and health problems due to contaminated air. Pollutants like ozone can disrupt vegetation, decreasing the availability of food and shelter for herbivores, which in turn can affect the complete food chain.

Another threatening part is the contribution of heavy metals applied as fertilizers in agriculture especially the use of cadmium which upon being applied through different farming procedures can cause aggregation of cadmium into the soil and on the growing plants; thus, when these plant wastes or agricultural residues turn into agricultural wastes, it can lead to soil contamination such as reducing soil health,

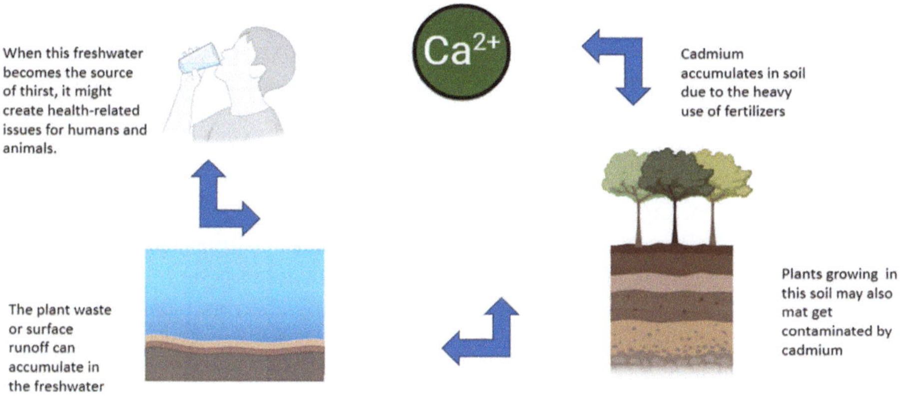

Fig. 2 Cadmium through agricultural wastes can contaminate soil and water and can cause health-related issues in humans

fertility, and productivity. Crops grown in contaminated soils can accumulate cadmium, leading to health damage in humans and animals. Additionally, surface runoff and leaching can spread cadmium into aquatic environments, ultimately creating a highly toxic environment. Figure 2 represents how cadmium produced through agricultural waste can cause health-related problems and affect soil fertility (Peana et al., 2022).

3 How Agricultural Waste Can Harm Human Beings Particularly?

Agricultural waste can affect human beings in various ways, primarily through its influence on public health. Inappropriate management of agricultural wastes, such as crop residues, animal manure, insecticides, and plastics, can cause the pollution of natural resources, which directly and indirectly has an impact on human beings. One significant route is through water pollution. Runoff from agricultural lands containing fertilizers, insecticides, and animal waste often pollutes rivers, lakes, and groundwater. The algal blooms release toxins that pollute drinking water and fish, causing severe health hazards like liver issues and gastrointestinal diseases. Furthermore, pathogens derived from untreated livestock manure can invade water sources, leading to outbreaks of waterborne illnesses such as cholera, typhoid, and diarrhea. Burning of crop leftovers, a common practice in many areas, produces particulate matter and poisonous gases such as carbon monoxide and volatile organic chemical compounds. These emissions destroy air quality and amplify the danger of respiratory problems, asthma, and cardiovascular issues in nearby areas. Longer exposure to contaminated air from agricultural waste burning has also been connected to long-term health hazards, including cancer. In terms of soil contamination, inappropriate disposal of agricultural waste, such as pesticides or cadmium-rich fertilizers, leads to the aggregation of toxic chemicals in the soil. These toxins can seep into crops, entering into the food chain and revealing humans to carcinogens and neurotoxins. Prolonged consumption of polluted food can result in chronic health problems, including developmental problems in children, reproductive issues, and kidney problems. Plastic waste from farm activities, such as mulch films and packaging materials, leads to environmental pollution. Over time, these plastics disintegrate into microplastics, which pollute soil and water. Microplastics can invade the food chain, posing unknown but potentially dangerous effects on human health. Additionally, poisonous chemicals from plastic disintegration may bioaccumulate in living organisms, increasing problems of endocrine disruption and other health complexity.

Lastly, the burning and decomposition of organic waste from agriculture can produce greenhouse gases like methane and nitrous oxide, contributing to climate change. Increasing temperatures and extreme weather events induce health challenges by amplifying the prevalence of vector-borne diseases, heat-related illnesses, and food insecurity.

Thus, agricultural waste, if not managed effectively, becomes a major source of pollution, adversely affecting water, soil, and ecosystems while emphasizing the need for sustainable agricultural practices.

This book chapter mainly emphasizes the importance of revolutionizing agricultural wastes into global policies of using clean energy and the environment pushing scientists to search for innovative ways to reuse farm waste. Advanced technologies are now being applied to turn waste residues into useful products, making farming more sustainable and decreasing environmental harm. These advanced applications adjust with global goals for sustainable development, like recycling resources better and reducing waste, and pollution. This has redesigned agriculture into a major player in the regenerative economy (reusing materials). It aims to promote the global economy (Duque-Acevedo et al., 2020).

To protect the environment from abundant agricultural waste, it should be reused instead of thrown it away. The focus is on a regenerative economy which explains recycling waste to make new and fresh valuable products. For example, Farm to Fork is a European plan that promotes sustainable food production and residue or leftover recycling. Green Deal aims to be carbon-neutral by 2050; these are all European plans that can be issued in India encouraging eco-friendly practices like reusing agricultural waste. Biowaste carries chemicals (phytochemicals) like vitamins, antioxidants, and natural pigments. These can be extracted and applied in preserving food products, making them healthier, or for skincare products or medicines to grow plants or produce clean energy—health-boosting products like supplements (Trabelsi et al., 2023). Another possibility is the use of bio-stimulants derived from agricultural wastes or leftovers or by-products, thus redesigning the concept of a green economy (Xu & Geelen, 2018).

4 Economic Approaches

4.1 Technologies Involved

A number of issues, including technological, climatic, demographic, environmental, socioeconomic, and regulatory aspects, are involved in garbage management. Because of their intrinsic complexity and nonlinearity, these systems are challenging to analyze, forecast, or optimize using traditional techniques. Artificial intelligence (AI) has recently surfaced as a promising substitute for solid waste management (SWM) problems. AI is particularly good at solving ambiguous problems, learning from mistakes, and dealing with data that is unclear or partial. Even though a lot of research has been done in this area, there are still few thorough evaluations evaluating AI's potential to address various SWM problems. This systematic review looks at 85 research that were published between 2004 and 2019 and emphasizes the application of AI in fields like planning, forecasting process parameters, monitoring waste bin levels, predicting waste characteristics, and optimizing vehicle routing (Abdallah et al., 2020).

However, organic waste management (OWM) has always been important for human societies, and the methods used need to be tailored to each population's particular traits. In this sense, the Kingdom of Saudi Arabia (KSA) is not an exception. The 3Rs—reducing waste, reusing, and recycling—are becoming more and more important in global sustainability trends, and KSA organizations are coordinating their efforts to achieve these objectives. However, a nation's economic situation frequently affects how far forward it is in implementing these sustainable practices. Richer countries can afford to invest in cutting-edge 3Rs technologies, but low-income developing nations find it difficult to carry out expensive "Zero Waste" projects. It is crucial to evaluate the viability of various OWM systems in Saudi Arabia in light of the nation's economic situation. This study examines pertinent facets of sustainability in waste management by analyzing OWM issues in the Kingdom of Saudi Arabia and comparing them to those in other nations. It suggests a comprehensive strategy for setting up a successful organic waste management system in Saudi Arabia, taking into account elements including institutional frameworks, appropriate technology, state policy, operational and financial management, and public awareness and involvement (Abdullah et al., 2022).

According to Rodić and Wilson (2017), solid waste management (SWM) is a cross-cutting issue that affects several aspects of growth in the environment, economy, and society—the three pillars of sustainability. According to Wilson (2007), the SDGs' guiding principles can be broadly classified as being similar to the objectives that have influenced the development of SWM practices over time. These objectives include resource value (SDG 11), environmental issues (SDGs 6 and 13), and public health (SDG 3). More recently, SDG 13 has been added to address climate change. For instance, achieving the objectives of sustainable SWM is necessary to achieve SDG 12.3 (cutting in half the amount of food waste produced worldwide per person at the retail and consumer sectors) and SDG 14 (proliferation of careless plastic waste usage and disposal that results in marine litter and issues connected to microplastics).

There are major negative effects on the environment and human health from uncollected and improperly disposed of garbage (SDG 6 and 13). Addressing these effects comes at a far larger expense than actually creating and running basic, sufficient waste management systems.

In order to achieve resilient and low-carbon economic growth, policymakers must consider more than just protecting national economies during times of crisis; they must also take important actions to create a more comprehensive structural shift that will be more resilient to future challenges from across the world (Masterson & Bugos, 2020). Countries would need to prepare for a different post-COVID economy by allowing capital, labor, expertise, and innovation to move into other industries and sectors in order to reverse the major setback and guarantee a robust and durable economic recovery (Sharma et al., 2021).

In the instance policymakers, practitioners, and researchers generally agree that digital technologies, like robotic automation, smart garbage bins, and artificial intelligence for item identification, are revolutionary tools for improving the efficacy and efficiency of waste management systems. Waste collection, sorting, and recycling procedures could be completely transformed by these developments, opening the door to a fully digitalized future. On the other hand, nothing is known about how widely these technologies are now being used in waste management companies. Current research frequently ignores the actual implementation levels in favor of focusing on organizations' ambitions to adopt digital technologies. Additionally, it leaves a crucial vacuum in our understanding of their practical use by failing to differentiate the level of technological integration across different stages of the waste management value chain (Borchard et al., 2022).

Modern technology combined with innovative business practices is transforming trash management through increased cost-effectiveness, sustainability, and efficiency. Sensor-equipped trash cans optimize routes, save labor costs, and use less fuel during collection operations. Cutting-edge technologies like robotic automation and AI-driven sorting systems increase recycling operations' accuracy, and waste-to-energy (WTE) systems reduce reliance on landfills by converting nonrecyclable waste into useful energy resources. In terms of the economy, programs such as Pay-As-You-Throw (PAYT) promote waste reduction, while Extended Producer Responsibility (EPR) encourages companies to invest in waste infrastructure and adopt eco-friendly designs. Furthermore, Material Recovery Facilities (MRFs) make money from recycled materials by recovering valuable resources from mixed garbage. These economic and technological strategies, which are supported by the ideas of the circular economy, reduce environmental damage while encouraging creative waste management solutions.

Agriculture has emerged as a key component of the world economy in recent years, helping to support the expanding population. Pressure on crop and animal systems has increased due to the growing desire for more food production, which frequently raises environmental issues. However, by increasing output, cutting expenses, and lessening the impact on the environment, the use of smart agricultural technologies is revolutionizing contemporary agriculture. These developments have major economic advantages in addition to addressing sustainability issues.

Precision agriculture (PA) optimizes the use of resources like water, fertilizer, and pesticides by using data and technology to track, evaluate, and cater to the unique requirements of particular fields and crops. Farmers' profitability is directly increased by this focused strategy, which maximizes yields while minimizing input expenditures. Similar to this, precision livestock farming (PLF) uses automated systems to monitor individual animals with an emphasis on behavior analysis, growth, milk supply, and disease prevention. PLF lowers operational inefficiencies and promotes economic gains through higher-quality products by improving animal productivity and health. When taken as a whole, these developments support a more profitable and sustainable agriculture sector (Monteiro et al., 2021).

4.2 *Previous Successful Case Studies*

In order to manage urban e-waste, WISE developed a comprehensive action plan centered on putting into practice practical tactics that support a circular economy. This program tackles the pressing issues brought on by the growing amount of e-waste, especially in urban regions of India.

The electrical and electronics industry in India has grown at an exponential rate, and by 2020, it is expected to reach a $400 billion market size (ASSOCHAM, 2018). However, the production of e-waste has increased dramatically due to faster technology improvements and more consumer purchasing power. After China, the USA, and Japan, India was the fourth-largest producer of e-waste in the world by 2016, with two million metric tons (MT) created. Maharashtra was the largest contributor in India, producing over 20% of all e-waste produced in the nation. This highlights the critical need for focused e-waste management interventions in the state.

Urban areas like Pune, the state's cultural and educational hub and one of India's eight biggest metropolises, are most affected by Maharashtra's problems. The Pune Municipal Corporation (PMC) reports that just 8.64 MT of the city's 16,000 MT of e-waste were appropriately recycled in 2016. Significant inefficiencies in the current e-waste disposal system are reflected in this discrepancy. Furthermore, the e-waste industry in Maharashtra is mostly dependent on unofficial recycling procedures, which frequently use dangerous techniques and present serious threats to the environment and public health.

Using Pune as a case study, WISE carried out an extensive investigation to evaluate the shortcomings and inefficiencies in Maharashtra's e-waste management system in response to these issues. Critical deficiencies were identified by the study, including inadequate formal recycling infrastructure, limited public awareness, and lax enforcement of regulations. These problems make it more difficult to handle e-waste sustainably and highlight the necessity of a multi-stakeholder coordinated strategy. In order to encourage efficient e-waste recycling and resource recovery, WISE suggested ways to close these gaps by promoting the integration of public organizations, commercial businesses, and unofficial recyclers into a unified framework. WISE aims to create a repeatable model for Maharashtra and beyond by using Pune as a reference case. It aims to establish Maharashtra as a leader in sustainable e-waste management practices, setting an example for other states to follow (Fiksel et al., 2021).

Countries all around the world have implemented a variety of rural solid waste management strategies that are suited to their unique requirements and systems of government.

The market-driven model is the most common in the USA, where waste treatment services are outsourced with government assistance. Numerous local businesses handle the collection and transportation of waste, guaranteeing a decentralized and competitive system. For waste management facilities, a "government-funded, privately operated" system is used, and it has worked well. When compared to direct government service supply, outsourcing waste management to private organizations lowers costs by about 25%, according to surveys done in more than 300 local municipalities.

In contrast, Germany manages rural garbage through a government-centric model, driven by worries about unregulated private company economic activity. The German model promotes a cooperative and open system by combining governmental oversight with active involvement from social groups and the broader public.

Japan takes a distinct strategy based on lean production concepts, emphasizing effective processing and recycling to optimize waste recovery. In Japan, farmers bear the majority of the costs associated with rural waste management; they pay for garbage services, and the government provides funding for recycling programs. This strategy ensures optimal resource utilization by emphasizing sustainability and precision.

The different priorities and difficulties that each nation faces in tackling rural solid waste management are reflected in these disparate approaches. Each of these case studies demonstrates successful waste management initiatives that are economically viable (Sun et al., 2024).

In contrast, Vietnam's informal sectors now handle a sizable amount of discarded electronics due to a lack of well-organized e-waste management and recycling infrastructure. Toxic compounds such as heavy metals (arsenic, manganese, nickel, lead, and zinc) and organic pollutants (flame retardants, PCBs, and dioxins) are released during these informal practices, which are frequently typified by antiquated techniques like manual disassembly, open burning, and unregulated plastic recycling. These contaminants seriously contaminate the environment and endanger the health of residents in the vicinity.

Furthermore, unofficial recycling workshops frequently release pollutants at levels on par with large-scale industrial operations in wealthy nations, underscoring the urgent need for organized waste management regulations. Creating a regulated e-waste management system would solve these problems and yield significant financial gains. For example, creating official recycling channels can lead to job possibilities in material recovery, processing, and garbage collection. In addition to lowering dependency on resource-intensive mining and imports, recovering valuable resources like gold, silver, copper, and rare earth metals from electronic waste can promote economic sustainability (Hoang et al., 2023).

By offering a consistent supply of recycled raw materials, an organized e-waste management system can help manufacturers reduce production costs and promote a circular economy. Formalizing the industry would also raise tax revenue for the government and lower medical and environmental cleanup expenses related to unregulated garbage processing.

Vietnam may accomplish long-term economic growth, enhance public health, and preserve the environment all at once by implementing a comprehensive plan. This strategy might be used as a template for other developing and emerging countries, highlighting the importance of organized e-waste management solutions on a worldwide scale (Abubakar et al., 2022).

4.3 Market Opportunities and Product Derivatives

Because it makes it possible to recover valuable products and energy from food waste, the circular bio-based economy has enormous potential for the food business. This policy supports sustainable food systems throughout Europe and the world under unified agriculture and fisheries policies, which is in line with the European Commission's Farm to Fork initiative. Aquaculture waste and other by-products of the fish and seafood industries hold great potential for the development of substitute fertilizers. By utilizing the high nutrient content of fisheries waste, these products can take the place of manufactured mineral fertilizers.

Research on the use of fishing by-products as fertilizers is still lacking, despite the fact that chances to extract valuable compounds from these materials have been investigated. In addition to reviewing current nutrient recovery systems and their results, this study looks at fishery waste The market potential of recovered items is significant, in addition to the environmental advantages. Fish waste fertilizers can help meet the increased demand for sustainable and organic agriculture inputs. Their adoption encourages less dependence on artificial fertilizers, which appeals to consumers who value eco-friendly and circular economy concepts. Value-added opportunities can also help the seafood processing industry by turning waste into items that can be sold, increasing profitability, and supporting sustainable practices. Streams refer to fish processing waste, fish sludge, and seafood by-products (Zhang et al., 2023).

However, the oil and gas sector contributes significantly to environmental waste by producing vast amounts of flue gases (including CO_2 and SO_2), oil sludge, and generated water. These waste streams can have detrimental effects on the environment if they are not disposed of properly. Recent developments, however, emphasize how crucial it is to combine waste treatment with resource recovery and reuse in order to adhere to the principles of the circular economy. By turning garbage into useful goods, this sustainable strategy not only lessens the ecological impact of the company but also promotes economic sustainability and supports international environmental goals.

There are significant financial advantages to these methods. For example, it is possible to reduce waste and generate extra income by processing and selling the hydrocarbons that are extracted from oil sludge. Natural water resources can be preserved and water acquisition costs reduced by reusing treated generated water in industrial processes. Furthermore, reducing greenhouse gas emissions and creating new market opportunities can be achieved by capturing flue gases like CO_2 and using them in enhanced oil recovery or turning them into chemicals and fuels.

By lowering expenses related to waste management and regulatory compliance, the financial benefits go even further. Additionally, this strategy creates opportunities for innovation, which results in the creation of fresh business models and technological advancements that promote sustainability. The three main waste streams examined in this study are gaseous emissions (flue gases), liquid waste (produced water), and solid waste (oil sludge). Their management approaches and possible economic benefits within the oil and gas sector are highlighted. Adopting such policies is essential to sustainable industrial practices since it guarantees both economic viability and environmental protection (Shahbaz et al., 2023).

In essence, since their commercialization in the 1950s, plastics have become an essential component of modern life, and their production has increased dramatically worldwide. The weight of plastic produced by 2015 was almost equivalent to the weight of all people. With 334 million metric tons of open waste dumping annually, 40 million of which is plastic, South Asia is the region with the worst plastic pollution crisis in the world.

Reducing pollution and promoting economic growth are two benefits of using plastic trash in infrastructure, such as road construction. Figure 3 represents the process of recycling plastics to roads. By substituting recycled plastics for virgin polymers in bitumen, enormous volumes of single-use plastics can be kept out of the ocean and landfills. One kilometer of plastic road, for example, could save 3 tons of carbon dioxide emissions from incineration and stop 57,803 food containers or 2.5 million straws from ending up in the trash.

This strategy also has financial advantages. Because recycled materials lessen dependency on pricey virgin resources, plastic highways are economical. In addition to boosting demand for recycling technology, they create jobs in the informal sector, particularly in the areas of waste collection, sorting, and processing. Furthermore, by boosting trade, connecting producers with markets, and expanding access to healthcare and education, better road infrastructure stimulates economic activity and eventually helps to reduce poverty. South Asia can simultaneously benefit the environment, society, and economy by turning plastic waste into necessary infrastructure. This will definitely benefit the environment and will improve the climatic health of our earth by reusing undegradable plastic wastes; on the other hand, it will also open new doors to economic progress toward the expansion of the economy and time which comes along with proper infrastructure and transportation.

Fig. 3 Different stages of the process of turning plastic waste into roads

4.4 Government Policies

One of the largest recognized biological economies is "agriculture" proving to be the largest producers of the biomass of significant amount. In the past 50 years, agriculture practices and production have tripled due to population growth, and agricultural land expansion with the green revolution. However, the rapid increase in the food production has raised considerable stress on the environment. Furthermore, this is compunded by the 23.7 million tons of food waste generated globally (Hamdan et al., 2022). These surges in agricultural output have strained the land water, and air resources which causes the adverse effect in health and sustainability of our environment. As the global demands are evolving under existing rules regulations, and policies, there was a speeding up in the exploitation of ordinary resources leading to massive amounts of excess production. When current tendencies remain waste production is expected to double by 2050 hypothetically reaching 4 billion tons. One of the important challenges confronted in farming waste management was the predominant rehearsal of growth burning in numerous regions; this process contributes to numerous ecological problems including air pollution and conservatory gas production that drive worldwide warming the agronomic sector alone books for around 21% of total GHG releases (Hamdan et al., 2022). Stubble burning not only promotes the concentration of injurious air effluences such as carbon dioxide (CO_2), ammonia, carbon monoxide, nitrogen oxide and sulfur oxide but likewise also leads to biodiversity damage and squalor of soil fertility which occured between 2008 and 2009 time period. The burning off with 98.4 million tons of ground waste produced around 141.15 million tons of carbon dioxide and other contaminants. Given these trials, there is a pressing need for vigorous policies and guidelines to guide maintainable agriculture practices. Effective constitutional policies frequently referred to as the agroecological policies are essential for dropping the environmental influences associated with cultivation while confirming food security for an increasing global population. These strategies can somehow help in targeting numerous outcomes connected to nourishments, pesticides, soil well-being, woods, and biodiversity; However, it is critical to recognize that simply having policies in place does not assure positive results in India which accounts for nearly 18% of the world's inhabitants. Agriculture plays the maximum vital role in nourishing its citizens and influence toward exports though open burning of crop remains a common repetition due to its suitability for clearing land rapidly before planning new crops (Govardhan et al., 2023). These performs significantly contribute to air contamination and health dangers. In an effort to mitigate the crop derived residue CDR scorching and strengthen waste management regulations, the Indian administration has introduced numerous projects over time. Government administrators and conservationists have worked together to encourage the sustainable expansion of several key legislative outlines governing environmental protection in India. The Environment Protection Act of 1986, the National Environment Tribunal Act of 1995, the National Environment Appellate Authority Act of 1997, and the Water Prevention and Control of Population Act 1974 along with the Water Prevention and Control of Pollution Cess Act 1977 are the main laws as mentioned in Table 1. Modifiable Conservational Quality and contamination control in India, the Ministry of Atmosphere Forest and Climate Alteration, and the Central Pollution Control Board are accountable for monitoring releases and air quality under these laws. In Addition to these guidelines, India recognized an environmental impact calculation EIA procedure in 1994 through MOF and CC to monitor air superiority effectively (Bhuvaneshwari et al., 2019).

The Supreme Court founded the (EPCA) Environmental Pollution (Prevention and Control) Specialist in 1998 especially to address transboundary air contamination issues in Delhi-NCR. An important step near addressing crop technique administration was taken with the overview of the national growth of appropriate farm equipment to minimize residue age group and promote supportable waste management practices (NPMCR) in 2014 (Sani et al., 2023). Contempt these efforts, tests remain due to insufficient enforcement of current regulations. The National Green Tribunal (NGT) has barred crop deposit burning in states corresponding to Haryana, Punjab, Rajasthan, and Uttar Pradesh from the time of 10 December 2015 though implementation remains feeble, and maritime farmers remain in this rehearsal due to economic weight and lack of viable replacements for sample contempt government enticements for alternative performs such as free power and diesel fundings (Koul et al., 2022). Intended at reducing growth burning, several farmers find these picks less appealing than burning residues. In answer to these trials, various state governments have applied stricter regulations on stubble burning for occurrence. On 8 October 2016, the Delhi High Court authorized states to tailor rules to address open burning at their levels. Accordingly states have imposed fines on farmers originate guilty of disrespectful these guidelines. The Indian government has also initiated schemes intended at promoting agriculture computerizations for the in situ organization of crop remains (Porichha et al., 2021). The central sector arrangement

Table 1 Government policies

No.	Policy	Year	Description	Outcome	Areas to focus	References
1.	Environment (protection) act	1986	The comprehensive legislation aims toward the protection and improvement of the environmental quality in India	A legal framework established for the protection of the environment	Control of the pollution, environmental standards with the regulation	Kalogiannidis et al. (2023)
2.	(EIA) Environmental Impact Assessment	1994	The environmental impact evaluation process established a project proposed before approval	It helps in the ensurement of the environmental consideration which is integrated into the project of development	Assessment of the air quality, biodiversity, and the management of resources	Kalogiannidis et al. (2023)
3.	EPCA Supreme Court	1998	This authority is established to address the transboundaries for air pollution problems in Delhi-NCR	Helps in implementing the measures to combat high air pollution in urban locations	Management of air quality and measures for pollution control	Gulia et al. (2022)
4.	(NGT) National Green Tribunal	2010	A more improved tribunal for handling environmental disputes and the enforcement of the environmental law in India	Strengthening the legal actions against the violations of the laws of the environment	Regulation of the environmental enforcement and the showing of the issues of pollution	Hernandez et al. (2024)
5.	(NPMCR) National Policy for the Management of Crop Residue	2014	Policies that aim toward the promotion of the sustainable management of the residues of the crop stubble burning	Encouraging alternative practices toward the minimization of burning	Machinery development and practices of sustainable waste management	Hernandez et al. (2024)

encourages using apparatuses like happy seeders and super straw administration systems that permit farmers to manage crop remains without burning them. Economic incentives include funding for purchasing these apparatuses 50% for individual agriculturalists and up to 80% for collectives. Moreover, innovative arrangements such as the biocomposting inventiveness below Rashtriya Krishi

Vikas Yojana have been thrown in to instruct farmers about maintainable practices providing financial funding for transitioning away from outdated methods that donate to pollution. In conclusion, there is a crucial need for complete policies and regulations on provision of sustainable farming practices while addressing the ecological challenges posed by farming. Waste management policymakers must consider trade-offs between the environmental impact on agriculture productivity to develop effective agro-environmental policies to promote sustainability without comprising food security by fostering partnership among governments, farmers, researchers, and societies through integrated approaches like Nexus thoughtful where agriculture interconnects with energy management and environmental guard. It is possible to create an additional sustainable agriculture sector that aids both people and plant (Erbaugh et al., 2024).

Agriculture waste is an unavoidable by-product of farming actions with approximately 140 billion metric tons produced globally each year. This supportable volume of waste presents both trials and an opportunity for agriculturalists as it can be transformed into appreciated energy and raw materials. Effective administration of agriculture waste is critical not only for the conservational sustainability but also for increasing the economic practicality of farming operations. Farm waste contains residues shaped from various agronomic processes such as crop agriculture and cattle rearing, providing a basis for the large production of agronomic waste. Common types of agricultural waste include physical crop residues (e.g., rice husk and maize stalks), fruit waste and farm waste (Fig. 4), and biological materials (e.g., tree trimming with grass clippings) (Cao et al., 2023). Organic waste can consist of up to 80% of total dense waste generated on farms. Cattle alone can produce important amounts of manure. On an average, approximately 527 KG of these residual substances are produced each day for every one thousand kilograms of live weight containing high value of these remaining substances. Many growers in developing regions like Nigeria often supervise their possible lack of drivable expertise and awareness leading to the underutilization of these capitals. This could also be converted into energy or raw materials for various economic possibilities. The idea of leftover to wealth highlights the potential for farmers to produce economic value from cultivation through advanced practices. Farm waste can be used into marketable products such as fertilizers, biofuels, and animal feed for occurrence. The Songhai farmhouse in Benin attends as a successful model by recycling leftover into biogas for cooking and energy, thereby achieving close zero waste. The farm remains such as cocoa pods with palm kernel shells can be converted into valued resources. Cocoa pods are often left in the grounds but can provide a rich foundation of potash fertilizer. Likewise palm kernel shells can generate over 750 MW of electricity if correctly utilized. Application of plant resources and animal excess as renewable energy sources is currently an alternative to fossil fuels, especially in rural groups where these resources are plentiful. In sustainable waste management practices, agriculturalists can adopt various maintainable practices to manage cultivation waste successfully. Composting process reprocesses nutrients back into the soil upholding or restoring organic material levels. Composting is a low-cost knowledge that helps farmers reduce expenses associated with biochemical fertilizers (Hu

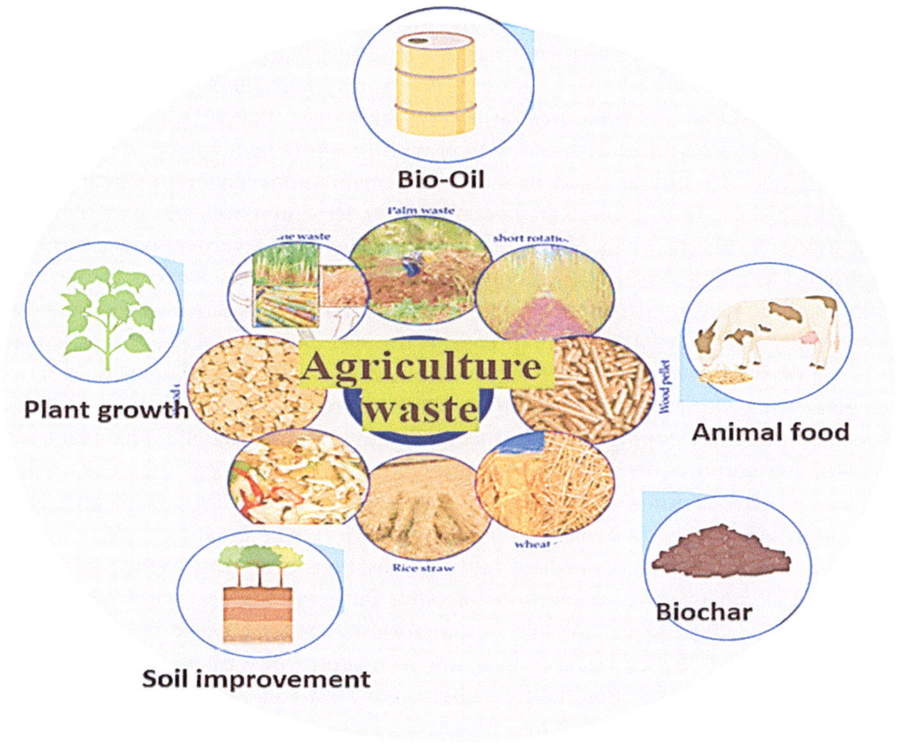

Fig. 4 Agriculture waste management in farms with the collection and processing of different types of waste like fruit waste, farm waste poultry, etc.

et al., 2023). Biomass conversion farming waste can be converted into bioenergy foundations like bioethanol and biodiesel dipping reliance on fossil fuels while providing additional revenue streams. Soil training livestock manure can aid as a soil conditioner or be transformed into biogas by the process of anaerobic digestion providing both energy and enhanced soil health. According to new studies, composting not only improves soil fertility but also contributes to carbon sequestration, and foods that battle climate change besides integrating composting with crop manufacturing systems have been shown to recover crop yields meaningfully while reducing chemical input costs. Trials and waste trials and waste application contempt the clear benefits of utilization of agricultural waste. Several trials hinder effective organization such as lack of information and inadequate admission to extension facilities, and meting out facilities many rural farmers lack awareness about the aids and approaches of waste application. Limited access to educational incomes avoids farmers from accepting groundbreaking perform and the nonappearance of services for dispensation agriculture waste bounds its operation potential to speech these trials. It is vital to enhance farmer teaching through postponement services that promote best practice and waste management. Inspiration cognition presentation and understanding agriculturalists' performance toward agriculture waste application can be framed using the motivation understanding behavior model. This theory posits that motivation cognition and performance of farmers' decisions are prejudiced by intrinsic motivations. Individual beliefs, extension incentives, and external pressures farmers' insights are shaped by personal norms and behavioral arrogances distressing their willingness to be involved in sustainable practices, and also ultimately these intellectual evaluations lead to practical actions concerning waste utilization (Kataya et al., 2023). Research indicates that advanced educational levels and years of agricultural experience positively influence farmers' likelihood to adopt waste utilization initiatives; for instance educated farmers are more likely to distinguish the economic benefits related to effective waste organization. Policy inferences and future directions to make the most of the potential benefits are derived from agriculture waste administration. Policymakers should focus on generating supportive environments that inspire innovations among farmers. This comprises developing financial incentives for adopting maintainable practices, establishing corporation between government agencies and local societies to simplify knowledge sharing, examining and research and growth for new technologies that improves waste processing competences. Moreover creating consciousness about successful case studies where farming waste has been transformed into gainful ventures can serve as inspiration for other farmers to adopt comparable practices. Farmer plays the critical role in transforming farming waste from a liability into an advantage that contributes to financial sustainability by adopting advanced performs such as composting biomass alteration. They can enhance their efficiency while minimizing environmental influences. However the knowledge gap and improving admission to resources are vital steps toward maximizing the possibility of agriculture waste as agriculturalists embrace maintainable practices. They not only recover their likelihoods but also donate significantly to global financial growth and ecological sustainability (Geng et al., 2024).

4.5 Local Communities Involved

Local communities play an essential role in the effective organization of agriculture waste converting what is often seen as a weight into variable resources that donate to environmental sustainability and financial resilience. Community-driven initiatives have emerged as a key approach for addressing the trials posed by agricultural waste, for example, the establishment of communal composting centers allows agriculturalists to together manage their biological waste. Producing nutrient-rich manure that can be refunded to the fields, this is not only enhances the soil richness but also reduces reliance on chemical composts. Promoting healthier farming practices, such creativities are often supported by local government, and NGOs throw consciousness programs that educate agriculturalists about the benefits of sustainable waste organization techniques, counting composting, and biogas productions in India where farming waste management is dangerous due to the scale of crop residue groups around 620,000,000 tons. Annually local groups have begun to adopt innovative performs like biogas groups from agriculture remains. Their approach not only offers farmers with a renewable energy source but also alleviates environmental issues associated with scorching crop residues which contribute knowingly to air pollution. Additionally, community engagement in waste organization fosters a sense of ownership and responsibilities among farmers. This can encourage them to participate actively in sustainable practices by collaborating with local governments and organizations (Ramos, 2024). Technical provision and resources are necessary for applying effective waste organization solutions. The incorporation of traditional information with modern techniques has proven beneficial, for example, performs such as grass cycling, where grass cutting are left on lawns to decay naturally, demonstrates how simple approaches can significantly reduce waste while enriching soil health. Furthermore, successful case studies from numerous rural communities have shown that selling fertilizer or biogas can expand their revenue watercourses as well as can enhance their economic stability. The founding of cooperative representations for composting allows small holder farmers to pool their possessions making it economically viable to devote in large-scale waste processing amenities that might otherwise be out of reach. For separate farmers involvement of local groups also plays a vital role in advocating for supportive government policies that facilitates the sustainable farming practices by participating in policy conversation and demonstrating the effectiveness of community-led creativities; these groups can influence guidelines that promote ecologically friendly waste management performs at a broader scale (Pornaro et al., 2022).

For instance, policies that provide financial incentives for adopting maintainable waste management knowledge or that support the formation of infrastructure for composting and biogas construction can significantly enhance the capacity of local groups to manage agriculture waste efficiently. In conclusion local groups are essential investors in the movement near maintainable agriculture waste association. The active influence not only helps mitigate environmental impact but also fosters financial chances and enhances the contributions. First job as these groups remain to innovate and collaborate with administrative bodies and NGOs, they covered the way for a more sustainable future in farming demonstrating that actual waste management is not just an ecological necessity but also a way to economic approval and resilience.

4.6 Contribution Toward the Environment

The agriculture-based organization is increasingly documented as a vital component of maintainable agriculture performance that can significantly donate to environmental guard and economic growth. Between 2020 and 2024, frequent studies have highlighted the multilayered benefits of effective agriculture waste organization particularly in the context of justifying environmental degradation and ornamental. The global economy one of the most persistent issues associated with agriculture waste is the indecorous disposal of crop remains which often leads to air contamination through burning performs. In addition to emitting harmful greenhouse gases, it also causes health problems and weather changes. Reports suggests that crop residue burning in republics like India leads to the emissions of significant quantities of carbon dioxide and particulate matter substance worsening global warming and air quality. Adapting of sustainable waste management techniques such as composting and bioenergy production by farmers can initiate dynamic transformations of the farming processes. Composting is a particularly effective technique for recycling (Fig. 5) organic leftover back into the soil, thus improving soil richness and reducing reliance on biochemical fertilizers. Educations have shown that composting can improve nutrient cycling, restore soil well-being, and promote biodiversity with agriculture ecologies. Furthermore these performs help justifying the adverse effects of nutrient overflow into water bodies which is meaningfully a concern in many agricultural regions (Afrasiabi et al., 2022).

The incorporation of composting and other waste organization strategies not only addresses environmental trials but also supports local economies by creating jobs and indorsing sustainable agriculture practices, for example, community composting creativities have been successful in various areas where they not only decrease waste but also provide agriculturalists with high quality organic composts that improve crop yields. Also the conversion of farming waste into bioenergy offers a dual advantage. It decreases waste while providing a renewable energy groundwork. The manufacture of biogas from biological waste can meaningfully decrease greenhouse gas productions associated with traditional fossil oils. In India government initiatives aimed at encouraging biogas plants have demonstrated how farming waste can be attached to produce clean energy thereby conducive to energy security and financial stability. These initiatives not only authorize farmers by providing them with alternate income sources but also align with worldwide sustainability goals by plummeting dependence on nonrenewable properties. The role of government strategies in facilitating these changeovers cannot be overstated. Effective policy frameworks that encourage sustainable agricultural performance are

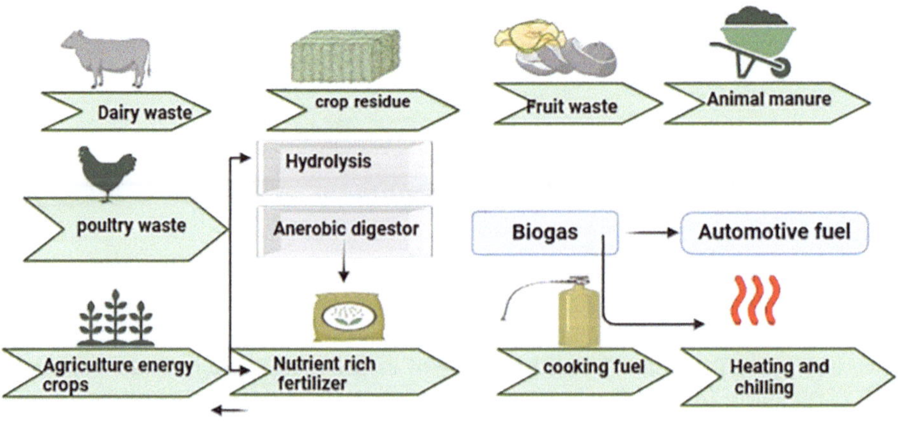

Fig. 5 Agriculture waste recycling

essential for inspiring farmers to approve innovative waste administration techniques. Government can provide economic incentives for composting services or biogas plants as well as instrument regulations that discourage harmful practices such as scorching crop residues. Awareness movements that educate farmers about the benefits of maintainable waste organization are also crucial in developing a culture of ecological stewardship within the agricultural societies. The involvement of effective agriculture waste organizations in ecological sustainability represents a significant leap toward encouraging a comprehensive economy (Kuamoah, 2020). By converting agricultural residues into appreciated resources through composting, biogas manufacture, and other innovative performances, farmers can recover soil health, decrease pollution, and generate economic occasions. Amid ongoing struggles with the trials posed by climate changes and supply depletion, the country started embracing sustainable cultivation waste administration will be pivotal in accomplishing long-term environmental goals while simultaneously clumping local economies. The collective labors of all local communities maintained by robust administration policies will be essential in understanding these objectives and promotion a more maintainable future of agriculture global (Kapoor et al., 2020).

5 Challenges

If environmental sustainability is our aim, waste management will be an important aspect of it, and to address this issue, various strategies can be used such as vermicomposting, aerobic composting (AC), and anaerobic composting (AnC) (Waqas et al., 2023). These technologies involve worms and microorganisms as the primary agents in composting for recycling nutrients, controlling soil processes, and preserving soil fertility by converting biodegradable waste into nutrient-rich compost (Mckenzie et al., 2022)· Vermicompost technology (VTT) is a multipurpose sustainable waste management strategy, but there are challenges. One of the main challenges is that it needs proper regulation and monitoring. In addition to that, adequate training and education is a must to ensure that it is being carried out safely and effectively (Waqas et al., 2023). We also need proper infrastructure and equipment to support a suitable environment for earthworm breeding and composting. Another challenge of VTT is the potential for contamination and health hazards for the people involved if the waste being processed contains hazardous materials or pathogens. Therefore, the waste being processed should be assessed and examined before processing. Though this technology has the potential to generate income by producing and selling compost, the infrastructure and equipment needed to support the process may be costly (Mckenzie et al., 2022). Therefore, it is important to carefully evaluate the costs and benefits of VTT to ensure that it is a sustainable and economically viable option for waste management. Proper regulation, monitoring, and training are needed to ensure that the process is being carried out safely and effectively. Additionally, there is a need to carefully assess the waste being processed, address potential contamination risks, and evaluate the economic viability of the technology. Addressing these challenges will be critical for successfully implementing VTT as a sustainable waste management strategy (Hajam et al., 2023).

Agricultural waste has a lot of organic matter and high levels of nitrogen, phosphorus, and potassium. Aerobic composting turns agricultural waste into organic fertilizer. Composting improves soil structure and enriches soil fertility levels (Nugroho et al., 2024). These traditional composting methods exhibit several drawbacks too. These include a substantial requirement for surface area, an extended fermentation period, significant emission of malodorous substances during the fermentation process, suboptimal product quality, and limited demand for substandard products. So, conventional aerobic composting is occasionally ineffective for decomposition due to the physiochemical characteristics of organic waste. These properties reduce composting temperature, which reduces decomposition and sanitation. NH_3, H_2S, and other pollutants can also occur from conventional aerobic composting (Nugroho et al., 2024). Additionally, the loss of organic nitrogen and carbon during the process is significant. However, by raising the treatment temperature above 70 °C, pathogens in animal manure can be destroyed in just 10–30 min.

Composting organic waste helps divert it from landfills, reducing methane (CH_4) emissions, a potent greenhouse gas typically produced through anaerobic decomposition in landfills (Manea et al., 2024). Several challenges are also there, including environmental problems (odors, bioaerosols, heavy metals, leachate, gas emissions), operational difficulties (segregation, pathogen detection, composting duration, low levels of adoption), and management constraints (toxic substances, lack of information, efficient pricing systems and regulations). Different types of organic waste may contain varying concentrations of nutrients and heavy metals, which further affects the composting process's intensity, the final compost's quality, and its suitability for reuse in agriculture (Manea et al., 2024). Bioaerosols are another environmental issue related to bacteria and fungi in the emissions which promotes health risks to workers; except this, odors from the composting process, management of heavy metals in compost, the release of low-level GHG, presence of pathogens, high infrastructure costs, operational difficulties, and low adoption rates are the challenges related to composting processes.

Development of sustainable biopolymer-based composites from agricultural waste biomass can also be challenging, due to their susceptibility to environmental degradation, particularly in moist conditions, which is caused by the hydrophilic nature of natural fibers used. They are vulnerable to water absorption and tend to degrade when exposed to moisture or water. They may be difficult to handle at high-temperature, and inconsistencies in natural fiber properties can cause dispute in the fabrication of biocomposites for structural applications due to the associated design uncertainties.

Agro-waste-based adsorbents, i.e., biochars, are now being used as sustainable alternatives to remove ibuprofen and carbamazepine from contaminated waters in wastewater treatment. Adsorption methods are used to remove pharmaceuticals from aqueous media here. However, it is associated with some limitations. Adsorption is a nondestructive technique involving only (Cintura et al., 2021) a transfer of contaminants, and therefore problems of sludge disposal are an issue. Hence, there is still some way to go for field-scale application from laboratory work.

A single adsorbent cannot be used to remove all the complex mixtures of pharmaceuticals given their diverse chemical structures. Therefore, we need to understand the chemical structures of both adsorbents and adsorbates and the mechanisms of their interaction, when developing adsorbents for enhanced removal of contaminants. Proper monitoring should be implemented to deliver accurate knowledge on the existing contamination levels, and further research on toxicity and health effects of pharmaceutical pollutants. Still and all, focusing only on end-of-pipe green technologies for water treatment is not sufficiently sustainable, and prevention remains the ultimate technique. As such, proper strategies to reduce the input of pharmaceuticals into the environment are the first step toward a sustainable environment. Moreover, a holistic green approach encompassing the full life cycle of pharmaceuticals from manufacturing to disposal offers a well-grounded, environmentally friendly passageway that ensures sustainable pharmaceutical production and mitigates their risks to the environment (Cintura et al., 2021).

Improper disposal of waste products could contaminate usable lands and cause problems for waste management authorities if they are dumped in landfills. These waste materials tend to contaminate water when they come into contact with water bodies. Recycled aggregate concrete has comparatively weak mechanical properties due to the existence of weak bonds between the old mortar and aggregate, alongside cross cracks and fractures in the recycled aggregates developed during the recycling process. The powdered form of waste products can combine with air and pollute the atmosphere. If the dumping ground is close to agricultural lands, this could affect cultivable lands. Therefore, it is of prime importance to get rid of or reuse waste materials properly, because it can affect the environment and human health in danger. As the global population increases, sustainable development should be crucial, and particularly the concrete industry should contribute to this development. Recycled aggregate concrete is now a promising industry but has comparatively weak mechanical properties due to weak bonds present between the old mortar and aggregate. Utilizing process, and coating recycled aggregates with cement slurry or admixture solution, can help mitigate the strength loss caused by incorporating recycled aggregates. To generate large quantities of recycled aggregate concrete, factory-scale production needs further studies (Nugroho et al., 2024).

Considering the current criteria for sustainable infrastructure and the associated environmental benefits, the use of agricultural wastes and their application is vast than we can think of. We need research and studies to generate more and more ideas like this.

6 Conclusion

This book chapter thoroughly explains agricultural waste and its management, emphasizing how crucial it is to advance economic development and environmental sustainability. Crop residues, poultry by-products, live-stock wastes and agro-industrial leftovers such as fruit peel, and sand oil-seed cakes are all considered forms of agricultural waste. When improperly managed, these pollutants significantly contribute to water, soil, and air pollution, which degrades the ecosystem and poses serious health hazards. For instance, burning crop waste releases greenhouse gases like carbon dioxide and methane, while nutrient-rich agricultural runoff eutrophies aquatic environments.

To overcome these obstacles, agricultural waste possesses immense capacity to boost the global economy through innovative technologies. Converting waste into biofuels, bioplastics, and organic fertilizers decreases environmental pollution, forms job opportunities in rural areas, and slows down dependency on fossil fuels. By fostering a circular economy, agricultural waste can drive innovation in industries, promote energy security, and support sustainable agricultural practices. Precision livestock farming and precision agriculture are covered in the chapter as ways to cut waste and increase resource efficiency. Anaerobic and aerobic composting, as well as the production of biogas from agricultural waste, are examples of sustainable alternatives. The importance of artificial intelligence (AI) in waste management is highlighted since it makes resource recovery, process improvement, and efficient monitoring possible.

The chapter also includes case studies that demonstrate the financial and environmental benefits of creative trash management techniques. These instances support global goals such as attaining carbon neutrality and promoting a circular economy.

Government policies and global initiatives are shown to play a critical role in agricultural waste management. The National Policy for Crop Residue Management, which was introduced in 2014, and India's Environment Protection Act of 1986 are among the important policies reviewed in this chapter. By implementing rules, offering incentives for sustainable technologies, and launching awareness campaigns, these policies aim to lessen the negative effects of agricultural practices on the environment. Notable projects include funding for technology like straw management systems and happy seeders, as well as the promotion of composting and biogas production through efforts like the Rashtriya Krishi Vikas Yojana.

To sum up everything, the application of agricultural waste is a transformative solution that corresponds to economic growth with environmental stewardship. By integrating waste management into global industries, we can unlock the economic value, mitigate climate change, and ensure a sustainable future for all.

References

Abdallah, M., Abu Talib, M., Feroz, S., Nasir, Q., Abdalla, H., & Mahfood, B. (2020). Artificial intelligence applications in solid waste management: A systematic research review. *Waste Management, 109*, 231–246. https://doi.org/10.1016/j.wasman.2020.04.057

Abdullah, N., Al-wesabi, O. A., Mohammed, B. A., Al-Mekhlafi, Z. G., Alazmi, M., Alsaffar, M., Anbar, M., & Sumari, P. (2022). Integrated approach to achieve a sustainable organic waste management system in Saudi Arabia. *Food, 11*(9), 1214. https://doi.org/10.3390/foods11091214

Abubakar, I. R., Maniruzzaman, K. M., Dano, U. L., AlShihri, F. S., AlShammari, M. S., Ahmed, S. M. S., Al-Gehlani, W. A. G., & Alrawaf, T. I. (2022). Environmental sustainability impacts of

solid waste management practices in the global south. *International Journal of Environmental Research and Public Health, 19*(19), 12717. https://doi.org/10.3390/ijerph191912717
Afrasiabi, A., Tavana, M., & Di Caprio, D. (2022). An extended hybrid fuzzy multi-criteria decision model for sustainable and resilient supplier selection. *Environmental Science and Pollution Research, 29*(25), 37291–37314. https://doi.org/10.1007/s11356-021-17851-2
ASSOCHAM (2018) Council on Climate Change & Environment study report - Delhi-NCR will generate 1,50,000 MT e-waste by 2020: ASSOCHAM study. https://www.assocham.org/news-detail.php?id=6677
Bhuvaneshwari, S., Hettiarachchi, H., & Meegoda, J. N. (2019). Crop residue burning in India: Policy challenges and potential solutions. *International Journal of Environmental Research and Public Health, 16*(5), 832. https://doi.org/10.3390/ijerph16050832
Borchard, R., Zeiss, R., & Recker, J. (2022). Digitalization of waste management: Insights from German private and public waste management firms. *Waste Management & Research: The Journal for a Sustainable Circular Economy, 40*(6), 775–792. https://doi.org/10.1177/0734242X211029173
Bracco, S., Calicioglu, O., Gomez San Juan, M., & Flammini, A. (2018). Assessing the contribution of bioeconomy to the total economy: A review of national frameworks. *Sustainability, 10*(6), 1698. https://doi.org/10.3390/su10061698
Briassoulis, D. (2024). Corrigendum to "Agricultural plastics as a potential threat to food security, health, and environment through soil pollution by microplastics: Problem definition" [Sci. Total Environ. 892 (2023) 164533]. *Science of the Total Environment, 954*, 176364. https://doi.org/10.1016/j.scitotenv.2024.176364
Cao, X., Williams, P. N., Zhan, Y., Coughlin, S. A., McGrath, J. W., Chin, J. P., & Xu, Y. (2023). Municipal solid waste compost: Global trends and biogeochemical cycling. *Soil & Environmental Health, 1*(4), 100038. https://doi.org/10.1016/j.seh.2023.100038
Cintura, E., Nunes, L., Esteves, B., & Faria, P. (2021). Agro-industrial wastes as building insulation materials: A review and challenges for Euro-Mediterranean countries. *Industrial Crops and Products, 171*, 113833. https://doi.org/10.1016/j.indcrop.2021.113833
Duque-Acevedo, M., Belmonte-Ureña, L. J., Cortés-García, F. J., & Camacho-Ferre, F. (2020). Agricultural waste: Review of the evolution, approaches and perspectives on alternative uses. *Global Ecology and Conservation, 22*, e00902. https://doi.org/10.1016/j.gecco.2020.e00902
Erbaugh, J., Singh, G., Luo, Z., Koppa, G., Evans, J., & Shyamsundar, P. (2024). Farmer perspectives on crop residue burning and sociotechnical transition in Punjab, India. *Journal of Rural Studies, 111*, 103387. https://doi.org/10.1016/j.jrurstud.2024.103387
Evans, A. E., Mateo-Sagasta, J., Qadir, M., Boelee, E., & Ippolito, A. (2019). Agricultural water pollution: Key knowledge gaps and research needs. *Current Opinion in Environmental Sustainability, 36*, 20–27. https://doi.org/10.1016/j.cosust.2018.10.003
Fiksel, J., Sanjay, P., & Raman, K. (2021). Steps toward a resilient circular economy in India. *Clean Technologies and Environmental Policy, 23*(1), 203–218. https://doi.org/10.1007/s10098-020-01982-0
Geng, W., Liu, L., Zhao, J., Kang, X., & Wang, W. (2024). Digital technologies adoption and economic benefits in agriculture: A mixed-methods approach. *Sustainability, 16*(11), 4431. https://doi.org/10.3390/su16114431
Govardhan, G., Ambulkar, R., Kulkarni, S., Vishnoi, A., Yadav, P., Choudhury, B. A., Khare, M., & Ghude, S. D. (2023). Stubble-burning activities in north-western India in 2021: Contribution to air pollution in Delhi. *Heliyon, 9*(6), e16939. https://doi.org/10.1016/j.heliyon.2023.e16939
Gulia, S., Shukla, N., Padhi, L., Bosu, P., Goyal, S. K., & Kumar, R. (2022). Evolution of air pollution management policies and related research in India. *Environmental Challenges, 6*, 100431. https://doi.org/10.1016/j.envc.2021.100431
Hajam, Y. A., Kumar, R., & Kumar, A. (2023). Environmental waste management strategies and vermi transformation for sustainable development. *Environmental Challenges, 13*, 100747. https://doi.org/10.1016/j.envc.2023.100747

Hamdan, M. F., Mohd Noor, S. N., Abd-Aziz, N., Pua, T.-L., & Tan, B. C. (2022). Green revolution to gene revolution: Technological advances in agriculture to feed the world. *Plants, 11*(10), 1297. https://doi.org/10.3390/plants11101297

Hernandez, D., Pasha, L., Arian Yusuf, D., Nurfaizi, R., & Julianingsih, D. (2024). The role of artificial intelligence in sustainable agriculture and waste management: Towards a green future. *International Transactions on Artificial Intelligence (ITALIC), 2*(2), 150–157. https://doi.org/10.33050/italic.v2i2.552

Hoang, A. Q., Tue, N. M., Tu, M. B., Suzuki, G., Matsukami, H., Tuyen, L. H., Viet, P. H., Kunisue, T., Sakai, S., & Takahashi, S. (2023). A review on management practices, environmental impacts, and human exposure risks related to electrical and electronic waste in Vietnam: Findings from case studies in informal e-waste recycling areas. *Environmental Geochemistry and Health, 45*(6), 2705–2728. https://doi.org/10.1007/s10653-022-01408-4

Hu, Z., Han, Y., Zalhaf, A. S., Zhou, S., Zhao, E., & Yang, P. (2023). Harmonic sources modeling and characterization in modern power systems: A comprehensive overview. *Electric Power Systems Research, 218*, 109234. https://doi.org/10.1016/j.epsr.2023.109234

Kalogiannidis, S., Kalfas, D., Giannarakis, G., & Paschalidou, M. (2023). Integration of water resources management strategies in land use planning towards environmental conservation. *Sustainability, 15*(21), 15242. https://doi.org/10.3390/su152115242

Kapoor, R., Ghosh, P., Kumar, M., Sengupta, S., Gupta, A., Kumar, S. S., Vijay, V., Kumar, V., Kumar Vijay, V., & Pant, D. (2020). Valorization of agricultural waste for biogas based circular economy in India: A research outlook. *Bioresource Technology, 304*, 123036. https://doi.org/10.1016/j.biortech.2020.123036

Kataya, G., Cornu, D., Bechelany, M., Hijazi, A., & Issa, M. (2023). Biomass waste conversion technologies and its application for sustainable environmental development—A review. *Agronomy, 13*(11), 2833. https://doi.org/10.3390/agronomy13112833

Koul, B., Yakoob, M., & Shah, M. P. (2022). Agricultural waste management strategies for environmental sustainability. *Environmental Research, 206*, 112285. https://doi.org/10.1016/j.envres.2021.112285

Kuamoah, C. (2020). Renewable energy deployment in Ghana: The hype, hope and reality. *Insight on Africa, 12*(1), 45–64. https://doi.org/10.1177/0975087819898581

Kumar Sarangi, P., Subudhi, S., Bhatia, L., Saha, K., Mudgil, D., Prasad Shadangi, K., Srivastava, R. K., Pattnaik, B., & Arya, R. K. (2022). Utilization of agricultural waste biomass and recycling toward circular bioeconomy. *Environmental Science and Pollution Research, 30*(4), 8526–8539. https://doi.org/10.1007/s11356-022-20669-1

Li, D., Tian, P., Shao, D., Hu, T., Luo, H., Dong, B., Khan, S., Cui, Y., & Luo, Y. (2022). Assessment of water pollution in the Tibetan Plateau with contributions from agricultural and economic sectors: A case study of Lhasa River Basin. *Environmental Science and Pollution Research, 29*(14), 20617–20631. https://doi.org/10.1007/s11356-021-17249-0

Manea, E. E., Bumbac, C., Dinu, L. R., Bumbac, M., & Nicolescu, C. M. (2024). Composting as a sustainable solution for organic solid waste management: Current practices and potential improvements. *Sustainability, 16*(15), 6329. https://doi.org/10.3390/su16156329

Masterson, J., & Bugos, S. (2020). Pandemic Disrupts Security Meetings. Arms Control Today, 50(4), 22–23.

Mckenzie, I., Diana, S., Jaikishun, S., & Ansari, A. (2022). Comparative review of aerobic and anaerobic composting for the reduction of organic waste. *Agricultural Reviews, 43*(2), 234–238. https://doi.org/10.18805/ag.R-191

Monteiro, A., Santos, S., & Gonçalves, P. (2021). Precision agriculture for crop and livestock farming—Brief review. *Animals, 11*(8), 2345. https://doi.org/10.3390/ani11082345

Nagendran, R. (2011). Agricultural waste and pollution. In *Waste* (pp. 341–355). Elsevier. https://doi.org/10.1016/B978-0-12-381475-3.10024-5

Nugroho, A. D., Ma'ruf, M. I., Nasir, M. A., Fekete-Farkas, M., & Lakner, Z. (2024). Impact of global trade agreements on agricultural producer prices in Asian countries. *Heliyon, 10*(2), e24635. https://doi.org/10.1016/j.heliyon.2024.e24635

Peana, M., Pelucelli, A., Chasapis, C. T., Perlepes, S. P., Bekiari, V., Medici, S., & Zoroddu, M. A. (2022). Biological effects of human exposure to environmental cadmium. *Biomolecules, 13*(1), 36. https://doi.org/10.3390/biom13010036

Porichha, G. K., Hu, Y., Rao, K. T. V., & Xu, C. C. (2021). Crop residue management in India: Stubble burning vs. other utilizations including bioenergy. *Energies, 14*(14), 4281. https://doi.org/10.3390/en14144281

Pornaro, C., Novello, A., Fidanza, M., & Macolino, S. (2022). Grasscycling: A key practice for sustainable turfgrass management. *Grasses, 1*(1), 45–52. https://doi.org/10.3390/grasses1010005

Ramos, A. (2024). Sustainability assessment in waste management: An exploratory study of the social perspective in waste-to-energy cases. *Journal of Cleaner Production, 475*, 143693. https://doi.org/10.1016/j.jclepro.2024.143693

Rodić, L., & Wilson, D. C. (2017). Resolving governance issues to achieve priority sustainable development goals related to solid waste management in developing countries. Sustainability, 9(3), 404. https://doi.org/10.3390/su9030404

Sani, M. N. H., Amin, M., Siddique, A. B., Nasif, S. O., Ghaley, B. B., Ge, L., Wang, F., & Yong, J. W. H. (2023). Waste-derived nanobiochar: A new avenue towards sustainable agriculture, environment, and circular bioeconomy. *The Science of the Total Environment, 905*, 166881. https://doi.org/10.1016/j.scitotenv.2023.166881

Searchinger, T. D., Malins, C., Dumas, P., Baldock, D., Glauber, J., Jayne, T., Huang, J., & Marenya, P. (2020). *Revising public agricultural support to mitigate climate change*. World Bank. https://doi.org/10.1596/33677

Seidavi, A., Azizi, M., Swelum, A. A., Abd El-Hack, M. E., & Naiel, M. A. E. (2021). Practical application of some common agro-processing wastes in poultry diets. *World's Poultry Science Journal, 77*(4), 913–927. https://doi.org/10.1080/00439339.2021.1960461

Shahbaz, M., Rashid, N., Saleem, J., Mackey, H., McKay, G., & Al-Ansari, T. (2023). A review of waste management approaches to maximise sustainable value of waste from the oil and gas industry and potential for the State of Qatar. *Fuel, 332*, 126220. https://doi.org/10.1016/j.fuel.2022.126220

Sharma, H. B., Vanapalli, K. R., Samal, B., Cheela, V. R. S., Dubey, B. K., & Bhattacharya, J. (2021). Circular economy approach in solid waste management system to achieve UN-SDGs: Solutions for post-COVID recovery. *Science of the Total Environment, 800*, 149605. https://doi.org/10.1016/j.scitotenv.2021.149605

Shyamsundar, P., Springer, N. P., Tallis, H., Polasky, S., Jat, M. L., Sidhu, H. S., Krishnapriya, P. P., Skiba, N., Ginn, W., Ahuja, V., Cummins, J., Datta, I., Dholakia, H. H., Dixon, J., Gerard, B., Gupta, R., Hellmann, J., Jadhav, A., Jat, H. S., … Somanathan, R. (2019). Fields on fire: Alternatives to crop residue burning in India. Science, 365(6453), 536–538. https://doi.org/10.1126/science.aaw4085.

Sun, D., Hao, L., & Xie, D. (2024). Governance of rural solid waste under a multi-subject governance model. *Scientific Reports, 14*(1), 27111. https://doi.org/10.1038/s41598-024-78732-5

Trabelsi, N., Mourtzinos, I., & Ramadan, M. F. (2023). Editorial: Valorization of bioactive compounds from bio-wastes of agro-food sector using green technologies. *Frontiers in Nutrition, 10*, 1334315. https://doi.org/10.3389/fnut.2023.1334315

Tripathi, N., Hills, C. D., Singh, R. S., & Atkinson, C. J. (2019). Biomass waste utilisation in low-carbon products: Harnessing a major potential resource. *npj Climate and Atmospheric Science, 2*(1), 35. https://doi.org/10.1038/s41612-019-0093-5

Wang, J., & Tao, J. (2020). An analysis of farmers' resource disposal methods for livestock and poultry waste and their determinants. *Chinese Journal of Population, Resources and Environment, 18*(1), 49–58. https://doi.org/10.1016/j.cjpre.2021.04.017

Waqas, M., Hashim, S., Humphries, U. W., Ahmad, S., Noor, R., Shoaib, M., Naseem, A., Hlaing, P. T., & Lin, H. A. (2023). Composting processes for agricultural waste management: A comprehensive review. *Processes, 11*(3), 731. https://doi.org/10.3390/pr11030731

Wilson, D. C. (2007). Development drivers for waste management. Waste Management & Research, 25(3), 198-207. https://doi.org/10.1177/0734242X07079149

Xu, L., & Geelen, D. (2018). Developing biostimulants from agro-food and industrial by-products. *Frontiers in Plant Science, 9*, 1567. https://doi.org/10.3389/fpls.2018.01567

Zhang, J., Akyol, Ç., & Meers, E. (2023). Nutrient recovery and recycling from fishery waste and by-products. *Journal of Environmental Management, 348*, 119266. https://doi.org/10.1016/j.jenvman.2023.119266

Waste: Challenges, Environmentally Friendly Handling, and Conversion into Goods of Value

Karpaga Swathi Marieswaran, Dharanidharan Murugan, S. Victor Vasanth Singh, M. Lavanya, Lakshmanan Muthulakshmi, and Nellaiah Hariharan

Abstract Transforming our world by 2030 through achieving substantial reduction in waste generation involves encouraging circular economic practices, including refuse, reduce, reuse, repurpose and, finally, recycle. Waste management is a critical global issue mainly affecting the environment, health, and economy by creating harmful pollution, loss of biodiversity, and serious diseases. Modern waste management emphasizes circular economic practices to minimize adverse impacts on the environment. Zero-waste initiatives rely on a combination of strategies to minimize waste generation and promote sustainable practices. Technologies such as recycling, composting, waste-to-energy (WtE) conversion, mechanical biological treatment (MBT), trickling filter processes, as well as aerobic and anaerobic digestion are being implemented to treat the waste generated. Waste provides raw materials for the manufacture of various products and by-products. Organic waste can be composted or digested anaerobically to produce biogas. Agricultural residues can be converted into biochar or bio-based plastics. Plastic waste is recycled or broken down into fuels and chemicals, while textile waste is upcycled. Municipal waste generates energy and recyclables, while wastewater is treated for nutrients or bioenergy. Agro-waste is used to manufacture products like nutraceuticals, biofuels, biofertilizers, and bio-nanoparticles. E-waste materials are processed to recover nanoparticles, alloys, and valuable metals. Waste cooking oil is used to produce bio-products such as lubricants and bio-asphalts. Additionally, conventional prod-

K. S. Marieswaran · L. Muthulakshmi (✉)
Biomaterials and Product Development Laboratory, Department of Biotechnology, Kalasalingam Academy of Research and Education, Krishnankoil, Tamil Nadu, India
e-mail: l.muthulakshmi@klu.ac.in

D. Murugan · S. Victor Vasanth Singh · M. Lavanya
Department of Biotechnology, Kalasalingam Academy of Research and Education, Krishnankoil, Tamil Nadu, India

N. Hariharan
Shelcel Biozeen Technologies Pvt. Ltd., Bangalore, Karnataka, India

S. Rajpoot, A. P. Das (eds.), *Innovative Technologies for Waste Management*, Sustainable Environmental Waste Management Strategies,
https://doi.org/10.1007/978-3-031-97184-6_14

ucts like notebooks and bags are produced by recycling or upcycling various waste materials. Waste valorization creates business opportunities, lowers raw material costs, promotes job creation, reduces pollution, conserves resources, and raises awareness about sustainability. The value of waste as raw materials for many useful products helps promote sustainable lifestyle practices in society.

Keywords SDG · Sustainability · Waste-to-energy · Zero waste · Biochar · Nutraceuticals · Bio-asphalts

1 Introduction

Wastes are considered materials with no known use that result from the manufacture of economically valuable or useful products. The world creates 2.01 billion tons of municipal solid waste annually, of which at least 33%, at a conservative estimate, is not managed in an environmentally safe manner. On average, every individual generates 0.74 kg of waste a day, with values ranging from 0.11 to 4.54 kg worldwide. At this rate, global waste generation is expected to reach 3.40 billion tons by 2050 (Maalouf & Mavropoulos, 2023). A country's gross domestic product (GDP) is always linked to its per capita waste generation rate, with high-income countries producing more waste than low-income countries. The mean per capita waste production in India is quantified at 370 g per day, in contrast to 2200 g in Denmark, 2000 g in the United States, and 700 g in China (Lebreton & Andrady, 2019). Waste management plays a crucial role in handling waste and has serious implications for the environment, humans, and the economy. Incorporating the 5R methodology in waste generation areas will minimize landfill waste and take the recycling program to the next level (Singh et al., 2022). This chapter endeavors to synthesize recent findings and insights into the scientific understanding of waste, its challenges, and potential solutions.

Waste can be categorized into different types based on its source of generation, including municipal solid waste (MSW), industrial waste, hazardous waste, electronic waste (E-waste), agricultural waste, sewage and wastewater, marine debris, biomedical waste, food waste, and construction and demolition waste (Mishra et al., 2023). Research results highlight the categorization of wastes according to criteria like physical state, biodegradability, chemical nature, and reusability. Based on biodegradability, wastes can be categorized into biodegradable and nonbiodegradable types (Das et al., 2024b). On the basis of reusability, wastes can be classified into reusable and nonreusable types. On the basis of chemical nature, wastes can be categorized as organic and inorganic wastes. Classifications of wastes according to these attributes, the types within each category, and examples for each type are provided in Table 1 (Demirbas, 2010). Understanding these classifications helps in developing strategies that are specific to effective waste management. Each type requires special handling procedures that might minimize environmental impact and boost recycling and reuse wherever possible.

Table 1 Different waste classification, types, and examples

Classification based on	Types	Examples
Physical state	Solid	Municipal solid waste, construction debris, hazardous wastes, medical wastes, and radioactive wastes
	Liquid	Wastewater, industrial effluents from tanneries and distilleries
	Gaseous	Emissions from vehicles, industrial gases
Biodegradability	Biodegradable	Food scraps, garden waste
	Non-biodegradable	Plastics, metals
Reusability	Reusable	Glass bottles, furniture
	Nonreusable	Disposable plastics
Chemical nature	Organic	Food waste, yard debris
	Inorganic	Solvents, heavy metals

Inadequate waste management systems significantly compromise environmental integrity and public health. Associated poor practices of open dumping and irregular collection lead to vectors like mosquitoes and rodents transmitting diseases like malaria and respiratory infections (Dey et al., 2023; Mishra et al., 2019). Environmental impacts include air and water pollution resulting from methane emission and leachates contaminating water in the ground that contributes to climate change. Besides, harmful disposal techniques destroy lands and environments, thereby increasing health hazards, especially for marginalized communities. Proper solid waste management (SWM) is indispensable to the conservation of public health and the environment. The composition of waste determines the waste management strategies to be used at a particular location (Pandit et al., 2021).

2 Waste Management Strategies

2.1 Bioremediation

It is a process in which organic pollutants or waste is biologically degraded under controlled conditions. Microorganisms are usually used to degrade organic contaminants by exploiting their metabolic or biochemical pathways to transform organic compounds (Das et al., 2015; Paul et al., 2023). For this process to be successful, microorganisms must enzymatically attack the contaminants in wastes and convert them to harmless substances. However, this works well only in environments that support the growth and activity of microbes (Das et al., 2024a; Pradhan et al., 2024). Therefore, this application of bioremediation involves manipulating ecological parameters, such that microbial growth and degradation proceed rapidly. Control and optimization of bioremediation processes is complex (Kanungo et al., 2024; Nag et al., 2024). These processes are influenced by factors such as degradation of pollutants by already existing microbial populations, availability of contaminants, and environmental conditions like the pH, temperature, soil type, presence of oxygen, and availability of nutrients (Abatenh et al., 2017).

Bioremediation strategies consist of two types: in situ (on-site) and ex situ (off-site). In situ bioremediation refers to the case wherein contaminants are dealt with on-site, while off-site bioremediation involves the segregation of contaminants followed by their treatment at a different location (Das et al., 2023; Dey et al., 2024). Biological methods are mostly used in situ since their on-site application, which involves minimal transport and digestion of pollutants, renders them more economical and less invasive. Off-site processes involve the removal of contaminants or excavation of contaminated soils prior to remediation. In situ and ex situ bioremediation involves different technologies such as bioventing, biosparging, bioreactor, composting, landfarming, bioaugmentation, and biostimulation.

2.1.1 Bioventing

Bioventing is a form of in situ bioremediation that utilizes microorganisms to degrade organic substances bound to soil in the unsaturated zone (the region extending from the earth's surface to the water table). Bioventing involves enhancing the activity of indigenous microorganisms as well as introducing air (or oxygen) and nutrients into the unsaturated zone in order to stimulate the in situ biodegradation of contaminants in soil (Lee et al., 2006).

2.1.2 Biosparging

Biosparging represents an in situ bioremediation methodology that employs indigenous microorganisms for the biodegradation of organic constituents within the saturated zone. In the biosparging process, air (or oxygen) along with nutrients (when deemed necessary) is introduced into the saturated zone to enhance the biological activity of the indigenous microbial population (Koçak & Ikizoğlu, 2020).

2.1.3 Bioaugmentation

Bioaugmentation constitutes an in situ technique whereby specifically selected standardized bacteria (microbes) are introduced into an area contaminated by undesirable substances. These introduced bacteria facilitate the degradation of the contaminant (Niu et al., 2009).

2.1.4 Biostimulation

Biostimulation pertains to the alteration of environmental conditions to promote the activity of bacteria capable of bioremediation. This enhancement may be achieved through the addition of various rate-limiting nutrients and electron acceptors, including phosphorus, nitrogen, oxygen, or carbon (Omokhagbor Adams et al., 2020).

2.1.5 Composting

Composting refers to the regulated decomposition of organic matter. This process involves the amalgamation of contaminated soil with nonhazardous organic materials, such as manure or agricultural residues. The inclusion of these organic materials fosters the establishment of a diverse microbial community, which subsequently facilitates the decomposition of organic contaminants (Osman et al., 2022).

2.1.6 Slurry Reactors

The ex situ remediation of contaminated soil and water is conducted utilizing slurry reactors or aqueous reactors. Bioremediation within these reactors encompasses the treatment of contaminated solid materials or water via an engineered containment system. A slurry bioreactor serves as a containment vessel and apparatus designed to create a triphasic (solid, liquid, and gas) mixing environment to augment the bioremediation rate of soil-bound and water-soluble pollutants. Bioremediation transpires in the aqueous slurry composed of contaminated soil and biomass (typically indigenous microorganisms) capable of degrading the target contaminants (Balseiro-Romero et al., 2019).

2.1.7 Landfarming

Landfarming is a simple technology that involves the excavation of contaminated soil, spreading it out on a prepared bed, and tilling the soil at regular intervals until the pollutants have been degraded. This methodology is aimed at stimulating indigenous biodegradative microorganisms to enhance the aerobic degradation of contaminants by these organisms. The practice is generally confined to the treatment of the uppermost 10–35 cm of soil. Given its potential to minimize monitoring and maintenance expenses, as well as clean-up liabilities, landfarming has garnered substantial attention as an alternative disposal strategy (Paul et al., 2021).

Bioremediation is a natural process, because of which it is employed as a permissible waste treatment method for a polluted medium such as soil. The microorganisms responsible for degrading the contaminant keep multiplying as long as the contaminant is available; the counts of these microbial populations begin to decrease when the contaminant has been degraded. The by-products resulting from the treatment process are generally non-toxic compounds, including carbon dioxide, water, and cell mass.

Bioremediation facilitates the degradation of specific pollutants without necessitating their relocation from one environmental medium to another, such as from terrestrial to aquatic or atmospheric environments. This process is frequently conducted in situ, thereby resulting in minimal interference with regular operational activities. Additionally, it negates the requirement for the off-site transportation of waste, along with the associated risks to human health and ecological systems that may arise from such transit. Ultimately, bioremediation proves to be more cost-effective compared to alternative methodologies employed for the remediation of hazardous wastes.

2.1.8 Phytoremediation

As part of natural ecosystems, plants act as filters and metabolizers of material produced by natural processes. In the context of phytoremediation, plants are utilized for extracting pollutants from soil and water (Teshale, 2019).

2.1.8.1 Phytoextraction

Phytoextraction (or phytoaccumulation) is the process in which plants remove contaminants from the soil and accumulate them (most usually heavy metals) in their roots and in the parts above the ground. Currently, this process is mainly used for extracting heavy metals and phytomining. Phytomining involves extracting metals from the soil by cultivating and harvesting specially selected hyperaccumulating plants (Napoli et al., 2019).

2.1.8.2 Phytostabilization

Phytostabilization, or phytoimmobilization, constitutes a methodology in which plants mitigate the mobility and migration of contaminants within contaminated soils. Leachable constituents are adsorbed and integrated into the plant structure, thereby forming a stable biomass from which the contaminants cannot reenter the environmental matrix (Li et al., 2021).

2.1.8.3 Phytotransformation

Phytotransformation, alternatively referred to as phytodegradation, pertains to the metabolization and degradation of organic contaminants that are absorbed by plants from the soil or, less frequently, the enzymatic breakdown of contaminants occurring outside the plant by enzymes secreted by the plant itself. In instances where phytodegradation transpires within plant tissues, contaminant uptake is contingent upon both the physicochemical characteristics of the contaminants, such as hydrophobicity, solubility, and polarity, and the specific traits of the plants involved (LeFevre et al., 2015).

3 Wastewater Treatment

Wastewater is classified as a by-product or end-product generated during municipal, agricultural, and industrial processes (Mishra & Das, 2023). It contains a wide combination of organic and inorganic materials, including those that are anthropogenic as well as naturally occurring (Mishra et al., 2022b; Mishra & Das, 2021). In general, the pollutants associated with wastewater include biodegradable organic matter, suspended solids, nutrients, heavy metals, and pathogens. These constituents of wastewater and their concentrations differ with the water source; consequently, characterization of wastewater is a difficult process. Wastewater treatment is a process adopted for the removal of contaminants before they are released into water bodies.

3.1 Quantification of Biodegradable Material in Wastewater

There are two ways of determining the content of biodegradable material in a wastewater sample, namely, BOD and COD. The measurement of BOD indicates the volume of oxygen that needs to be supplied so that aerobic microorganisms can oxidize the biodegradable material within a specific time-frame (usually 5 days) at a constant temperature of 20 °C and upon incubation in the dark. A sample of wastewater is saturated with oxygen and then charged with an inoculum that represents a broad diversity of microbes. After incubation for 5 days, the concentration of oxygen in the sample is assayed (before and after incubation) and the concentration reported as milligrams of oxygen per liter of wastewater. COD quantitates the amount of oxygen required to chemically oxidize any oxidizable organic matter contained in the wastewater. The organic matter content is determined by applying reaction stoichiometry following oxidation of the organic compounds by a potent chemical oxidant. This analytical procedure involves the addition of a known volume of the sample to a mixture of oxygen-rich potassium dichromate and concentrated sulfuric acid. It is important to take note that almost all the organic compounds present in wastewater are oxidized when subjected to strong chemical oxidants. Thus, in a waste-water sample, the value of COD normally exceeds the measured BOD. The BOD-COD ratios for sewage normally lie between 0.2:1 and 0.5:1 (Jayasinghe et al., 2024).

4 Treatment Process

Waste materials present in wastewater can be categorized into three principal classifications: industrial wastes, domestic or municipal wastes, and agricultural wastes (Kumar et al., 2023a, 2023b, 2023c; Mishra & Das, 2022). Each category of waste possesses distinct characteristics, thereby necessitating the application of varied treatment methodologies. The methodologies for waste treatment encompass physical, chemical, and biological treatments.

4.1 Physical Treatment

Physical treatment encompasses a range of processes including screening, sedimentation (clarification), aeration, flotation, skimming, degasification, and equalization. These treatment methodologies do not induce significant chemical or biological alterations in the wastewater, as they are confined to physical processes. In the sedimentation process, physical phenomena related to the gravitational settling of solids are enacted. Typically, this process involves the temporary retention of wastewater within a tank under quiescent conditions, allowing for the sedimentation of denser solids, followed by the extraction of the clarified effluent. An additional physical treatment technique is aeration, which involves the introduction of air to facilitate the oxygenation of the wastewater. Filtration is yet another physical operation employed in the treatment process, where wastewater traverses a filter medium that segregates solids from the liquid. For instance, a sand filter is deployed to eliminate entrained solids within the treated wastewater. The treatment process frequently incorporates the allowance of greases or oils to ascend to the surface, followed by skimming or the physical removal of these substances from the wastewater. In specific industrial wastewater treatment scenarios, substantial or undesirable wastes may be generated in excessive amounts for brief durations. Given that such "slugs" or intermittent inflows of waste could disrupt a biological treatment process, these wastes are occasionally stored, amalgamated with other wastewaters, and progressively released to avert "shocks" to the treatment facility. This procedure is termed equalization. A secondary form of "equalization" is employed to mitigate extreme fluctuations in flow rates. For example, the wet well of a pumping station may be subjected to significantly large volumes of wastewater at highly variable rates, subsequently pumping the waste onward at more stable rates (Amresh et al., 2022).

4.2 Chemical Treatment

Chemical methods of treatment encompass chlorination, ozonation, neutralization, coagulation, adsorption, and ion exchange. Chemical treatment pertains to the application of one or multiple chemical reactions to render water potable. Chlorination stands as the most prevalent chemical treatment technique employed. Chlorine serves as an oxidizing agent and is notably efficient in eradicating bacteria while inhibiting the decomposition of wastewater. Another effective oxidizing agent employed as a disinfectant is ozone. A chemical process frequently utilized in numerous industrial wastewater treatment operations is neutralization, which involves the introduction of either an acid or a base to restore pH levels to neutrality. Given that lime constitutes a base, it is often utilized in the neutralization of acidic wastes. Coagulation entails the addition of a chemical that, through a chemical reaction, generates an insoluble end-product that aids in the removal of substances from wastewater. Polyvalent metals are conventionally employed as coagulating agents in wastewater treatment, with common coagulants including lime (which may also serve in neutralization), various iron-containing compounds (such as ferric chloride or ferric sulfate), and alum (Amresh et al., 2022).

4.3 Biological Treatment

Biological treatment methodologies employ microorganisms, predominantly bacteria, in the biochemical degradation of waste materials in wastewater into stable end-products (Mishra et al., 2022a). An increase in the population of microorganisms, or sludge, occurs, while a portion of the waste is transformed into carbon dioxide, water, and other resultant products. Broadly, biological treatment methodologies can be delineated into aerobic and anaerobic categories. This classification hinges on the presence of dissolved oxygen. In these techniques respiration serves as the basic principal mechanism through which the treatment is carried out. The techniques encompass activated sludge treatment methods, trickling filters, oxidation ponds, and lagoons. The fundamental process involved is the oxidation of organic matter.

5 Anaerobic Methods

In this category, anaerobic digestion constitutes the primary mechanism for breakdown. These methodologies incorporate septic tanks and lagoons, wherein a significant proportion of the carbon contained in organic compounds is reduced to hydrocarbons. The solids that are extricated are chiefly organic in nature, although they may also encompass certain inorganic solids. Furthermore, the sludge, which comprises both solids and liquids, necessitates treatment. It is frequently imperative to administer treatment for the management of odors, inhibition of biological activity, or eradication of pathogenic organisms. Wastewater frequently harbors elevated concentrations of nitrogen and phosphorus. Consequently, the excessive release of these substances into the environment engenders a nutrient accumulation phenomenon, termed eutrophication, which subsequently stimulates rampant growth of weeds, algae, and cyanobacteria (blue-green algae). This phenomenon may catalyze an algal bloom, characterized by a rapid proliferation of algal biomass, a substantial fraction of which ultimately succumbs. A considerable quantity of oxygen is depleted during the decomposition process within the aquatic environment, leading to the mortality of a majority, if not all, animal life. This, in turn, generates additional organic matter for bacterial decomposition. In addition to inducing deoxygenation, certain algal species can generate toxins that compromise potable water supplies. Distinct treatment processes must be employed to eliminate nitrogen and phosphorus (Amresh et al., 2022).

6 Waste Collection and Environment-Friendly Management

Waste collection and management of waste (Fig. 1) are crucial components of sustainable environmental conservation methodologies, representing interdisciplinary strategies that incorporate aspects of engineering, environmental science, and social policy. The increasing generation of waste poses a contemporary challenge to ecological integrity and public health, primarily driven by elevated rates of urbanization and industrialization (Yadav & Karmakar, 2020). Fundamental waste collection systems play a vital role in reducing the ongoing pollution of the environment via

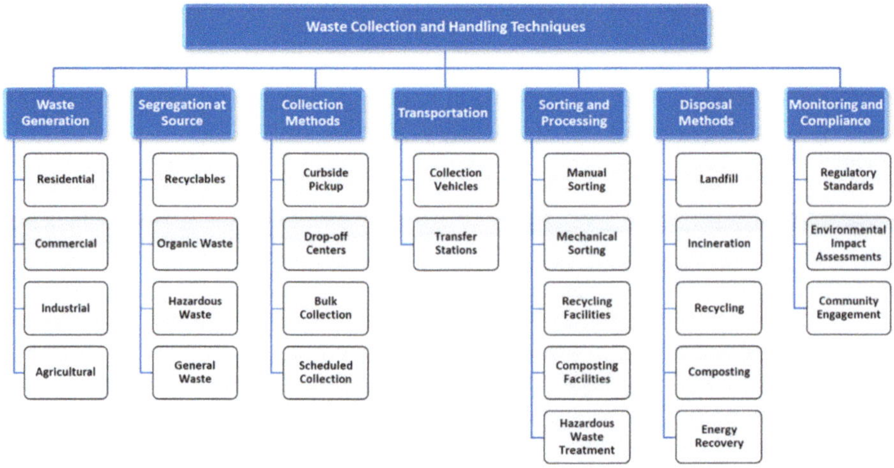

Fig. 1 Waste collection and handling techniques

land, water, and air. Other innovations include source separation, recycling, and composting to recover the material content of waste. Another development, the waste-to-energy conversion with the automation of sorting systems for wastes, has opened up new prospects in more effective and sustainable waste management (Naveen & Sivapullaiah, 2020). Socio-economic determinants of the patterns of waste generation and disposal must inform policy measures and educational efforts. This more holistic approach is fundamentally integrative but addresses the most crucial logistical challenges involved in waste collection, while fostering environmental stewardship. Incorporating scientific research with community engagement will be necessary, with the global crisis intensifying in terms of waste management, with the solutions expected from it to be more aligned with a circular economy approach, while sustaining the environment (Wang et al., 2021a, 2021b). These new trends thus call for more intensive and eco-friendly waste collection and disposal practices. Good waste management is essential not only for resource recovery but also for ensuring good health of the people as well as safeguarding the environment (Edirisinghe et al., 2022).

7 Waste Collection Techniques

Waste collection techniques are crucial elements of urban waste management systems and comprise a variety of approaches based on specific contexts and the needs of communities:

1. The most common approach, door-to-door, means generators take their personally produced wastes to collectors who come to their very doorsteps. Specialized vehicles have scheduled slot times at prearranged collection points where they collect waste and transport it, and are often accompanied by a specific ringing device or bell that marks their arrival (Perrot & Subiantoro, 2018).
2. Curbside collection system: This is the most used technique as it is practiced in almost all urban locations. With the possibility of being both manual and automated, the curbside system favors inhabitants and addresses the waste collection problems of densely populated areas and littering.
3. Drop-place is a community place where people come and deposit recyclables and other hazardous waste items, illustrative of a collective contribution toward environmentalism; this initiative can prove successful only if the public starts working enthusiastically on this project (Panainte-Lehadus et al., 2022).
4. Modern technological innovations, such as pneumatic techniques and automated waste collection systems (AWCS), make use of pneumatic tubes to collect waste without the involvement of humans, but with a heavy investment in infrastructural development (Rossi et al., 2022).
5. Involvement of local communities has encouraged community-based collection approaches, which have increased recycling rates. Intelligent waste collection

technologies have maximized the use of Internet of Things (IoT) in real-time monitoring, which has improved operational efficiency (Acharya et al., 2021).
6. Targeted special collection for hazardous waste offers the means of reaching underserved regions so that safe management of dangerous materials can help prevent undesirable environmental effects. This diversity in approaches, supplemented with ongoing innovation and public education, is a precondition for the progress of efficient, sustainable, and, therefore, "fit-for-today-waste-stream" management concepts (Mazzei & Specchia, 2023).

7.1 Waste Handling Techniques

Waste management techniques are based mainly on waste handling practices in which possible adverse environmental impacts are minimized. Several techniques have been applied at all stages of waste management, which comprise collection, separation, treatment, and disposal:

1. Source separation reduces contamination and makes recycling processes better by separating recyclable materials from organic waste as well as toxic products at the source level (Panainte-Lehadus et al., 2022).
2. Concerning landfill curbing, it is within the management and treatment of waste materials by a landfill site. On a very wide scale, this affects the environment by giving rise to problems like groundwater contamination and generation of greenhouse gases, among other things, with this practice being considered an example of the inept use of land resources. For that matter, managing organic waste by turning a biodegradable into highly prized compost improves soil fertility, composting will effectively curb contributions to landfills and subsequently result in lower emissions of methane gases (Farooqi et al., 2021).
3. Recycling plays a central role in recovery processes that aim at recovering materials, including plastics, metals, and paper. The process is typical of conserving natural resources while extracting and processing raw material (Labra Cataldo et al., 2024).
4. The use of color-coded containers is one very effective strategy in the implementation of waste management schemes as it allows for source separation during segregation of various types of waste. This means that public education is augmented, appropriate practices are emphasized, and contamination levels in streams are minimized. In addition, these color-coded containers are highly useful for the waste management teams and allow for more efficient transportation and processing operations (Seay & Ternes, 2022).
5. The implementation of zero-waste strategies is aimed at rethinking resources life cycles, in reducing waste production, by improving repair and reuse initiatives.
6. Public education and the facilitation of public participation are the major steps taken to empower citizens to adopt a culture of sustainability and responsible

waste management. Electric and alternative fuel collection vehicles minimize the harmful environmental by-products of transporting waste. Proper management and treatment of hazardous wastes ensure that there is no emission of harmful substances, thereby eliminating any harm to public health and environmental integrity (Acharya et al., 2021).

7. Mechanical Biological Treatment (MBT): processes are a recent innovation in waste management techniques. MBT combines the technologies of mechanical separation and biological treatment, making it feasible to treat mixed waste effectively. The process separates recyclables from organic material, which can be stabilized through biological treatment, thereby minimizing the volume sent to landfills and potential environmental hazards (Gadaleta et al., 2022).
8. Composting is based on the activity of natural microorganisms that facilitate the aerobic breakdown of organic materials. Organic waste management by composting involves the decomposition of organic wastes, such as kitchen and yard wastes, through aerobic decomposition, resulting in nutrient-rich compost (Widyatmika & Bolia, 2023).
9. Incineration is a thermal waste management method that burns organic wastes in high temperatures with a massive volume reduction and possible energy recovery. In this process, waste is burnt to ash and flue gases, with the liberation of heat (Das et al., 2021).
10. Waste to Energy (WtE): specifically refers to burning nonrecyclable waste to give usable energy. The majority of waste-to-energy (WtE) methods entail the incineration of a combination of municipal solid waste, resulting in the production of energy, which can be utilized as electricity, steam, or for district heating purposes (Alao et al., 2022).
11. Anaerobic digestion refers to the biological process that decomposes organic materials without the presence of oxygen. This method produces biogas, which primarily consists of methane and carbon dioxide, in addition to a by-product known as digestate. This technique is used to decrease waste and produce renewable energy (Alao et al., 2022).
12. Trickling filter methods are biological processes used in the treatment of wastewater where fixed media provide the substrata for microbial growth. As the wastewater passes through the media, breakdown of organic polymers in the wastewater through the agency of microorganisms takes place, with oxygen being consumed in the process. This results in the rapid reduction of BOD and suspended solids (Apaydin & Akçay Han, 2023).
13. The management of hazardous waste encompasses the processes of identifying, treating, and disposing of materials that present risks to human health and the environment. This category of waste includes chemical by-products, lead-acid batteries, biomedical waste, and electronic waste, commonly referred to as e-waste (Kleppinger, 2022).

The above exemplifies the concept of a circular economy in waste management and explains that resources are used ceaselessly through recycling, reuse, and refurbishment. The ultimate goal is to have as little waste as possible from the extraction of natural resources and production. With these sustainable practices, there should be a thrust on the design of durable products, an effective recycling system, and sensitization of communities on consumption patterns that produce less waste.

8 Limitations of Present Waste Management and Disposal Techniques

Global waste production has increased significantly and poses a problem to current waste management practices and disposal methods. Some of the major concerns in the case of waste management are environmental sustainability, public health-related issues, and economic development, among others. Though technology has greatly advanced and policy-making significantly improved throughout the world, it is not possible for modern waste management techniques to reduce the ill effects of wastes (Farooqi et al., 2021).

8.1 Limitations

- Environmental Impact: Improper waste management might lead to leachate generation, a process that leads to massive contamination of soil and groundwater by landfills. Moreover, one of the major environmental risks that result from landfills is the generation of methane, a potent greenhouse gas (Kumar et al., 2023a, 2023b, 2023c).
- Spatial Problems (space): Urban lands with a shrinking space in addition to an increased number of people lead to unavailability of land to erect new landfill sites. Such conditions result in high costs and environmental degradation in various other areas. Some landfills pose persistent environmental threats that require constant monitoring or even observation to prevent various kinds of risks. The consequences of this are increased liabilities that municipalities experience over several years (Naveen, 2021).
- Air Pollution: The incineration process, which is carried out in the processing plant, emits countless harmful pollutants, such as dioxins and furans, and raises numerous health concerns among citizens living close by while also degrading the air quality.
- Capital Investment: The installation and running expenses of incineration plants are cost-prohibitive, which makes them only feasible in high-income regions.
- Public Opposition: Considering health and environmental reasons, and the fact that communities are vigorously opposing incineration facilities, choosing sites and operations is quite challenging (Widyatmika & Bolia, 2023).
- Contamination: High levels of contaminants in recycling streams often result in the loss of valuable material, which is degraded during the processing stage.

- Economic Feasibility: The volatile market prices for recyclable materials provide no stable ground for cost-efficient recycling programs, and thus result in more landfill disposal (Bajaj et al., 2024).
- Infrastructure: In most regions, a lack of infrastructure and technological equipment for recycling significantly impedes material recovery operations. This activity is essentially scale-dependent and requires available open space, which is always scarce in highly populated areas (Rafiquee & Shabbiruddin, 2024).
- Contamination Risks: The presence of nonbiodegradable contaminants within the compostable streams diminishes the quality of compost.
- Odour and Pest Problems: In unmanaged composting, odors develop, causing infestation, posing a problem to the public (Cirik & Gocer, 2020).
- Regulatory and Policy Obstacles: The lack of uniformity in legislation and its enforcement mainly prevents effective waste management practices, which in turn affects the effectiveness of both the implementation of these practices and the regulations enforcing waste management (Chevillard et al., 2018).
- Technological Constraints: The waste management technologies currently available are poorly developed or very expensive to acquire and implement, and so are restricted in their usage (Maureira et al., 2023).
- Sociocultural and behavioral factors: Laying down public education and community contribution toward segregating waste and minimizing its generation may discourage an effective system of waste management.
- Resource Scarcity: Financial and technical constraints could be critical factors, hindering effective waste management. These constraints are more evident in developing countries (Pires & Martinho, 2019).

The limitations inherent in the waste management and disposal techniques used call for comprehensive, integrated, and adaptive approaches. Waste management systems that will advance into the future promise improvements in sustainability and resilience when tackling technological, economic, and social obstacles. The needed changes in this crucial field will be better supported through cooperative efforts from governmental bodies, industries, and local communities.

9 Challenges from Evolving Waste and Their Management

The unprecedented escalation in waste generation across the globe has occurred as a direct consequence of accelerated urbanization, industrialization, and technological innovations. This new situation has completely altered the landscape of waste management, giving rise to new difficulties in managing waste streams at their traditional places of handling, that is, beyond the waste generation site. Often these outgrow the capabilities of modern systems for waste management as these systems are inadequate to manage modern waste streams given their complexity and diversity (Naveen et al., 2021; Maureira et al., 2023).

9.1 Problems

- Diverse Waste Streams: Historically, the management of waste had solely dealt with solid wastes, mainly consisting of household and industrial components. Shifts in consumer behavior have given rise to multiple forms of waste. Notably, electronic waste, which encompasses obsolete electronic products, is currently the fastest-growing category of waste worldwide. The United Nations estimates this waste to have increased by 21% by the year 2024. The issue of plastic pollution has escalated to concerning levels, with estimates indicating that approximately 300 million tons of plastic are produced annually, a significant portion of which is either disposed of in landfills or released into the natural environment. Estimates also predict that the global generation of e-waste is likely to reach 74 million metric tons by the year 2030 (Maureira et al., 2023).
- Change in waste composition: The nature of waste has changed dramatically, which has made its handling even more difficult. This change includes significantly higher levels of hazardous materials like lead and cadmium, present in electronic wastes and volatile additives commonly linked to plastics. The presence of many hazardous constituents requires specific treatment technologies that are often not present in the current structures of waste management (Wikurendra et al., 2023).
- One of the principal issues of managing new categories of waste lies in the general lack of holistic regulatory and policy frameworks. Most countries do not have uniform laws on how emerging types of waste should be disposed of and recycled. Take the example of e-waste where the existing laws are fractured and cause rampant dumping as well as inadequate recycling methods. Such a situation arises because present policies can lag behind fast-changing technology environments, while in other places, newly found materials have yet to be addressed by lawmakers (Pires & Martinho, 2019).
- Technological Issues: This is the most significant existing condition. Waste management technologies should thus be invented to mitigate the occurrence of varied wastes appropriately. Traditional methods of recycling fail to do well in handling wastes that happen to be electronic, whose material composition is diversified and will require different kinds of processes. There are superior ways and technologies in sorting as well as recycling for the attaining of rare resources without compromising the ecosystem.
- Economic Factors: Economic elements are very crucial to policy development regarding waste management. High costs in the adoption of sophisticated technologies in waste management make it difficult for most developing countries to manage this effectively. Fluctuating market prices, uncertainty, and economic downturn in terms of selling recycled materials render most recycling ventures impossible in most urban areas. Hence, there are mostly dump sites or landfills which only contribute to the exacerbation of the crisis in managing waste (Naveen et al., 2021).
- Public Enlightenment and Involvement Public enlightenment and community involvement, above all else, are basic requirements for proper waste manage-

ment. The critical role these activities play in waste disposal and recycling is unknown to most members of communities. Behavioral characteristics, which range from simple access to facilities for the deposit of recyclables and the seeming effectiveness of recycling initiatives, are factors that greatly influence a participating level. There will, then, be no effective waste management initiatives absent for public participation (Naveen & Sivapullaiah, 2020; Acharya et al., 2021).

- Inadequate waste management has profound detrimental effects on environmental quality and human health. The presence of landfills and open dumping sites serves as a primary source of pollution for both soil and groundwater. It is said that the burning of waste produces an emission of harmful pollutants into the air. An important issue is related to health hazards from electronic waste due to health risks for individuals working in informal recycling processes, since such situations expose workers to dangers from these harmful substances. These consequences call for an immediate intervention in the proper management of waste. Issues related to waste generation and proper management are very complex, demanding multidimensional approaches considering elements of technology and regulation as well as social ones (Kumar et al., 2023a, 2023b, 2023c; Pheakdey et al., 2022).

Improvements to waste management technology will doubtless be a response to the diversity of types and classes of waste. The various challenges inherent in waste management will be overcome through the enhancement of practical implementation, elevated public involvement, and improved regulatory structures. Transitioning communities worldwide into sustainable waste practices is expected to have environmental, economic, and social development benefits.

10 Waste Conversion Methods

In sustainable waste management, waste conversion methods play a major role by transforming the different types of waste into value-added products (Appels et al., 2008). These methods include physical, biological, and thermochemical processes, and are depicted in Fig. 2. Each of these approaches offers unique advantages depending on the specific type of waste and the desired value-added product.

10.1 Physical Conversion

Refuse-derived fuel (RDF) is the outcome of physically converting waste, particularly municipal solid waste (MSW). This fuel ignites rapidly in combustion boilers. The production of RDF involves eliminating noncombustible elements such as glass and metal from MSW. The remaining flammable components are then crushed,

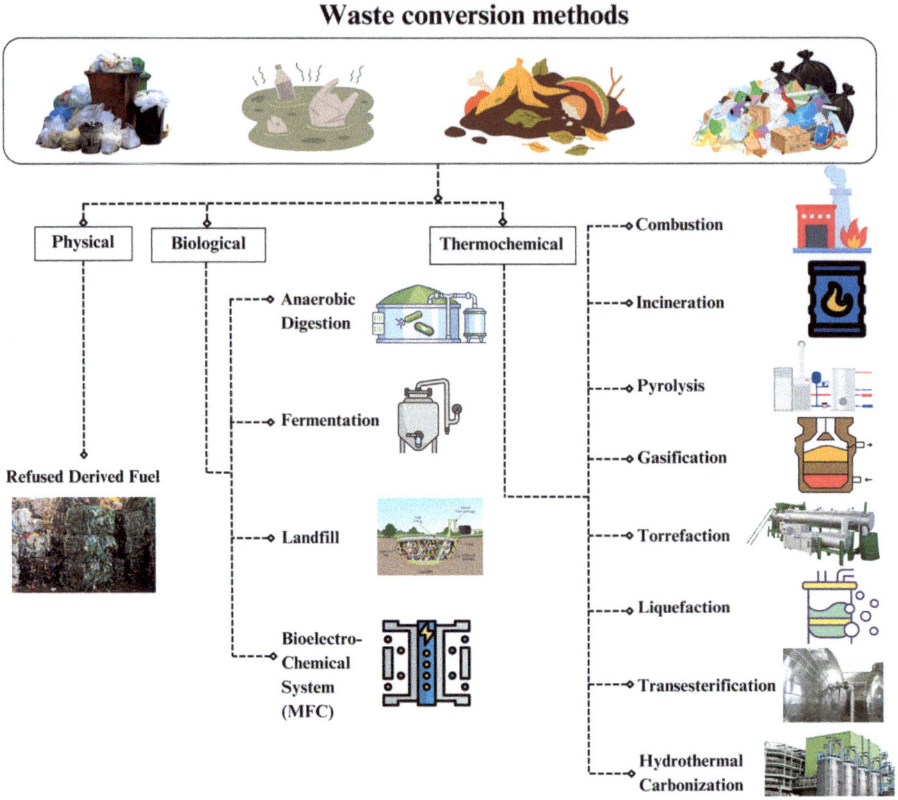

Fig. 2 Physical, biological, and thermochemical waste conversion methods

sorted, and compressed into pellets, which are marketed as fuel. Initially, this technique was developed to create a fuel that could be mixed with coal for use in coal-powered electricity plants (Bhowmik et al., 2023).

10.2 Biological Conversion

The decomposition of organic waste content through microbial activity is known as biological conversion. This process is widely used in many countries as it is environmentally suitable for energy recovery from waste. This conversion process is preferred for wastes with high moisture content and a large fraction of organic biodegradable matter. The commonly used biochemical conversion processes are anaerobic digestion, fermentation, and electrochemical methods (Bhatia et al., 2018).

10.2.1 Anaerobic Digestion

Anaerobic digestion or bio-methanation is the process in which the organic biodegradable matter of waste is degraded by a group of microorganisms to produce biogas in an oxygen-free environmentand stabilize the sludge (Sahota et al., 2018). The resulting sludge can be utilized as a soil enhancer or organic supplement in agriculture (Kumar & Samadder, 2017). The composition of the substrate and process parameters determines the quality of the generated biogas. Typically, biogas consists of 75% CH_4, 25–50% CO_2, and 1–15% other gases (including water vapor, NH_3, H_2S, etc.) (Sahota et al., 2018; Surendra et al., 2014). There are four significant stages in anaerobic processing performed by various microbial groups, in particular, hydrolysis, acidogenesis, acetogenesis, and methanogenesis. During hydrolysis, microorganisms produce hydrolases that convert the complex polymeric materials of the waste biomass into simpler substances (sugars, amino acids, and fatty acids) (Aryal et al., 2018). Subsequently, acidogenic bacteria ferment these simpler substances into various volatile fatty acids and gaseous elements (H_2 and CO_2). Acetogenic bacteria further reduce these components to acetic acid. Methanogenic bacteria then utilize the intermediate products from the various steps to ferment them into methane, carbon dioxide, and water (Bhatia et al., 2017). This process can be applied to all types of biological waste, including food, agro-industry, animal, and municipal waste (Parthiba Karthikeyan et al., 2018). Landfill waste disposal generates greenhouse gases that endanger the environment. However, utilizing anaerobic digestion allows for the capture of methane, which can serve as an energy source (Wang, 2014). This process is strictly anaerobic and is influenced by various process parameters such as pH (for methanogenic bacteria 6.5–7.2, other fermentative bacteria 4.0–8.5) and temperature (Appels et al., 2008). Anaerobic digestion can be categorized based on substrate moisture content (wet or dry), digestion temperature (mesophilic or thermophilic), and feeding method (continuous, fed batch, or batch). To enhance solid waste anaerobic digestion yield, it is often combined with other substrates (Huang et al., 2016). This establishes a positive synergism in the medium and the co-substrate supplies nutrients that are missing in the waste (Khalid et al., 2011). This process can address waste management and energy challenges while potentially generating revenue through side products like biofertilizers.

10.2.2 Fermentation

Fermentation is a bioprocessing or biochemical technology to convert carbohydrates into liquid fuels and value-added products using microbial cultures (Richard et al., 2021,). Commonly used microorganisms include bacteria (such as *Escherichia coli*, *Clostridium acetobutylicum*, *Synechocystis*, and *Clostridium tyrobutyricum*), yeast (*Saccharomyces cerevisiae*), and fungi (*Rhizopus*) (Ramamurthy et al., 2021). For biofuel or energy production, fermentation involves various process designs and configurations, which include simultaneous saccharification and fermentation

(SSF), separate hydrolysis and fermentation (SHF), co-fermentation of pentoses and hexoses, and consolidated bioprocessing (CBP) (Morales-Martinez et al., 2017). In SHF, biomass saccharification and fermentation are sequentially for biofuel conversion. The major disadvantages to its commercialization are the extended processing time and high costs. SSF reduces overall processing time and cost by combining biomass saccharification and fermentation in a single unit operation. But SSF requires higher enzyme loads to compensate for the optimal conditions, such as temperature and pH, to achieve a better yield.

To overcome this, simultaneous saccharification and co-fermentation (SSCF) and consolidated bioprocessing are alternative approaches for energy or biofuel production. Genetically modified microorganisms have been introduced to increase the yield of bio-products. In sequential fermentation, *Saccharomyces cerevisiae* is used for hexose utilization in the first phase, while *Candida shehatae* is used for pentose utilization in the second phase (Kumar et al., 2023a, 2023b, 2023c). Other genetically engineered microorganisms (*Saccharomyces cerevisiae*, *Escherichia coli*) have been developed to simultaneously uptake glucose and xylose, enhancing process efficiency and ethanol yield (Wang et al., 2019). Consolidated bioprocessing (CBP) combines cellulase production, biomass hydrolysis, and sugar fermentation into biofuels in a single process. For efficient ethanol production directly from biomass, this process uses a monoculture or co-culture microorganisms, such as the bacterium *Clostridium thermocellum* bacteria and the fungus *Trameteshirsuta*, both of which are naturally cellulolytic (Akinosho et al., 2014). During CBP, 86% of sugars in the waste are converted into ethanol, and value-added products like biopolymers and propanediol are obtained from unutilized lignocellulosic biomass and organic waste effluent. To increase yield, as well as the economic and environmental benefits, CBP integration and optimization are necessary (Kumar et al., 2023a, 2023b, 2023c).

11 Landfills

Landfilling remains a widely used and economical approach for converting municipal solid waste (MSW) into energy through biological processes. This method is used globally in countries such as Brazil, Malaysia, Bangladesh, Iran, and Nigeria, among others (Kaur et al., 2021). Sanitary landfilling involves the controlled disposal of waste on land, aiming to minimize environmental impact by recovering biogas and managing leachate. The decomposition of the organic matter in landfills produces landfill gas (LFG), which contains CO_2, N_2, CH_4, and H_2S and can be utilized to produce heat and electricity (Yong et al., 2019). LFGs released in the environment have multiple harmful effects, including global warming, endangering human life, and potential hazards like fires and explosions.

The Intergovernmental Panel on Climate Change (IPCC) method can be employed to estimate the methane content in LFG produced by anaerobic biodegradation of MSW in landfills (Bhowmik et al., 2023). The landfill disposal process includes waste dumping, compaction, and burial. In waste dumping, waste is collected and deposited with protective measures like leachate and gas collection systems, implemented to minimize environmental impact (Wang et al., 2021a, 2021b), and then evenly spread and compacted to enhance stability and land recovery. In waste burial, the waste is covered with soil to control odors, limit excessive biodegradation, reduce leachate formation, deter vermin, and protect the local biodiversity (Bhowmik et al., 2023).

12 Bioelectrochemical Systems (BECS)

Bioelectrochemical conversion is one of the emerging waste conversion technologies wherein wastewater is converted to bioenergy. Microbial fuel cells (MFCs) are a type of bioelectrochemical cell where one electrode, typically the anode, interacts with microorganisms (usually respiring bacteria) to transfer electrons from organic substrates to the electrode (Seelam et al., 2018). In MFCs, the breakdown of waste organic matter in the anodic compartment generates electrons, which are transported to the cathode via an external circuit. Simultaneously, protons move from the anode to the cathode through a proton exchange membrane (PEM). MFCs require an oxidative agent, such as oxygen, to produce electricity (Lee et al., 2019). MFC is an eco-friendly technique which generates energy from wastewater; however, this technology is still in its early stages, and the electricity produced is currently insufficient for large-scale application.

13 Thermochemical Conversion

Thermochemical conversion is the process in which waste is chemically dissociated into value-added products and fuels under controlled conditions (i.e., at a defined temperature, pressure, residence time, etc.). Commonly used thermochemical processes are direct combustion, incineration, pyrolysis, gasification, liquefaction, and torrefaction (Kartikesh et al., 2024).

13.1 Combustion

Combustion is a popular technology, used in most of the waste-to-energy plants which accompanies subprocesses such as drying, gasification, devolatilization, and gas- and solid-phase combustion (Perera et al., 2021). Direct combustion is employed to convert chemical energy into mechanical energy, electricity, and heat; however, the release of NO and CO_2 must be accounted for (Ong et al., 2019). This process is less common in developing countries due to an increase in environmental pollution.

However, it is utilized in advanced nations like the United States, Malaysia, and China, primarily to reduce excess waste by condensing it up to 90%. This is achieved by utilizing the waste to generate electricity and heat (Lu et al., 2017).

13.2 Incineration

Incineration is the primary thermal conversion method, which involves the controlled combustion of solid waste at elevated temperatures. This process reduces waste volume up to 80–90% through complete combustion in an incinerator. In 2003, the US Environmental Protection Agency (US EPA) recognized municipal solid waste incineration as a sustainable electricity source due to significant reductions in emissions. China, the world's largest producer of MSW, relies heavily on incineration and landfills. The major benefits of the incineration of solid waste include the waste volume reduction, on-site waste processing, minimized air emissions, obtaining of aseptic ashes, and reduced space requirement compared to landfills (Bhowmik et al., 2023). The significant advantage of solid waste incineration is the complete elimination of all biological contaminants and conversion of organic materials into safer by-products (Brunner & Rechberger, 2015).

13.3 Pyrolysis

Pyrolysis involves the irreversible thermochemical breakdown of organic matter in waste biomass to generate biofuels (Rahimi et al., 2022). The process heats waste biomass to 300–800 °C without air or oxygen, yielding solid fuels (coal, biochar), liquid fuel (bio-oil), and syngas. Depending on parameters like resistance time, heating rate, temperature, and lignocellulosic composition of the residue, the amount of energy produced varies. Pyrolysis has three types: slow, intermediate, and fast. In slow pyrolysis, the slow heating rate (temperature is less than 450 °C) and longer vapor residence time result in greater production of biochar due to slower carbonization of the waste biomass. Fast pyrolysis occurs at 450–800 °C, leading to rapid condensation of volatile hydrocarbon vapors into bio-oil because of quick heating and brief vapor residence times (Wang et al., 2020; Rahimi et al., 2022). Bio-oil comprises an organic phase (tar and hydrocarbon-rich heavier compounds) and a watery phase (water and other biochemicals) (Singh et al., 2021). The organic phase can be upgraded through catalytic and non-catalytic processes to remove nitrogen, oxygen, and sulfur compounds to obtain clean transportation fuels. This may reduce heating value and potentially lead to nitrogen oxide (NO_x) and sulfur oxide (SO_x) emissions during combustion (Kumar & Strezov, 2021). The gas phase, particularly H_2 and CO, can be transformed to liquid hydrocarbons via catalytic Fischer-Tropsch synthesis. In intermediate pyrolysis, the reaction temperature is the range of 300–500 °C which produces liquid products with lower tar yield and viscosity compared to fast pyrolysis (Dhyani & Bhaskar, 2018).

13.4 Gasification

In gasification, the waste biomass is the partially oxidized at high temperatures (500–1600 °C) in the presence of a gasifying agent (i.e., air, oxygen, CO_2, or steam), which results in gas, tar, and solid products. This method is the preferred thermochemical conversion method for waste biomass with high moisture content (up to 35%) (Molino et al., 2016). During gasification the biomass is heated at high temperatures, i.e., >800 °C, which results in production of syngas. This gas mixture primarily consists of combustible H_2 and CO, along with varying amounts of CO_2 and CH_4, and trace quantities of benzene, toluene, aliphatic hydrocarbons, and tars. Syngas can be used to generate electricity in turbines and internal combustion engines (Nicodème et al., 2018). In developed countries, gasification is the most prevalent thermochemical conversion method because of the lower production of CO_2 compared with combustion. When compared with gasification, the fuel formed from the waste biomass pyrolysis is preferred because of its higher heat content and low transportation cost (Dhyani & Bhaskar, 2018).

13.5 Torrefaction

Torrefaction is also called mild pyrolysis in which the waste biomass is heated at a low temperature between 200 and 300 °C in an oxygen-free environment for 60 min. The process is categorized based on temperature ranges: light (200–235 °C), mild (235–275 °C), and severe (275–300 °C) (Chen et al., 2015). The objective of this process is to pretreat biomass by creating torrefied pellets or briquettes, enhancing homogeneity and heating value while reducing moisture content and low-weight organic volatile components, similar to charcoal production through carbonization (Fernandez et al., 2016). During torrefaction volatile matter is converted to hot vapors, which then condensed. The resulting liquid contains water and carboxylic acids along with traces of aldehydes, ketones, ethers, esters, and alcohols (Acharya et al., 2015). The briquettes obtained have properties comparable to coal, and hence can be used as its green alternative. Charcoal/coal can be used for domestic heating and improves biomass energy potential in combustion, pyrolysis, and gasification processes (Sukiran et al., 2017). The advantage of this process is that it does not emit any greenhouse gases.

13.6 Liquefaction

Liquefaction, also called hydrothermal liquefaction (HTL), is a process in which the polymeric structure of waste biomass is converted into liquid components (fuels) at 250–374 °C and 4–22 MPa for less than an hour (Yang et al., 2019). The hot compressed water acts as both a catalyst and reactant, making it similar to pyrolysis

(Toor et al., 2011). Liquefaction predominately produces bio-crude oil. Compared to pyrolysis, liquefaction offers advantages such as higher energy efficiency, lower operating temperatures, and reduced tar yield (Ong et al., 2019). The mechanism of liquefaction is that waste biomass is hydrolyzed to monomers, then to smaller compounds through cleavage and decarboxylation. The smaller compounds thus obtained are recombined to give new products by condensation and polymerization. Commonly used solvents in liquefaction are ethanol and subcritical water. The breakdown of hemicellulose, cellulose, and lignin, present in waste, into volatile matter is achieved using solvents (Yang et al., 2019). Like pyrolysis two phases are formed during liquefaction: a tar phase and a watery phase. The watery phase contains water and certainbiochemicals that can be reused as a solvent in the process (Hu et al., 2018). Bio-crude oil produced in this process requires minimal upgrading by hydrotreating because of its reduced oxygen content, increased hydrocarbon yields, enhanced energy density, and flowability (Rahimi et al., 2022).

14 Transesterification

In the transesterification process, nonedible oils undergo reaction with alcohols to produce biodiesel. Ethanol and methanol are commonly used in this process because they are easily available and inexpensive. In this process, the viscosity of nonedible oils is decreased, and this facilitates the conversion of triglycerides into esters (Avhad & Marchetti, 2015). Waste cooking oils and waste fats are combined with alcohol to produce methanol and methyl esters. The methyl esters can be catalytically transformed into biodiesel and glycerol, and the biodiesel-methanol mixture subsequently separated and reused (Veljković et al., 2015). Phosphoric acid, sulfuric acid, organic sulfonic acids, and hydrochloric acid are some of the acid catalysts used in the transesterification process. Alkali catalysts such as NaOH, KOH, carbonates, sodium ethoxides, and sodium methoxide are also employed. Alkali-catalyzed transesterification is more efficient than acid-catalyzed processes (Rahimi et al., 2022).

15 Hydrothermal Carbonization

Hydrothermal carbonization (HTC) is a thermal process first developed in 1913 for converting cellulose into coal, used to treat high moisture-containing (80%–90%) waste (Ali et al., 2021). This method subjects moisture-rich waste to various reactions including dehydration, decarboxylation, hydrolysis, and condensation at relatively low temperatures (180–350 °C) and autogenous pressures (Le Man et al., 2010). HTC offers several significant advantages, such as requiring minimal space for treatment, odor-free processing, and achieving substantial waste volume reduction within hours (Ali et al., 2021). HTC's process temperatures are sufficient to eliminate pathogens and neutralize other potentially harmful organic contaminants which results in hygienic, easily stored, and transportable energy-rich products. Furthermore, HTC allows for energy recovery from the liquid, further enhancing its efficiency (Lohri et al., 2017), and produces nitrogen-rich liquids that could be utilized as a fertilizer (Libra et al., 2011). This process generates highly carbonized material, which is highly functional and suitable for valuable end-use applications.

16 Waste-to-Value-Added Products

Waste valorization involves the converting waste residues into valuable products for promoting a sustainable economy and enhanced efficiency in resource utilization. Transforming waste (such as agro-waste, electronic waste, toxicants) into valuable products is a sustainable approach as it repurposes discarded materials and reduces environmental harm in the process. Through the biological conversion of several types of waste, including agricultural residuals, food waste, industrial waste, and electronic waste, into biofuels, bioplastics, biofertilizers, and nanomaterials and other products, waste valorization plays a promising role in minimizing pollution and promoting a circular economy. A list of waste-to-value added products and their conversion techniques are presented in Table 2.

17 Agricultural Waste to Bio-products

17.1 Vermicompost

Vermicompost is a dense mass derived from organic matter rich in earthworms; tiny organisms and their respective growth hormones; decomposing enzymes such as chitinase, cellulase, lipase, amylase, and proteases; and vitamins that are necessary for growth. Earthworms help in synthesizing certain enzymes (e.g., nitrate reductase) in the waste biomass that have the potential to boost the yield of crops while simultaneously enhancing their physiochemical quality and soil health (Ahmad et al., 2022).

For the partial fermentation, various types of feed materials, such as crop residues and animal manure in proper ration, are used. Vermicomposting uses locally available materials such as wheat straw, maize stalks, legumes, grasses, and red worms, of which the latter is used to decompose the feed. Wheat, soybean residue, haricot bean, maize stalks, chick pea, finger millet, groundnut residues, and rhodes grass have been used in vermicomposting (Elema, 2021).

The organic content of potato peel and its high biodegradability make it rapidly decomposable by earthworms or microorganisms. Besides, the abundance and easy biodegradation of potato waste suggest the possibility of use of leftovers of the

Table 2 Value-added products from waste and their conversion techniques

Type of waste	Conversion methods	Value-added products
Agricultural waste (peels of fruits and vegetables, molasses, rice and wheat straw, etc.)	Biodegradation, fermentation	Biofertilizers
	Enzymatic hydrolysis	Biofuels (bioethanol, biobutanol)
	Anaerobic digestion	Biogas
	Biopolymer extraction and chemical modification	Bioplastics
E-waste (PCB, batteries, LCD screens)	Pyrometallurgy, hydrometallurgy, bioleaching	Recovered metals (Au, Ag, Pd, Co, etc.)
	Calcination, acid leaching, solvothermal method, sol-gel method	Photocatalytic materials (TiO_2, ZnO)
	Solvent extraction, electrowinning, dialysis, green synthesis (using microbes, plant derivatives)	Nanoparticles (Ag, Au, Pt, Co, etc.)
Waste cooking oil	Transesterification, pyrolysis	Biodiesel
	Epoxidation, transesterification	Biolubricants
	Microbial fermentation	Bioemulsifiers, biosurfactants
Plastic waste	Pyrolysis	Bio-oil Plastic oil
	Hydrothermal liquefaction	Bio-crude oil
	Gasification	Synthetic gas
Textile waste	Enzymatic hydrolysis, fermentation	Bio-based chemicals (bioethanol, lactic acid, sorbitol)
	Mechanical recycling, blending	Recycled yarn, composites

vermicomposting process. Local management of potato peel waste is by feeding it to animals, conversion into compost, or disposal in dustbins. A considerable portion is eventually delivered to waste disposal sites, thereby emitting methane, a more potent GHG than CO_2. Potato peel has a large amount of starch, cellulose, hemicellulose, polyphenols, lipids, and fermentable sugars, which makes it more suitable for vermicomposting (Canales-Gutiérrez et al., 2023).

17.2 *Biofertilizers*

Present efforts are focused on the production of biofertilizers from farm residues using residues of fruits such as melons, pineapples, oranges, bananas, and papayas. These biofertilizers are produced via solid-state fermentation and are often used in farms and gardens. The peels of papaya, banana, and carrot have been used to produce an economical and renewable biofertilizer for possible use in some farming operations. They consist of metallic ions, phenolic compounds, beta-carotene, vitamins, glucose, fructose, and volatile fatty acids which are categorized as carbohydrates. Banana skin also contains lipids, proteins, crude fiber, and carbohydrates (simple sugars such as glucose and fructose), phosphorus, vitamin E, malic acid, and trace acids. Carrot peels are constituted of water, carbohydrate elements, vitamin C, niacin, phytochemicals, and metallic ions.

The crop waste of tubers can be used in the preparation of biofertilizers through by fermentation. The tuber crop residues utilized include cassava, potato, sweet potato, carrot, and beetroot. Waste fertilizer from tuber crops exhibited the highest efficiency compared with others (Darch et al., 2019).

17.3 Biofuels

Agro-waste from sugarcane (such as sugarcane bagasse and bark) and maize (such as corncob, corn stalk, and corn husk) are subjected to acid hydrolysis to remove lignin for the production of biofuels. Besides, ethanol can be produced from non-food items such as corncob, cornstalk, corn husk, sugarcane bagasse, and sugarcane bark, as these contain a lot of sugar and serve as good feedstocks for ethanol production (Braide et al., 2016).

Natural resources for biofuel production include agricultural residues like corncob, corn stover, sugarcane bagasse, fruit peel, and organic waste from plantations. Bioethanol is an excellent substitute for gasoline; a lot of waste products are available, and the raw material is lignocellulosic biomass, which is abundantly available. Cassava pulp is highly rich in starch and also contains a small amount of fiber, which makes it very ideal for conversion into ethanol. *Saccharomyces cerevisiae* is the widely used yeast in the manufacturing of ethanol as it is resistant to acidic pH, thereby minimizing chances of contamination. They also have a higher ethanol tolerance than other microorganisms and rapid fermentation rates, besides being unaffected by fluctuations in temperature or substrate concentration (Hossain et al., 2020).

Biobutanol can be produced from different agricultural residues such as corn, wheat, cassava peels, molasses, and sugarcane bagasse. Since these materials are of biological origin, they contribute toward making the biomanufacturing process cost-effective. This form of biological fuel is an upgraded type with a heat of vaporization higher than that of gasoline. Thus, as a motor fuel, biobutanol emits fewer toxic gases and can be easily transported through the existing pipeline infrastructure, which reduces transportation costs (Awogbemi & Von Kallon, 2022).

Biobutanol finds use in the manufacture of products as diversified as paints and coatings, cosmetics, detergents, chemical intermediates, and herbicides. However, it is also used in resins and plasticizers, and especially in the food and healthcare sector. Biobutanol has been produced from synthetic gas or biogas and the new technology using cheaper raw material like wood logs, agricultural stover, crop residues, sewage, and other degraded biomass used for green methanol synthesis. Bio-methanol made from these feedstocks is a green fuel with a high-octane number and a reasonable price. It is to be used as an additive to gasoline or ethanol in internal combustion engine (ICE) vehicles (Awogbemi & Von Kallon, 2022).

17.4 Bioplastics

Banana peel can be used as a raw material in the development of biodegradable plastics, thereby generating environmentally friendly plastic products. The bioplastic synthesized from banana peel also has several desirable characteristics such as low cost, easy availability, and biodegradability. A study in Packaging Bulletin has shown that banana starch provides a fundamental and appropriate factor for the production of biodegradable plastic.

Biodegradable plastic obtained from maize is one of the most widely used in its class. Starch derived from corn is another ingredient used in the synthesis of polylactic acid (PLA), making this crop an ideal source of most polymers that were in the past produced from petroleum. Bioplastic derived from maize is recyclable, reversible, and degradable, with the carbon chains in the starch closely resembling those of polymers from oil (Rai & Choure, 2023).

Potato is also used for manufacturing biodegradable plastics which are primarily produced from potato peel and spoiled potatoes. Some plastics can be made of compostable biodegradable materials that can be processed from potato starch. Furthermore, plastic made from wheat straw is roughly ten times as efficient as that made from common materials. Lignin, which is a major component of wheat straw, can be used to make plastics (Rai & Choure, 2023).

18 E-Waste to Renewable Materials

18.1 Extraction of Metals

Gold (Au), palladium (Pd), and silver (Ag) are among the most preferred metals obtained from the electronic wastes, including PCBs, due to their high economic value resulting from their high conductivity and excellent corrosion resistance. Gadolinium (Gd), praseodymium (Pr), dysprosium (Dy), and neodymium (Nd) are essential for high-performance magnets and speakers, and are not replaceable in electronics and other industrial goods.

A type of scrap originating from indium is used in LCD screens, while other scrap from computers, laptops, mobile phones, and flat screen TVs contain other recyclable materials. Lanthanum, terbium, and gallium are some components used in the manufacture of electric light sources, particularly LED lights and fluorescent lamps.

Some batteries may include yttrium (Y), europium (Eu), ceramics (Ce), or cobalt (Co) for enhancing their performance and durability. The recycling of these metals facilitates their conversion into pieces of jewelry appealing to the eye. Further, they are also used in the manufacture of connectors found in electronics and circuit boards.

In addition, both platinum and palladium are essential for the manufacturing of cardiac pacemakers and many surgical instruments. The renewal of e-waste ensures that a steady supply of new applications is provided. These metals are especially used in the aerospace industry because of their superior performance, and platinum is used in catalytic converters to reduce vehicle emissions. Overall, the process of collection and recycling of these beneficial materials puts into practice a closed-loop model, emphasizing resource optimization and overall decrease in losses (Aboghrbia & El-Emam, 2024).

18.2 Photocatalytic Materials

There are many photocatalytic materials that can be acquired from electronic wastes. Recycled materials from electrical and electronic equipment (EEE) include different metal oxides such as TiO_2 and ZnO, sulfides, carbonaceous materials, materials derived from biomass templates, and carbon quantum dots (CQDs). Moreover, metals such as platinum, gold, silver, etc., extracted from e-waste, may be employed in photocatalytic processes, including water splitting and CO_2 reduction, due to their high electron mobility, optical transparency, and abundance of active sites. The potential recovery of TiO_2 from multilayer ceramic capacitors (MLCC) with $BaTiO_3$, Pd, Ag, Ni, and Sn has been investigated through synthesis of Nb-Pb-co-doped and Pd-loaded $TiO_2/BaTiO_3$ nanocomposites with chlorination-leaching processes (Bahadoran et al., 2022).

ZnO has been suggested as a photocatalyst rather than TiO_2. ZnO nanostructures can be regarded as a viable candidate for use as a photocatalyst in the photodegradation of POPs under solar light. The findings of a study on the role ZnO nanostructures in solar photocatalysts for photodegradation that ZnO are preferable to TiO_2 on account of its relative inexpensiveness non-toxicity, and better absorption of the solar spectrum (Ong et al., 2018).

Sulfides are also deemed suitable photocatalysts due to their ability to absorb visible light. It is more essential to mention that some of the samples containing metal sulfides and/or transition metal sulfides such as CdS and ZnS can exhibit photocatalytic behavior. Furthermore, WS_2, Cu_2S, In_2S_3, MoS_2, and CdS and complex structures and ternary systems such as $CdIn_2S_4$, $CdLa_2S$, and $ZnIn_2S_4$ exhibit good photocatalytic performance (Bahadoran et al., 2022).

Indium tin oxide (ITO) thin-film electrodes have photocatalytic activity and can be used for the degradation of dyes and for H_2 production; The indium content of the ITO films of discarded LCDs is fairly high, which makes them a rather promising secondary source of indium (Bahadoran et al., 2022).

18.3 Nanoparticles

Computer circuit boards, smartphones, laptops, vehicles, and super-capacitor electronic wastes have nanoparticles of metals such as Fe, Hg, Cu, Pb, and other precious metals such as Au, Ag, Rh, Pd, and Pt, which can be recovered. Apart from used batteries, other categories of e-waste also provide very rich sources of recoverable materials. Cu nanoparticles have been recovered using acidic $CuCl_2$ etchants that arise in printed circuit board fabrication. Moreover, $CuSO_4$ solutions extracted from PCBs are chemically reduced to obtain Cu nanoparticles, which are further stabilized organically. Common metals like Fe, Au, Cu, Hg, and Pb have been extracted from printed circuit board wastes during recycling (Abdelbasir et al., 2020).

Silver nanoparticles (AgNPs) are recovered because of their antimicrobial properties, UV resistance, and optical and electrical properties coupled with high thermal and electrical conductivity (Mondal et al., 2023).

Gold found in PCBs is more than that found in gold-rich ores. In small qualitative laboratories, it would be possible to recover very small quantities of gold from e-scrap, sufficient for synthesizing nanoparticles. Old computers can be used to acquire more gold, especially from high gold-density pieces like CPU pins and wires, and RAM chips and contacts (Do Nascimento et al., 2021).

18.4 Utilization of Waste Cooking Oil (WCO)

Valorization and recycling of WCO is highly recommended for reducing its negative effects on environment. For example, lubricants, pyrolytic oil, soaps, polyurethane, biohydrogen, biodiesel, polyhydroxybutyrate, and electricity can be produced from WCO. A large amount of WCO is disposed of untreated, causing damage, which can be minimized by using the WCO for biodiesel production (Kim et al., 2021). A lot of other value-added products like polyhydroxyalkanoates, lipases, microbial lipids (microbial oils, single-cell oils), and bioemulsifiers can be recovered (Lopes et al., 2020).

Using WCO as a feedstock in the manufacturing of nonfuel products, such as biolubricants, bio-asphalts, bioplasticizers, biosurfactants, and 3D printing resin, will significantly lower their production cost. This is because the cost of the feedstock comprises of more than 50% of the total production cost of these commodities. Therefore, the use of WCO as a feedstock will have the total effect of reducing the prices of these products at their point of sale. Conversion of WCO into nonfuel products is not only economically viable but also extremely environmentally friendly (Hosseinzadeh-Bandbafha et al., 2022).

18.5 Plastic Waste to New Products

Plastic waste is considered to be the major fraction of solid waste across nations. The management of plastic waste is a universal issue, specifically because most of the plastics are not biodegradable. This issue is very critical in developing countries due to shortage of technologies and policies for management of plastic waste. The different types of plastics that contribute to waste are polypropylene (PP), polyvinyl chloride (PVC), polyethylene terephthalate (PET), polystyrene (PS), high-density polyethylene (HDPE), low-density polyethylene (LDPE), and several others.

Plastics can be utilized in different ways by converting them to oil, bio-crude oil, hydrogen, and aromatic char and synthesis gas through pyrolysis, liquefaction, and/or gasification. Plastic pyrolysis oil can be recommended for use as an effective substitute for fossil diesel in diesel engines (Choi et al., 2021).

Depending on their type, plastics can be used as raw materials in manufacturing a number of materials and fuels. For example, PET, PP, and PS are used in making liquid fuels, while LDPE, PP, and PS are used in construction materials. Printable paper is made from linear low-density polyethylene (LLDPE). High-demand fibers involve PVC, PET, and PP, and manufacturing of new polymers involves LDPE, PET and high-density polyethylene (HDPE), PVC, and PS. PP is used to produce polar waxes, and both PP and PET are crucial materials for the synthesis of carbon nanoparticles (Abdelbasir et al., 2020).

Carbon nanotubes (CNTs) can be synthesized using many carbon precursors, of which waste plastics form a favorable carbon source. Use of waste plastics for the production of CNTs is not only useful in addressing the issue of plastic waste management but also helps in obtaining valuable outputs. CNTs can be produced from various types of plastics, including PP, PE, polyvinyl alcohol (PVA), and PET (Nižetić et al., 2019).

18.6 Upcycling Textile Waste

Textile waste has been transformed into valuable products like composites for automotive application, microbial fuel cells (MFCs) for energy production, potassium ion-exchange materials, and biochar for ecological management. It is also used for sound absorption, insulation, and as electromagnetic shields. By-products from the cotton industry, including cotton silver waste, cotton oil, and short staple cotton waste and cotton gin residue, have been used to synthesize cellulose nanocrystals (CNCs) that have important uses in medicines and hydrogels. Besides, a wide range of mixed fiber by-products is used for the preparation of mixtures of composites for textile fiber-reinforced composites (TRFC), a low carbon, and non-toxic building material for premium structures (Subramanian et al., 2022).

A novel resource recovery method utilizing textile waste is proposed for the valorization of three categories: cotton, coloured cotton, and cotton-polyester-blended textiles. Through enzymatic hydrolysis and fermentation, these different kinds of textile wastes are transformed into valuable bio-based chemicals such as sorbitol, terephthalic acid (TPA), bioethanol, ethylene glycol (EG), and lactic acid (LA) (Cho et al., 2023).

Various blends of fibers sourced from yarn waste, rags, and fabric clippings have been transformed into open-end yarns. The resulting yarn from these blended materials demonstrates enhanced tensile strength. These recycled yarns can be used to produce a superior textile product that includes denim, chino fabric for pants, and towels (Jamshaid et al., 2021).

19 Conclusion

The transformation of waste into valuable goods is the foundation for achieving sustainability and reducing environmental degradation. Future focus on zero-waste initiatives hinges on strengthening circular economic practices, fostering innovation, and adopting cutting-edge technologies such as waste-to-energy (WtE) systems, advanced recycling, and bio-conversion techniques. Enhancing waste segregation and decentralized waste management systems will empower local communities while minimizing logistical challenges. The integration of AI and IoT in waste tracking and treatment can further optimize resource recovery and reduce inefficiencies.

Future prospects include scaling up bio-based solutions such as biochar and bionanoparticles for industrial and agricultural applications, along with the development of biofuels and bioplastics from agro-waste. Expanding the upcycling of textile and plastic wastes into high-value products will contribute to a reduction in virgin resource extraction. E-waste management holds the potential for the recovery of critical metals, alloys, and electronic components, driving advancements in green technology.

As the demand for eco-friendly products grows, investments in research, entrepreneurship, and education on waste valorization can drive a shift in global culture toward sustainability. With concerted efforts, waste can evolve from an environmental burden to an economic asset, creating a resilient, resource-efficient, and sustainable world by 2030 and beyond.

References

Abatenh, E., Gizaw, B., Tsegaye, Z., & Wassie, M. (2017). The role of microorganisms in bioremediation—A review. *Open Journal of Environmental Biology, 2*(1), 38–46. https://doi.org/10.17352/ojeb

Abdelbasir, M., McCourt, K. M., Lee, C. M., & Vanegas, D. C. (2020). Waste-derived nanoparticles: Synthesis approaches, environmental applications, and sustainability considerations. *Frontiers in Chemistry, 8*, 1–18. https://doi.org/10.3389/fchem.2020.00782

Aboghrbia, A., & El-Emam, D. A. (2024). E-waste valorization and management for precious metals mining: Towards a circular economy. *Springer Proceedings in Earth and Environmental Sciences, Part F3499*, 199–210. https://doi.org/10.1007/978-3-031-71555-6_17

Acharya, B., Dutta, A., & Minaret, J. (2015). Review on comparative study of dry and wet torrefaction. *Sustainable Energy Technologies and Assessments., 12*, 26–37. https://doi.org/10.1016/j.seta.2015.08.003

Acharya, A., et al. (2021). The impact of COVID-19 outbreak and perceptions of people towards household waste management chain in Nepal. *Geoenvironmental Disasters, 8*(1), 14. https://doi.org/10.1186/s40677-021-00188-w

Ahmad, A., et al. (2022). Rice straw vermicompost enriched with cellulolytic microbes ameliorate the negative effect of drought in wheat through modulating the morpho-physiological attributes. *Frontiers in Environmental Science, 10*, 902999. https://doi.org/10.3389/fenvs.2022.902999

Akinosho, H., Yee, K., Close, D., & Ragauskas, A. (2014). The emergence of Clostridium thermocellum as a high utility candidate for consolidated bioprocessing applications. *Frontiers in Chemistry, 2*, 66. https://doi.org/10.3389/fchem.2014.00066

Alao, A., Popoola, O. M., & Ayodele, T. R. (2022). Waste-to-energy nexus: An overview of technologies and implementation for sustainable development. *Cleaner Energy Systems, 3*, 100034. https://doi.org/10.1016/j.cles.2022.100034

Ali, F., et al. (2021). Sustainable management of municipal solid waste to fuel: An overview for a better tomorrow. In *Advanced technology for the conversion of waste into fuels and chemicals: Volume 2: Chemical processes* (pp. 289–314). Elsevier. https://doi.org/10.1016/B978-0-323-90150-5.00008-X

Apaydin, Ö., & Akçay Han, G. S. (2023). Analysis of municipal solid waste collection methods focusing on zero-waste management using an analytical hierarchy process. *Sustainability (Switzerland), 15*(17), 13184. https://doi.org/10.3390/su151713184

Appels, L., Baeyens, J., Degrève, J., & Dewil, R. (2008). Principles and potential of the anaerobic digestion of waste-activated sludge. *Progress in Energy and Combustion Science, 34*(6), 755–781. https://doi.org/10.1016/j.pecs.2008.06.002

Aryal, T., Kvist, F., Ammam, D. P., & Ottosen, L. D. M. (2018). An overview of microbial biogas enrichment. *Bioresource Technology, 264*, 359–369. https://doi.org/10.1016/j.biortech.2018.06.013

Avhad, M. R., & Marchetti, J. M. (2015). A review on recent advancement in catalytic materials for biodiesel production. *Renewable and Sustainable Energy Reviews, 50*, 696–718. https://doi.org/10.1016/j.rser.2015.05.038

Awogbemi, O., & Von Kallon, D. V. (2022). Valorization of agricultural wastes for biofuel applications. *Heliyon, 8*, e11117. https://doi.org/10.1016/j.heliyon.2022.e11117

Bahadoran, A., et al. (2022). Photocatalytic materials obtained from E-waste recycling: Review, techniques, critique, and update. *Journal of Manufacturing and Materials Processing, 6*(4), 69. https://doi.org/10.3390/jmmp6040069

Bajaj, H., Sharma, A., Arora, D., Yadav, M., Sharma, D., & Bajwa, P. S. (2024). Challenges in E-waste management. In *Sustainable management of electronic waste*. Wiley. https://doi.org/10.1002/9781394166923.ch10

Balseiro-Romero, M., et al. (2019). Modelling the ex situ bioremediation of diesel-contaminated soil in a slurry bioreactor using a hydrocarbon-degrading inoculant. *Journal of Environmental Management, 246*, 840–848. https://doi.org/10.1016/j.jenvman.2019.06.034

Bhatia, K., Kim, S. H., Yoon, J. J., & Yang, Y. H. (2017). Current status and strategies for second generation biofuel production using microbial systems. *Energy Conversion and Management, 148*, 1142–1156. https://doi.org/10.1016/j.enconman.2017.06.073

Bhatia, K., Joo, H. S., & Yang, Y. H. (2018). Biowaste-to-bioenergy using biological methods—A mini-review. *Energy Conversion and Management, 177*, 640–660. https://doi.org/10.1016/j.enconman.2018.09.090

Bhowmik, D., et al. (2023). Multitudinous approaches, challenges and opportunities of bioelectrochemical systems in conversion of waste to energy from wastewater treatment plants. *Cleaner and Circular Bioeconomy, 4*, 100040. https://doi.org/10.1016/j.clcb.2023.100040

Braide, W., Kanu, I., Oranusi, U., & Adeleye, S. (2016). Production of bioethanol from agricultural waste. *Journal of Fundamental and Applied Sciences, 8*(2), 372. https://doi.org/10.4314/jfas.v8i2.14

Brunner, H., & Rechberger, H. (2015). Waste to energy—Key element for sustainable waste management. *Waste Management*, 37, 3–12. https://doi.org/10.1016/j.wasman.2014.02.003

Canales-Gutiérrez, Á., Belizario-Quispe, G., Chui-Betancur, H. N., Huaquisto-Ramos, E., Calatayud-Mendoza, A., & Gutierrez-Flores, I. R. (2023). Abundance and mortality of Eisenia fetida (earthworm) using potato peel as a substrate in vermicomposting process. *International Journal of Agricultural Resources, Governance and Ecology, 19*(2), 123. https://doi.org/10.1504/IJARGE.2023.136645

Chen, H., Peng, J., & Bi, X. T. (2015). A state-of-the-art review of biomass torrefaction, densification and applications. *Renewable and Sustainable Energy Reviews, 44*, 847–866. https://doi.org/10.1016/j.rser.2014.12.039

Chevillard, G., Lucas-Gabrielli, V., & Mousquès, J. (2018). "Medical deserts" in France: Current state of research and future directions. https://doi.org/10.3917/eg.474.0362

Cho, E. J., Lee, Y. G., Song, Y., Kim, H. Y., Nguyen, D. T., & Bae, H. J. (2023). Converting textile waste into value-added chemicals: An integrated bio-refinery process. *Environmental Science and Ecotechnology, 15*, 100238. https://doi.org/10.1016/j.ese.2023.100238

Choi, H. S., Choi, Y. S., & Park, H. C. (2021). Fast pyrolysis characteristics cellulosic biomass with varying reaction conditions. *Renewable Energy, 14*(10), 131–135. https://doi.org/10.1016/j.renene.2011.08.049

Cirik, K., & Gocer, S. (2020). Performance of anaerobic membrane bioreactor treating landfill leachate. *Journal of Environmental Health Science and Engineering, 18*(2), 383. https://doi.org/10.1007/s40201-019-00376-9

Darch, T., et al. (2019). Fertilizer produced from abattoir waste can contribute to phosphorus sustainability, and biofortify crops with minerals. *PLoS One, 14*(9). https://doi.org/10.1371/journal.pone.0221647

Das, A. P., Ghosh, S., Mohanty, S., & Sukla, L. B. (2015). Advances in manganese pollution and its bioremediation. In L. B. Sukla, N. Pradhan, S. Panda, & B. K. Mishra (Eds.), *Environmental microbe biotechnology* (1st ed., pp. 313–328). Springer. https://doi.org/10.1007/978-3-319-19018-1_16

Das, K., Islam, M. N., Billah, M. M., & Sarker, A. (2021). COVID-19 and municipal solid waste (MSW) management: A review. *Environmental Science and Pollution Research International, 28*(23), 28993–29008. https://doi.org/10.1007/s11356-021-13914-6

Das, A. P., Mohanty, A. K., van Hullebusch, E. D., Figueiredo, G., Das, A. P., Mohanty, A. K., van Hullebusch, E. D., & Figueiredo, G. (2023). Bioremediation as an emerging technology for the removal of synthetic microplastic pollutants from marine ecosystem. *Marine Pollution Bulletin, 194*, 115297. https://doi.org/10.1016/J.MARPOLBUL.2023.115297

Das, A., Dey, S., Bala, K., Gnanasekaran, R., & Das, A. P. (2024a). Microbial bioremediation technology for sustainable treatment and management of synthetic microfiber waste. In *Sustainable microbial technology for synthetic and cellulosic microfiber bioremediation* (pp. 221–238). Springer. https://doi.org/10.1007/978-3-031-62600-5_12

Das, A., Ray, S. K., Mohanty, M., Mohanty, J., Dey, S., & Das, A. P. (2024b). Ecotoxicity of microplastic wastes and their sustainable management: A review. *Environmental Chemistry and Ecotoxicology, 6*, 144–152. https://doi.org/10.1016/J.ENCECO.2024.05.003

Demirbas, A. (2010). Waste management, waste resource facilities and waste conversion processes. *Energy Conversion and Management, 52*(2), 1280–1287. https://doi.org/10.1016/j.enconman.2010.09.025

Dey, S., Das, A., Mallick, K., Sahu, A., & Das, A. P. (2023). *Environmental petroleum waste: Pollution, toxicity, sustainable remediation* (pp. 159–175). Springer. https://doi.org/10.1007/978-3-031-48220-5_7

Dey, S., Dash, A., & Das, A. P. (2024). Bioremediation as a sustainable and circular solution for mining waste management and bioenergy generation. In *Sustainable management of mining waste and tailings: A circular economy approach* (pp. 107–119). Routledge. https://doi.org/10.1201/9781003442455-6

Dhyani, V., & Bhaskar, T. (2018). A comprehensive review on the pyrolysis of lignocellulosic biomass. *Renewable Energy, 129*, 695–716. https://doi.org/10.1016/j.renene.2017.04.035

Do Nascimento, D. R. V., Wohnrath, K., & Garcia, J. R. (2021). Synthesis of gold nanoparticles using recovered gold from electronic waste. *Orbit, 13*(2), 153–159. https://doi.org/10.17807/orbital.v13i2.1455

Edirisinghe, G. L. M., Wijayasundara, M., & De Alwis, A. A. P. (2022). Waste generation and characteristics in Sri Lankan textile and apparel sector: Case study of the Biyagama Industrial Export Processing Zone, Sri Lanka. *Nature Environment and Pollution Technology, 21*(2), 697. https://doi.org/10.46488/NEPT.2022.V21I02.030

Elema, R. (2021). Preparation and characterization of vermicompost made from different sources of materials. *Open Journal of Plant Science*. https://doi.org/10.17352/ojps.000031

Farooqi, U. R., Kareem, A., Rafi, F., & Ali, S. (2021). Solid waste, treatment technologies, and environmental sustainability: Solid wastes and their sustainable management practices. In *Handbook of research on waste diversion and minimization technologies for the industrial sector*. IGI Global. https://doi.org/10.4018/978-1-7998-4921-6.ch003

Fernandez, A., Saffe, A., Pereyra, R., Mazza, G., & Rodriguez, R. (2016). Kinetic study of regional agro-industrial wastes pyrolysis using non-isothermal TGA analysis. *Applied Thermal Engineering, 106*, 1157–1164. https://doi.org/10.1016/j.applthermaleng.2016.06.084

Gadaleta, G., De Gisi, S., Todaro, F., & Notarnicola, M. (2022). Environmental comparison of different mechanical–biological treatment plants by combining life cycle assessment and material flow analysis. *Clean Technologies, 4*(2), 380–394. https://doi.org/10.3390/cleantechnol4020023

Hossain, M., et al. (2020). Production of bioethanol from agricultural wastes utilizing fermentation potential of Saccharomyces cerevisiae. *Journal of Bangladesh Agricultural University, 18*, 1. https://doi.org/10.5455/jbau.8647

Hosseinzadeh-Bandbafha, H., et al. (2022). Managing the hazardous waste cooking oil by conversion into bioenergy through the application of waste-derived green catalysts: A review. *Journal of Hazardous Materials, 424*(Pt C), 127636. https://doi.org/10.1016/j.jhazmat.2021.127636

Hu, Y., Feng, S., Bassi, A., & (Charles) Xu, C. (2018). Improvement in bio-crude yield and quality through co-liquefaction of algal biomass and sawdust in ethanol-water mixed solvent and recycling of the aqueous by-product as a reaction medium. *Energy Conversion and Management, 171*, 618–625. https://doi.org/10.1016/j.enconman.2018.06.023

Huang, X., Yun, S., Zhu, J., Du, T., Zhang, C., & Li, X. (2016). Mesophilic anaerobic codigestion of aloe peel waste with dairy manure in the batch digester: Focusing on mixing ratios and digestate stability. *Bioresource Technology, 218*, 62–68. https://doi.org/10.1016/j.biortech.2016.06.070

Jamshaid, H., Hussain, U., Mishra, R., Tichy, M., & Muller, M. (2021). Turning textile waste into valuable yarn. *Cleaner Engineering and Technology, 5*, 100341. https://doi.org/10.1016/j.clet.2021.100341

Jayasinghe, J. A. W. W., Kiriella, K. G. T. N., Jayarathne, L., Priyadarshani, C., Shiromani, M. K. S., & Rajapakshe, R. M. A. O. A. (2024). Study of BOD: COD ratio as an indicator for wastewater of rubber industry sector. In *Proceedings of International Forestry and Environment Symposium* (Vol. 27). https://doi.org/10.31357/fesympo.v27.6573

Kanungo, J., Sahoo, T., Bal, M., & Behera, I. D. (2024). Performance of bioremediation strategy in waste lubricating oil pollutants: A review. *Geomicrobiology Journal, 41*(4), 360–373. https://doi.org/10.1080/01490451.2023.2245395

Kaur, A., Bharti, R., & Sharma, R. (2021). Municipal solid waste as a source of energy. *Materials Today: Proceedings, 81*, 904–915. https://doi.org/10.1016/j.matpr.2021.04.286

Khalid, A., Arshad, M., Anjum, M., Mahmood, T., & Dawson, L. (2011). The anaerobic digestion of solid organic waste. *Waste Management, 31*(8), 1737–1744. https://doi.org/10.1016/j.wasman.2011.03.021

Kim, H., et al. (2021). Valorization of waste-cooking oil into sophorolipids and application of their methyl hydroxyl branched fatty acid derivatives to produce engineering bioplastics. *Waste Management, 124*, 195–202. https://doi.org/10.1016/j.wasman.2021.02.003

Kleppinger, E. W. (2022). Remedial options in hazardous waste management—An overview. In *Hazardous waste management* (Vol. 1). Elsevier.

Koçak, E., & Ikizoğlu, B. (2020). Types of waste in the context of waste management and general overview of waste disposal in Turkey. *International Journal of Agriculture Environment and Food Sciences, 4*(4), 520–527. https://doi.org/10.31015/jaefs.2020.4.17

Kumar, A., & Samadder, S. R. (2017). A review on technological options of waste to energy for effective management of municipal solid waste. *Waste Management, 69*, 407–422. https://doi.org/10.1016/j.wasman.2017.08.046

Kumar, R., & Strezov, V. (2021). Thermochemical production of bio-oil: A review of downstream processing technologies for bio-oil upgrading, production of hydrogen and high value-added products. *Renewable and Sustainable Energy Reviews, 135*, 110152. https://doi.org/10.1016/j.rser.2020.110152

Kumar, K., Ding, L., Zhao, H., & Cheng, M. H. (2023a). Waste-to-energy pipeline through consolidated fermentation–microbial fuel cell (MFC) system. *Processes, 11*(8), 2451. https://doi.org/10.3390/pr11082451

Kumar, A., et al. (2023b). A critical review on sustainable hazardous waste management strategies: A step towards a circular economy. *Environmental Science and Pollution Research, 30*, 105030. https://doi.org/10.1007/s11356-023-29511-8

Kumar, M., Pandit, S., Patel, V., Khanna, N., Nag, M., Lahiri, D., Ray, R. R., Das, A., & Das, D. (2023c). Maximization of energy recovery from starch processing wastewater by thermophilic dark fermentation coupled with microbial fuel cell technology. *Geomicrobiology Journal, 41*, 392. https://doi.org/10.1080/01490451.2023.2209555

Labra Cataldo, M., Oyinlola, S., Sigdel, D. N., & Gallego-Schmid, A. (2024). Waste management in Nepal: Characterization and challenges to promote a circular economy. *Circular Economy and Sustainability, 4*, 439. https://doi.org/10.1007/s43615-023-00283-0

Le Man, H., Behera, S. K., & Park, H. S. (2010). Optimization of operational parameters for ethanol production from Korean food waste leachate. *International Journal of Environmental Science and Technology, 7*(1), 157–164. https://doi.org/10.1007/BF03326127

Lebreton, L., & Andrady, A. (2019). Future scenarios of global plastic waste generation and disposal. *Palgrave Communications, 5*(1), 6. https://doi.org/10.1057/s41599-018-0212-7

Lee, T. H., Byun, I. G., Kim, Y. O., Hwang, I. S., & Park, T. J. (2006). Monitoring biodegradation of diesel fuel in bioventing processes using in situ respiration rate. *Water Science and Technology, 53*(4–5), 263–272. https://doi.org/10.2166/wst.2006.131

Lee, H., Terbish, N., Holder, S. L., Popuri, S. R., & Nalluri, L. P. (2019). A study on development of alternative biopolymers based proton exchange membrane for microbial fuel cells and effect of blending ratio and ionic crosslinking on bioenergy generation and COD removal. *Journal of Polymer Research, 26*(12), 285. https://doi.org/10.1007/s10965-019-1957-4

LeFevre, H., Müller, C. E., Li, R. J., Luthy, R. G., & Sattely, E. S. (2015). Rapid phytotransformation of benzotriazole generates synthetic tryptophan and auxin analogs in Arabidopsis. *Environmental Science & Technology, 49*(18), 10959–10968. https://doi.org/10.1021/acs.est.5b02749

Li, J., et al. (2021). Effect of manure and compost on the phytostabilization potential of heavy metals by the halophytic plant wavy-leaved saltbush. *Plants, 10*(10), 2176. https://doi.org/10.3390/plants10102176

Libra, A., et al. (2011). Hydrothermal carbonization of biomass residuals: A comparative review of the chemistry, processes and applications of wet and dry pyrolysis. *Biofuels, 2*(1), 71. https://doi.org/10.4155/bfs.10.81

Lohri, R., Diener, S., Zabaleta, I., Mertenat, A., & Zurbrügg, C. (2017). Treatment technologies for urban solid biowaste to create value products: A review with focus on low- and middle-income settings. *Reviews in Environmental Science and Bio/Technology, 16*, 81. https://doi.org/10.1007/s11157-017-9422-5

Lopes, M., Miranda, S. M., & Belo, I. (2020). Microbial valorization of waste cooking oils for valuable compounds production—A review. *Critical Reviews in Environmental Science and Technology, 50*(24), 2583–2616. https://doi.org/10.1080/10643389.2019.1704602

Lu, J. W., Zhang, S., Hai, J., & Lei, M. (2017). Status and perspectives of municipal solid waste incineration in China: A comparison with developed regions. *Waste Management, 69*, 170–186. https://doi.org/10.1016/j.wasman.2017.04.014

Maalouf, A., & Mavropoulos, A. (2023). Re-assessing global municipal solid waste generation. *Waste Management and Research, 41*(4), 936. https://doi.org/10.1177/0734242X221074116

Maureira, D., Romero, O., Illanes, A., Wilson, L., & Ottone, C. (2023). Industrial bioelectrochemistry for waste valorization: State of the art and challenges. *Biotechnology Advances, 64*, 108123. https://doi.org/10.1016/j.biotechadv.2023.108123

Mazzei, H. G., & Specchia, S. (2023). Latest insights on technologies for the treatment of solid medical waste: A review. *Journal of Environmental Chemical Engineering, 11*(2), 109309. https://doi.org/10.1016/j.jece.2023.109309

Mishra, S., & Das, A. P. (2021). Current treatment technologies for removal of microplastic and microfiber pollutants from wastewater. In *Wastewater treatment: Cutting-edge molecular tools, techniques and applied aspects* (pp. 237–251). Elsevier. https://doi.org/10.1016/B978-0-12-821881-5.00011-8

Mishra, S., & Das, A. P. (2022). Treatment of the wastewater polluted with synthetic microfiber released from washing machine. In B. B. Das, H. Hettiarachchi, P. K. Sahu, & S. Nanda (Eds.), *Lecture notes in civil engineering* (Vol. 207, pp. 109–117). Springer Science and Business Media Deutschland GmbH. https://doi.org/10.1007/978-981-16-7509-6_9/COVER

Mishra, S., & Das, A. P. (2023). Microbial remediation of synthetic microfiber contaminated wastewater. In M. P. Shah (Ed.), *Microbial technologies in industrial wastewater treatment* (pp. 337–350). Springer. https://doi.org/10.1007/978-981-99-2435-6_17

Mishra, S., Rout, P. K., & Das, A. P. (2019). Solar photovoltaic panels as next generation waste: A review. *Biointerface Research in Applied Chemistry, 9*(6), 4539.

Mishra, S., Dash, D., & Das, A. P. (2022a). Detection, characterization and possible biofragmentation of synthetic microfibers released from domestic laundering wastewater as an emerging source of marine pollution. *Marine Pollution Bulletin, 185*, 114254. https://doi.org/10.1016/J.MARPOLBUL.2022.114254

Mishra, S., Singh, R. P., Rout, P. K., & Das, A. P. (2022b). Membrane bioreactor (MBR) as an advanced wastewater treatment technology for removal of synthetic microplastics. In *Development in wastewater treatment research and processes* (pp. 45–60). Elsevier. https://doi.org/10.1016/B978-0-323-85583-9.00022-3

Mishra, S., Ghosh, S., van Hullebusch, E. D., Singh, S., & Das, A. P. (2023). A critical review on the recovery of base and critical elements from electronic waste-contaminated streams using microbial biotechnology. *Applied Biochemistry and Biotechnology, 2023*, 1–30. https://doi.org/10.1007/S12010-023-04440-X

Molino, A., Chianese, S., & Musmarra, D. (2016). Biomass gasification technology: The state of the art overview. *Journal of Energy Chemistry, 25*(1), 10–25. https://doi.org/10.1016/j.jechem.2015.11.005

Mondal, M. S., Paul, A., & Rhaman, M. (2023). Recycling of silver nanoparticles from electronic waste via green synthesis and application of AgNPs-chitosan based nanocomposite on textile material. *Scientific Reports, 13*(1), 1–15. https://doi.org/10.1038/s41598-023-40668-7

Morales-Martinez, T. K., et al. (2017). Assessment of different saccharification and fermentation configurations for ethanol production from Agave lechuguilla. *Bioresources, 12*(4), 8093–8105. https://doi.org/10.15376/BIORES.12.4.8093-8105

Nag, M., Lahiri, D., Ghosh, S., Sarkar, T., Pati, S., Das, A. P., Ram, D. K., Bhattacharya, D., & Ray, R. R. (2024). Application of microorganisms in biotransformation and bioremediation

of environmental contaminant: A review. *Geomicrobiology Journal, 41*, 374. https://doi.org/10.1080/01490451.2023.2261443
Napoli, M., Cecchi, S., Grassi, C., Baldi, A., Zanchi, C. A., & Orlandini, S. (2019). Phytoextraction of copper from a contaminated soil using arable and vegetable crops. *Chemosphere, 219*, 122–129. https://doi.org/10.1016/j.chemosphere.2018.12.017
Naveen, P. (2021). Scenarios of waste management nexus in Bangalore. *Energy Nexus, 1*, 100004. https://doi.org/10.1016/j.nexus.2021.100004
Naveen, B. P., & Sivapullaiah, P. V. (2020). Solid waste management: Current scenario and challenges in Bengaluru. In *Sustainable sewage sludge management*. IntechOpen. https://doi.org/10.5772/intechopen.90837
Nicodème, T., Berchem, T., Jacquet, N., & Richel, A. (2018). Thermochemical conversion of sugar industry by-products to biofuels. *Renewable and Sustainable Energy Reviews, 88*, 151–159. https://doi.org/10.1016/j.rser.2018.02.037
Niu, G. L., Zhang, J. J., Zhao, S., Liu, H., Boon, N., & Zhou, N. Y. (2009). Bioaugmentation of a 4-chloronitrobenzene contaminated soil with Pseudomonas putida ZWL73. *Environmental Pollution, 157*(3), 763–771. https://doi.org/10.1016/j.envpol.2008.11.024
Nižetić, S., Djilali, N., Papadopoulos, A., & Rodrigues, J. J. P. C. (2019). Smart technologies for promotion of energy efficiency, utilization of sustainable resources and waste management. *Journal of Cleaner Production, 231*, 565. https://doi.org/10.1016/j.jclepro.2019.04.397
Omokhagbor Adams, G., Tawari Fufeyin, P., Eruke Okoro, S., & Ehinomen, I. (2020). Bioremediation, biostimulation and bioaugmention: A review. *International Journal of Environmental Bioremediation & Biodegradation, 3*(1), 28–39. https://doi.org/10.12691/ijebb-3-1-5
Ong, B., Ng, L. Y., & Mohammad, A. W. (2018). A review of ZnO nanoparticles as solar photocatalysts: Synthesis, mechanisms and applications. *Renewable and Sustainable Energy Reviews, 81*, 536–551. https://doi.org/10.1016/j.rser.2017.08.020
Ong, H. C., Chen, W. H., Farooq, A., Gan, Y. Y., Lee, K. T., & Ashokkumar, V. (2019). Catalytic thermochemical conversion of biomass for biofuel production: A comprehensive review. *Renewable and Sustainable Energy Reviews, 113*, 109266. https://doi.org/10.1016/j.rser.2019.109266
Osman, A. I., et al. (2022). Biochar for agronomy, animal farming, anaerobic digestion, composting, water treatment, soil remediation, construction, energy storage, and carbon sequestration: A review. *Environmental Chemistry Letters, 20*, 2385–2485. https://doi.org/10.1007/s10311-022-01424-x
Panainte-Lehadus, M., et al. (2022). Study on the method of household waste collection: Case study. *Applied Sciences (Switzerland), 12*(15), 7490. https://doi.org/10.3390/app12157490
Pandit, Y., Nakagawa, R. R., Timilsina, K. K., & Saijo, T. (2021). Taking the perspectives of future generations as an effective method for achieving sustainable waste management. *Sustainable Production and Consumption, 27*, 1526. https://doi.org/10.1016/j.spc.2021.03.019
Parthiba Karthikeyan, O., Trably, E., Mehariya, S., Bernet, N., Wong, J. W. C., & Carrere, H. (2018). Pretreatment of food waste for methane and hydrogen recovery: A review. *Bioresource Technology, 249*, 1025–1039. https://doi.org/10.1016/j.biortech.2017.09.105
Paul, O., Jasu, A., Lahiri, D., Nag, M., & Ray, R. R. (2021). In situ and ex situ bioremediation of heavy metals: The present scenario. *Journal of Environmental Engineering and Landscape Management, 29*(4), 454–469. https://doi.org/10.3846/jeelm.2021.15447
Paul, A., Dey, S., Ram, D. K., & Das, A. P. (2023). Hexavalent chromium pollution and its sustainable management through bioremediation. *Geomicrobiology Journal, 41*, 324–334. https://doi.org/10.1080/01490451.2023.2218377
Perera, S. M. H. D., Wickramasinghe, C., Samarasiri, B. K. T., & Narayana, M. (2021). Modeling of thermochemical conversion of waste biomass—A comprehensive review. *Biofuel Research Journal, 8*(4), 1481–1528. https://doi.org/10.18331/BRJ2021.8.4.3

Perrot, F., & Subiantoro, A. (2018). Municipal waste management strategy review and waste-to-energy potentials in New Zealand. *Sustainability (Switzerland), 10*(9), 3114. https://doi.org/10.3390/su10093114

Pheakdey, D. V., Van Quan, N., Khanh, T. D., & Xuan, T. D. (2022). Challenges and priorities of municipal solid waste management in Cambodia. *International Journal of Environmental Research and Public Health, 19*(14), 8458. https://doi.org/10.3390/ijerph19148458

Pires, A., & Martinho, G. (2019). Waste hierarchy index for circular economy in waste management. *Waste Management, 95*, 298. https://doi.org/10.1016/j.wasman.2019.06.014

Pradhan, G., Tripathy, B., Ram, D. K., Digal, A. K., & Das, A. P. (2024). Bauxite mining waste pollution and its sustainable management through bioremediation. *Geomicrobiology Journal, 41*, 335–344. https://doi.org/10.1080/01490451.2023.2235353

Rafiquee, A., & Shabbiruddin. (2024). Optimal selection and challenges of municipal waste management system using an integrated approach: A case study. *Energy Sources, Part A: Recovery, Utilization and Environmental Effects, 46*(1), 1996. https://doi.org/10.1080/15567036.2023.2298285

Rahimi, Z., Anand, A., & Gautam, S. (2022). An overview on thermochemical conversion and potential evaluation of biofuels derived from agricultural wastes. *Energy Nexus, 7*, 100125. https://doi.org/10.1016/j.nexus.2022.100125

Rai, P. K., & Choure, K. (2023). Agriculture waste to bioplastics: A perfect substitution of plastics. In *Value-addition in agri-food industry waste through enzyme technology*. Elsevier. https://doi.org/10.1016/B978-0-323-89928-4.00011-0

Ramamurthy, C., et al. (2021). Microbial biotechnological approaches: Renewable bioprocessing for the future energy systems. *Microbial Cell Factories*. https://doi.org/10.1186/s12934-021-01547-w

Richard, E. N., Hilonga, A., Machunda, R. L., & Njau, K. N. (2021). Life cycle analysis of potential municipal solid wastes management scenarios in Tanzania: The case of Arusha City. *Sustainable Environment Research, 31*, 1. https://doi.org/10.1186/s42834-020-00075-3

Rossi, M., Papetti, A., & Germani, M. (2022). A comparison of different waste collection methods: Environmental impacts and occupational risks. *Journal of Cleaner Production, 368*, 133145. https://doi.org/10.1016/j.jclepro.2022.133145

Sahota, S., et al. (2018). Review of trends in biogas upgradation technologies and future perspectives. *Bioresource Technology Reports, 1*, 79. https://doi.org/10.1016/j.biteb.2018.01.002

Seay, J., & Ternes, M. E. (2022). A review of current challenges and legal advances in the global management of plastic waste. *Clean Technologies and Environmental Policy, 24*, 731–738. https://doi.org/10.1007/s10098-022-02289-y

Seelam, J. S., Maesara, S. A., Mohanakrishna, G., Patil, S. A., Ter Heijne, A., & Pant, D. (2018). Resource recovery from wastes and wastewaters using bioelectrochemical systems. In *Waste biorefinery: Potential and perspectives* (pp. 535–570). Elsevier. https://doi.org/10.1016/B978-0-444-63992-9.00018-5

Singh, A., Nanda, S., Guayaquil-Sosa, J. F., & Berruti, F. (2021). Pyrolysis of Miscanthus and characterization of value-added bio-oil and biochar products. *Canadian Journal of Chemical Engineering, 99*(S1), S55–S68. https://doi.org/10.1002/cjce.23978

Singh, N., Ogunseitan, O. A., & Tang, Y. (2022). Medical waste: Current challenges and future opportunities for sustainable management. *Critical Reviews in Environmental Science and Technology, 52*(11), 2000. https://doi.org/10.1080/10643389.2021.1885325

Subramanian, K., et al. (2022). An overview of cotton and polyester, and their blended waste textile valorisation to value-added products: A circular economy approach–research trends, opportunities and challenges. *Critical Reviews in Environmental Science and Technology, 52*(21), 3921–3942. https://doi.org/10.1080/10643389.2021.1966254

Sukiran, A., Abnisa, F., Wan Daud, W. M. A., Abu Bakar, N., & Loh, S. K. (2017). A review of torrefaction of oil palm solid wastes for biofuel production. *Energy Conversion and Management, 149*, 101–120. https://doi.org/10.1016/j.enconman.2017.07.011

Surendra, C., Takara, D., Hashimoto, A. G., & Khanal, S. K. (2014). Biogas as a sustainable energy source for developing countries: Opportunities and challenges. *Renewable and Sustainable Energy Reviews, 31*, 846–859. https://doi.org/10.1016/j.rser.2013.12.015

Teshale, E. (2019). Review on utilization of plants for the phytoremidiation purpose on heavy metal contaminated soil. *Journal of Natural Sciences Research.* https://doi.org/10.7176/JNSR

Toor, S. S., Rosendahl, L., & Rudolf, A. (2011). Hydrothermal liquefaction of biomass: A review of subcritical water technologies. *Energy, 36*, 2328–2342. https://doi.org/10.1016/j.energy.2011.03.013

Veljković, B., Banković-Ilić, I. B., & Stamenković, O. S. (2015). Purification of crude biodiesel obtained by heterogeneously-catalyzed transesterification. *Renewable and Sustainable Energy Reviews, 49*, 500–516. https://doi.org/10.1016/j.rser.2015.04.097

Wang, J. (2014). Decentralized biogas technology of anaerobic digestion and farm ecosystem: Opportunities and challenges. *Frontiers in Energy Research, 2.* https://doi.org/10.3389/fenrg.2014.00010

Wang, L., York, S. W., Ingram, L. O., & Shanmugam, K. T. (2019). Simultaneous fermentation of biomass-derived sugars to ethanol by a co-culture of an engineered Escherichia coli and Saccharomyces cerevisiae. *Bioresource Technology, 273*, 269–276. https://doi.org/10.1016/j.biortech.2018.11.016

Wang, D., Jiang, P., Zhang, H., & Yuan, W. (2020). Biochar production and applications in agro and forestry systems: A review. *Science of the Total Environment, 723*, 137775. https://doi.org/10.1016/j.scitotenv.2020.137775

Wang, J., Qin, C., Qu, X., Ran, C. L., & Chen, B. (2021a). A smart municipal waste management system based on deep-learning and internet of things. *Waste Management, 135*, 20. https://doi.org/10.1016/j.wasman.2021.08.028

Wang, Y., Levis, J. W., & Barlaz, M. A. (2021b). Life-cycle assessment of a regulatory compliant U.S. Municipal Solid Waste Landfill. *Environmental Science & Technology, 55*(20), 13583–13592. https://doi.org/10.1021/acs.est.1c02526

Widyatmika, M. A., & Bolia, N. B. (2023). Understanding citizens' perception of waste composting and segregation. *The Journal of Material Cycles and Waste Management, 25*(3), 1608. https://doi.org/10.1007/s10163-023-01636-5

Wikurendra, A., Achmad, S., Herdiani, N., & Nurika, G. (2023). Forecast of waste generated and waste fleet using linear regression model. *Polish Journal of Environmental Studies, 32*(2), 1867. https://doi.org/10.15244/pjoes/158779

Yadav, V., & Karmakar, S. (2020). Sustainable collection and transportation of municipal solid waste in urban centers. *Sustainable Cities and Society, 53*, 101937. https://doi.org/10.1016/j.scs.2019.101937

Yang, J., (Sophia) He, Q, & Yang, L. (2019). A review on hydrothermal co-liquefaction of biomass. *Applied Energy, 250*, 926–945. https://doi.org/10.1016/j.apenergy.2019.05.033

Yong, J., Bashir, M. J. K., Ng, C. A., Sethupathi, S., Lim, J. W., & Show, P. L. (2019). Sustainable waste-to-energy development in Malaysia: Appraisal of environmental, financial, and public issues related with energy recovery from municipal solid waste. *Processes, 7*(10), 676. https://doi.org/10.3390/pr7100676

Plastic Waste, Microplastic Pollutions, Impacts, and Management

Saji Baby, Jenan Bahzad, Himangi Neve, and Ganesh Deshmukh

Abstract Plastic is a nonbiodegradable synthetic substance. It poses a significant threat owing to its wide usage and harmful consequences upon degradation. As it breaks down, harmful substances are released, leading to pollution and adversely affecting the environment, humans, and wildlife. The non-decomposing nature of plastic, which requires high-energy ultraviolet light for breakdown, causes a concerning increase in plastic waste (PW) in land and oceans. Yearly, 380 million tons of plastic are produced globally, contributing to environmental pollutants. This problem is particularly severe in developing countries. To address the associated pollutants, it is essential to understand plastic, PW, its fate, and biodegradation. In this chapter, numerous synthetic plastics, their properties, significance, and waste quantification are investigated, and topics, such as bioplastics, micro- and nanoplastics, pollution crises, persistent organic pollutants (POPs), dioxins, toxicity, potential substitutes, and waste management (WM) efficacy are explored. Recycling technologies and case studies are also discussed in this chapter, rendering insights for better options, management principles, regulations, and policies that aim to address the environmental impact of PW.

S. Baby (✉)
Centre of Excellence in Sustainability, Environment and Climate Change, Dr. Vishwanath Karad MIT World Peace University, Pune, India
e-mail: saji.baby@mitwpu.edu.in

J. Bahzad
Department of Sciences, College of Basic Education, Public Authority for Applied Education and Training, Kuwait, Kuwait

H. Neve
School of Chemical Engineering, Dr. Vishwanath Karad MIT World Peace University, Pune, India

G. Deshmukh
School of Economics and Public Policy, Dr. Vishwanath Karad MIT World Peace University, Pune, India

S. Rajpoot, A. P. Das (eds.), *Innovative Technologies for Waste Management*, Sustainable Environmental Waste Management Strategies,
https://doi.org/10.1007/978-3-031-97184-6_15

Keywords Forever chemicals · Greenhouse gases · Endocrine-disrupting chemical · Kuznets curve · Circular economy · 5Rs

1 Introduction

Plastic is a major component of municipal solid waste (MSW). Owing to factors like population growth, urbanization, and industrial development, plastic poses significant environmental challenges (Kibria et al., 2023). Although plastic is nonbiodegradable and derived from petroleum, its unique properties make it indispensable with over 50 years of ubiquitous use in various sectors (Eagle et al., 2016; Wang et al., 2016; Landon-Lane, 2018; Mazhandu et al., 2020). Plastic production has consistently increased globally, averaging 8.7% annually from 1950 to 2012 and reaching almost 300 million tons today. In 2013, plastic production saw a 3.9% expansion over 2012, reaching 299 million tons (Plastics—The facts 2014, 2014). After that, annual plastic production in 2014 exceeded 300 million metric tons (Kara, 2017).

Plastics, which are composed of basic polymers, solvents, and various additives, such as plasticizers and pigments, are shaped via heat and pressure for several applications (UNEP, 2023). There are two types of polymers: (1) synthetic, which is derived from petroleum by scientists and engineers, and (2) natural, which occurs in nature and is extractable. The first synthetic plastic was bakelite, which was created in 1907 by Leo Baekeland, coining the term "plastics" from the Greek word *plastikos*, meaning moldable. Alexander Parkes patented the first synthetic material, Parkesine in 1862, which serves as a cost-effective substitute for ivory or tortoiseshell (Science Museum, 2019). The Plastics Age is aptly labeled in the twentieth and twenty-first centuries, acknowledging various usages of plastics. Worldwide, over 400 million metric tons of numerous plastics were manufactured in 2015 alone. Figure 1 indicates the 2015 comparison of primary waste production as well as generation by the type of plastic utilized.

Plastic materials that constitute 50% of production are typically nondegradable and derived from petrochemical sources. This reliance on crude oil production results in an increasing trend in PW, comprising hydrocarbons (HCs) with significant chemical energy, and potentially recoverable to meet energy demands (Antelava et al., 2021).

The global urban population that is heavily reliant on plastics reached 4.2 billion in 2019 (Kedzierski et al., 2020). For PW management, several strategies comprising landfilling, recycling, microbial degradation, incineration, and energy recovery (ER) have been implemented (Ilyas et al., 2018). This review especially concentrates on thermal cracking and conversion (TCC) processes as more environmentally sustainable alternatives to incineration. The longer polymer chains are broken down into short ones by the TCC processes, producing products similar to those found in nature. In this chapter, the energy potential of products acquired from pyrolysis along with gasification is highlighted, thus discussing real-life (RL)

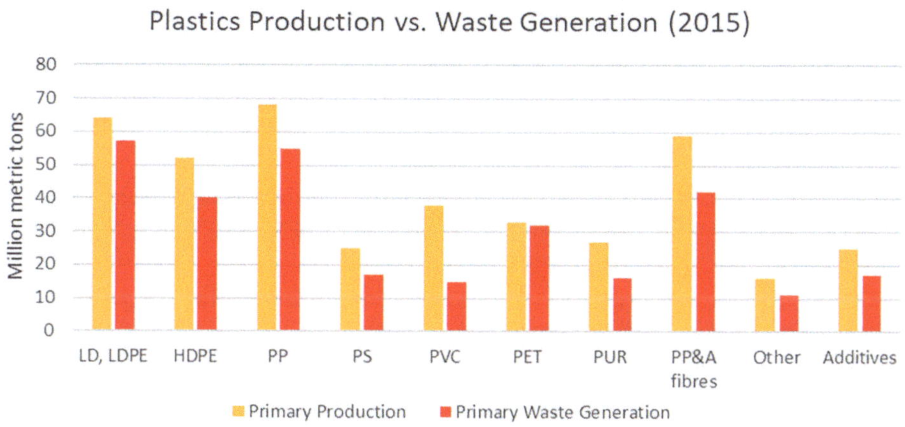

Fig. 1 Plastic production vs. waste generation by type of plastic (Antelava et al., 2021)

applications in industry. TCC methods, their influencing factors, and operational standards with a particular concentration on the energy potential of products from pyrolysis and gasification are explored in the chapter, thereby offering a condensed review for environmental authorities and stakeholders.

A steady increase in global energy-related CO_2 emissions is highlighted by the International Energy Outlook report (US Energy Information Administration, n.d.), emphasizing the significance of CO_2 utilization pathways. For CO_2 use and conversion of PW into clean fuel, thermal treatment of PW is proposed as an efficient method (Singh et al., 2019; Alli et al., 2018). TCC emerges as a promising longer-term solution for PW accumulation (Fox & Stacey, 2019). In the article, current work and product yields from PW pyrolysis along with gasification are investigated, offering opportunities for integrating such energy into industrial applications.

2 Structure, Properties, and Characteristics

The plastic's manufacturing starts with distillation in an oil refinery or from natural gas processing, encompassing the separation of heavy crude oil into lighter fractions. Naphtha is an intermediate liquid hydrocarbon (C5–C10). Also, it is essential in fractional distillation and an important step in plastic production. Over 13,000 chemicals are linked to plastics with more than 3200 having potentially hazardous properties, comprising carcinogenicity and environmental persistence (UNEP, 2023).

Ten groups of chemicals related to plastics are recognized as major concerns in this report owing to known toxicity and potential release. These encompass flame retardants, UV light stabilizers, PFASs, phthalates, bisphenols, alkylphenols, biocides, metals, PAHs, and non-intentionally added substances. The plastics' physical properties depend on the polymer chain's arrangement. In Table 1, factors that determine the plastic's properties are exhibited.

Table 1 Factors that might determine the properties of plastic

Factors	Effect
Reaction conditions	The polymer chain's length as well as branching is affected by the pressure, temperature, and catalysts
Monomer	The type of forces betwixt polymer chains is affected by the type of monomer utilized
Additives	Additives could "lubricate" polymer chains, join them along with cross-links, preserve them from decay, or contribute to physical properties to a great extent

Centered on plastics' properties, they are extensively categorized into two types: (1) thermoplastics and (2) thermosets.

Thermoplastics, which are composed of linear chains of molecules held by weak van der Waals forces, are also known as thermo-softening plastics. Acrylic, polyamide, polyester, polyethene, polystyrene (PS), polypropylene (PP), acrylonitrile butadiene styrene (ABS), polycarbonate (PC), nylon, Teflon, natural rubber, polyvinyl chloride (PVC), et cetera are some of the examples. A high-strength polymer utilized in pipe and medical machines as well is called PVC, which is also an extensively utilized polymer globally. uPVC ("u" stands for unplasticized) is strong, rigid, light, and durable and utilized for window frames.

Thermosetting plastics or thermosets are fabricated by irreversible polymerization. Due to strong intermolecular forces, the polymer chains can't slide past each other (cross-links). Instances comprise urea-formaldehyde, vulcanized rubber, bakelite, epoxy resin, duroplast, silicone, polyimides, cyanate esters, polyurethane (PUR), and phenolic resins. These materials, once molded, cannot be stretched or melted.

Elastomers and plastomers are a third type of plastic. Elastomers are found in two types, such as thermoplastic and thermoset elastomers. Most elastomers like natural rubber show unique properties of regaining their original shape subsequent to stretching. Plasticity and elasticity are merged by plastomers like ethylene-alpha olefin copolymers, offering improved properties. Elastomers are selected for flexibility, whereas plastomers are preferred when both flexibility and toughness are required. In Table 2, the molecular structure of different plastics is exhibited, and in Table 3, the properties of thermoset as well as thermoplastic are displayed.

The thermoplastic polymers' ideal properties, namely, lower density, corrosion resistance, user-friendly design, and higher strength, are accountable for the plastic's great demand. It is possible to further divide thermoplastic into two subcategories, which is shown in Table 4.

Thermoplastics are the most extensively utilized plastics. Among them, PE 29.6%, PP 18.9%, PUR 7.4%, PVC 10.4%, PS 7.1%, and PET 6.9% are the most common and exhibited in Table 5. The purpose of the standard resin identification code in Fig. 2 is to guide consumers and recyclers toward proper disposal methods.

1. PET: Lightweight and durable, used in soda, juice, bottled water, food packaging, and cooking oil.
2. HDPE: Strong and rigid, common in liquid soap and shampoo bottles and shopping bags.

Table 2 Molecular structure of different plastics

Types of plastic	Thermosets	Elastomer/plastomer	Thermoplastics
Molecular structure	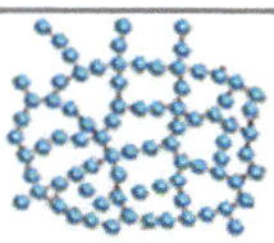	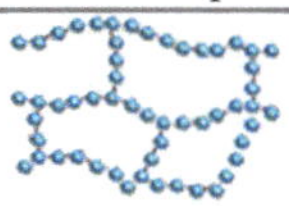	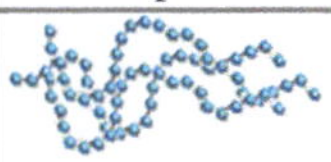
General characteristics	A hard and tightly meshed, branched molecular structure is displayed by the thermosets. They form a three-dimensional cross-linked structure during curing. Reshaping by heating is not possible after shaping, and further modifications can only be achieved via machining	Elastomers have a cross-linked structure. They have a looser mesh compared to thermosets, which gives rise to a degree of elasticity. Weak intermolecular forces hold together the elastomer molecules. Also, heating cannot reshape elastomers after shaping	Thermoplastics have a linear or unbranched and branched or irregular molecular structure that finds out their thermal behavior and strength. At ordinary temperatures, thermoplastics are flexible. Also, they have a relatively lower melting point (MP), resistant to chemicals. The service temperature range for thermoplastics is significantly low compared to thermosets

3. PVC: Hard plastic for electrical cables, glass cleaning bottles, as well as water pipes.
4. LDPE: Easily molded, hard, and strong, often in bags, bottles, storage boxes, and toys.
5. PP: Easily molded, flexible, hard, and fat-resistant, utilized in food containers, drug bottle caps, plant pots, and straws.
6. PS: Easily molded and very stiffer at room temperature, found in CS boxes, plastic tools, and cups
7. Label seven plastics: Unknown resin mixture, utilized in the beverage and food industry (Chiba et al., 2018).

3 Use of Plastic

Plastics are lightweight and inexpensive; have strength, durability, and resistance to corrosion; and offer an extensive range of applications that contribute to medical, technological, and societal advancements. Plastic polymer production has doubled globally from 2000 to 2019, thus reaching 460 million tons (Mt) yearly, and is expected to nearly triple by 2050 (OECD(x), 2022; OECD(z), 2016). Thirty-five percent of global plastic consumption is constituted by the packaging, followed by building and construction (23%), electrical and electronics (8%), furniture and

Table 3 Properties of thermoset and thermoplastic

Properties/ parameters	Thermosets	Thermoplastics
Elasticity	Very poor elasticity	Good elasticity, but relies on the type
Remolding ability	Highly intractable cross-link, i.e., cannot be remolded	Easily reshaped on heating
Attractive forces	Stronger attractive forces (covalent bond) between polymer chain	Weak intermolecular attractive forces between polymer chains
Solubility	Not soluble in organic compounds	They are soluble in organic solvent
Melting point	Higher MP	Lower MP
Molecular weight (MW)	Are higher in MW	Have lower MW when compared to thermoset plastics
Tensile strength	Have high tensile strength and are brittle and inelastic	Have low tensile strength. They are flexible and elastic
Resistance	Superior flexural as well as input resistance	Super abrasion along with dimensional stability
Flexibility	They are not flexible; however, owing to the network structure formed by cross-link, they are rigid	They are flexible and not rigid
Reversible change	Undergoes irreversible transformation that causes it to harden or set	Soften without chemical changes when heated
Synthesis	The process of condensation polymerization can be utilized to synthesize thermoset plastics	The process of addition polymerization can be utilized to synthesize thermoplastics
Processing	Processed utilizing reaction injection molding and compression molding	The thermoforming process, rotational molding, injection molding, blow molding, and extrusion process are all utilized
Curing process	Comprises polymers, which cross-link together in the course of the curing process, for forming an irreversible chemical bond	Softens while heated and turns out to be more fluid while additional heat is applied, the reversible curing process is reversible, and no chemical bonding
Features and benefits	Enhances the material's mechanical properties and chemical resistance, structural integrity, and heat resistance	Higher strength, shrink performance, and easy bendability
High-temperature mechanical performance	Excellent	Poor to good
Coefficient of friction	Moderate to very higher	Very lower to moderate
Elongation	Good to excellent	Poor to good
Colorability	Poor to excellent	Good to excellent
Heat deformation	Good to excellent	Poor to good

(continued)

Table 3 (continued)

Properties/ parameters	Thermosets	Thermoplastics
Cold temperature compatibility	Good to excellent	Poor to good
Advantages	Lower resin viscosity Good fiber wetting More resistant that thermal stability is excellent after polymerized Chemically resistant/corrosive resistant Mechanical creep resistance Heat resistant High flexible design Excellent aesthetic appearance Cost-effective	Highly recyclable Easy for repair by welding along with solvent bonding Reformable capabilities, i.e., post-formable Flexibility Chemical resistant Shrink resistant Aesthetically superior finishes Eco-friendly manufacturing
Disadvantages	Brittle Nonrecyclable through standard approaches Not post-formable or remolded	Poorer melt flow (melt when heated) Required to be heated above MP for processing; generally, more expensive

Table 4 Thermoplastic categories

Crystalline thermoplastics	Amorphous thermoplastics	Semicrystalline thermoplastics
Materials with highly ordered, crystal-like structures or molecular chains, displaying translucent properties and higher mechanical impact resistance. They also comprise amorphous elements	Materials with randomly arranged molecular chains, thus leading to transparency. They lack a specific molecular structure and typically form solid, clear, stiff materials with minimal shrinkage	A combination of crystalline and amorphous properties, often tougher and less fragile compared to amorphous thermoplastics
Instances: high-density polyethylene (HDPE), low-density polyethylene (LDPE), and PP	Examples: polymethyl methacrylate (PMMA), PVC, PC, ABS, PS, PC, and PMMA	Instances: polyamide-imide (PAI), polyethylene (PE), PP, polybutylene terephthalate (PBT), polyethylene terephthalate (PET), and polyamides (PA) (nylon)

housewares (8%), transport (8%), agriculture (8%), toys and sports (3%), mechanical engineering (2%), medical (2%), and other sectors (3%). The augmentation of plastic usage in automotive, electrical and electronics, and construction industries, particularly in emerging markets, namely, Brazil, China, India, and Mexico, contributes to overall industry growth. The largest market share is held by PE, surpassing 25.3% in 2022, with Asia Pacific dominating at 45.1% owing to rapid industrialization and a growing skilled workforce, specifically benefiting the manufacturing sector. Plastic polymer types along with waste material groups are taken into account in this study, which are shown in Table 6 (Milbrandt et al., 2022).

Table 5 Plastic polymer types and waste material groups considered in this study (Milbrandt et al., 2022)

Resin identification code	Polymer name	Common product
01 PET	PET	Bottles, containers, jars, trays, and carpet
02 HDPE	HDPE	Bottles, bags, milk jugs, containers, and toys
03 PVC	PVC	Pipes, pool liners, siding, bags, shoes, and tile
04 LDPE	LDPE (another form is linear LDPE or LLDPE)	Bags, wraps, flexible container lids, cable coating, squeezable bottles, and agricultural films
05 PP	PP	Yogurt tubs, tupperware plastics, hangers, straws, and diapers
06 PS	PS or expanded PS (PS/EPS)	Disposable cups as well as plates, take-out containers, and peanut packing
07 OTHERS	PC, ABS, nylon, polylactide as PLA, and acrylic	CDs, safety glasses, medical storage, and baby bottles
PW materials	**Descriptions**	
Missed plastic packaging #3–#7	Jars, plastic bottles (PBs), and containers that are made of polymers other than PET or HDPE; items have symbols 3 via 7 when marked for identification	
Film/wraps/bags	Grocery bags, storage bags, shrink wrap, trash bags, sheet film plastic, bubble wrap, and so on	
Durable plastic products	Items that last for more than a single usage (for instance: furniture, appliances, toys, along with sporting goods) have symbols betwixt one and seven	
Nondurable plastic products	Items consumed in a shorter period like plastic plates, cups, and other disposable products have symbols betwixt 1 and 7	
Remainder/ composite plastics	These items can be durable, nondurable, or packaging, and cannot be categorized in any other material type (for instance: disposable diapers, clothing, footwear, etc.)	

4 Impacts of Plastics and Microplastics on Environment

This higher amount of PW accordingly affects the ecosystem via marine pollution, soil pollution, and air pollution through ocean dumping, landfilling, as well as open dumping, respectively (Kibria et al., 2023; Mourshed et al., 2017; Paletta et al., 2019). Plastic consists of toxic pollutants, which damage the environment and lead to land, water, and air pollution. Most of the PW ends up in landfills, affecting the environment ultimately. Moreover, most plastics don't degrade.

After that, microplastics (MPs) are defined as pieces smaller than 5 mm with a low size boundary recommended at 333 mm. Also, they are a rising environmental concern. Worldwide, MPs have been recognized from the Arctic to the Antarctic, in

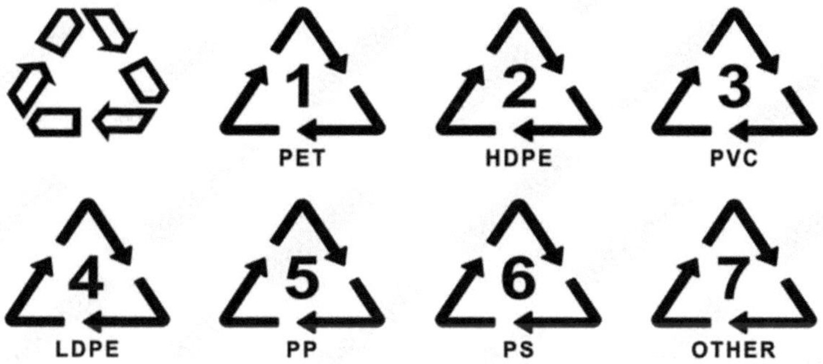

Fig. 2 Plastic recycling codes

rivers, lakes, agricultural soils, sediments, and even in the atmosphere. MPs that are extensively distributed in numerous ecosystems render ingestion risks for various organisms, thus impacting digestive tracts and resulting in internal injuries. MPs' uptake is influenced by their properties, comprising shape and color. MPs can be transferred by earthworms and mites via food webs. Plastic additives that are also known as plasticides pose biochemical risks with some of the plastics releasing toxic monomers linked to reproductive abnormalities and cancer in several species. The plastic pollution's impact extends beyond physical presence, encompassing biochemical effects from plastic additives.

4.1 Plastic Pollution and Impact on the Environment

Important environmental challenges are posed by plastic pollution with a lack of recycling and the accumulation of enormous amounts of plastics in oceans along with landfills yearly. Harmful effects on aquatic life are caused by plastic pollution in marine environments, leading to ingestion-related problems and releasing toxic chemicals into the oceans (Ilyas et al., 2018).

The presence of toxic elements like phthalates and BPA in plastics raises concerns regarding human exposure and environmental degradation. Plastic degradation, especially in landfills, contributes to methane release and further environmental difficulties. Globally, less than 10% of PW is recycled, thus intensifying its environmental threat, non-degradability, and potential health risks. The rise in plastic usage directly relates to the increase in PW, which are categorized as municipal and industrial, with distinct difficulties in recycling heterogeneous PW. Pollution is caused by the mismanagement of PW, influencing climate, biodiversity, and social dimensions, with annual costs ranging from USD 300 to 1500 billion (Purcell et al., 2023).

In aquatic environments, dust, and air, phthalates and BPA have been identified, which raises concerns about their potential adverse effects on both wildlife and humans. The plastic products' wide usage contributes to the extensive presence of

Table 6 Polymer type and its plastic applications

Polymer type	Plastic application
Thermoplastic polymers	
PE	Packaging, fibers and textiles, consumer goods, pipes, as well as fittings
PP	Packaging, automotive applications, and consumer goods
PVC	Building and construction, electronics, healthcare, automobile, other sectors, blood bags and tubing, piping and siding, wire and cable insulation, and windshield system components
PET	PBs, packaging, tape applications, cosmetic jars that are rigid, microwavable containers, and transparent films
PS	Protective packaging, lids, containers, bottles, trays, disposable cutlery, and tumblers
Expanded polystyrene (EPS)	Building materials, surfboards, and white foam packaging
ABS	Drain-waste-vent pipe systems, musical instruments, golf club heads, automotive trim components, bumper bars, luggage and protective carrying cases, medical devices for blood access, small kitchen appliances, toys (comprising Lego and Kre-O bricks), household and consumer goods
Styrene-acrylonitrile resin (SAN)	Food containers, kitchenware, bottles, computer products, packaging material, battery cases, and plastic optical fibers
PA	Textiles, carpets, automotive industry, kitchen utensils, and sportswear
PC	Electronic components, data storage, construction materials, automotive, railway, aircraft, security components, phones, and medical applications
Polymethyl methacrylate (PMMA)	Transparent glass substitute, medical technologies and implants, daylight redirection, and artistic and aesthetic usages
Thermoplastic elastomers (TPE)	Automotive and household appliances sectors
Polyarylsulfone	Membranes and fuel cells
Fluoropolymers and polytetrafluoroethylene (PTFE)[a]	Automotive and aircraft bearings and seals, cookware coatings, flame retardants, linings of piping and chemical tanks, implants, packing for lithium-ion batteries, and catheters for biomedical applications
Polyether ether ketone (PEEK)	Bearings, pumps, piston parts, higher-performance liquid chromatography columns, electrical cable insulation, compressor plate valves, ultrahigher vacuum applications, and medical implants
Polyoxymethylene (POM)	In precision parts that require higher stiffness, lower friction, and excellent dimensional stability
PBT	Housings in electrical engineering, showerheads, in automotive construction as plug connectors and irons, processed into fibers in toothbrushes and false eyelashes, and in the keycaps of some high-end computer keyboards

(continued)

Table 6 (continued)

Polymer type	Plastic application
Polylactic acid (PLA)	Plastic film, cutlery bottles, disposable tableware, 3D printing, monofilament fishing line, and medical implants
Polyhydroxyalkanoates (PHA)	Packaging and medical implants
Polybutylene succinate (PBS)	Packaging, mulching films, tableware, and release materials for pesticide and fertilizer
Polybutylene adipate terephthalate (PBAT)	Cling wrap for food packaging, compostable plastic bags, and water-resistant coatings like on paper cups
Polycaprolactone (PCL)	Medical implants and drug delivery device
Thermosetting polymers	
Polyurethane (PUR)	Cleaning products, automobile seats, furniture, hoses, decorations, sculptures, and filling of spaces and cavities
Unsaturated polyester	Sheet and bulk molding compounds and the toner of laser printers, as overlays on roads and bridges
Epoxy resins	Paints and coatings, industrial tooling and composites, adhesives, petroleum and petrochemical, and consumer and marine applications
Melamine resin	Construction material, cabinet and furniture making, and kitchen utensils and plates
Vinyl ester	As an alternative to polyester and epoxy materials and the thermoset polymer matrix in composite materials, the manufacture of fiberglass-reinforced plastics (FRP), vessels, and tanks
Silicone	Automotive, coatings, aerospace, cookware, dry cleaning, defoaming, lubricants, mold making, medicine, ophthalmology, personal care, plumbing and building construction, and toys and hobbies
Phenol-formaldehyde resin	Molded product production, comprising billiard balls, coatings and adhesives, and laboratory countertops
Urea-formaldehyde	Adhesives, particle boards, finishes, molded objects, and medium-density fiberboard
Acrylic resins	Ingredient in paints

[a] PTFE can't be remolded owing to its higher melt viscosity

plastic particles in numerous environmental components, namely, air, rain, soil, snow, water, and food. While pure plastics are commonly insoluble and nontoxic, the presence of toxic additives like phthalates and the potential leaching of these substances into the environment raise environmental and health concerns.

BPA's environmental concentrations could vary to reach 21 mg L^{-1} in freshwater (FW), and sediment concentrations are orders of magnitude higher compared in the water column. Phthalates and BPA could bioaccumulate with high concentration factors in invertebrates compared to vertebrates. Further research is required for population-level impacts in the natural environment, while adverse effects are observed in laboratory studies at environmentally related concentrations. So, understanding the long-term effects, exposure mixtures' effects, and plastics' role as sources of contaminants is important.

4.2 Impact on Marine and Coastal Environment

Plastic debris (PD) has important environmental consequences that are initially studied in marine settings. Aesthetic problems and hazards for maritime activities, namely, fishing as well as tourism, are well-documented. Fishing nets that are discarded contribute to ghost fishing, leading to commercial fisheries losses. Marine organisms can be harbored by the floating PD, enabling the non-native species' transport.

For debris ingestion, seabirds are widely studied, rendering data on temporal and spatial patterns. A special concern is provided for small plastic fragments or MPs with sizes below current detection levels. The accumulation of PW in the ocean forms plastic soups and garbage patches, namely, the Great Pacific Garbage Patch, which poses a risk to aquatic animals.

Through entanglement and ingestion, marine litter harms fish, seabirds, turtles, and marine mammals. Only limited knowledge prevails regarding the implications for lesser charismatic invertebrates vital to marine ecosystems. MPs that are extensively present in numerous environments raise concerns about their influence on marine life, even in remote areas like the deep seafloor and Arctic sea ice.

4.2.1 Impact on Coastal Environment

Coastal areas usually suffer from litter accumulation, resulting in decreased attractiveness for tourists and potential harm to marine ecosystems. MPs are almost ubiquitous and are seen on ocean surfaces near river mouths along coasts and even in deep-sea sediments and are easily swallowed by mussels, fish, and other marine animals that are linked to the transfer of persistent chemicals (e.g., DDT, PCBs) via the environment. Surface runoff, rivers, and sewage outfalls create issues.

4.3 Impact on Soil and Land

Contrarily, data on PD accumulation in terrestrial and FW habitats are specifically inadequate. Contamination of plastic that has been reported since the 1960s has grown worldwide, affecting oceans from the poles to the Equator and constituting a considerable portion of shoreline debris. Gyres, oceanic convergences, and enclosed seas, namely, the Mediterranean, are particularly contaminated. In some European locations, reports suggest seabed debris quantities exceeding 10,000 items per hectare. For PW management, urgent action is warranted, which estimates that 60–80%

of marine debris contains plastic. Furthermore, toxic elements are absorbed by the plastics, influencing marine and inland species, with recent evidence of significant plastic pollution affecting both ocean and terrestrial ecosystems and emphasizing the critical need for comprehensive management strategies (Villarrubia-Gomez et al., 2018; Chiba et al., 2018).

4.4 *Biodiversity Impacts*

The plastics' toxicity poses a threat to biodiversity as chemicals leach into waterways and soils, affecting aquatic and terrestrial organisms. MPs accumulate in soil and water due to this microbial community, and nutrient availability are altered and processes like nutrient cycling and carbon sequestration are negatively affected (Purcell et al., 2023). In natural environments, plastics could block waterways, reduce food and nesting sites, and lead to habitat destruction. Plastics might be ingested by the species, comprising birds and marine life, causing injuries, choking, and death (Knowlton, 2016). Owing to plastic pollution's durability, abundance, and persistence, it poses a considerable threat to biodiversity. Over 90% of debris encounters involve plastic in oceans, resulting in different consequences for organisms, encompassing entanglement, ingestion, habitat creation, and species dispersal. When compared to ingestion, entanglement is more fatal (79%). Plastic pollution is also faced by FW systems and agricultural soils with potential ingestion by animals. Landfills make a contribution to groundwater (GW) contamination, GHG emissions, and leachate, thus posing risks to nearby water sources (Olofsson & Halvarsson, 2020).

4.5 *Toxicity and Human Health*

4.5.1 Pollution and Toxicity

An important invisible aspect of plastic pollution is toxic chemical additives, comprising POPs, endocrine-disrupting chemicals (EDCs), and heavy metals. Increasing exposure is faced by humans via production and recycling, which hinders mechanical recycling (MR) and creates higher-volume waste residues, which are hazardous, from chemical recycling (CR). Toxic substances are released into the air by burning plastics, thus posing risks to human health, causing cancer, heart disease, headaches, nausea, and damage to the nervous and reproductive systems (Verma et al., 2016). Drainages can be blocked by mismanaged PW, which could also create breeding grounds for disease-carrying mosquitoes and flies (Kehinde et al., 2020). Hazardous chemicals can be emitted throughout the plastics' lifecycle (LC), affecting ecosystems and human exposure. Plastic-related chemicals that are released into soil, air, or water pose risks to humans and biota via inhalation or consumption of contaminated food as well as water. Hazardous chemicals are released by the plastic dumped in landfills, affecting soil quality and surrounding GW over a prolonged period. Air pollution is caused by the incineration of PW, releasing harmful chemicals and carcinogenic substances, namely, PAHs, dioxins, and nitro-PAHs.

4.5.2 Health Impacts

The plasticizer group mainly phthalates within plastics could pose considerable health risks as they leach or migrate from the polymer structure, influencing humans during numerous phases of plastic life. Bisphenol-A (BPA) is an extensively utilized plasticizer and known EDC. BPA migrates into food from plastic baby bottles, heat-affected food can coatings, and several food packaging types.

BPA's exposure may persist from different sources in spite of efforts to produce "BPA-free" food containers, rendering BPA's prevalence in plastics. Toxic additives leave plastics via emissions, releases, degradation, leaching, and migration, affecting the environment at numerous life stages.

Diverse plastics show varied leaching rates for additives impacted by factors, such as permeability and physical structure. Phthalates' higher concentration in flexible PVC, especially in medical equipment, increases concerns owing to endocrine disruption potential, with intravenous medical bags containing phthalates of up to 40% by weight. PVC, which is generally utilized in medical tubing, poses additional concerns, as lead stabilizers readily leach into drinking water from PVC pipes.

PVC, which depends on vinyl chloride monomer (VCM) as its building block, is a major pollutant during production, use, and end-of-life stages. Substantial mercury emissions and waste are caused by China's unique coal-based VCM production, affecting air quality and environmental health.

4.5.3 Effects on Humans: Epidemiological and Experimental Evidence

Human exposure to toxic plastic chemicals is exposed by biomonitoring. Potential health risks are recommended by laboratory studies on animals. Chemicals from plastics relate to human body burdens and reproductive abnormalities. To study endocrine disruptors' impact, the conventional toxicological approach is inadequate, emphasizing the need for a paradigm shift in chemical testing. Through geography and age, phthalates and BPA in human biomonitoring exhibit variations, with house dust and oral drugs' major uptake pathways. Phthalates' relations to adverse health effects in humans are reported, and BPA displays adverse effects at lower doses, particularly during development. Adverse responses in animals for lowering BPA doses are associated with human disease trends. In toxicity studies, the focus on individual chemicals is insufficient. The cumulative effects of chemical mixtures from common household products must be considered by the research. Evolving production patterns, confidential industrial specifications, and the need for collaboration between industry and independent scientists are the challenges.

4.6 Social Impact

Threats to human health are posed by harmful plastic pollution that surpasses safe limits by entering food and water supplies. Studies recommend individuals could ingest up to 5 g of plastics weekly, equivalent to a credit card. Open burning of plastics makes a contribution to air pollution, influencing fertility and hormonal activity and posing risks to vulnerable groups. Plastic pollution relates to premature births, neurodevelopmental disorders, reproductive defects, and numerous health issues. Island nations are disproportionately affected by marine plastic pollution. Plastic pollution's annual cost is evaluated at a minimum of USD 300 billion (Purcell et al., 2023; United Nations (UN) Human Rights, 2022).

5 Plastic Footprint: Global Warming and Climate Change

Plastics, mainly as of fossil fuels (FFs), considerably make a contribution to climate transformation that exceeds carbon budgets as well as jeopardizes the 1.5 °C limit (Hamilton et al., 2019). Plastic emissions, which are 0.8–1.8 Gt CO_2e yearly, surpass 189 coal power plants, thus hindering climate goals (Hamilton & Vallette, 2019). Ten to 32% of the global carbon budget is encompassed in plastics, hindering temperature control (Zheng & Suh, 2019; Hann, 2022; Chen et al., 2023). Littering and plastic pollution lasted 500–1000 years, thus exacerbating environmental problems (Olofsson & Halvarsson, 2020). Plastics' carbon footprint equals 1500 coal-fired plants, thus stressing their climate impact. By 2060, the plastic lifecycle will emit 3.1 Gt CO_2e, emphasizing the issue (Organization for Economic Cooperation and Development (OECD), 2022). The carbon footprint of 25,000 PBs is mirrored by landfilling food waste. Environmental impact is further highlighted by the fossil fuel combustion, 90% of 2011 CO_2 emissions.

5.1 Climate Change

A substantial carbon footprint is posed by the plastic lifecycle, which is from extraction to disposal, thus escalating with augmented production. A major plastic precursor is methane released from natural gas extraction, which makes a significant contribution to global emissions. Annually, plastic incineration and production add over 850 million metric tons of GHGs, exceeding coal power plants' emissions. By

2030, 1.34 gigatons of CO_2e could be emitted by projected plastic growth, challenging the Paris Agreement's goals.

The main source of CO_2 emissions is plastic incineration. The proposed CR makes a significant contribution to the plastic lifecycle's environmental impact. Landfill emissions via lower have detrimental impacts on GW and surface water. MR is limited in its efficacy while having low emissions with renewable energy. Owing to cost disparities, CR might divert plastic to fuel despite being energy-intensive. CO_2 emissions are exacerbated by incinerating plastic as fuel, from CR.

Global climate agreements may restrict virgin plastic production after 2050 in spite of industry expansion. Gas extraction for plastic production remains with petroleum usage decreasing; however, it poses potential CO_2 emissions. Owing to sustainability and toxicity concerns, the fossil-carbon-centric virgin plastic era might phase out by 2050.

6 Biodegradable and Compostable Plastic

Efforts for creating economically viable biodegradable plastics are incomplete. Unluckily, there have not been many decisive actions toward solving the prevailing issue of the huge dumps of PW into the water bodies and onto the land.

6.1 Bioplastic

An alternative to fossil-based plastics is offered by the bioplastics derived from renewable sources, thus potentially decreasing usage.

Nevertheless, they signify over 0.5% of worldwide plastic utilization and might face challenges, comprising cost, efficiency, and ethics. By 2060, bioplastics are projected to encompass only a mid-single digit of plastic utilization even in aggressive policy scenarios. The environmental influence is ambiguous as microbial degradation is resisted by certain bioplastics. Negative environmental interactions could be caused by improper waste disposal. Higher costs than fossil-based peers, equivalent or higher carbon footprints, and ethical concerns associated with first-generation biomass usage, potentially competing with food production and contributing to biodiversity issues, are the challenges (Rosenboom et al., 2022; Naser et al., 2021).

6.2 Durability

Plastic demand is reduced by the durability impacts PW and its relation with GDP intensity as longer-lasting products. Packaging, which has a half-year lifespan, constitutes over 40% of yearly PW. Common polymers like PET, LDPE, PP, and HDPE in packaging must be aligned by household recycling.

Recycling is affected by the short lifespan of packaging, which also contrasts with the longer lifecycle of global plastic stock. Counterintuitively, augmented durability might challenge mismanaged PW by slowing degradation owing to thicker structures. This could exacerbate negative biodiversity results in mismanaged PW. Bioplastic is an eco-friendly alternative that utilizes renewable resources; however, it faces challenges in cost and marine degradability (Sagnelli et al., 2017).

6.3 The Building Blocks of Bioplastics

Most bioplastics depend on polysaccharides, which are carbohydrates with monosaccharide chains linked by glycosidic bonds. Both stability and biodegradability are offered by polysaccharides, such as starches, glycogen, and cellulose, thus making them suitable for diverse applications in industries, namely, food, pharmaceuticals, and medicine. Starches, which are common in human diets from grains and seeds, comprise amylose and amylopectin polymers. A structure is shared with amylopectin by glycogen stored in animal livers and skeletal muscles. Cellulose, which is abundant in plants, is extensively utilized in paper manufacturing. These natural polysaccharides replace synthetic polymers in conventional plastics, promoting biodegradability (Shafqat et al., 2020).

6.4 The Issues with Bioplastics

Biodegradable plastics pose difficulties in their disposal, which takes months to break down and risk contamination of recyclable plastic. During degradation, some bioplastics still generate MPs. By creating an enzyme-activated compostable bioplastic, an experiment at the Department of Energy's Lawrence Berkeley National Laboratory and UC Berkeley (2021) addressed these problems. Embedding enzymes into PLA and PCL plastics enabled rapid degradation in weeks devoid of producing MPs. The broken-down components could be reconstructed into new bioplastic products, offering a cost-effective solution.

6.5 Cross-linking Bioplastics

Starch has emerged as a key material as bioplastics have gained popularity as an alternative to traditional plastics, replacing petroleum-based polymers with natural ones. When compared to regular plastics, cross-linked bioplastics, particularly those incorporating starch, show accelerated degradation (Matusow, 2021). By addressing stability concerns, Sagnelli et al. (2017) experimented with cross-linking starches and lipids complex to amylose at a specific temperature, thus strengthening the starches for plastic-like applications. Mechanical analysis, concentrating on stress and strain levels (Fig. 2), exposed augmented strain at break for cross-linked control and AO starches. The flexibility and cohesion of the bioplastic are impacted alone by the cross-linking, presenting a promising alternative to prevailing plastics.

Another approach, such as mixing biodegradable plastic with straw, involves blending different starches in bioplastics. Rossi et al. (2020) carried out experiments to show that elasticity is improved by adding straw to prevailing bioplastics, making the resulting bioplastic easier to compost, enhancing performance, and increasing cost-effectiveness.

6.6 *Biodegradable Plastic Packaging*

Yang et al. (2021) presented a novel environmentally friendly solution—a biodegradable packaging bag designed to decompose after usage. This innovation encompasses two layers: an inner PE glycol film and an outer layer of degradable paper, unlike nondegradable pesticide packaging. The packaging offers a nontoxic alternative that is harmless to animals, humans, as well as the environment, joined with degradable glue and coated with a moisture-proof layer. This prototype showcases a practical application of bioplastic with the ability for large-scale production:

- Nanoparticles: These particles under 100 nm offer a potential solution for improving bioplastic properties. Their addition enhances gas barrier properties, improving package integrity and decreasing contaminants. In spite of this, the safety of nanoparticles remains under-tested and debated, limiting their immediate usage in cosmetic packaging (Cincinelli et al., 2019).
- Seaweed-Based Bioplastic: Seaweed emerges as an eco-friendly alternative being biodegradable, waste-free, lightweight, and suitable for single-use packaging. Although not widely adopted, the brand Ooho implements seaweed in "notpla," a material utilized for sustainable plastic bottle alternatives (Drobac et al., 2020).

7 Macro- to Nanoplastic

Micro- and nanoplastics as global pollutants pose an ecosystem threat owing to persistence. Plastic pollution spans macro to nano sizes, including numerous items. But, consensus on size-centric categorization is inadequate.

Bergmann et al. (2017) proffered a section on expanding microplastic research in the wake of increased public awareness. The section is introduced by Richard C. Thompson, emphasizing the necessity for hypothesis-driven methods. Sources and pathways of MPs are outlined by Mark A. Browne, whereas Martin G.J. Löder and Gunnar Gerdts critically assess methods that highlight potential misevaluation risks. Global distribution and its effects on marine organisms are summarized by Amy Lusher.

Albert A. Koelmans models environmental contaminant transfer to marine organisms. Concerns about cellular impact are raised by nanoparticles, which are smaller (<1 μm). LCA and economic analysis help in WM decisions, which favor recycling for lower environmental impacts, with variations described by material types and energy consumption. The economic literature on MSW plastic fraction is limited, necessitating further standardization. According to the Waste Framework Directive (WFD), prevailing tools support the PW prevention hierarchy.

Based on MPs' deliberate production (primary) or breakdown of larger particles (secondary), their (<5 mm) posturing environmental threats are classified. Primary MPs found in household items and hygiene products raise concerns owing to their prevalence. Inadequate handling, processing, and residues contribute to their accumulation with potential mitigation actions. Secondary MPs result from environmental processes and sources, such as clothing fibers and tire wear. MPs could be removed by advanced wastewater treatment; however, less developed facilities face challenges, particularly in developing countries. Classification challenges are presented by nanoplastics (<1 μm). The EU Commission recommends <100 nm; however, debates persist. Secondary ones are from the fragmentation of larger particles, with mechanisms still unclear, whereas primary nanoplastics come from numerous products. It is important to understand the effects of plastics, such as nano and micro on the environment, comprising biota.

8 Plastics and Medical Wastes

Kibria et al. (2023) described the substantial generation of medical plastic waste (MPW) owing to advanced medical technologies and healthcare service expansion. Urgent measures, encompassing plastics from gloves, personal protective equipment (PPE), syringes, cannula pipes, and medicine coverings are needed for the safe disposal of MPW. Approximately, 817,000 tons of MPW were generated in 200 medium and large-sized Chinese cities in 2018. A surge in single-use plastics (SUPs), such as hospital suits, gloves, face masks, hand sanitizer bottles, and more, has been caused by the plastic-centric PPE in COVID-19 protection's role. The five nations—India, Brazil, the USA, Nigeria, and the UKA—could create over 1.6 Mt of PP facemasks daily that take centuries to decompose and release MPs if each person discarded one mask daily (Benson et al., 2021). This contributes to an estimated 22.4–53.6 tons of CO_2 emissions daily from waste treatment, facemask production, as well as transportation.

The disposal of corona-related plastics, particularly 14,500 tons of pandemic waste in March–April 2020 poses a significant challenge in Bangladesh. This comprises 3076 tons from Dhaka City alone, with 5877 tons from discarded hand gloves, contributing to plastic pollution. The outstanding values of 1592 tons originate from utilized medical masks and 900 tons from hand sanitizer bottles (Graulich et al.,

2021). The steps like introducing biodegradable masks and promoting mask reuse after decontamination are necessitated by the microparticles' release from masks, amounting to 1246.62 items.

9 Plastic Waste Management and Disposal Solutions

Arranging proper collection and transfer of PW becomes challenging in developing and poor nations owing to limited land space, causing a significant portion to remain uncollected and mixed with MSW. With 7 cities in India contributing 500 kT yearly, Indonesia, India, Thailand, Malaysia, Iran, and Bangladesh collectively produce approximately 1 Mt of PW.

With higher-income countries like Europe, the USA, and Canada leading in PW generation owing to factors, such as income, standard of living, and consumer awareness, the per capita generation of PW varies with income levels (Kaza et al., 2018; Masud et al., 2019). Through better waste disposal practices, PD accumulation in the environment and its consequences can be significantly mitigated. Littering, which is a behavioral problem, has augmented with the usage of disposable products, emphasizing the need for enhanced waste handling. Disposal to landfill raises concerns about the longevity of plastics and the leaching of harmful chemicals while landfill space is diminishing.

For PW management, the 3Rs—reduce, reuse, and recycling—are crucial, with the additional considerations of ER and molecular redesign. Efforts to decrease raw material use, encourage reuse, and recover energy via incineration are constant. Nevertheless, recycling remains an important aspect with considerable regional variations and difficulties in recycling composite items and mixed wastes. Incorporating design and LC analysis of plastics is the aim of the emerging concept of molecular redesign (the fifth R) in green chemistry. This comprises establishing plastics with decreased toxicity, improved recyclability, and the usage of renewable feedstocks.

9.1 Waste Management Infrastructure

Key methods to manage PW considered solutions for PW issues are:

- Mechanical recycling (MR)
- Chemical recycling (CR)
- Downcycling (e.g., use in road surfacing)
- Plastic to fuel
- Plastic to energy (waste incineration)
- Reuse
- Landfill

The determined environmental presence of PD, which is illustrated by debris from a plane ingested by an albatross 60 years after a crash, is highlighted by Thompson. Concerns regarding finite resources for plastic production and the additive effects of diverse chemicals contribute to global WM issues. Concerns about waste disposal capacity are raised by current plastic usage, including PWs from electrical equipment, packaging, and end-of-life vehicles. Plastic production and disposal practices are recognized as reasons for concern.

As per Takada and Bell, neither mechanical nor CR can effectively reduce plastic pollution without regulating and minimizing plastic production. In 2050, projected trends indicate substantial plastic production, incineration, and landfilling, emphasizing the need for changes in current practices. An important factor in reducing mismanaged PW is enhancing WM infrastructure, comprising collection, sanitary landfills, as well as incineration with or devoid of ER. Plastic recycling, which has been evolving since the 1970s, appears promising to meet future demands. But, the waste accumulation in landfills and the natural environment is caused by the challenges arising from plastics' ubiquity (Law et al., 2020). An extensive range of solutions is essential to augment resource recovery (RR), providing a lower percentage of worldwide plastic packaging recycled in 2018 (Markit, 2019).

For waste prevention and management policies, WFD establishes a hierarchy, thus highlighting prevention, reuse, recycling (mechanical or feedstock), other recovery (e.g., ER), and disposal. LC thinking, life cycle assessment (LCA), and economic assessment techniques play crucial roles in assessing the economic and environmental influences of alternative PW management structures.

Waste plastic management remains a major environmental problem, encompassing strategies like recycling, microbial degradation, landfilling, conversion into useful materials, and incineration. Yet, global statistics expose challenges, with only 9% of PW recycled and 27% mismanaged. COVID-19 impacts, incorporating reduced plastic usage in 2020 and augmented littering, further compound the PW problem exhibited in Table 7.

As per a survey, over 40% of manufactured plastic, encompassing food containers, baby bottles, as well as medical materials, serves as single-use packaging. This trend also exposes vulnerabilities to pollution in spite of the acknowledged value of plastics in public domains.

With Table 4 highlighting key techniques (28), effective waste plastic management is crucial for proper disposal or utilization in new product manufacturing. As per (10), recycling technologies are actively developed, and efforts extend to improving biodegradability, namely, including oxo-biodegradable (OBD) additives at just 1% during plastic production. This solution has gained traction in Europe, Latin America, South Asia, Africa, and the Middle East, thus addressing the environmental impact of plastic escaping collection. When independently tested, OBD plastics prove to be ultimately biodegradable on land or in the sea.

Although current methods like recycling, landfilling, and incineration are commonly used for PW management, they fall short in preventing by-products and toxic chemicals. To address this challenge, plastic recycling and environmentally friendly

Table 7 Share of plastic treated by waste management category, after disposal of recycling residues and collected litter, 2019

Countries	Mismanaged and uncontrolled litter (%)	Landfilled (%)	Incinerated (%)	Recycled (%)
Canada	7	82	4	6
USA	4	73	19	4
Other OECD America	20	69	0	10
OECD EU countries	5	37	44	14
Other OECD Europe	8	61	23	8
OECD Oceania	7	75	12	7
OECD Asia	2	14	72	12
China	27	36	24	13
India	46	36	4	13
Latin America	42	47	1	10
Middle East & North Africa	40	54	1	5
Other Africa	64	30	1	6
Other EU	11	80	3	6
Other Eurasia	40	54	1	5
Other non-OECD Asia	45	42	4	9
World	22	49	19	9

technologies are important. Improved pyrolysis, which with careful management and advanced technology, emerges as a promising solution.

9.1.1 Landfill

Olofsson and Halvarsson (2020) emphasize that 50–80% of waste is only gathered primarily through open dumping in developing countries, causing plastic leakage into the ecosystem. The importance of exploring circular economy (CE) opportunities is underscored by inadequate proper waste collection systems. The CE's principles, including waste reduction, reuse, recycling, and material recovery, are highlighted by Khan and Kabir (2020), offering a holistic approach to WM. Globally, landfilling remains the dominant method for MSW disposal owing to cost considerations, especially in developing countries. The continuous demand for plastics makes a contribution to the accumulation of PW in landfills, thus posing environmental risks and pollution issues. When compared to other types of waste, plastics, taking over a thousand years to degrade, present long-term environmental challenges. The urgency of addressing landfilling problems is highlighted by the alarming increase in MSW generation, with plastics comprising 7%–12%. Environmental problems, such as unpleasant odors, methane release, leachate production, and the pathogenic microorganisms' spread, are caused by landfilling. With methane having

20–23 times higher global warming potential compared to CO_2, open landfills contribute significantly to greenhouse (GH) gas emissions. Babaremu et al. (2022), in spite of its environmental impact, landfilling remains the most adopted WM method globally. Plastics utilized in landfills could persist for up to 500 years, highlighting the requirement for alternative strategies. Certain countries, such as Denmark, Sweden, Austria, Belgium, Netherlands, Germany, as well as Switzerland, have successfully limited waste disposal in landfills to about 5% compared to nations, namely, Mexico in which 78.54% of waste is sent to areas of disposal. Globally, the pressing environmental problems related to landfilling underscore the imperative for sustainable WM practices.

9.1.2 Incineration

Incineration is an alternative to landfilling. Also, it is a general approach for MSW management. However, it poses environmental challenges. Plastics in MSW can produce carcinogenic substances like nitro-PAHs, polycyclic aromatic HCs (PAHs), and dioxins during incineration. To mitigate these pollutants, large-scale plants utilize gas-cleaning systems, producing gaseous fuel for electricity generation or chemical feedstock. Pyrolysis is a thermal decomposition process. Also, it is another option with varying product characteristics and yields.

Incineration adversely affects the environment despite continuous operation benefits, releasing toxic substances that pollute the atmosphere (Jeswani et al., 2021). Incineration methods comprise gasification and combustion. However, the release of harmful gases and matter, including CO, VOCs, semi-VOCs, and hazardous materials, during waste plastics' incineration raises environmental concerns. Plastic additives, which are categorized as EDCs, pose further risks.

Solid waste (SW) emits climate-relevant gases, while its incineration is attractive for the generation of thermal energy and reduction of waste volume, significantly contributing to the GH effect (Shah et al., 2023). The high GH gas emissions make incineration less environmentally friendly than other plastic recycling approaches like gasification along with pyrolysis.

9.1.3 Plastic to Energy: Fuel

Due to plastics' abundance and higher energy content, they are attractive for energy applications. In this article, the management of PW through pyrolysis and gasification, two thermochemical conversion (TCC) techniques are reviewed, highlighting product distribution, yields, energy content, as well as integration into energy systems. Pyro-oils are produced by pyrolysis with energy content (41.10–46.16 MJ kg^{-1}) nearly conventional fuels, making them potential candidates for fuel applications. To meet market standards, post-conversion treatments maintain oil quality and remove sulfur content. Under specific conditions, syngas (SG) from gasification, with lower fuel potential (20 MJ kg^{-1}), could be improved to over 40 MJ kg^{-1}. SG serves as a precursor for valuable chemicals and could be transformed into commercial products like liquid fuels via Fischer-Tropsch synthesis.

Commercial-scale plants for PW TCC and RL applications of acquired products are discussed, showing the technical and economic feasibility of integration into energy systems. In cooking stoves, internal combustion engines, gas turbines, and direct-injection diesel engines, the fuels generated were used.

With thermochemical and catalytic conversion being the primary methods, fuel production from plastics is a globally significant research area. Among these, TCC, especially pyrolysis treatment, dominates.

9.1.4 Thermochemical

In an inert environment at elevated temperatures, large polymers are broken into smaller HCs by thermochemical treatment:

(a) **Thermal Pyrolysis**: With studies recommending temperatures as high as 700–900 °C for optimal yields, polyolefins' non-catalytic thermal pyrolysis is a higher-energy procedure necessitating temperatures of at least 350–500 °C.
(b) **Catalytic Cracking**: A catalyst's addition improves the conversion as well as fuel quality in polyolefin pyrolysis than purely thermal pyrolysis.

9.1.5 Gasification

Gasification, which utilizes air rather than oxygen, is a cost-efficient and straightforward approach for PW management. The nitrogen's presence during the process poses a notable drawback despite gasification's simplicity (Babaremu et al., 2022). Gasification encompasses the chemical as well as thermal change of carbon-centric materials into a primarily gaseous output, using air, oxygen, or steam as gasification agents. The process, typically exothermal, operates within a temperature range of 800–1100 °C with air and oxygen (up to 1500 °C) as the oxidant. The resulting products comprise solids (nonvolatile metals), liquids (oil and tar), and gases with varying heating values. The gas yield, ranging from 30% to 60% by weight, relies on factors, namely, temperature, waste input, and process framework. Owing to PW disposal's low degradability, particularly in marine environments, it is a pressing environmental concern. There has been a considerable reduction in PW sent to landfills in Europe, accompanied by increased energy valorization and recycling, thus signifying a shift toward more sustainable practices. To transform waste plastics into chemicals and fuels, pyrolysis is widely acknowledged as an efficient method. Thus, a promising solution for PW management is offered.

9.1.5.1 Gasification Technology

Gasification is also called "indirect combustion." It is an endothermic thermal conversion procedure that converts carbonaceous fuels into SG, chiefly composed of hydrogen (H_2) and carbon monoxide (CO) with a lower oxidant amount compared to stoichiometric combustion. For numerous valuable chemicals and fuels, SG serves as a versatile building block. Hence, potential applications are offered in fertilizers, transport fuels, and chemicals, or as natural gas substitutes (Klavins et al., 2018). This makes gasification an appealing alternative for energy production as of PW. Waste gasification is a complex procedure that happens at temperatures exceeding 600 °C, with variations centered on the oxidizing agent and heat source, categorized as oxygen, steam, air-blown, autothermal, direct, and allo-thermal, or indirect gasifiers. Air/oxygen, CO_2, steam, electrically generated plasma, and, notably, supercritical water (SCW) gasification technology are included in the choice of gasification agent (Bai et al., 2019, 2020). SCW, which is known for its higher efficiency and clean energy attributes, renders a conducive reaction environment for hydrocarbon decomposition. Research efforts like that by Bungay have explored utilizing ultrahigh purity N_2, CO_2, and air for oxidizing remaining char and combustibles in samples. There are four main stages, emphasizing its potential as an advanced technology for efficient and sustainable energy production from PW.

9.1.6 Recycling

PW is a significant contributor to environmental degradation. Also, it can be transformed into useful items, namely, toys and bags via recycling efforts (Angeles-Hurtado et al., 2021). The CE policies establishment is crucial for recycling household waste and mitigating pollution effects by acknowledging the plastics' environmental influence. Generating employment opportunities and the resulting proliferation of new plastic products poses challenges in classification while steps in industrialization have directed to innovative recycling initiatives (Babaremu et al., 2022). Recycling remains an essential approach, thus emphasizing the significance of CE policies for effective PW management.

9.1.6.1 Mechanical Recycling

MR, which is also termed secondary recycling, is a procedure of recovering and reusing PW via mechanical means, primarily applicable to plastics with a single polymer. The MR's efficiency becomes compromised in the presence of contaminants, especially green waste particles (Al-Salem et al., 2017). Waste collection, separation, cleaning, crushing, and pelletizing for conversion are involved in the conventional MR process, leading to new product production devoid of changing the material's chemical composition. But, the process might induce considerable changes in plastics particularly when wetting agents are utilized, impacted by substances encountered during treatment, thus potentially changing polymer compositions.

The issues, such as heterogeneity and degradation of PW during the recycling process due to polycondensation and polymerization, are highlighted by Arun Kumar, causing chemical reactions. In spite of its challenges, MR remains a key approach for recovering and reusing plastics, particularly those with a single-polymer composition.

Recycling stands out as the superior method for effective PW management over incineration and landfilling. Generally, recycled polymers offer economic advantages over new plastics with price variations reliant on polymer categories and quantities ranging from 20% to 100% (Okan et al., 2019). The process of recycling is classified into primary, secondary, tertiary, and quaternary recycling.

MR is a crucial facet. It involves two types, such as upcycling and downcycling. PW is reprocessed solely via mechanical approaches, including grinding, washing, separating, drying, granulating, and compounding in MR (Ragaert et al., 2017). Recyclates produced through these processes can replace virgin polymers in new plastic production, especially for thermoplastics, which can be remelted and processed utilizing several techniques. Nevertheless, since thermoset plastics do not remelt, they pose challenges.

MR operates below capacity despite being a mature industry owing to technical, financial, and policy challenges. In 2016, significant amounts of plastic were burned or landfilled in the EU, which highlights the industry's underutilization. Technical challenges comprise the complexity of plastics, multilayered structures, and additives hindering processing. Industry growth is further impeded by competition with cheap virgin plastics and a limited market for recyclate. Moreover, the inadequate mandatory recycled plastic content in products and policy support makes a contribution to the industry's struggles.

Additionally, in many developed countries, dependence on waste export models has hindered the development of domestic recycling infrastructure, exemplified by Australia's crisis following China's policy changes. This overdependence on exports decreases the incentive to develop robust domestic MR industries for PW in higher-income countries. For the advancement of MR in PW management, efforts to address these challenges and promote CE principles remain imperative.

9.1.6.2 Chemical Recycling

9.1.6.2.1 Pyrolysis and Gasification Come Under Chemical Recycling

Major petrochemical corporations view CR as a strategy to maintain the status quo while meeting recycling targets instead of limiting the exponential development in the production of plastics. This technique doesn't aim to decrease plastic production; however, it concentrates on regulatory compliance and serves as a marketing strategy to sidestep potential plastic bans. CR is presented as a solution for handling plastic scrap types that MR struggles with, comprising contaminated and mixed polymers, potentially contributing marginally to the CE. Yet, significant concerns are raised by the scale-up, energy intensity, pollution, and residues related to CR, thus lacking comprehensive industry data. The potential problems, especially concerning chemical/solvent "purification" leading to hazardous residue, cannot be overlooked, while techniques like pyrolysis provide some insights. It is emphasized that the plastic pollution problem could be fundamentally resolved by neither CR nor MR, especially as petrochemical plastic production escalates to offset carbon liabilities in FFs. Plastic pollution will persist as long as petrochemical companies perceive plastic production as a safer refuge. The key to addressing this challenge lies in global regulations that limit plastic production within bounds that are ecologically sustainable and promote higher rates of both MR and CR. CR remains a public relations distraction until such regulations are in place.

Depolymerizing polymers, via CR and decomposing plastic into valuable raw materials, produces monomers. In the polymers' decomposing, various chemical reactions are adopted; however, methanolysis and glycolysis are the most commercialized ones.

Pyrolysis stands out as an extensively adopted technology among the various techniques. It involves crushing sorted plastics, melting plastics for precise refinement, and converting them into useful products, such as diesel and other petrochemical applications (Kumagai et al., 2020). Gasification is another important approach that transforms waste into energy by melting plastics in the presence of oxygen as well as higher-temperature steam (1200–1500 °C).

9.1.6.2.2 Pyrolysis

Pyrolysis is a thermal degradation process that involves heating plastics without oxygen, and breaking down PW into gas and oil. Plastic polymers are effectively reduced into small molecules by pyrolysis, producing viscous liquids at lower temperatures (400 °C or less) and small gas molecules at higher temperatures (600 °C or more). Reduced CO_2 emissions, minimized landfilling, accelerated product commercialization, and the production of materials useful for electricity generation and heat are the advantages of pyrolysis. Another valuable process that discourages excessive landfilling and reduces costs related to the incineration of PW is gasification. To address environmental degradation, pyrolysis technology emerges as an efficient solution. Thus, economic viability is presented on a large scale. Also, pyrolysis technology serves as an alternative energy source, thus recycling municipal PW into a potential hydrocarbon fuel.

PW is transformed into a valuable fuel by numerous types of pyrolysis processes that are done under controlled conditions at temperatures of 370–420 °C. This environmentally friendly technology encompasses heating plastics in an inert atmosphere over uncontrolled incineration and landfilling, thus yielding a hydrocarbon distillate containing aliphatic, cyclic aliphatic, and aromatic HCs. The pyrolysis products rely on multiple factors, comprising reactor type, temperature, pressure, catalyst, and feedstock composition. Pyrolysis is an important process for recovering fuel oil from PW that mainly utilizes HDPE, LDPE, PP, and PS. When PC,

PVC, and PET are commonly utilized, the latter two present challenges owing to hazardous chlorinated gas emissions, particularly at lower temperatures. As per the research, PVC and PET yield lower liquid outputs than other plastics during pyrolysis. More technological advancements and research are required to enhance the process potentially by including PET or PVC with other biomasses, thus showing coactive benefits. Hence, the pyrolysis of PW could recover 50–90% of fuel oil by weight, generating liquid oils akin to gasoline or diesel fractions. While PC, PET, and PVC produce additional gases like CO, CO_2, and HCl, the gases retrieved from pyrolysis vary with polyolefins generating mainly C1–C4 compounds.

A study by Khazaal and Abdulaaima (2023) explored municipal PW recycling for hydrocarbon fuel production through semi-batch pressurized thermal (Th-PP) and catalytic pyrolysis (Cat-PP). Promising oil recovery percentages are revealed by Th-PP with PP > LDPE > HDPE > PS and a considerable portion of PP pyro-oil falling within the gasoline fraction (C6–C12). Augmented cracking reactions are exhibited by Cat-PP utilizing a Ni-Mo-w/zeolite-based catalyst, thus favoring diesel formation. PP pyro-oil from thermal pyrolysis showed the highest selectivity toward conventional gasoline, while LDPE > PP > PS > HDPE was the order of oil yields. In this study, a promising source of energy and an efficient recycling solution for the plastic pollution predicament are signified; however, challenges remain, and optimization of high temperature and retention time is essential for improving the pyrolysis process's economic and environmental viability.

Landfilling is the most traditional method implemented for WM in the majority of countries. The addition of additives like antioxidants as well as stabilizers supports the PW polymers that are thrown away in landfills for slowing down the plastic biodegradation process extensively. Hence, for PW management, landfilling seems not to be an appropriate technique. Also, landfilling's usage must be limited owing to:

(a) The plastics' longer-term firmness.
(b) Their slower kinetics for biodegradation.
(c) Limited availability of areas for landfilling.
(d) Starting materials' complete usage.
(e) Health problems arise owing to GW contamination by leeching and the toxic chemicals' inclusion and their tendency for leaching from landfill areas to GW.

The procedure of thermal degradation (tertiary recycling) of SW plastic at the temperature range from 300 to 900 °C in a non-oxygenated environment for generating gaseous fuel as well as liquid is termed pyrolysis. Polymeric substances that are made up of monomers are called plastics. Also, a method of conversion of utilized and discarded plastics into valuable products in the form of fuel, monomers, and other useful resources is pyrolysis. Pyrolysis is a beneficial technique of WM than other traditional methods in the subsequent ways:

1. Most of the PW downcycled in the course of recycling generate articles of the lower process and poor-quality products since plastic losing its salient features, such as clarity, flexibility, and strength in the course of multiple cycles is avoided in pyrolysis.

2. From the pyrolysis technique, the higher cost related to sorting, blending, and washing prior to waste plastics' MR is completely omitted.
3. The melting and remolding of discarded or utilized plastics into new articles are comprised in MR. Hence, in this technique, only thermo-PW articles could be utilized efficiently. Contrarily, both thermoplastic and thermosetting plastics are used in the pyrolysis process as feedstock.

   (a) The pyrolysis approach is highly proficient in the new material composites' treatment, especially in the developing technology for replacing conventional materials with polymeric composites in the majority of applications. Finally, the reaction conditions like reacting gas, temperature, heating rate, catalytic process (in the presence of a catalyst), or thermal pyrolysis (in the absence of a catalyst), and the final product, yield, and composition could be different reliant on the composition of the feedstock (Qureshi et al., 2020).

In road construction, the plastic-coated mixture, combined with hot hydrocarbon, is used. This method, including waste plastic and rubber, serves as a partial replacement for traditional bitumen, improving specific road mix mechanical characteristics. The PW's usage in the mix decreases the bitumen requirement by approximately 10%, enhances road strength and performance, removes the need for anti-stripping agents, prevents PW disposal via incineration and landfilling, and promotes an eco-friendly technology. Plastic roads act as a preventive measure against the deteriorating effects of augmented traffic conditions, ultimately serving as a cost-saving and resource-effectual solution for construction (Kalpana & Surendaran, 2018). Gasification as well as pyrolysis, which convert carbonaceous substances into ash, tar, coke, gas, and char, are thermal processes. Char, tar, and gas are produced by pyrolysis, whereas carbon-containing products are transformed into a primarily gaseous form by gasification. The products' composition and quantities rely on the complete process configuration and input fuel. Gasification is generally utilized to explain the entire technology, even though pyrolysis processes can happen before gasification. However, pyrolysis can be independently implemented without undergoing the gasification process.

In the future, pyrolysis and waste gasification are expected to gain more prominence, which is driven partly by public resistance to waste incineration in some countries and the superior ability of gasification and pyrolysis for conserving waste chemical energy.

The thermal decomposition of organic matter devoid of oxidizing agents, such as oxygen, CO_2, and steam, typically occurring at temperatures between 300 and 850 °C is pyrolysis. Pyrolysis processes are commonly endothermic, which requires energy input; the resulting products' energy content and composition rely on the waste input. The products include:

1. Solid: a char-like substance containing residual solid products (10–35 MJ/kg, 20–50% char content).
2. Liquid: a complex mixture of HCs, namely, phenols, organic acids, PAHs, and alcohols (5–15 MJ/kg, 30–50% liquid amount).

3. Gas: a mixture of CH_4, CO, H_2, CO_2, and other volatile constituents (3–12 MJ/ Nm^3, 20–50% gas yield).

The pyrolysis products' heating values and mass yields vary across diverse processes and rely on waste input composition. Mixed plastics often lead to higher inorganic residues and char, while higher-quality PW and rubber help higher oil and gas ratios. Plastics, with higher volatile matter and lower ash content, show considerable potential for pyrolysis to generate liquid oil.

Pyrolysis, performed in an inert atmosphere or pyrolysis gaseous atmosphere, involves heating PW for degrading polymers into small-chain HCs and monomers (alkenes), thus making it suitable for the manufacture of fuel and chemicals. The process applies to mixed PW and could be utilized for polyolefin and polyester plastics. In a continuous supply chain process, pyrolysis products are often re-polymerized into plastics, thereby bringing it closer to a cyclic process with higher carbon efficiency. Pyrolysis, which requires three capital equipment pieces, is comparatively easier to operate compared to gasification, which demands at least five pieces of capital equipment.

As per a survey notes, over 40% of produced plastic serves as single-use packaging, thus contributing to plastic particles' widespread presence in several environments. Owing to the augmented usage of PPE, hygiene products, and single-use items, the COVID-19 pandemic intensifies PW challenges. Thus, a surge is caused in PW generation. During lockdowns, the reliance on online food distribution further compounds the issue. This situation highlights susceptibility to pollution while emphasizing the societal value of plastics (Khazaal & Abdulaaima, 2023). The merits and demerits of diverse SW management techniques are exhibited in Table 8.

9.1.6.2.3 Toxic Additives Challenge Mechanical and Chemical Recycling.

As highlighted by Takada and Bell, the postconsumer polymeric materials' effective recycling faces numerous technical challenges. These challenges include the following.

1. **Thermo-Degradation During the Recycling Process**:
   (a) Recycling processes, involving heat and mechanical sheer, lead to thermo-degradation of polymers, posing an important issue.
   (b) Waste plastics may undergo prior degradation (heat, light, oxygen, and moisture exposure) before reaching recycling facilities, influencing the recyclate product's quality.
2. **The Immiscibility of Polymers in Remelting Phase**:
   (a) The primary higher-volume polymers' immiscibility during the remelting phase can lead to the final recyclate product's degradation.
   (b) Although efforts minimize cross-polymer contamination, achieving 100% contamination-free mono-streams remains a challenge.

Table 8 Pyrolysis and gasification advantages in comparison with other common waste management techniques

S. n.	Solid waste management technique	Advantages	Disadvantages
01	Landfilling	Cheap	Toxic chemicals' production
		Deal with a larger volume of waste	Occupy larger volume
		Decreased transportation distance	Heavy metal's leeching into the ground and soil pollution
		Applicable to non-recycling waste	Deforestation
		Lower operating and labor costs	Nitrogen compounds and heavy metal contaminants are contained in leachate generation (Chua et al., 2019) A larger land area occupation (Bai et al., 2019) Longer decomposition cycle and environmental pollution (Bai et al., 2020) Land restoration is not possible always
02	Incineration	Waste volume reduction	Higher cost
		Combustible toxins' devastation	Higher energy output
		ER	Lead to air pollution
		Minimal area of land occupation	Not appropriate for all types of materials
		Appropriate for commingled with excessive contamination, separation problems, or polymer property deterioration	Nitrogen oxide (NO_x) and sulfur oxide (SO_x) emission
03	Pyrolysis	Ideal for PW management	Technology in the development process
		Low GHG emission	Relative cheap
		Procession of the larger variety of feedback	Comparatively more costly than landfilling
		Higher ER	Thermoplastic treated efficiently
		Higher conversion efficacy and polluting gaseous emissions Appropriate for commingled PW An extensive range of products, liquid, gas, and solid could be generated from PWs	The yield is affected by the moisture content of the feedstock Higher cost of operation

(continued)

Table 8 (continued)

S. n.	Solid waste management technique	Advantages	Disadvantages
04	Gasification	Flexibility to utilize diverse plastics and plastics mixed with other feedstocks, i.e., co-gasification (Lopez et al., 2018) Fuel SG, which is produced by gasification, is easier for handling, meter, control, and burn compared to solid MSWs Gasification needs a significantly small fraction of the stoichiometric amount of oxygen essential for combustion. Gasification takes place in a lower-oxygen environment that limits the formation of dioxins, SO_x, and NO_x The gasifier's decreasing environment could enhance the solid residues' quality, especially metals	Needs dry feedstock, augmenting processing costs To utilize the generated SG in highly effectual internally fired cycles (e.g., gas turbines and combined cycles, Otto engines) or to produce higher-quality fuels (e.g., diesel, gasoline, hydrogen) and chemicals, the SG needs to be treated—cleaned properly, and conditioned, which is very expensive

3. **Management of Additives**:
   (a) The extensive usage of chemical additives in plastics introduces challenges in both CR and MR.
   (b) Degradation products, toxic chemical additives, and "side products" are carried via the recycling chain, thus affecting the plastic recyclate's quality.
   (c) Highly toxic additives like brominated dioxins could persist via the recycling phases and even enter the environment after disposal.
4. **Exposure Risks for Workers**:
   (a) During numerous processing phases, workers in both MR and CR may face exposure to toxic substances and EDCs.
   (b) During MR, workers, especially during shredding and extrusion, may be exposed to substances, such as metals, VOCs, phthalates, PAHs, PBDEs, PAE, and PBDD/F.
   (c) During different processing phases, CR workers may also encounter volatile gases as well as particulate matter from toxic additives.
5. **Externalization of Costs to the Mechanical Recycling Sector**:
   (a) The current plastic pollution paradigm involves virgin plastic's cheap production, externalizing the costs to the MR sector.
   (b) For downstream mechanical recyclers, additives, especially toxic and incompatible ones, added to polymers during production pose economic challenges.

6. **Accountability and Regulation**:
   (a) It is crucial to hold virgin plastic producers accountable for additives and unrecyclable products via regulations, taxes, or policy measures.
   (b) Addressing the economic viability of MR necessitates regulatory measures to ensure accountability for plastic products' entire LC.
7. **Transition to Circular Economy**:
   (a) Addressing the exponential increase in virgin plastic production requires more robust measures while MR contributes to a CE.
   (b) Freezing or decreasing recent plastic production levels and mandating recycled polymer content are recognized as crucial prerequisites to transition to a CE for plastics.

9.1.6.3 Downcycling of Plastic Waste

Downcycling is a method that is suitable for lower-quality plastics like PET, LDPE, and HDPE since they could be remelted for new products. Nevertheless, recycled plastics lose certain properties and functionality, unlike metals and glass, thus making them unsuitable for original applications. Collection, isolation, washing, cleaning, desiccation, chipping/sizing, filling, dyeing, accumulation, and pelletization are the physical steps. After successive cycles, constraints include deterioration and property variations owing to reduced MW. Thermoplastics are best suited for downcycling, whereas thermosetting plastics and higher-MP viscous polymers are not suitable.

Downcycling is a common issue in MR, which involves the polymer's degradation and mixing, causing the loss of crucial characteristics of the original polymer. As outlined by Takada and Bell, this phenomenon is related to numerous challenges.

1. **Characteristic Loss in Recycling**:
   (a) The degradation of plastic qualities is caused by MR processes, making it challenging to return the material as input to new plastics.
   (b) Degraded polymers are often utilized to create less valuable, limited-application plastic products, thus marking the transition into downcycling.
2. **Uncertainty with Chemical Recycling**:
   (a) The CR's application to downcycled products for the recovery of monomers or polymers is unclear.
   (b) Concerns arise concerning the integration of postconsumer plastics into road surfaces in which the material's durability and the management of maintenance or demolition remain uncertain.

3. **Environmental and Health Implications**:
    (a) Limited studies have analyzed the environmental and human health implications of blending postconsumer plastics into bitumen or paving utilizing a melting process.
    (b) Concerns comprise perpetuating a linear extraction-production-disposal cycle, risks to road workers from inhaling toxic emissions during construction, and the release of additives and MPs into the environment from weathering and abrasion.
4. **Concerns with Plastic Road Construction**:
    (a) Specific concerns associated with plastic road construction involve risks to workers from inhaling toxic emissions, incorporating chlorinated and brominated dioxins, carbon monoxide, acrolein, and other gases.
    (b) Environmental concerns are raised by the potential release of chemical additives and MPs into the environment owing to weathering and abrasion from vehicle tires.
5. **Worker Risks in the Manufacturing Process**:
    (a) The downcycling process that involves shredding and extrusion into lower-value products poses risks to workers.
    (b) Volatile compounds or particulates are released by heating waste during manufacturing, thereby exposing workers to emissions, such as dioxins, phthalates, VOCs, and particulate-adsorbed toxic monomers.

Addressing these challenges is essential to establish sustainable recycling practices and ensure the safety of both workers and the environment.

9.1.6.4 Upcycling of Plastic Waste

Upcycling involves creatively transforming waste plastics into higher-quality products for both artistic and environmental purposes like making flower pots, garden sprinklers, and more from PBs. Even though impurities in mixed plastics limit chemical processing's application, it is a cost-effective approach. Specific steps, namely, segregating centered on resins and colors, thorough washing, and re-extrusion, are essential to enhance the process. For example, PBs made from a blend of recycled PET and fresh PET showcase this approach's efficiency (Yu et al., 2016).

9.1.7 Conversion of Waste Plastic to Carbon-Based Compounds

A valuable source for producing carbon-based products is waste plastic, with carbon being the most prized. Owing to the carbon materials' unique properties, finding applications in biomedical, catalysis, environmental, analytical, energy, and electronic fields, the interest in them has surged. The conversion of waste plastic to carbon-centric compounds is facilitated by the synthesis schemes, including catalytic pyrolysis and the CVD approach. In numerous applications, such as fuel cells, solar cells, water purification, supercapacitors, bioimaging, drug delivery, thermal and structural materials, packaging, fabrics, fibers, and polymer industries, materials like graphite, activated carbon, graphene, carbon nanotubes, and fullerene are utilized.

10 Concept of Circular Economy

A CE's concept to manage PW raises a vital question about its viability in comparison with conventional recyclable materials. Burning PW not only generates toxic by-products but also eliminates the extracted petrochemical resources and the embedded energy within the plastic item, while materials, such as paper, glass, and metal, can be endlessly recycled. This embedded energy includes the energy expended in oil extraction, petrochemical refinement, polymer manufacturing, and transportation throughout the production process. Unluckily, this entire energy investment is lost within seconds in the incinerator's combustion chamber, thus yielding only minimal calorific energy. An important difficulty is sustainable WM. An attractive solution is presented by the CE's concept. To sustainably reduce the PW's accumulation, the 4R concept (reduce, reuse, recycle, and recover) within the CE methodology offers a promising technique. In cases where the first three Rs are impractical-grounded on plastic-type or processing costs, RR could be attained via a waste-to-fuel-and-energy (WtE) technique.

The CE theory, which concentrates on preventing plastics from becoming waste, still confronts challenges as plastics persist as an important environmental and social concern, often termed the "lubricant of globalization" (Babaremu et al., 2022; Qu et al., 2019). Critics argue that the CE might divert attention from plastic pollution's root causes, leading global organizations to highlight the reduction and phasing out of plastics instead of solely depending on recycling efforts. Employing CE solutions like waste separation, resource awareness, and small-scale recycling initiatives shows promise in low- and middle-income countries (LMICs). The CE model rejects the linear approach in favor of a restorative and regenerative cycle, including both technical and biological cycles by addressing marine plastic pollution (Foundation EM, 2017). The biological cycle involves anaerobic and composting digestion, whereas the technical cycle employs strategies, such as repair, reuse, remanufacture, recycling, and refurbishment.

The CE model's industrial operationalization could be categorized into resource switch, emphasizing renewable and non-harmful inputs; resource circularity, concentrating on the reuse and recovery of end-of-life materials; and RR, ensuring efficient plastic use. Material input regulations, bans on nondegradable and nonrecyclable plastics, encouragement of bio-based materials, influence and packaging regulations in product design, resource efficacy in manufacturing and service provision, and end-of-life material management policies, promoting circular practices via recycling targets, are the policies supporting the CE.

11 Future of the Plastic Age

Looking ahead, the "plastic age" envisioned by Yarsley and Couzens in the 1940s seems far from its conclusion, with plastics poised to continue contributing significantly to societal advancements. With plastics playing a crucial role in scientific, medical, and environmental domains, Andrady anticipates a rapidly evolving technological landscape by 2030. The potential applications comprise medical breakthroughs, lightweight components for fuel-efficient transportation like the Boeing 787, contributions to renewable energy generation, insulation for carbon emission reduction, and smart plastic packaging for monitoring perishable goods' quality.

In conclusion, current production, usage, and disposal practices raise sustainability concerns, while plastics hold substantial promise for the future, influencing wildlife and human health. Comprehensive solutions are imperative through recognizing the environmental and health hazards related to PW accumulation. The significant measures are individual responsibility, industry adoption of green chemistry and material reduction, product design for reusability and recyclability, and government interventions through standards and funding academic research. For informed decision-making, lifecycle analysis, encompassing all stages of plastic production, usage, and disposal, is advocated.

Yearly, production continues to rise at around 9% despite growing awareness of environmental issues and health concerns linked to plastics. To address the environmental hazards related to PW accumulation, urgent actions are necessary. Thus, the importance of proactive measures and interdisciplinary collaboration is highlighted.

12 Action

To address the challenges related to plastic pollution, containing fragmented policies, unsustainable production and consumption patterns, inadequate information on chemicals in plastics, and insufficient capacity to manage PW, global action is imperative. The subsequent actions are proposed to reduce chemical-associated impacts:

- ***Reduce Production and Consumption***: Initiate reductions in plastic consumption and production, thus prioritizing non-crucial plastics.
- ***Design Safer Plastics***: Design and produce plastics free of hazardous chemicals, thereby prioritizing the phaseout of internationally restricted chemicals.
- ***Avoid Regrettable Substitutions***: Prevent regrettable substitutions and burdens shifting by performing chemical alternatives evaluations and LCAs.

- ***Enhance Transparency***: By disclosing the quantity and identity of all chemicals utilized in plastic products, ensure transparency along the plastics value chain, enabling informed decision-making.
- ***Update Regulatory Testing Guidelines***: Revise regulatory testing guidelines to incorporate rapid and cost-efficient methodologies like bioassays and computational tools, addressing knowledge gaps.
- ***Develop Waste Management Frameworks***: Develop robust WM frameworks aligned with the waste hierarchy, rendering specific attention to the protection of the informal sector.

Capacity-building efforts, encompassing legal frameworks, chemical substitution competencies, supply chain management, chemical assessments, WM technology, and the development of local and international networks, are crucial, particularly in developing countries. For a successful transition to safe and sustainable material cycles, the collaborative engagement of stakeholders across the plastic value chain, comprising regulatory authorities, industries, waste managers, scientists, and the public, is vital. Safeguarding human health as well as the environment from plastic pollution's influences is the aim of the collective effort.

12.1 Epilogue and Recommendation

In conclusion, three key recommendations for the public are offered to address environmental concerns associated with PW:

(a) **Consider Necessity**

- Prioritize the necessity of items before obtaining them, emphasizing the reduction of unnecessary consumption instead of a blanket switch to alternatives.

(b) **Choose Sustainable Alternatives**

- Explore reusable or disposable alternatives with a concentration on recyclability and natural biodegradability. Assess sustainability factors like water, carbon, and land use footprints to make environmentally conscious choices.

(c) **Recycle and Dispose Properly**

- Ensure proper disposal in designated bins or recycle when feasible for items challenging to replace or recycle, namely, food wrappers and cigarette butts.

In this report, the comprehension of the flow of plastic items into rivers and oceans is contributed, emphasizing actionable steps for consumers. Also, insights for businesses and policymakers to support sustainable practices, encompassing waste minimization, carbon sequestration, soil enhancement, energy supply, and the creation of value-added products are rendered in this report. Encouraging responsible choices that address environmental challenges related to plastic pollution is the overarching goal.

References

Alli, R. D., Kannan, P., Al Shoaibi, A., & Srinivasakannan, C. (2018). Performance prediction of waste polyethylene gasification using CO2 in a bubbling fluidized bed: A modelling study. *Chemical and Biomedical Engineering Quarterly, 32*(2), 349–358.

Al-Salem, S. M., Antelava, A., Constantinou, A., Manos, G., & Dutta, A. (2017). A review on thermal and catalytic pyrolysis of plastic solid waste (PSW). *Journal of Environmental Management, 197*, 177–198.

Angeles-Hurtado, L. A., Rodríguez-Reséndiz, J., Salazar-Colores, S., Torres-Salinas, H., & Sevilla-Camacho, P. Y. (2021). Viable disposal of post-consumer polymers in Mexico: A review. *Frontiers in Environmental Science, 9*, 749775.

Antelava, A., Jablonska, N., Constantinou, A., Manos, G., Salaudeen, S. A., Dutta, A., & Al-Salem, S. M. (2021). Energy potential of plastic waste valorization: A short comparative assessment of pyrolysis versus gasification. *Energy & Fuels, 35*(5), 3558–3571.

Babaremu, K. O., Okoya, S. A., Hughes, E., Tijani, B., Teidi, D., Akpan, A., Igwe, J., Karera, S., Oyinlola, M., & Akinlabi, E. T. (2022). Sustainable plastic waste management in a circular economy. *Heliyon, 8*(7), e09984. https://doi.org/10.1016/j.heliyon.2022.e09984

Bai, B., Liu, Y., Wang, Q., Zou, J., Zhang, H., Jin, H., & Li, X. (2019). Experimental investigation on gasification characteristics of plastic wastes in supercritical water. *Renewable Energy, 135*, 32–40.

Bai, B., Wang, W., & Jin, H. (2020). Experimental study on gasification performance of polypropylene (PP) plastics in supercritical water. *Energy, 191*, 116527.

Benson, N. U., Fred-Ahmadu, O. H., Bassey, D. E., & Atayero, A. A. (2021). COVID-19 pandemic and emerging plastic-based personal protective equipment waste pollution and management in Africa. *Journal of Environmental Chemical Engineering, 9*(3), 105222.

Bergmann, M., et al. (2017). Vast quantities of microplastics in Arctic Sea Ice—A prime temporary sink for plastic litter and a medium of transport. In *Fate and impact of microplastics in marine ecosystems*: *From the coastline to the open sea* (pp. 75–76). ISBN 9780128122716. https://doi.org/10.1016/B978-0-12-812271-6.00073-9.

Chen, X., McDonald, K., & Rose, M. (2023). Stemming the plastic-climate crisis. Pacific environment.

Chiba, S., Saito, H., Fletcher, R., Yogi, T., Kayo, M., Miyagi, S., Ogido, M., & Fujikura, K. J. M. P. (2018). Human footprint in the abyss: 30 year records of deep-sea plastic debris. *Marine Policy, 96*, 204–212.

Chua, H., Bashir, M., Tan, K., & Chua, H. (2019) A sustainable pyrolysis technology for the treatment of municipal solid waste in Malaysia. In *6th International Conference on Environment (ICENV2018)*: *Empowering Environment and Sustainable Engineering Nexus Through Green Technology*.

Cincinelli, A., et al. (2019). A potpourri of microplastics in the sea surface and water column of the Mediterranean Sea. *Trends in Analytical Chemistry, 110*, 321–326.

Drobac, J., Alivojvodic, V., Maksic, P., & Stamenovic, M. (2020) Green face of packaging—Sustainability issues of the cosmetic industry packaging. In *MATEC Web of Conferences* (Vol. 318, p. 01022). https://doi.org/10.1051/matecconf/202031801022

Eagle, L., Hamann, M., & Low, D. R. (2016). The role of social marketing, marine turtles and sustainable tourism in reducing plastic pollution. *Marine Pollution Bulletin, 107*(1), 324–332.

Foundation EM. (2017). *The new plastics economy*: *Rethinking the future of plastics & catalysing action*. Retrieved August 2022.

Fox, J., & Stacey, N. (2019). Process targeting: An energy-based comparison of waste plastic processing technologies. *Energy, 170*, 273–283.

Graulich, K., Kohler, A., Low, C., Sutter, D. W., Mehlhart, G., Egebæk, K. R., Bilsen, V., Bley, F., & Manshoven, S. (2021). *Impact of COVID-19 on single-use plastics and the environment in Europe* (p. 22). European Environment Agency.

Hamilton, & Vallette, J. (2019). The new coal: Plastics & climate change, (beyond plastics, October 2021),

Hamilton, L. A., Feit, S., Muffett, C., Kelso, M., Rubright, S. M., Bernhardt, C., Schaeffer, E., Moon, D., Morris, J., & Labbe-Bellas, R. (2019, May). *Plastic & climate: The hidden costs of a plastic planet* (Center for International Environmental Law). OECD, Global Plastics Outlook. https://www.ciel.org/plasticandclimate/

Hann, S. (2022). *Is net zero enough for the materials production sector?* Zero Waste Europe. https://www.eunomia.co.uk/reports-tools/is-net-zero-enough-for-the-materials-production-sector/

Ilyas, M., Ahmad, W., Khan, H., Yousaf, S., Khan, K., & Nazir, S. (2018). Plastic waste as a significant threat to environment—A systematic literature review. *Reviews on Environmental Health, 33*(4), 383–406. https://doi.org/10.1515/reveh-2017-0035

Jeswani, H., Kruger, C., Russ, M., Horlacher, M., Antony, F., Hann, S., et al. (2021). Life cycle environmental impacts of chemical recycling via pyrolysis of mixed plastic waste in comparison with mechanical recycling and energy recovery. *Science of the Total Environment, 769*, 144483.

Kalpana, & Surendaran, D. (2018). Utilisation of waste plastic in bituminous roads—A review. *International Journal of Pure and Applied Mathematics, 119*(17), 1143–1156.

Kara, L. L. (2017). Plastics in the marine environment. *Annual Review of Marine Science, 9*, 205–229.

Kaza S, Yao L, Bhada-Tata P, Van Woerden F (2018) *What a waste 2.0: A global snapshot of solid waste management to 2050, urban development.* © World Bank. http://hdl.handle.net/10986/30317

Kedzierski, M., Frere, D., Le Maguer, G., & Bruzaud, S. (2020). Why is there plastic packaging in the natural environment? Understanding the roots of our individual plastic waste management behaviours. *Science of the Total Environment, 740*, 139985.

Kehinde, O., Ramonu, O. J., Babaremu, K. O., & Justin, L. D. (2020). Plastic wastes: Environmental hazard and instrument for wealth creation in Nigeria. *Heliyon, 6*(10), e05131.

Khan, I., & Kabir, Z. (2020). Waste-to-energy generation technologies and the developing economies: A multi-criteria analysis for sustainability assessment. *Renewable Energy, 150*, 320–333. https://doi.org/10.1016/j.renene.2019.12.132

Khazaal, R. M., & Abdulaaima, D. A. (2023). Valuable oil recovery from plastic wastes via pressurized thermal and catalytic pyrolysis. *Energy Conversion and Management, 20*, 100430. https://doi.org/10.1016/j.ecmx.2023.100430

Kibria, M. G., Masuk, N. I., Safayet, R., Nguyen, H. Q., & Mourshed, M. (2023). Plastic waste: Challenges and opportunities to mitigate pollution and effective management. *International Journal of Environmental Research, 17*, 20. https://doi.org/10.1007/s41742-023-00507-z

Klavins, M., Bisters, V., & Burlakovs, J. (2018). Small scale gasification application and perspectives in circular economy. *Environmental and Climate Technologies, 22*, 42–54.

Knowlton, A. R. (2016). Effects of fishing rope strength on the severity of large whale entanglements. *Conservation Biology, 30*(2), 318–328.

Kumagai, S., Nakatani, J., Saito, Y., Fukushima, Y., & Yoshioka, T. (2020). Latest trends and challenges in feedstock recycling of polyolefinic plastics. *Journal of the Japan Petroleum Institute, 63*, 345–364. https://doi.org/10.1627/jpi.63.345. ISSN 1346-8804.

Landon-Lane, M. (2018). Corporate social responsibility in marine plastic debris governance. *Marine Pollution Bulletin, 127*, 310–319.

Law, K., Starr, N., Siegler, T., Jambeck, J., Mallos, N., & Leonard, G. (2020). The United States contribution of plastic waste to land and ocean. *Science Advances, 6*(44), eabd0288. https://doi.org/10.1126/sciadv.abd0288

Lopez, G., Artetxe, M., Amutio, M., Alvarez, J., Bilbao, J., & Olazar, M. (2018). Recent advances in the gasification of waste plastics. A critical overview. *Renewable and Sustainable Energy Reviews, 82*, 576–596.

Markit IHS. (2019). *Plastic recycling and sustainability*. Retrieved June 2024.

Masud, M. H., Akram, W., Ahmed, A., Ananno, A. A., Mourshed, M., Hasan, M., & Joardder, M. U. H. (2019). Towards the effective E-waste management in Bangladesh: A review. *Environmental Science and Pollution Research, 26*(2), 1250–1276.

Matusow, J. (2021). A growing responsibility for sustainable packaging [Review of a growing responsibility for sustainable packaging]. *A growing responsibility for sustainable packaging; beauty packaging*. https://www.beautypackaging.com/issues/2021-04-01/view_features/a-growing-responsibility-for-sustainable-packaging/

Mazhandu, Z. S., Muzenda, E., Mamvura, T. A., & Belaid, M. (2020). Integrated and consolidated review of plastic waste management and bio-based biodegradable plastics: Challenges and opportunities. *Sustainability, 12*(20), 8360.

Milbrandt, A., Coney, K., Badgett, A., & Beckham, G. T. (2022). Quantification and evaluation of plastic waste in the United States. *Resources, Conservation and Recycling, 183*, 106363.

Mourshed, M., Masud, M. H., Rashid, F., & Joardder, M. U. H. (2017). Towards the effective plastic waste management in Bangladesh: A review. *Environmental Science and Pollution Research, 24*(35), 27021–27046.

Naser, A. Z., Deiab, I., & Darras, B. M. (2021). Poly (lactic acid) (PLA) and polyhydroxyalkanoates (PHAs), green alternatives to petroleum-based plastics: A review. *RSC Advances, 11*(28), 17151–17196.

OECD(x). (2022). *Global plastics outlook: Economic drivers, environmental impacts and policy options*. OECD Publishing. https://doi.org/10.1787/de747aef-en

OECD(z). (2016). Global plastics outlook; World Economic Forum, the new plastics economy: Rethinking the future of plastics.

Okan, M., Aydin, H. M., & Barsbay, M. (2019). Current approaches to waste polymer utilization and minimization: A review. *Journal of Chemical Technology & Biotechnology, 94*, 8–21.

Olofsson, F., & Halvarsson, H. (2020). *Small scale energy conversion of plastic waste. Degree project in industrial engineering and management*. Malardalen University. ERA402/ERA403.

Organization for Economic Cooperation and Development (OECD). (2022). *Global plastics outlook: Economic drivers, environmental impacts and policy options*. OECD.

Paletta, A., Leal Filho, W., Balogun, A. L., Foschi, E., & Bonoli, A. (2019). Barriers and challenges to plastics valorisation in the context of a circular economy: Case studies from Italy. *Journal of Cleaner Production, 241*, 118149.

Plastics—The facts 2014: An analysis of European Plastics Production, Demand and Waste Data, 2014. Plastics Europe.

Purcell, J., Sayyad, K., Heinzmann, L., Awramenko, S., & Dorner-Müller, S. (2023). *Plastic pollution: Pathways to net zero*. Credit Suisse Center of Sustainability 53.

Qu, S., Guo, Y., Ma, Z., Chen, W., Liu, J., Liu, G., et al. (2019). Implications of China's foreign waste ban on the global circular economy. *Resources, Conservation and Recycling, 144*, 252–255.

Qureshi, M. S., Oasmaa, A., Pihkola, H., Deviatkin, I., Tenhunen, A., & Mannila, J. (2020). Pyrolysis of plastic waste: Opportunities and challenges. *Journal of Analytical and Applied Pyrolysis, 152*, 104804.

Ragaert, K., Delva, L., & Geem, K. V. (2017). Mechanical and chemical recycling of solid plastic waste. *Waste Management, 69*, 24–58.

Rosenboom, J. G., Langer, R., & Traverso, G. (2022). Bioplastics for a circular economy. *Nature Reviews Materials, 7*(2), 117–137.

Rossi, G., Conti, L., Fiorineschi, L., Marvasi, M., Monti, M., Rotini, F., Togni, M., & Barbari, M. (2020). A new eco-friendly packaging material made of straw and bioplastic. *Journal of Agricultural Engineering, 51*(4). https://doi.org/10.4081/jae.2020.1088

Sagnelli, D., Hooshmand, K., Kemmer, G., Kirkensgaard, J., Mortensen, K., Giosafatto, C., Holse, M., Hebelstrup, K., Bao, J., Stelte, W., Bjerre, A. B., & Blennow, A. (2017). Cross-linked amylose bio-plastic: A transgenic-based compostable plastic alternative. *International Journal of Molecular Sciences, 18*(10), 2075. https://doi.org/10.3390/ijms18102075

Science Museum. (2019). *The age of plastic: From parkesine to pollution*. Retrieved September 3, 2023.

Shafqat, A., Tahir, A., Mahmood, A., Tabinda, A. B., Yasar, A., & Pugazhendhi, A. (2020). A review on environmental significance carbon foot prints of starch based bio-plastic: A substitute of conventional plastics. *Biocatalysis and Agricultural Biotechnology, 27*, 101540. https://doi.org/10.1016/j.bcab.2020.101540

Shah, H. H., Amin, M., Iqbal, A., Nadeem, I., Kalin, M., Soomar, A. M., & Galal, A. M. (2023). A review on gasification and pyrolysis of waste plastics. *Frontiers in Chemistry, 10*, 960894. https://doi.org/10.3389/fchem.2022.960894

Singh, P., Déparrois, N., Burra, K., Bhattacharya, S., & Gupta, A. (2019). Energy recovery from cross-linked polyethylene wastes using pyrolysis and CO2 assisted gasification. *Applied Energy, 254*, 113722.

UNEP. (2023). *Chemicals in plastics: A technical report*. United Nations Environment Programme and Secretariat of the Basel, Rotterdam and Stockholm Conventions.

United Nations (UN) Human Rights. (2022). *UN human rights experts urge treaty to address 'plastic tide.'* Retrieved May 26, 2023.

US Energy Information Administration. (n.d.). *International energy outlook 2019*. Retrieved May 12, 2020, from http://www.eia.gov/ieo

Verma, R., Vinoda, K. S., Papireddy, M., & Gowda, A. N. S. (2016). Toxic pollutants from plastic waste: A review. *Procedia Environmental Sciences, 35*, 701–708.

Villarrubia-Gomez, P., Cornell, S., & Fabres, J. (2018). Marine plastic pollution as a planetary threat: The drifting piece in the sustainability puzzle. *Marine Policy, 96*, 213–220.

Wang, J., Tan, Z., Peng, J., Qiu, Q., & Li, M. (2016). The behaviors of microplastics in the marine environment. *Marine Environmental Research, 113*, 7–17.

Yang, H. (2021). Eco-friendly composite packaging bag, US20210094267A1.

Yu, J., Sun, L., Ma, C., Qiao, Y., & Yao, H. (2016). Thermal degradation of PVC: A review. *Waste Management, 48*, 300–314.

Zheng, J., & Suh, S. (2019). Strategies to reduce the global carbon footprint of plastics. *Nature Climate Change, 9*, 374–378. https://doi.org/10.1038/s41558-019-0459-z

Waste Management: Tackling Challenges, Sustainable Practices, and Transforming Waste into Value

Noorie Shama, Subhadra Rajpoot , and Aditi Nayak

Abstract Waste management has become a global concern as the total amount of waste generated is overwhelming, including municipal, industrial, and hazardous waste, which all contribute to the problem. Waste streams get contaminated with hazardous materials, making recycling more challenging. Encouraging individuals and businesses to reduce, reuse, and recycle and its management is challenging. To manage biomedical waste, there is a need for cost-effective, eco-friendly, and less contaminating approaches for a greener and safer environment. The eco-friendly handling of waste includes the 3Rs, composting organic waste, waste-to-energy technologies, and zero-waste system to eliminate waste. The last portion of the chapter is the conversation of waste into goods with additional values like the use of waste to produce bioactive compounds like carotenoids and the conversation of food waste to carbon dots contributing to the concept of circular economy. Feedstocks from dairy industry are used for the generation of biofuels like bioethanol and biogas. Pyrolysis is a microbial technology used for the formation of bio-oil and biochar. Bioenergy production utilizing biowaste can open new avenues in the field of waste valorization and energy research. Overall, the chapter aims to provide an overview of current challenges, eco-friendly waste handling, and conversion of waste into goods with additional values. By implementing these strategies and promoting public awareness, we can have more sustainable waste management practices that benefit both the environment and society.

Keywords Waste to energy technologies · Biomedical waste · Carotenoids · Carbon dots · Biochar

---

N. Shama · A. Nayak (✉)
Department of Life Science, Guru Nanak Institute of Pharmaceutical Science and Technology, Kolkata, India
e-mail: aditi.nayak@gnipst.ac.in

S. Rajpoot
Amity University, Greater Noida, Greater Noida, Uttar Pradesh, India

S. Rajpoot, A. P. Das (eds.), *Innovative Technologies for Waste Management*, Sustainable Environmental Waste Management Strategies,
https://doi.org/10.1007/978-3-031-97184-6_16

1 Introduction

Waste is a discarded substance that is no longer in use. Waste management refers to the collection, transportation, processing, and disposal of this waste in a manner that reduces the harmful effects of this waste on human health and the ecosystem. Waste is generated by almost all the activities of humans. In the past large amounts of waste production have been a source of concern (Chen et al., 2022). In prehistoric times as the population was low, waste management meant just the disposal of useless substances, as vast land and resources were available (Amasuomo & Baird, 2016). The land at that time was potential enough to mitigate the waste generated without its degradation. The population increased as people migrated to cities during the Industrial Revolution which led to a population explosion in turn upsurging both the volume and the variety of waste produced which became a challenging issue for the people.

Municipal solid waste (MSW) is an unavoidable by-product of human activity which is a major issue globally. SW includes the kitchen, household, municipality, and industrial sector waste. The rise in household waste during the COVID-19 pandemic has contributed to the issues related to MSW, for this waste is disinfected before sorting and stored for 9 days so that harmful viruses can be inactivated contributing to the safety measures of MSW management (Das et al., 2021a).

Industrial hazardous wastes including chemicals, solids leftovers, effluents, heavy metals, etc. all contribute to dangerous environmental concerns that require key strategies to promote industrial waste management. Reducing resource use, waste treatment before discarding, technological innovations, and circular economy steps are some solutions that are adopted.

Textile waste are synthetic and non-synthetic materials discarded from the textile industry making their decomposition a problem. Strategies for recovery of valuable polymers and monomers include dilute acid hydrolysis and enzymatic hydrolysis, fiber extraction, biochemical recycling techniques, and bio-based textile waste (Wojnowska-Baryła et al., 2022). Agricultural waste is generated mainly from crop residue, livestock, aquaculture, and agro-industries. Cellulose and lignocellulose are major wastes that can be used to produce biofuels and other value-added products (Koul et al., 2022). Sources of biomedical waste include infectious material, syringes, pharmaceuticals, pipes, needles, and hazardous substances, which require proper handling as they can pose health risks to patients in clinical settings and other places where they are discarded (Dehal et al., 2022). Strategies include waste sorting, inactivation using chemical disinfectants, sterilization, etc. Plastic waste increased with the increase in population; there are many adopted solutions and technologies for mitigating plastic pollution like property management landfills, recycling, pyrolysis, liquefaction, construction of roads, and bricks making among which tar, concrete, and road construction are major solutions (Huang et al., 2022). Sanitary waste items including sanitary pads, baby diapers, tampons, etc. are non-biodegradable waste and pose environmental issues if not properly handled. Public awareness programs to aware people of the need for sanitary waste management, methods of incineration, waste-to-energy systems, and infrastructure policies are necessary (Kaur et al., 2018). There is an urgent need for electronic waste (e-waste) management strategies due to their short lifespan, which contributes to their high accumulation. For this the principles of reduction and reuse need to be implemented. Key approaches include promoting producer responsibility in which producers recycle devices, encouraging people not to follow trends of buying new products before the old ones are used up, implementing techniques for extracting valuable metals from discarded electronics, and recycling devices before discarding (Islam et al., 2024). For eco-friendly handling of waste, some steps that can be taken are waste minimization and sorting in different categories (e.g., organic, recyclable, and other types), composing (through microbial activity to make waste into useful humus that enhances soil health), recycling technologies, safe disposal methods (sanitary landfills, incineration, and energy production methods), biomedical waste (sterilization, disinfection), and public awareness programs, dramas, and campaigns to reduce harmful environmental effects (Roulia, 2024; Abubakar et al., 2022).

Waste conversion to valuable products is an essential and innovative idea to manage waste, reduce its harmful effects, and utilize it. Household and kitchen waste can be converted to carbon dots, biogas, and biofertilizers through microbial degradation, which enriches soil fertility (Kumar et al., 2020). Through pyrolysis, hazardous wastes like masks which increase the amount of waste produced during COVID-19, and industrial, and municipal waste can be converted to high-value products like biochar, reducing environmental impacts (Luo et al., 2021). Pyrolysis and co-pyrolysis convert waste to usable products along with carbon sequestration (Usmani et al., 2021). Biomass can be converted to biofuels and bioplastics, offering a renewable alternative to fossil fuels that require millions of years to renew and reducing carbon footprint. Waste is converted to usable chemicals by processes like fermentation, hydrolysis, etc. (Antonetti et al., 2017). Valuable resources are also products of fish waste which served as an important step for a sustainable economy (Coppola et al., 2021). Nanocellulose which is used in different packaging, biocomposites, and as construction materials is extracted from agricultural waste (Ahmad Khorairi et al., 2023). Biodiesel production gives out glycerol as a by-product which can be converted to bioethanol which decreases the amount of waste as well as contributes to an alternative source of energy (Elsayed et al., 2024).

2 Strategies to Overcome Challenges of Waste Management

Waste management strategies are employed to overcome the difficulties faced during waste handling; the primary strategy employed is the differentiation of waste into different categories to facilitate management processes, recyclable waste (paper, glass bottles, electronics, juice cans), compostable (organic wastes like spoiled food waste, fruits and vegetable peel, agro waste), hazardous waste (paints, syringes, medical knives, gloves), and nonrecyclables (sanitary waste, tampons, diapers, plastic bags, bottles, thermocol boxes) (Fig. 1).

Fig. 1 Differentiating waste into disposable categories like recyclables, compostable, hazardous, and nonrecyclable

2.1 Municipal Solid Waste

A significant challenge is faced in terms of MSW management. Solid waste produced has tremendously increased, recycling reduced during the pandemic, and 20–30% of household waste has increased due to online shopping, consumption, and increased gardening during lockdown (Das et al., 2021a). Waste generated by infected people further contributed to the spread of the virus through the waste (Mol & Caldas, 2020). Limited landfills available, improper collection, sorting infrastructure, and improper disposal are major issues faced by the environment which can cause pollution, global warming, and health risks (Ayilara et al., 2020).

Biosolids absorb infectious agents and thus have a high concentration of pathogens (Abdel-Shafy & Mansour, 2018).

Among the composition of MSW plastics and biosolids can be recycled and organic household waste can be used to produce fertilizers (Ragazzi et al., 2020). Implementing the principles of a circular economy helps waste management and their transformation into usable products (Elleuch et al., 2018). Sorting waste into different bins according to their categories is the first step in a safe handling strategy. Methods like chemical disinfection of waste before sorting and storage for around 9 days before recycling can help in preventing the spread of diseases and pathogens (Sharma et al., 2020).

Incineration is the process by which waste is subjected to high temperatures to kill harmful pathogens. Waste is loaded in a bunker using a using overhead crane to avoid contamination, and using a feeding chute, the waste is then incinerated. Pyrolysis is another way to treat MSW rather than using landfills (Perrot & Subiantoro, 2018).

Used electronics, gloves, papers, and masks from infected people need to be separated as hazardous waste from normal waste to reduce the spread of pathogens and viruses during pandemics (Lin et al., 2020b).

For sewage waste anaerobic digestion, aerobic digestion, chlorine treatment, and dewatering can reduce pathogenic agents. These processed biosolids can then be used in thermal hydrothermal liquefaction. This process uses high pressure and moderate temperature to mimic the natural process of fossil fuel production and bio-crude oil from biomass in a few hours rather than in a million years.

2.2 *Industrial Waste*

Wastes are generated in industries like the food industry (leftover seeds, peels, and inedible materials) and the agro-industry. These industries consist of waste produced during harvest, and processing of raw products like crops, cereals, meat, and poultry (sawdust, rice straw, wastewater, fiber of coconut, skin, horns, features, blood, hairs, bones, shells, etc.).

Paper and pulp industry waste includes metals, glass, fibers, primary paper sludge, and large amounts of solid paper sludge. When these wastes are degraded and discarded without proper management, they cause high risk of accumulation of hazardous gases in air and contamination of the aquatic ecosystem (Mohammad et al., 2021). To overcome this problem, different strategies can be employed.

Recycling and reuse of waste by feeding them to animals is an efficient way to utilize these by-products; the compounds present in these like gluten, fiber, corn feed, beet pulp, and sunflower meal have high nutrition value and can be used in the animal feed industry (Gaur et al., 2020). Composting waste is another important way to recycle this waste. Using organic agro- and food industrial waste as biofertilizer can overcome the problem of ammonia toxicity due to using manure as fertilizer (Correddu et al., 2020).

The traditional method of in situ burning which produces smoke and CO_2 can be changed to four-phase composting process. This process consists of the mesophilic, thermophilic, cooling, and maturation phases; this will sanitize the compost and reduce the C: N ratio. Though solid waste management requires the policy of 3R (reduce, reuse, and recycle), there is an addition of the fourth R, i.e., "recovery"; this mentions about the by-products produced which can be further used for other useful purposes contributing to a circular economy (Schieber, 2017).

Valorization is another significant process for disposal and remediation, to produce value products from discarded residues (Das et al., 2019).

Waste hydrolysis is followed by saccharification to produce bioethanol, a commercial operational process to use waste from agro-industry (Guan et al., 2016). It was observed that solid-state fermentation produces diesel with reduced potassium (<90%) and increased cellulose (>57%) and other superior properties. Wastes from the paper and pulp industry are treated through fermentation and saccharification to produce liquid ethanol, and through dark fermentation and thermomechanical pulping to produce biohydrogen. Wastewater were reported to contain fatty acids, residual gums, and aromatic compounds; they are employed to produce soaps and candles; biosurfactants are produced by wastewater from sugarcane and oil mills (Welz, 2019).

2.3 Textile Waste

The rising textile industries have doubled their production rate, though there has been 40% deduction in the age of the clothes, among which 73% of the discarded waste is burned on land or landfilled, 12% is recycled, and 1% is reused to produce new garments. Among the textile fibers, cotton is globally used (90%) cellulosic fiber. Types of fibers include plant, animal, and mineral fibers, among which natural fiber is cotton and man-made fiber is viscose. Recycling of textile waste can provide a new phase toward waste management, but sorting issues due to different colors and fibers make the process challenging.

Sorting is mainly unreliable as it is done manually on the basis of dye, fiber type, and quality of the fiber; to overcome this issue, a complement could enhance the sorting procedure which will use Fourier-transform infrared spectroscopy (FTIR) (Nayak et al., 2015). RFID tags and 2D barcodes can be used for particularly texting identification and near-infrared spectroscopy (NIRS) for sorting fibers into cotton, nylon, polyester, wool, etc. (Girfanova & Abdyrasulova, 2021).

Pretreatment—Concentrated acid hydrolysis, dilute acid hydrolysis, and enzymatic hydrolysis are treatments to obtain sugars and polymer by-products from textiles. Using 85% phosphoric acid, 79.2% sugar was recovered by hydrolysis under standard conditions. Sodium hydroxide when combined with urea at low temperature efficiently dissolves cellulose from waste textiles.

The *n*-methylmorpholine *N*-oxide (NMMO) process separates cellulose from fibers and facilitates the production of biogas by increasing enzymatic hydrolysis. For wool fiber waste, the combination of enzymes, urea, and alkaline hydrolysis are steps of wool recycling.

Biodegradability is influenced by factors such as crystallinity, hydrophobicity, degree of polymerization, and the type of microbes present in the soil; accordingly, natural fibers are degraded faster than synthetic fibers. Fungi degrade fibers faster from the inside out, bacteria degrade the outer layer first, and cellulose found in cotton is more susceptible to microbial degradation. Natural dye-used textiles are degraded more easily than synthetic due to their complexity.

Biodegradation of synthetic fibers like PET and nylon has been challenging for which special microbes and enzymes have been identified for breaking down these polymers.

Organic recycling of textile waste properties:

1. Cellulose a major component of cotton (30–40%) as feedstock.
2. Efficient pretreatment methods to increase product yield.

3. Different fermentation techniques and usages of microbes like *Saccharomyces cerevisiae*.
4. Viscose textile was reported to show higher ethanol yield.
5. Anaerobic digestion yields biogas.
6. Optimum substrate-to-inoculum ratio yields higher methane.
7. Production of biofuel, methanol, ethanol, cellulose enzyme, and glucose recovery (Nikolić et al., 2017).

2.4 Agricultural Waste

Agricultural waste management faces challenges due to the high scale of waste production, pollution, health risks, and lack of infrastructure for proper dumping and disposal. In overcoming challenges waste minimization should be practiced, for this precision farming can be employed, and for removing organic waste and utilizing it to improve soil health, organic waste should be converted to biofertilizer and compost (Gontard et al., 2018). Anaerobic digestion converts agriculture waste to biogas which can be used as an alternative to fossil fuels to reduce global warming; pyrolysis is an emerging technology that can be used to produce high-value biochar and bio-oil from waste; this is a valuable conversation (Sohn et al., 2018). Agricultural residues are mainly converted to biogas and biofertilizers by anaerobic digestion; technical limitations due to certain issues like low yield by lignin-rich materials, digestate handling, problems in handling feedstocks, and low cost are major challenges. To overcome these strategies implemented are as follows: (1) to enhance digestion certain selective enzymes can be used, and chemical and thermal pretreatment methods can be adopted for better yield (20–30%). (2) Upgrading biogas to remove CO_2 and produce high-quality biomethane and the emergence of techniques like microbial electrolysis cells (MEC) that use electroactive microorganisms on organic waste to generate biomethane and biohydrogen (Obi et al., 2016).

For converting agro waste to building blocks and usable material, certain challenges are faced, restricted amount of biochemicals, more cost, low quality, and requirement of a standardized valorization process. For these problems certain steps can be taken, such as advanced extraction processes yield high-quality bioactive compounds, and polyphenols can be recovered from agro waste. In place of harmful bisphenol, aromatic lignin can be used as a biopolymer subunit for bioplastics, a sustainable alternative to plastics. Polyhydroxyalkanoates can be produced from agricultural waste as a substitute for conventional plastics (Bolzonella et al., 2018).

2.5 Biomedical Waste

Biomedical waste is waste generated in healthcare settings; managing this waste is a significant challenge due to the health risks and concerns regarding patients, doctors, nurses, cleaners, etc.; it involves training workers in clinical settings how to deal with biomedical waste, management infrastructure, and public awareness (Vanapalli et al., 2021). Human health and safety are prioritized over the environment during the time of pandemic; if this waste is not properly disposed of, it leads to environmental health hazards in the future.

People should follow special guidelines in healthcare settings; colored dustbins should be used for sorting waste and collecting and transporting waste (Das et al., 2021b).

Mobile incinerators have been an innovative idea to manage biomedical waste during the pandemic. Larger capacity incinerators could be introduced with the capacity of 50–200 kg/h which could increase the treatment ability of the incinerators.

For the safety purpose of people during the pandemic, this is needed for the development of systems where workers are fewer in number and maximum work is automated and should include AI to reduce manual burden (Shammi et al., 2021).

Hydrogen peroxide vapors, UV lamps, dry heat, moist heat, 75% ethanol, gamma irradiation, and ethylene oxide are some disinfectant approaches for recycling and reuse of PPEs; this would reduce waste generated and encourage decontamination techniques, recycling, and reuse. Reusable masks can also be used for waste minimization.

The method of co-processing biomedical waste in cement kilns is an effective method to decrease the usage of nonrenewable energy resources such as coal and limestone, in this method high temperature of 1450 °C is used to burn waste that leaves no residue, the inorganic portion of the waste gets mixed in cement, acidic gases produced during this process are neutralized by the alkaline environment in the kiln (Liu et al., 2021). Future generations, i.e., students, should be aware of the environmental impacts of BMW, and programs should be launched to make the public aware of health hazards and risk posed by BMW on the environment.

2.6 Plastic Waste

Plastic waste is a growing concern due to rising consumption by people, production by industries, and its usage in almost all the sectors of packaging, healthcare, and electronics. Plastics are the most dangerous form of waste as they are not biodegradable and break down into microplastics. Due to their shape, size, resistance, and other factors, they come in contact with pollutants, become more hazardous, and accumulate in land, air, and water. Globally 15% of plastic waste is recyclable while 85% is still incinerated (Gazzotti et al., 2022). As these are nonbiodegradable, they pose long-term environmental risks to animals, humans, and aquatic animals. Adopting strategies of 4R, i.e., reducing, reusing, recycling, and recovering, helps in minimizing waste production and sustainability managing waste. Many areas lack the facilities for waste recycling and management which causes the accumulation of large debris of waste; strategies include the production of bioplastics to reduce the usage and reliance on traditional plastics. Despite efforts taken to encourage people to waste management, recycling and sorting activities remain slow for public awareness programs, and policies like deposit return schemes can encourage active recycling behavior (Huang et al., 2022).

Recent reports have classified plastic waste management strategies into six forms:

1. Landfills—The process includes digging soil and burning waste in the soil, the process is often considered the least good among other strategies due to their contribution to land (soil), air, and water, and this could be hazardous if they leach out in the soil contaminating the soil and the plants growing in that area. Landfills can also contaminate water bodies and air by methane emissions (Huang et al., 2022).
2. Pyrolysis—This process converts waste material into valuable products like gas, oil, and char by thermal decomposition in high temperatures and anaerobically.
3. Recycling—Mechanical and chemical processing breaks down plastic waste into molecules and monomers for reformation into new reusable material.
4. Road construction and tar—Plastic waste is converted to tar that can be used for road construction for high resistance and durability.
5. Concrete production—Plastic waste is utilized for the formation of concrete with better properties like reduced density and corrosion resistance which are better alternatives for traditional aggregates for reducing environmental pollution. Plastic-infused concrete has been shown to have low thermal conductivity making it useful for construction of thermal insulating rooms (Gopinath et al., 2020).
6. Liquefaction—Liquefaction uses moderate temperature and pressure to convert plastic waste into liquid hydrocarbons which are used for the formation of fuels. Hazardous and more contaminated plastics are mainly used for liquefaction (Zhang et al., 2018).

2.7 Sanitary Waste

Sanitary waste refers to sanitary pads, baby diapers, tampons, and hygienic absorbents of bodily fluids. Super absorbent polymers are used in the formation of these sanitary products which are not managed properly due to various reasons like social stigma, shyness among girls and women, lack of awareness, inadequate infrastructure, and lack of policy for handling sanitary waste.

The following strategies can be employed (Kaur et al., 2018):

1. Companies should write about the details of the composition of the product so that its disposal and appropriate technology can be used for its management.
2. Easily manageable and cost-effective materials should be used for pad production.
3. To facilitate decomposition eco-friendly chemicals should be used in the production of sanitary products.
4. Toilets should be properly designed and maintained with a sanitary napkin-vending machine.

5. Sanitary waste should be disposed of away from other types of waste in different color bins.
6. New policies and awareness programs should be implemented for free pad distribution in schools and colleges; the government should not impose any GST on sanitary products, with safe and proper disposal of sanitary waste.

2.8 E-Waste

Electronic waste or e-waste are discarded devices like televisions, mobile phones, refrigerators, discarded cells, etc. E-waste management strategies include reducing and discarding electronic devices before time, reusing them, and recycling them to reduce their harmful impacts (Islam et al., 2024). Formal e-waste management strategies include collecting them, sorting them according to their category and testing them, refurbishing, or reusing them, disassembly, recycling, and sanitary landfills (Borthakur et al., 2019).

Other management strategies include extended producer responsibilities like design enhancement, product quality enhancement, consumer responsibilities, take-back system (TBS), and other facilities for collecting e-waste, processing, and disposal. Take-back systems include flow from the consumer to the retailer or secondhand shops, collection, plant, and material recycling. Informal e-waste management includes collection, reuse, disassembly, recycling, and landfilling or open burning (Ranasinghe & Athapattu, 2020).

Way forward for sustainable waste management strategies include (Borthakur, 2023; Giri & Adhikari, 2020):

1. Environment-friendly handling by reducing their consumption by re-evaluation for new products.
2. Implementing recycling bins for e-waste.
3. Enhancing the durability of electronic devices.
4. Strict laws and policies for electronics generation and processing.
5. Motivating people to differential collections of organic waste and e-waste.
6. Reduce, reuse, reject, design, reconsider, recycle, and recover are the 7Rs system for significant management components.

3 Green Handling of Waste

3.1 Reduce, Reuse, Recycle (3Rs)

Reducing the production of waste is the primary step under eco-friendly methods. It has been reported that preventive strategies can decrease the amount of waste produced in both the industrial and consumer sectors (Kabirifar et al., 2020). Products need to be designed for better durability and resistance so that their shelf life can be increased and waste generation decreased; also packing methods should be modified to reduce any unnecessary packaging to reduce plastic consumption. Awaring customers about the harmful impacts of disposing of single-use products, and other goods should be implemented. Awareness programs should be held to encourage people to reduce food waste, use reusable containers, and eliminate the trend of buying new products before the old ones are exhausted.

Reuse is another way to eliminate waste production, remanufacturing, and refurbishing techniques to extend the shelf life of devices like refurbishing mobile phones, laptops, and other electronic devices to reuse them rather than discarding them (Nie et al., 2023). Used products are upcycled to transform them into higher value better products, such as the transformation of old wood into new furniture not only extends its life but also reduces the demand for raw material, thus minimizing consumption. Promoting donations of onetime used items and online platforms that facilitate the system of exchanging, swapping, renting, or reusing clothes, furniture, and electronic devices significantly enhances the reuse process. Recycling is the process of turning waste into valuable goods. A closed-loop system for the total recycling of waste is needed.

3.2 Waste Minimization and Sorting

Waste is an environmental problem along with a huge amount of economic loss. Some methods employed are (Mostaghimi & Behnamian, 2023):

1. Manufacturing products with better durability and resistance have long lifespans.
2. Product generation should be cleaner and safer so that nothing extra is wasted and harmed during the production process.
3. Improved designs of machines, products, or their covers.
4. Making the most use of the resources.
5. Toxic materials should be eliminated as they make the waste management and eco-friendly handling process more difficult.
6. Recycling should be facilitated by using only those materials that can be later broken down without much harm to nature.
7. Awareness programs should be conducted; consumption style and pattern should be changed to minimize waste.

Waste sorting is a crucial process in the management of waste before treatment or disposal. It includes different separation techniques such as (Pinto & Stecca, 2021):

1. Manual separation—bulky and large items like large rocks, wood, and cloth etch are handpicked and removed from the waste debris; the process requires time consumption and is labor intensive.
2. Air separation—the method uses large moving fans to remove light objects like paper and plastic.

3. Mills and shredders—hanging hammers and rollers are used to reduce the size of the waste; waste materials are broken down into small pieces.
4. Magnetic and nonmagnetic separation—this technique removes metals by using electromagnets, although all metals do not get separated in magnetic separators like copper, and aluminum; for them, current separators are used.

3.3 Composting

Composting is the process of breaking down organic waste into nutrient-rich compost by microorganisms under controlled conditions of oxygen, moisture, and pH to increase soil fertility and growth. There are different techniques of composting (Manea et al., 2024; Rao & Parsai, 2023). There are different types of composting techniques followed according to the type, nature, and amount of waste generated (Fig. 2):

1. Onsite composting—This method employs composting of small amounts of organic waste like waste food generated in small quantities; in this method the waste food is kept in a yard, and layers of waste are covered with soil and turned occasionally; this cannot be employed for large amounts of animal and municipal waste.
2. Anaerobic composting—This method includes a pit or a sealed container from which oxygen is removed and sealed and waste is loaded with anaerobic micro-

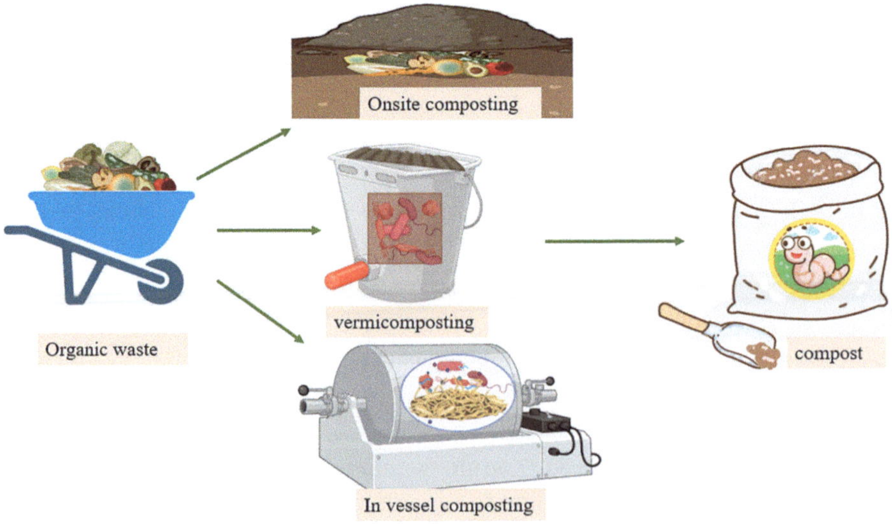

Fig. 2 Organic waste is converted to compost by strategies such as on-site (aerobic and anaerobic), vermicompost, and in-vessel composting

organisms. It breaks down the waste anaerobically naturally over time and produces biogas.
3. Aerobic composting—It is also known as turning composting and is suitable for large amounts of waste generated in communities and municipalities. The method includes forming 14–16 ft piles or windrows of waste which are aerated periodically by turning them; this method produces a significant amount of compost. Another form of aerated composting is static pile composting in which piles are not turned, and piles of wood chips and newspapers are used in making the layers for aeration purposes.
4. Vermicomposting—In this method, red worm bins are used having pounds of red worm wigglers which break down the organic waste material to form nutrient-rich worm castings. The castings are used for increasing fertility in the garden and can also be used as potting soil.
5. In-vessel composting—In this method any type of waste (animal, yard, kitchen) can be used, and is a faster process than other methods. It uses a rolling drum in which waste material is fed along with microbes and optimum temperature, pH, air, and moisture are maintained; the method is fast, better, and odorless.

3.4 Solid Waste Management

Solid waste includes all the organic, inorganic, and electronic waste produced in consumer and industrial settings. Eco-friendly handling techniques include disposal in different color bins, landfilling, incineration, recovery, and gasification. Disposal of wastes like paper, glass, organic, plastics, and electronics in five different color bins helps in eco-friendly handling of waste; landfilling is the process in which a large amount of waste debris is collected and buried in the ground; different engineering techniques are employed to improve their efficiency and cause less harm to the area and the groundwater; area method involves layering the waste on the ground and covering it with earth on regular intervals; trenches are dug in low-lying areas and is space efficient. For hilly and mountainous areas, the ramp method is employed, where waste is layered and compacted along the slope progression; each of these methods is applied according to its area of need. Incineration is the method in which a large amount of waste is burned at a very high temperature to reduce its volume and harmful nature, in the presence of controlled conditions in which heat produced during combustion is used to generate electricity (Mohod & Bagal, 2023). The incinerators are engineered with filtration systems to reduce the toxic pollutants; also, scrubbers and electrostatic precipitators are attached to remove harmful gaseous particles. This method reduces waste volume by up to 90% and also produces energy.

Waste recovery is another method to exact useful components from it; there are two methods, first one is primary recovery in which the materials are used without much change like the recovery of cans, and the second method includes melting of plastic to make new items from it; organic recovery and energy recovery are also another component. Advanced technologies like plasma gasification are used for syngas production from waste using plasma torches at temperatures 4000–10,000 °C; due to this extremely high-temperature, syngas is produced with the least amount of air pollution and nontoxic residues are obtained which are further used for construction (Kaur et al., 2023).

3.5 *Liquid Waste Management*

Eco-friendly handling of liquid waste is the management, handling, treatment, and disposal of liquid waste; it includes domestic sewage waste, industrial discharge, municipality wastewater, and toxic substances. Steps under eco-friendly handling include (Sniatala et al., 2023):

1. Liquid waste is characterized into different forms so that proper treatment methods can be chosen.
2. Liquid waste mainly hazardous waste is prone to harm workers if there are leaks or spills. Wastewater should be segregated according to its type (hazardous and nonhazardous), stored in leakproof containers, labeled, and transported using a vacuum truck with proper monitoring.
3. Treatment of wastewater includes primary, secondary, and tertiary treatment stages. These are processed in improved water treatment plants with nine-stage processors (bar screen, grit chamber, settling tank, aeration tank, sludge digestor, sludge pump, sludge dryer, settling tank, and chlorination tank). They are divided into three stages: the first is in which filtration, sedimentation, and coagulation take place, the second stage includes removing biological organic contaminations, breaking BOD (biochemical oxygen demand), and TSS (total suspended solids), and in the third stage, the water is disinfected using chlorine, UV, or ozone (Fig. 3).

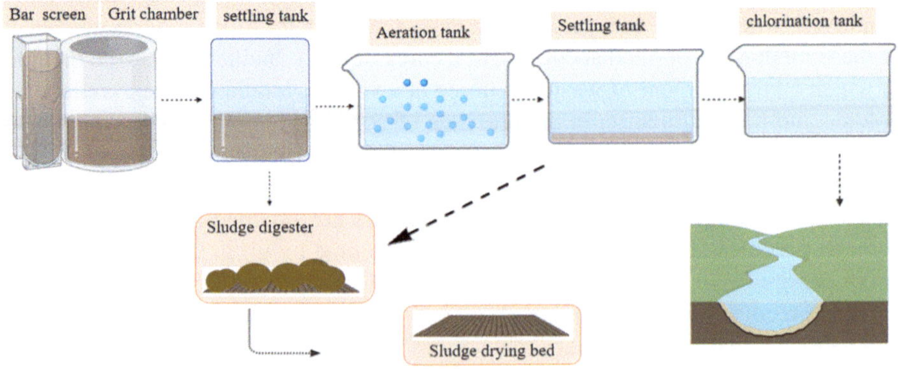

Fig. 3 Steps of wastewater treatment technique

4. Disposal of the treated water in the municipality, for agriculture purposes, injection wells, further treatment methods to produce energy or extract macronutrients, and other uses.

4 Waste-to-Value Conversion

Different categories of waste are utilized for the production of value-added products. An account of types of waste used, sources or the subtype, valuable products obtained, and their wide applications are given in Table 1.

4.1 Kitchen Waste Value Conversion

Household and kitchen waste can be converted into a wide range of valuable products such as (Sindhu et al., 2019; Sharma et al., 2022; Wu et al., 2021):

1. Production of activated carbon absorbent used in the grape industry, with additional zinc chloride activation.
2. Antioxidant production from organic waste like potato peel, grape seeds, and fatty food used in the food and cosmetic industry for preservation.
3. Fish and shellfish waste used for bioactive production acts as a beneficial agent for antimicrobial, anticancer, and antidiabetic properties.
4. Kitchen waste like pineapple peel for bioethanol production, starchy waste for bioethanol, castor seeds, and pepper seeds for biodiesel formation, organic waste for biogas, and orange peels for the production of bioelectricity.
5. Biomaterials are of significant importance, urban food waste and potato waste conversion to biopolymer. Bio-nanocomposite like epoxy resin blends is produced from waste vegetable oils. Shrimp shell waste is used for the production of chitosan which is a polymer of *N*-acetylglucosamine.
6. Tomato waste from the canning industry is used for the production of corrosion inhibitors.
7. Potato chip industries were reported as a suitable source of docosahexaenoic acid (DHA) which acts as antidiabetic, antihypertensive, and anticancer.
8. Kitchen waste is used for the production of enzymes like amylase, cellulase, protease, pectinase, and xylanase.
9. Organic acids are produced mainly from fruit product processing industries such as fumaric acid from apple waste, citric acid, succinic acid, lactic acid from bakery waste, propionic acid, and gluconic acid.

Table 1 Types of wastes, sources, valuable product outputs, and their applications

Sl. no	Type of waste	Source	Product output	Application	References
1.	Kitchen waste	Fruit peel waste Fish waste Pineapple peel Potato waste Tomato waste	Carbon absorbent Bioactive compound Bioethanol Biopolymer Corrosion inhibitors	Grape industry Antimicrobial, anticancer, and antidiabetic As fuel Production of chitosan Used on corrosive substances as a protectant	Sindhu et al. (2019)
2.	Hazardous waste	Nonbiodegradable masks Plastics	HHV biochar 3DGFs Aromatic oils	Used in electronic and filtration devices due to its lightweight Used in the formation of filter film Used for fragrance	Luo et al. (2021)
3.	Biomass	Sweet potato peel and corn residue Lignocellulose biomass	Citric acid, aspartic acid Sugarcane bagasse, sorghum	Heat, fuel, char Biogas, syngas	Usmani et al. (2021)
4.	Fish waste	Fish bones, scales, gills, skin, and internal organ waste	Collagen Peptides Chitin Oil Enzyme	Beverage, food, and cosmetic industry Antihypertensive, neuroprotective properties, etc. Antibacterial, antifungal derivatives Animal feed Production of fish sauce and silage cod uracil glycosylase	Mishra et al. (2023)
5.	Agro waste	Agricultural residues, straws, husks, stalks Paper and pulp from the paper industry Forest by-products	Nanocellulose	Used in sustainable packaging, and biomedical use for drug delivery, wound dressing, coating, and film-making purposes	Yu et al. (2021)
6.	Glycerol	The by-product of biodiesel production	1,3-propanediol Hydrogen Propanoic acid SCO Butanol Glyceric acid Citric acid	Manufacture of polymer Fuel As a pesticide and solvent Source of lipids Polymer, solvent, lubricant Flavoring agent, preservative Acidulant and emulsifier	Chilakamarry et al. (2021)

4.2 *Hazardous Waste High-Value Utilization*

Hazardous materials like face masks which have made up most of the garbage during the COVID-19 pandemic can be used to produce high HHV biochar, high-quality three-dimensional graphene films (3DGFs), and aromatic bio-oil (Luo et al., 2021). Pyrolysis of face masks forms fewer pollutants than normal combustion. Co-pyrolysis of HB and mask produces gaseous products for the formation of 3DGFs. Masks can be converted to biochar by pyrolysis which in turn can be used as fuels and for improving the health of the land and carbon sequestration. Bio-oil an aromantic high-quality product is used as a renewable energy resource that can be formed into different types of fuels. Chemical treatment for the conversion of hazardous materials like plastic and masks to lightweight graphene foam which is used in electronic and filtration devices due to its lightweight (Jung et al., 2021).

4.3 *Pyrolysis for Value Production*

Waste generated creates impacts toward harmful environmental stresses by both production and disposal of waste; this can be utilized for the production of value-added products by using these raw waste materials. Pyrolysis is a thermochemical process by which wastes can be converted to valuable products (Papari et al., 2021; Ge et al., 2021):

1. Thermochemical pyrolysis converts organic waste to high-quality bio-oil; these bio-oils contain phenols, aldehydes, and other compounds which can be further used in a wider range of different pharmaceutical products in industries.
2. Biochar is a valuable resource with a high carbon content. It is formed by processing biomass through pyrolysis. It serves as a good-quality fertilizer and carbon sequester agent that improves the water retention capacity of the soil and is used for bioremediation of land.
3. Gases produced during pyrolysis of agricultural waste can be used as fuels in turbines or engines, they are called carbon monoxide syngas, and also microwave pyrolysis forms biofuels from these agro wastes.
4. Carbon nanotubes can be produced by pyrolysis of plastic waste, which is a valuable material in electronics, material science, and nanotechnology due to its flexibility and conductivity.
5. 3D graphene foam is formed from plastic waste through thermochemical pyrolysis. It is lightweight and highly conductive.

4.4 *Biomass to Biofuel and Bioplastic*

Biomass mostly contains agricultural and forestry by-products which account for major significant share of annually produced renewable sources (Usmani et al., 2021). These are agricultural biomass residues and livestock materials that are daily produced materials of the world. These waste biomasses can be converted into some valuable products:

1. Chemicals and organic acids—waste biomass is used in the production of succinic acid, a biobased chemical from waste wheat residues and livestock; citric acid and aspartic acid from sweet potato peel and corn residue; methanol from forest biomass; and lignin from wood chips. Production of these compounds needs a pretreatment process in which biomass is converted to hydrolysate and then to organic acids and chemicals (Dessie et al., 2020).
2. Lignocellulose biomass is the major significant resource for the generation of heat, fuel, and char; agricultural waste biomass like sugarcane bagasse, corn, and sorghum biomass can be used for the production of ethanol in large amounts in industrial settings which reduces global warming (10% ethanol consumption decreases 12–19% emission of greenhouse gas). Biogas and synthetic gas (syngas) are also formed from gasification of biomass waste (Dessie et al., 2020).
3. Biomass waste is used for the formation of homopolymers like poly-3-hydroxybutyrate (PHB) which has a high-quality potential of bioplastic. Alkaline pretreatment of rice straw produced high-quality, better yield of sugars and homopolymer for bioplastic production (Al-Battashi et al., 2019).

4.5 Chemicals from Waste

The synthesis of eco-friendly, nonhazardous chemicals from waste is known as green chemistry. It uses processes like aerobic digestion, anaerobic digestion, and enzymatic hydrolysis to produce ethanol from agricultural residues, biogas from livestock waste, and organic acids; these are sustainable resources and a good alternative to fossil fuels (Mishra et al., 2023).

Production of green solvents for chemical production from waste expresses the goal of waste management and reduction of environmental impacts. Supercritical fluids (SCFs) produced from waste materials is a green solvent containing water, CO_2, neoteric, deep eutectic, bio-based solvents, and supramolecular solvents.

SCFs are used in industries for dying fibers, downstream processing, and analysis of contaminated soil. These solvents are used for the formation of phenolic compounds from fruit peels like grapes, papaya, and oranges; phytochemicals and phenols from soya bean and olive waste; and solanesol and saponins from tobacco and *Agave salmiana* bagasse (Mishra et al., 2023). Ionic liquids (ILs) can be used for extraction of bioactive compounds from agro waste including cellulose, lactic acid, levulinic acid from the husk of coconut, wheat, and rice. Acetone-butanol-ethanol fermentation converts lignocellulose residue of agricultural waste to butanol having wide applications as fuel and chemical. The glycerol by-product of diesel can be converted to succinic acid which can be further used in food preservation, resin, and polymer industries (Antonetti et al., 2017).

4.6 *Fish Waste to the Valuable Resource*

The development of fisheries techniques has given rise to massive increments in aquaculture along with the accumulation of large quantities of fish wastes. It has been reported that two-thirds of fisheries are discarded as waste posing significant environmental concerns along with huge economic loss. Fish waste needs to be managed properly along with valuable conversation into useful products; these are done by using different extraction processes like acid alkaline hydrolysis, enzymatic hydrolysis, fermentation, chemical extraction, and deacetylation, followed by further processing by ultrafiltration, nanofiltration, gel filtration, chromatography, mass spectrometry, FTIR, and X-ray diffraction, to obtain collagen, peptides, chitin, oil, and enzymes (Coppola et al., 2021):

1. Collagen derived from fish waste can be used in the beverage, food, and cosmetic industries; collagen is partially hydrolyzed or denatured to obtain gelatin. Fish collagen contains some superior qualities like high biocompatibility, unique amino acid residues, hydroxyproline, omega-3 fatty acids, antioxidant properties, wound healing properties, low cytotoxicity, and inflammatory responses (Xu et al., 2020).
2. Fish peptides obtained from fish waste have antimicrobial activity and antihypertensive and neuroprotective properties. Fish peptides are used in cosmetic industries due to their wide application in skin care and anti-photoaging properties. These peptides have been reported to show enhanced blood glucose levels, benefit bone health, and have therapeutic effects in oncology (Lin et al., 2020a).
3. Fish chitin from fish scales contains unique properties like antifungal and antibacterial; chitosan derivatives of chitin having flexibility and resistance are used for fruit and vegetable preservation. Chitin is used as a cleaning agent in the water filtration system due to its effective function in removing contamination from water (Yadav et al., 2019).
4. Fish oil extracted from the processing of fish wastes is used as animal feed as they have a high quantity of omega-3 fatty acids which provides improved health. They are used as nutraceuticals for managing diseases like asthma, diabetes, and cardiovascular disease. Fish oil from fish liver mainly contains antioxidant properties containing phenolics and carotenoids. Fish oil is a good fat-liquoring agent for leather formation (Alfio et al., 2021).
5. Fish enzymes and protease states for 60% of global industrial enzyme market. Salt-activated fish gastric proteases are used in the production of fish sauce, and silage cod uracil glycosylase is a fish enzyme Atlantic cod having high molecular biology applications. Fish proteases are used in laundry as a cleaning agent due to their ability to break down proteins making them better cleaning agents. Fish waste from sea bass is converted to fish lipase for the production of structured lipids and specific ethers (Coppola et al., 2021).

4.7 Nanocellulose: Agro Waste to Value

Wastes like agricultural residues, straws, husks, and stalks; paper and pulp from the paper industry; and food and forest by-products are all major sources for extraction of nanocellulose. Extraction methods include (Kamel et al., 2020; Yu et al., 2021):

1. Chemical treatment removes lignin and hemicellulose for its isolation and further processing.
2. Mechanical milling and grinding breakdown of cellulose fiber into nano dimensions.
3. Enzymatic hydrolysis selectively degrades non-cellulosic components to process nanocellulose; all these steps prove the purity and yield. Nanocellulose is used in sustainable packaging, and biomedical use for drug delivery, wound dressing, coating, and film-making purposes.

Other improvement modifications include:

1. Tempo oxidation method including steps like selective oxidation, chemical treatments, energy efficiency, and mechanical defibrillation.
2. Enzymatic pretreatment by the combination action of endoglucanases and exoglucanases.
3. Preparation of cellulosic nanocrystals like NFC and CNC; surface modifications of nanocellulose.

4.8 Glycerol to Bioethanol

Glycerol forms 10% of the by-product of biodiesel production, containing salt, methanol, and fatty acids. This glycerol is a waste and can be utilized for the production of valuable products such as (Chilakamarry et al., 2021; Anitha et al., 2016):

1. Individual and industrial purposes widely use ethanol; it is used as a solvent and for medical purposes for antiseptic and antidote for poisons. Ethanol is used for cleaning purposes and blended in gasoline for vehicle fuel.
2. 1,3-Propanediol is used in the manufacture of polymers and the cosmetic industry.
3. Hydrogen is produced by microbial fermentation of waste glycerol, used as fuel, and reduces pollutants associated with fossil fuels.
4. Propanoic acid is a preservative and solvent used as a solvent, for pesticides, and pharmaceuticals. This conversion highlights the promise of glycerol as an excellent feedstock contributing to sustainability.
5. Single-cell oil produced from waste glycerol is a valuable source of lipids. Fed-batch fermentation process is used for better yield of SCO.

6. Butanol is produced from glycerol and used as a valuable alternative fuel, used in further production of plastics, polymers, solvents, lubricants, etc. It is also used as a flavoring agent.
7. Waste glycerol is processed to form glyceric acid used in flavoring, preservation, and skincare.
8. Citric acid is produced in the biodiesel industry by using the by-products which are further used as an acidulant, emulsifier, and in the beverage industry; strains of *Yarrowia lipolytica* are used for more yield.

5 Conclusion

Waste management is essential for reducing pollution, harmful effects of waste on human health, and also their conversation into value-added resources. There are many strategies to overcome the challenges faced during waste management such as municipal solid waste, industrial waste, textile waste, agricultural waste, biomedical waste, plastic waste, sanitary waste, and e-waste. For eco-friendly handling of waste, better-designed strategies include 3Rs (Reduce, Reuse, Recycle), waste minimization and sorting, composting, solid waste management, and liquid waste management techniques.

Waste-to-value conversion is the conversion of waste sources into valuable resources. Kitchen waste is utilized for the production of organic acids and antioxidants; hazardous waste like masks which have made up much garbage during the pandemic can be utilized for fuels. An account of pyrolysis is given; it is a process by which many valuable products are made from waste, biomass to biofuel conversion; waste-to-green chemical conversion is the utilization of waste to make non-toxic chemicals. Fish waste is valuable waste that can be fully utilized for the production of a list of products like collagen, peptides, oils, enzymes, etc. Also, nanocellulose and bioethanol are produced from agricultural waste and waste glycerol from diesel production plants.

References

Abdel-Shafy, H. I., & Mansour, M. S. M. (2018). Solid waste issue: Sources, composition, disposal, recycling, and valorization. *Egyptian Journal of Petroleum, 27*, 1275–1290. https://doi.org/10.1016/j.ejpe.2018.07.003

Abubakar, I. R., Maniruzzaman, K. M., Dano, U. L., et al. (2022). Environmental sustainability impacts of solid waste management practices in the global south. *International Journal of Environmental Research and Public Health, 19*, 12717. https://doi.org/10.3390/ijerph191912717

Ahmad Khorairi, A. N. S., Sofian-Seng, N.-S., Othaman, R., et al. (2023). A review on agro-industrial waste as cellulose and nanocellulose source and their potentials in food applications. *Food Reviews International, 39*, 663–688. https://doi.org/10.1080/87559129.2021.1926478

Al-Battashi, H. S., Annamalai, N., Sivakumar, N., et al. (2019). Lignocellulosic biomass (LCB): A potential alternative biorefinery feedstock for polyhydroxyalkanoates production. *Reviews in Environmental Science and Biotechnology, 18*, 183–205. https://doi.org/10.1007/s11157-018-09488-4

Alfio, V. G., Manzo, C., & Micillo, R. (2021). From fish waste to value: An overview of the sustainable recovery of Omega-3 for food supplements. *Molecules, 26*, 1002. https://doi.org/10.3390/molecules26041002

Amasuomo, E., & Baird, J. (2016). The concept of waste and waste management. *Journal of Management and Sustainability, 6*, 88. https://doi.org/10.5539/jms.v6n4p88

Anitha, M., Kamarudin, S. K., & Kofli, N. T. (2016). The potential of glycerol as a value-added commodity. *Chemical Engineering Journal, 295*, 119–130. https://doi.org/10.1016/j.cej.2016.03.012

Antonetti, E., Iaquaniello, G., Salladini, A., et al. (2017). Waste-to-chemicals for a circular economy: The case of urea production (waste-to-urea). *ChemSusChem, 10*(912), 920. https://doi.org/10.1002/cssc.201601555

Ayilara, M., Olanrewaju, O., Babalola, O., & Odeyemi, O. (2020). Waste management through composting: Challenges and potentials. *Sustainability, 12*, 4456. https://doi.org/10.3390/su12114456

Bolzonella, D., Battista, F., Cavinato, C., et al. (2018). Recent developments in biohythane production from household food wastes: A review. *Bioresource Technology, 257*, 311–319. https://doi.org/10.1016/j.biortech.2018.02.092

Borthakur, A. (2023). Design, adoption and implementation of electronic waste policies in India. *Environmental Science and Pollution Research International, 30*, 8672–8681. https://doi.org/10.1007/s11356-022-18836-5

Borthakur, A., Govind, M., & Singh, P. (2019). Inventorization of E-waste and its disposal practices with benchmarks for depollution: The global scenario. In *Electronic waste management and treatment technology* (pp. 35–52). Elsevier.

Chen, D., Zhang, Y., Xu, Y., et al. (2022). Municipal solid waste incineration residues recycled for typical construction materials—A review. *RSC Advances, 12*, 6279–6291. https://doi.org/10.1039/d1ra08050d

Chilakamarry, C. R., Sakinah, A. M. M., Zularisam, A. W., & Pandey, A. (2021). Glycerol waste to value added products and its potential applications. *Systems Microbiology and Biomanufacturing, 1*, 378–396. https://doi.org/10.1007/s43393-021-00036-w

Coppola, D., Lauritano, C., Palma Esposito, F., et al. (2021). Fish waste: From problem to valuable resource. *Marine Drugs, 19*, 116. https://doi.org/10.3390/md19020116

Correddu, F., Lunesu, M. F., Buffa, G., et al. (2020). Can agro-industrial by-products rich in polyphenols be advantageously used in the feeding and nutrition of dairy small ruminants? *Animals, 10*, 131. https://doi.org/10.3390/ani10010131

Das, S., Lee, S.-H., Kumar, P., et al. (2019). Solid waste management: Scope and the challenge of sustainability. *Journal of Cleaner Production, 228*, 658–678. https://doi.org/10.1016/j.jclepro.2019.04.323

Das, A. K., Islam, M. N., Billah, M. M., & Sarker, A. (2021a). COVID-19 and municipal solid waste (MSW) management: A review. *Environmental Science and Pollution Research, 28*, 28993–29008. https://doi.org/10.1007/s11356-021-13914-6

Das, A. K., Islam, M. N., Billah, M. M., & Sarker, A. (2021b). COVID-19 pandemic and healthcare solid waste management strategy—A mini-review. *Science of the Total Environment, 778*, 146220. https://doi.org/10.1016/j.scitotenv.2021.146220

Dehal, A., Vaidya, A. N., & Kumar, A. R. (2022). Biomedical waste generation and management during COVID-19 pandemic in India: Challenges and possible management strategies. *Environmental Science and Pollution Research International, 29*, 14830–14845. https://doi.org/10.1007/s11356-021-16736-8

Dessie, W., Luo, X., Wang, M., et al. (2020). Current advances on waste biomass transformation into value-added products. *Applied Microbiology and Biotechnology, 104*, 4757–4770. https://doi.org/10.1007/s00253-020-10567-2

Elleuch, B., Bouhamed, F., Elloussaief, M., & Jaghbir, M. (2018). Environmental sustainability and pollution prevention. *Environmental Science and Pollution Research International, 25*, 18223–18225. https://doi.org/10.1007/s11356-017-0619-5

Elsayed, M., Eraky, M., Osman, A. I., et al. (2024). Sustainable valorization of waste glycerol into bioethanol and biodiesel through biocircular approaches: A review. *Environmental Chemistry Letters, 22*, 609–634. https://doi.org/10.1007/s10311-023-01671-6

Gaur, V. K., Sharma, P., Sirohi, R., et al. (2020). Assessing the impact of industrial waste on environment and mitigation strategies: A comprehensive review. *Journal of Hazardous Materials, 398*, 123019. https://doi.org/10.1016/j.jhazmat.2020.123019

Gazzotti, S., De Felice, B., Ortenzi, M. A., & Parolini, M. (2022). Approaches for management and valorization of non-homogeneous, non-recyclable plastic waste. *International Journal of Environmental Research and Public Health, 19*, 10088. https://doi.org/10.3390/ijerph191610088

Ge, S., Yek, P. N. Y., Cheng, Y. W., et al. (2021). Progress in microwave pyrolysis conversion of agricultural waste to value-added biofuels: A batch to continuous approach. *Renewable and Sustainable Energy Reviews, 135*, 110148. https://doi.org/10.1016/j.rser.2020.110148

Girfanova, L. R., & Abdyrasulova, R. R. (2021). Improvement of textile waste sorting processes. *IOP Conference Series: Earth and Environmental Science, 666*, 022027. https://doi.org/10.1088/1755-1315/666/2/022027

Giri, J., & Adhikari, R. (2020). Urgency of proper E-waste management plan in Nepal: An overview. *Nepal Journal of Science and Technology, 19*, 107–118. https://doi.org/10.3126/njst.v19i1.29790

Gontard, N., Sonesson, U., Birkved, M., et al. (2018). A research challenge vision regarding management of agricultural waste in a circular bio-based economy. *Critical Reviews in Environmental Science and Technology, 48*, 614–654. https://doi.org/10.1080/10643389.2018.1471957

Gopinath, K. P., Nagarajan, V. M., Krishnan, A., & Malolan, R. (2020). A critical review on the influence of energy, environmental and economic factors on various processes used to handle and recycle plastic wastes: Development of a comprehensive index. *Journal of Cleaner Production, 274*, 123031. https://doi.org/10.1016/j.jclepro.2020.123031

Guan, W., Shi, S., Tu, M., & Lee, Y. Y. (2016). Acetone–butanol–ethanol production from Kraft paper mill sludge by simultaneous saccharification and fermentation. *Bioresource Technology, 200*, 713–721. https://doi.org/10.1016/j.biortech.2015.10.102

Huang, S., Wang, H., Ahmad, W., et al. (2022). Plastic waste management strategies and their environmental aspects: A scientometric analysis and comprehensive review. *International Journal of Environmental Research and Public Health, 19*, 4556. https://doi.org/10.3390/ijerph19084556

Islam, M. K., Khatun, M., & Mourshed, M. (2024). An in-depth analysis and review of management strategies for E-waste in the south Asian region: A way forward towards waste to energy conversion and sustainability. *Heliyon, 10*, e28707. https://doi.org/10.1016/j.heliyon.2024.e28707

Jung, S., Lee, S., Dou, X., & Kwon, E. E. (2021). Valorization of disposable COVID-19 mask through the thermo-chemical process. *Chemical Engineering Journal, 405*, 126658. https://doi.org/10.1016/j.cej.2020.126658

Kabirifar, K., Mojtahedi, M., Wang, C., & Tam, V. W. Y. (2020). Construction and demolition waste management contributing factors coupled with reduce, reuse, and recycle strategies for effective waste management: A review. *Journal of Cleaner Production, 263*, 121265. https://doi.org/10.1016/j.jclepro.2020.121265

Kamel, R., El-Wakil, N. A., Dufresne, A., & Elkasabgy, N. A. (2020). Nanocellulose: From an agricultural waste to a valuable pharmaceutical ingredient. *International Journal of Biological Macromolecules, 163*, 1579–1590. https://doi.org/10.1016/j.ijbiomac.2020.07.242

Kaur, R., Kaur, K., & Kaur, R. (2018). Menstrual hygiene, management, and waste disposal: Practices and challenges faced by girls/women of developing countries. *Journal of Environmental and Public Health, 2018*, 1730964. https://doi.org/10.1155/2018/1730964

Kaur, A., Bharti, R., & Sharma, R. (2023). Municipal solid waste as a source of energy. *Materials Today Proceedings, 81*, 904–915. https://doi.org/10.1016/j.matpr.2021.04.286

Koul, B., Yakoob, M., & Shah, M. P. (2022). Agricultural waste management strategies for environmental sustainability. *Environmental Research, 206*, 112285. https://doi.org/10.1016/j.envres.2021.112285

Kumar, H., Bhardwaj, K., Sharma, R., et al. (2020). Fruit and vegetable peels: Utilization of high value horticultural waste in novel industrial applications. *Molecules, 25*(12), 2812. https://doi.org/10.3390/molecules25122812

Lin, P., Hua, N., Hsu, Y.-C., et al. (2020a). Oral collagen drink for antiaging: Antioxidation, facilitation of the increase of collagen synthesis, and improvement of protein folding and DNA repair in human skin fibroblasts. *Oxidative Medicine and Cellular Longevity, 2020*, 8031795. https://doi.org/10.1155/2020/8031795

Lin, P.-T., Ni, T.-Y., Chen, T.-Y., et al. (2020b). Reducing the consumption of personal protective equipment by setting up a multifunctional sampling station in the emergency department to screen for COVID-19 infection in Taiwan. *Environmental Health and Preventive Medicine, 25*, 34. https://doi.org/10.1186/s12199-020-00874-5

Liu, J., Vethaak, A. D., An, L., et al. (2021). An environmental dilemma for China during the COVID-19 pandemic: The explosion of disposable plastic wastes. *Bulletin of Environmental Contamination and Toxicology, 106*, 237–240. https://doi.org/10.1007/s00128-021-03121-x

Luo, Z., Zhu, X., Deng, J., et al. (2021). High-value utilization of mask and heavy fraction of bio-oil: From hazardous waste to biochar, bio-oil, and graphene films. *Journal of Hazardous Materials, 420*, 126570. https://doi.org/10.1016/j.jhazmat.2021.126570

Manea, E. E., Bumbac, C., Dinu, L. R., et al. (2024). Composting as a sustainable solution for organic solid waste management: Current practices and potential improvements. *Sustainability, 16*, 6329. https://doi.org/10.3390/su16156329

Mishra, K., Siwal, S. S., Nayaka, S. C., et al. (2023). Waste-to-chemicals: Green solutions for bioeconomy markets. *Science of the Total Environment, 887*, 164006. https://doi.org/10.1016/j.scitotenv.2023.164006

Mohammad, A., Goli, V. S. N. S., & Singh, D. N. (2021). Discussion on 'challenges, opportunities, and innovations for effective solid waste management during and post COVID-19 pandemic, by Sharma et al. (2020)'. *Resources, Conservation & Recycling, 164*, 105175. https://doi.org/10.1016/j.resconrec.2020.105175

Mohod, A. V., & Bagal, M. V. (2023). Technological developments in the energy generation from municipal solid waste (landfill gas capture, combustion, pyrolysis and gasification). In *360-Degree waste management, Volume 1* (pp. 139–157). Elsevier.

Mol, M. P. G., & Caldas, S. (2020). Can the human coronavirus epidemic also spread through solid waste? *Waste Management & Research, 38*, 485–486. https://doi.org/10.1177/0734242X20918312

Mostaghimi, K., & Behnamian, J. (2023). Waste minimization towards waste management and cleaner production strategies: A literature review. *Environment, Development and Sustainability, 25*, 12119–12166. https://doi.org/10.1007/s10668-022-02599-7

Nayak, R., Singh, A., Padhye, R., & Wang, L. (2015). RFID in textile and clothing manufacturing: Technology and challenges. *Fashion and Textiles, 2*, 9. https://doi.org/10.1186/s40691-015-0034-9

Nie, Y., Wang, Y., Li, L., & Liao, H. (2023). Literature review on power battery echelon reuse and recycling from a circular economy perspective. *International Journal of Environmental Research and Public Health, 20*, 4346. https://doi.org/10.3390/ijerph20054346

Nikolić, S., Lazić, V., Veljović, Đ., & Mojović, L. (2017). Production of bioethanol from pretreated cotton fabrics and waste cotton materials. *Carbohydrate Polymers, 164*, 136–144. https://doi.org/10.1016/j.carbpol.2017.01.090

Obi, F., Ugwuishiwu, B., & Nwakaire, J. (2016). Agricultural waste concept, generation, utilization and management. *Nigerian Journal of Technology, 35*, 957. https://doi.org/10.4314/njt.v35i4.34

Papari, S., Bamdad, H., & Berruti, F. (2021). Pyrolytic conversion of plastic waste to value-added products and fuels: A review. *Materials, 14*, 2586. https://doi.org/10.3390/ma14102586

Perrot, J.-F., & Subiantoro, A. (2018). Municipal waste management strategy review and waste-to-energy potentials in New Zealand. *Sustainability, 10*, 3114. https://doi.org/10.3390/su10093114

Pinto, D. M., & Stecca, G. (2021). Optimal planning of waste sorting operations through mixed integer linear programming (pp 307–320).

Ragazzi, M., Rada, E. C., & Schiavon, M. (2020). Municipal solid waste management during the SARS-COV-2 outbreak and lockdown ease: Lessons from Italy. *Science of the Total Environment, 745*, 141159. https://doi.org/10.1016/j.scitotenv.2020.141159

Ranasinghe, W. W., & Athapattu, B. C. L. (2020). Challenges in E-waste management in Sri Lanka. In *Handbook of electronic waste management* (pp. 283–322). Elsevier.

Rao, J. N., & Parsai, T. (2023). A comprehensive review on the decentralized composting systems for household biodegradable waste management. *Journal of Environmental Management, 345*, 118824. https://doi.org/10.1016/j.jenvman.2023.118824

Roulia, M. (2024). Sustainable utilization of humic substances and organic waste in green agriculture. *Agriculture, 14*, 115. https://doi.org/10.3390/agriculture14010115

Schieber, A. (2017). Side streams of plant food processing as a source of valuable compounds: Selected examples. *Annual Review of Food Science and Technology, 8*, 97–112. https://doi.org/10.1146/annurev-food-030216-030135

Shammi, M., Behal, A., & Tareq, S. M. (2021). The escalating biomedical waste management to control the environmental transmission of COVID-19 pandemic: A perspective from two south Asian countries. *Environmental Science & Technology, 55*, 4087–4093. https://doi.org/10.1021/acs.est.0c05117

Sharma, H. B., Vanapalli, K. R., Cheela, V. S., et al. (2020). Challenges, opportunities, and innovations for effective solid waste management during and post COVID-19 pandemic. *Resources, Conservation and Recycling, 162*, 105052. https://doi.org/10.1016/j.resconrec.2020.105052

Sharma, A., Kuthiala, T., Thakur, K., et al. (2022). Kitchen waste: Sustainable bioconversion to value-added product and economic challenges. *Biomass Conversion and Biorefinery, 15*, 1749–1770. https://doi.org/10.1007/s13399-022-02473-6

Sindhu, R., Gnansounou, E., Rebello, S., et al. (2019). Conversion of food and kitchen waste to value-added products. *Journal of Environmental Management, 241*, 619–630. https://doi.org/10.1016/j.jenvman.2019.02.053

Sniatala, B., Kurniawan, T. A., Sobotka, D., et al. (2023). Macro-nutrients recovery from liquid waste as a sustainable resource for production of recovered mineral fertilizer: Uncovering alternative options to sustain global food security cost-effectively. *Science of the Total Environment, 856*, 159283. https://doi.org/10.1016/j.scitotenv.2022.159283

Sohn, J., Vega, G. C., & Birkved, M. (2018). A methodology concept for territorial metabolism—Life cycle assessment: Challenges and opportunities in scaling from urban to territorial assessment. *Procedia CIRP, 69*, 89–93. https://doi.org/10.1016/j.procir.2017.10.005

Usmani, Z., Sharma, M., Awasthi, A. K., et al. (2021). Bioprocessing of waste biomass for sustainable product development and minimizing environmental impact. *Bioresource Technology, 322*, 124548. https://doi.org/10.1016/j.biortech.2020.124548

Vanapalli, K. R., Sharma, H. B., Ranjan, V. P., et al. (2021). Challenges and strategies for effective plastic waste management during and post COVID-19 pandemic. *Science of the Total Environment, 750*, 141514. https://doi.org/10.1016/j.scitotenv.2020.141514

Welz, P. J. (2019). Edible seed oil waste: Status quo and future perspectives. *Water Science and Technology, 80*, 2107–2116. https://doi.org/10.2166/wst.2020.043

Wojnowska-Baryła, I., Bernat, K., & Zaborowska, M. (2022). Strategies of recovery and organic recycling used in textile waste management. *International Journal of Environmental Research and Public Health, 19*, 5859. https://doi.org/10.3390/ijerph19105859

Wu, M., Hu, J., Shen, F., et al. (2021). Conceptually integrating a multi-product strategy for the valorization of kitchen waste towards a more sustainable management. *Journal of Cleaner Production, 306*, 127292. https://doi.org/10.1016/j.jclepro.2021.127292

Xu, Y., Duan, L., Li, Y., et al. (2020). Nanofibrillar decellularized Wharton's jelly matrix for segmental tracheal repair. *Advanced Functional Materials, 30*, 1910067. https://doi.org/10.1002/adfm.201910067

Yadav, M., Goswami, P., Paritosh, K., et al. (2019). Seafood waste: A source for preparation of commercially employable chitin/chitosan materials. *Bioresources and Bioprocessing, 6*, 8. https://doi.org/10.1186/s40643-019-0243-y

Yu, S., Sun, J., Shi, Y., et al. (2021). Nanocellulose from various biomass wastes: Its preparation and potential usages towards the high value-added products. *Environmental Science and Ecotechnology, 5*, 100077. https://doi.org/10.1016/j.ese.2020.100077

Zhang, K., Shi, H., Peng, J., et al. (2018). Microplastic pollution in China's inland water systems: A review of findings, methods, characteristics, effects, and management. *Science of the Total Environment, 630*, 1641–1653. https://doi.org/10.1016/j.scitotenv.2018.02.300

Vermicompost Use in Sustainable Agriculture: Impact on Soil Fertility and Plant Growth

Rahul Gogoi, Madhurjya Protim Borah, Anshu, Madhurjya Ranjan Sharma, Madhumita Barooah, and Sudipta Sankar Bora

Abstract Vermicomposting, the use of earthworms to decompose organic waste into quality compost, is an important component of sustainable agriculture because it improves soil health and plant performance. This chapter discusses the benefits of vermicomposting approach toward soil health and agricultural productivity. The chapter opens with a consideration of vermicomposting practices including the selection of organic wastes and the decomposition of biological matter by earthworms. The chapter explains the ways in vermicompost improve soil features, addressing soil structure, fertility, and water-holding capacity. Vermicompost improves soil aeration, and aggregation of soil particles, which promotes root development and less potential for erosion. Vermicompost improves the availability of important macronutrients including nitrogen, phosphorus, and potassium, and improves water-holding capacity of soils, which can be particularly beneficial in drought conditions. In terms of plant performance, vermicomposting has demonstrated to improve seed germination, rooting structure, and overall plant health and yield. Crops grown in vermicompost soils also frequently demonstrate greater resistance to disease and other stressors. The chapter includes illustrative case studies in terms of the valuable functions of vermicomposting around the world and in different farming applications, with a focus on demonstrating the versatility and capacity of vermicomposting. The chapter also considers the challenge of scaling up vermi-

R. Gogoi · M. R. Sharma · M. Barooah
Department of Agricultural Biotechnology, Assam Agricultural University, Jorhat, Assam, India

M. P. Borah
Department of Biosciences and Bioengineering, Indian Institute of Technology Jammu, Jammu, Jammu and Kashmir, India

Anshu
Department of Medical Oncology, All India Institute of Medical Sciences, Delhi, India

S. S. Bora (✉)
Silchar Regional Forensic Science Laboratory, Ghungur, Silchar, Assam Under the Directorate of Forensic Science, Kahilipara, Guwahati, Assam, India

S. Rajpoot, A. P. Das (eds.), *Innovative Technologies for Waste Management*, Sustainable Environmental Waste Management Strategies,
https://doi.org/10.1007/978-3-031-97184-6_17

composting and optimizing processes associated with composting earthworms. The chapter concludes by reiterating the constructive agronomic and ecological aspects of vermicomposting with regard to agricultural sustainability through soil health, crop productivity, and involvement with ecological sustainability.

Keywords Vermicomposting · Nutrient cycling · Soil structure · Agricultural productivity · Environmental sustainability

1 Introduction

Sustainable agriculture includes farming practices that meet present demands for food and textiles while protecting future generations' capacity to provide for themselves. Earthworms and other microbes break down organic materials in a natural process called vermicomposting (Fig. 1), like food scraps and agricultural residues, producing a nutrient-rich product called vermicompost (Rehman et al., 2022). In recent decades, this method has garnered a lot of interest for its ability to convert

Fig. 1 Schematic representation of vermicomposting

biological waste into valuable resources. Vermicomposting is a natural process where organic materials are broken down by earthworms and other bacteria. Given the escalating challenges of environmental degradation, waste management, and food security, vermicomposting emerges as a practical and environmentally responsible substitute for conventional trash disposal techniques and synthetic fertilizers, in accordance with sustainable development goals (Mohite et al., 2024).

The primary function of vermicomposting comprises, turning organic waste into a substance that resembles humus substance that is filled with helpful bacteria, needed plant nutrients, and compounds that encourage growth. The favorable qualities of vermicompost render it an extraordinary amendment for soil, as it helps to improve the structure and fertility of the soil, subsequently facilitating improved plant growth and resilience against diseases (Lim et al., 2015). In addition, using vermicompost has been linked to a decrease in reliance on pesticides and chemical-based fertilizers, both of which are usually linked to environmental contamination and soil degradation. By leveraging the inherent capacity of earthworms to digest organic wastes, vermiculture offers a feasible alternative for waste management that encourages the return of nutrients to the soil environment (Cruz et al., 2024).

A fundamental difficulty confronting modern agriculture is the excessive dependence on synthetic fertilizers. While these fertilizers are successful in improving immediate crop yields, they are damaging to soil quality and contribute to environmental damage over time. The runoff from fertilized fields may result in nutrient accumulation in aquatic settings, leading to eutrophication and severe consequences on aquatic ecosystems. Additionally, the misuse of chemical fertilizers has the capacity to disrupt soil microbial populations, which are essential for maintaining soil health and encouraging the cycling of nutrients (Pahalvi et al., 2021). In contrast, vermicompost acts as a slow-release nutrient source that not only stimulates plant growth but also fosters microbial diversity and activity inside the soil. Consequently, the incorporation of vermicompost into farming practices helps the long-term sustainability of farming systems by boosting soil health and lowering dependency on synthetic fertilizers (Pathma & Sakthivel, 2012).

Vermicomposting fulfils a dual role, enhancing not only soil fertility but also numerous critical soil qualities that promote sustainable agriculture operations. This process enhances soil structure, leading to greater aggregation and porosity. These enhancements promote deeper root penetration and better water infiltration. The altered soil structure enhances moisture retention, lowering irrigation requirements and boosting plant resistance to drought (Ullah et al., 2024). Moreover, vermicomposting encourages biological activity within the soil by providing an environment receptive to beneficial bacteria. These microbes perform a key role in digesting organic debris, fixing nitrogen, and limiting the growth of plant diseases. These benefits collectively lead to increased soil fertility, higher crop yields, and enhanced plant health (Nazli et al., 2020).

Furthermore, vermiculture plays a decisive role in supporting sustainable waste management techniques through the recycling of organic waste, which would otherwise contribute to increased landfill buildup and environmental deterioration (Lirikum et al., 2022). Organic waste, covering agricultural by-products, food remains, and yard detritus, forms a large component of the garbage generated in both urban and rural environments. By directing this trash into vermiculture systems, the organic material is turned into a useful resource, thereby minimizing greenhouse gas emissions and other environmental problems associated with landfill disposal. Consequently, vermiculture aligns with circular economy principles, wherein resources are repurposed and waste creation is avoided (Bhatia & Sindhu, 2024).

Vermicomposting procedures are the key to successful conversion of organic waste by earthworms into compost that is rich in nutrients. This chapter addresses the essential stages, materials, and methodologies connected with successful vermicomposting and highlight its adaptability to diverse scales and contexts.

2 Vermicomposting Practices

Vermiculture is used in many contexts than merely agriculture. The use of vermicomposting as a waste management technique in both municipal and industrial settings has drawn increased interest in recent years. Nowadays, large-scale vermiculture systems are being used to handle a variety of organic wastes, including food waste, sewage sludge, and animal dung (Sharma & Garg, 2018). Additionally, in urban environments, vermiculture has proven to be a highly successful technique of composting organic waste generated by both residential and commercial entities, providing a feasible substitute for traditional landfill-based waste management practices.

2.1 Selection of Organic Wastes

Selecting the organic waste is one of the most crucial processes in the vermicomposting process. Fruit and vegetable peels, crop leftovers, yard debris, coffee grounds, and animal manure are some examples of suitable materials. These elements provide the nutrition and organic matter required for earthworm metabolism and reproduction (Maharjan et al., 2022). In contrast, some substrates should be avoided since they may be harmful to earthworm survival and could interfere with the vermicomposting process as a whole. These include citrus peels, onions, garlic, and food waste that has too much acidity or salinity (Wang et al., 2024a). Additionally, organic wastes from industrial sources, such as paper sludge, food processing leftovers, and brewery waste, are being used more and more. Pre-composting these materials helps to stabilize their composition while reducing the dangers of toxins, heavy metals, and dangerous microorganisms. Furthermore, pre-composting improves earthworm digestive efficiency and reduces ammonia toxicity, creating a condition that is favorable for earthworm activity (Gaur et al., 2020).

2.2 Role of Earthworms

Earthworms are the key biological agents used in the vermicomposting practice; thus, choosing the right species is essential. Because of their quick reproduction and capacity to adapt well to the breakdown of organic materials, epigeic species—in particular, *Eisenia fetida*, sometimes known as red wigglers, and *Lumbricus rubellus*—are preferred. After consuming organic materials, these earthworms process them in their digestive systems to produce castings that are rich in humus (Walia & Kaur, 2024d). The resultant castings, sometimes referred to as vermicast, are distinguished by their high content of nutrients that are available to plants and their microbial activity. Studies reveal that vermicast has greater levels of nitrogen, phosphorous, potassium, and other vital micronutrients than regular compost (Chatterjee et al., 2020). Additionally, earthworms aid in the vermicompost's inoculation with advantageous bacteria, strengthening its favorable effects on soil health.

2.3 Methods and Techniques

A variety of techniques that are adapted to particular needs can be used to implement vermicomposting in a range of resource conditions and scales. The windrow system is a cost-effective technique that is especially well-suited for large-scale applications like urban trash management and substantial agricultural activities. It involves placing organic waste in long rows on the ground and adding earthworms (Alshehrei & Ameen, 2021). On the other hand, bed systems—which are made of stationary materials like concrete, wood, or other long-lasting materials—are frequently used in medium-sized operations because they are simple to monitor. Bin systems, usually composed of wood or plastic, provide a small and controllable solution for domestic or small-scale applications, which makes them ideal for urban environments (Shukla et al., 2024).

Maintaining ideal conditions is essential to promoting successful vermicomposting, regardless of the approach selected. In particular, the moisture content should be maintained between 60% and 80% because the earthworms may suffer from both too much dryness and too much saturation. For active earthworm activity, temperatures between 15 and 25 °C were optimum. Adequate aeration is also necessary to avoid anaerobic conditions and the development of disagreeable odors (Walia & Kaur, 2024c).

2.4 Processing Time and Harvesting

Several factors influence the vermicomposting process's duration, including the type of earthworm used, the type of organic waste, and the environmental conditions. In order to produce mature vermicompost, the procedure usually takes

2–3 months to complete (Yatoo et al., 2020). The final compost is separated from the earthworms during the harvesting stage. Common manual separation techniques used in small-scale operations include exposing the compost to light, which encourages the earthworms to burrow deeper and makes it easier to collect them. In contrast, in order to increase productivity, larger enterprises frequently use mechanical sifters or separators (Walia & Kaur, 2024b).

2.5 Quality Control

Providing high-quality vermicompost is crucial to maximizing its use in agriculture. The main criteria for assessing quality include NPK (nitrogen, phosphorus, and potassium) in a balanced ratio that corresponds to the nutritional requirements of different crops; a high degree of microbial diversity and activity that indicates compost maturity and biological efficacy; mature vermicompost that is free of harmful pathogens, which guarantees its safety for use in food production; and maintaining an ideal pH and low electrical conductivity because a neutral pH and lower salinity make vermicompost suitable for a variety of crops (Gill et al., 2023).

2.6 Integration into Farming Systems

Many sustainable agriculture strategies can be used with vermicomposting. By increasing soil organic matter and providing a natural source of nutrients, it supports organic farming. Additionally, vermicompost's ability to improve soil structure and water retention benefits agroforestry practices. According to Reddy (2016), vermicompost can be successfully added to integrated nutrient management systems along with other organic and inorganic inputs to achieve balanced crop nutrition. Vermicomposting's adaptability to different agricultural settings demonstrates how useful it is as a tool for increasing sustainability and productivity.

3 Impact on Soil Health

As a natural and efficient way to restore and maintain productive soils, vermicomposting significantly improves the fertility of soil, which has an effect on soil health and structural integrity and fostering a healthy soil ecosystem, both of which are consistent with the tenets of sustainable agriculture.

3.1 Improvement in Soil Structure

By improving aggregation and reducing compaction, vermicompost significantly enhances the configuration of soil constituents, such as organic matter, silt, clay, and sand. The excreta produced by earthworms, commonly referred to as earthworm castings, function as natural agents for binding soil. According to Singh et al. (2020), these castings help create stable micro- and macroaggregates, which create a friable and aerated environment that promotes root penetration and water infiltration. Additionally, vermicompost reduces soil compaction, especially in clay-rich soils that are prone to hardpan formation. As Blouin et al. (2019) point out, this happens through lowering bulk density and enhancing aggregate stability, which makes the soil more cultivable. Furthermore, vermicompost produces stable aggregates that strengthen the soil's resilience to erosion, reducing sediment displacement and surface runoff. The preservation of the topsoil layer, which is necessary to sustain agricultural productivity, depends on this process.

3.2 Improvement of Fertility in the Soil

The fertility of the soil determines the availability of nutrients required for plant development. By augmenting macronutrients and micronutrients in bioavailable forms, vermicompost contributes significantly to increased soil fertility. The three primary nutrients needed for plant growth are potassium (K), phosphorus (P), and nitrogen (N) which are rich in this organic amendment (Shrivastav et al., 2020). Vermicompost releases these vital nutrients gradually, in contrast to synthetic fertilizers that could cause nutrient leaching or runoff. For instance, the nitrogen included in vermicompost is easily absorbed by plant roots as nitrates and ammonium (Saleem et al., 2023). Vermicompost offers essential micronutrients like boron, copper, manganese, and zinc in addition to macronutrients. Although these micronutrients are essential for enzymatic processes and plant health in general, intensively farmed soils often lack them. Additionally, vermicompost enhances the cation exchange capacity (CEC) of the soil, improving the exchange and retention of nutrients. This enhancement guarantees better fertilizer availability for plants and lowers nutrient losses from leaching (Assunção et al., 2022).

3.3 Increase in Water-Holding Capacity

A major barrier to agricultural productivity is water scarcity, particularly in regions that are arid or semiarid. Nonetheless, it has been demonstrated that adding vermicompost greatly improves the soil's capacity to hold onto water. Similar to a sponge, the high organic matter content—particularly the humus component—absorbs and stores water before gradually releasing it to plant roots (Ingrao et al., 2023). This feature is particularly helpful in sandy soils that are prone to quick drainage; in these situations, vermicompost serves as a water reservoir, reducing the need for frequent watering and encouraging the preservation of vital water resources. Additionally, improved water retention makes it possible for crops to endure drought conditions better. Compared to those using conventional techniques, farmers who integrate vermicompost into their farming practices frequently report improved yields during times of water scarcity (Razzaghi et al., 2020).

3.4 Promotion of Soil Microbial Activity

The decomposition of organic matter, the cycling of nutrients, and the avoidance of disease all depend on the high microbial life abundance seen in healthy soils. Vermicompost is a powerful microbial inoculant that makes it easier to introduce and sustain healthy microorganisms. This material includes a broad range of bacteria, fungi, and actinomycetes that support vital soil functions such as organic matter breakdown, phosphate solubilization, and nitrogen fixation (Minhas et al., 2024). Studies reveal that vermicompost-treated soils exhibit noticeably higher microbial biomass and activity levels than untreated soils (Uz & Tavali, 2014). Furthermore, vermicompost contains enzymes such as phosphatase and dehydrogenase, which are required for breaking down complex organic compounds into simpler forms that plants can consume. Vermicompost not only aids in the cycling of nutrients but also suppresses disease by introducing helpful microorganisms like *Pseudomonas* and *Bacillus*. According to Zhang and Cheng (2022), these bacteria have the ability to create antibiotics and other metabolites that inhibit soilborne pathogens, hence reducing the occurrence of illnesses.

3.5 Carbon Sequestration and Organic Matter Enrichment

Vermicomposting provides a key function in the practice of carbon sequestration by preserving organic carbon inside the soil and limiting its emission as carbon dioxide. The humus produced during vermicomposting integrates effortlessly into the organic carbon reservoir of the soil, which helps to the enhancement fertility of soil and structure over the long term (Raza et al., 2024). Furthermore, this technique contributes in the mitigation of climate alteration by trapping atmospheric carbon. In addition, the enhanced organic matter created from vermicompost increases soil fertility by fostering an appropriate atmosphere for microorganisms, consequently boosting availability of nutrient and improving the retention of soil moisture (Bera et al., 2024).

3.6 *Mitigation of Soil Degradation*

Vermicompost is a useful tool in the battle against soil deterioration, tackling important issues such erosion, salinisation, and nutrient loss. Vermicompost application is crucial for re-establishing soil health in areas that have degraded as a result of intense agricultural methods, overgrazing, or deforestation (Mishra & Dash, 2022). Restoring organic matter and vital nutrients is how this restoration is accomplished. Vermicompost application significantly increases yields on degraded fields, according to empirical data gathered from case studies carried out in Africa and India (Walia & Kaur, 2024a). Additionally, by stabilizing pH levels, vermicompost efficiently lowers soil salinity, making it a useful supplement for saline soils where conventional fertilizers usually don't produce good results (Liu et al., 2025).

3.7 *Global Applications*

Vermicompost's benefits are apparent across a range of geographic locations and farming methods, making it a flexible strategy for encouraging sustainable farming. Vermicompost enhances nutrient retention and lowers runoff in tropical regions where nutrient leaching is common. Examples of this can be found among Southeast Asian rice farmers who have reported increased crop yields and reduced use of synthetic fertilizers after using vermicomposting methods (Sharma & Garg, 2023). Additionally, even with little rainfall, the water-retentive qualities of vermicompost have helped to cultivate drought-resistant crops like millet and sorghum in dry and semiarid areas like Rajasthan, India. Furthermore, vermicompost has gained popularity in urban agriculture, especially for rooftop gardens and small-scale vegetable farming, because of its odorless and lightweight nature, which makes it an ideal soil amendment for constrained urban spaces (Saha et al., 2017).

4 Cycling of Nutrients and Their Availability

Nutrient cycling is a basic method in soil ecosystems where nutrients are continually recycled and then made available to plants for growth. In organic farming systems, vermicomposting plays a crucial role in nutrient cycling, mainly because of earthworm activity, which greatly increases soil fertility and results in farming methods that are more sustainable (Nair et al., 2021). Vermicomposting involves earthworms converting organic resources into a nutrient-dense substance called vermicompost from a variety of waste products, such as plant residues, kitchen scraps, and manure. As a result, the soil becomes more accessible to macronutrients like potassium, phosphorus, and nitrogen as well as micronutrients and helpful microbes, leading to the best possible plant performance and overall health (Karimi et al., 2024).

4.1 *Macronutrient Enhancement: Nitrogen, Phosphorus, and Potassium*

Three key macronutrients for plant health are included in the vermicomposting process: nitrogen (N), phosphorus (P), and potassium (K). These nutrients are necessary for numerous physiological functions, such as plant growth, root development, and possibly disease resistance. By mineralizing organic matter, vermicomposting releases these nutrients, making them easily absorbed by plants (Sinha & Tandon, 2020).

4.1.1 Nitrogen (N)

Nitrogen serves as a limiting nutrient in several soil types, consequently affecting plant growth and total production. In the process of vermicomposting, organic nitrogen generated from waste products undergoes conversion into ammonium and nitrate forms. These chemicals are quickly digested by plants, contributing to their development and output. Furthermore, vermicomposting indicates a lower loss of nitrogen compared to alternative composting processes due to diminished volatilization and leaching (Farzadfar et al., 2021).

4.1.2 Phosphorus (P)

Phosphorus plays a significant function in energy transmission, photosynthesis, as well as root formation. Phosphorus bound in organic matter is liberated during vermicomposting as a result of microbial decomposition and made more available by earthworm activity. In order to help break down organic phosphates and increase their availability to plants, earthworms create enzymes. Vermicompost has been shown to have more plant-available phosphorus than traditional compost (Malhotra et al., 2018).

4.1.3 Potassium (K)

By decomposing organic debris and minerals high in potassium in the soil, earthworms help to maintain this nutrient cycle by releasing potassium in an amount that plants can readily absorb. Potassium is necessary for numerous physiological functions in plants, for example, controlling water, activating enzymes, and improving disease resistance. This nutrient is important for the general health of plants, especially with regard to their capacity to endure stress and produce their best yield (Sardans & Peñuelas, 2021).

4.2 *Micronutrient Availability*

Vermicomposting not only raises the amounts of macronutrients but also makes micronutrients like iron, copper, zinc, and manganese more readily available. These micronutrients are nevertheless very important for plant health even if they are needed in comparatively smaller amounts. These micronutrients are made more accessible to plants by earthworms' solubilization and mineralization of them (Manzoor et al., 2024). Additionally, because microbes aid in the subsequent breakdown of micronutrients from soil particles, the increased microbial activity linked to vermicompost significantly adds to nutrient cycling (Jacoby et al., 2017). For example, research has indicated that adding vermicompost to plants increases their iron bioavailability, which is essential for the production of chlorophyll and overall plant development (Atiyeh et al., 2000). Additionally, the copper and zinc found in vermicompost are in forms that plants may easily absorb, providing the nutrients required for cellular processes and enzymatic reactions.

4.3 *Soil Microbiome and Nutrient Cycling*

Earthworms are essential to the cycle of nutrients in vermicomposting by contributing to the soil microbiota. Earthworms aid in the organic molecules' breakdown while also encouraging the growth of a healthy microbial population in the soil. This microbial community, which includes bacteria, fungus, and actinomycetes, is essential for nutrient cycling, decomposition, and the creation of humus (Edwards & Arancon, 2022). Earthworms break down organic debris, creating a soil matrix that promotes microorganism growth. Furthermore, the activity of these microbes causes a surge in nutrient mineralization. Bastida et al. (2021) found that the amount of microbiological biomass in soils treated with vermicomposting is substantially higher than in soils maintained using traditional composting methods. Increased microbial variety leads to enhanced cycling of key elements like nitrogen, phosphorous, and sulfur, expanding the soil's total nutrient availability. Furthermore, several earthworm species contain digestive enzymes that aid in the breakdown of complicated organic materials, increasing their accessibility inside organic waste (Khaldoon et al., 2022).

4.4 *Improvement in Soil Organic Matter and Humus Formation*

Humus, formed by the breakdown of organic waste, is essential for nitrogen cycling in soil ecosystems. Vermicompost, with its high humic content, enhances the availability of nutrients, soil structure, and water retention. Vermicompost contains high quantities of humic acids, which improve soil aggregation, increase soil porosity, and promote root penetration. This improved soil structure, combined with increased microbial activity, leads to more nutrient availability for plant uptake (Ravi et al., 2019). Additionally, humus can chelate and bind important minerals such as calcium, magnesium, and trace elements, reducing the risk of leaching and guaranteeing plants a consistent flow of vital nutrients during the growth season. As a result, this procedure not only increases soil nutrient reserves, but it also helps to maintain a balanced nutrient cycle, minimizing reliance on synthetic fertilizers and promoting long-term soil fertility (Trevisan et al., 2010).

5 Plant Performance in Vermicomposting

The practice of using earthworms to break down organic materials into compost that is rich in nutrients is known as vermicomposting, greatly affecting plant performance. This technology not only gives needed nutrients to the soil but also optimizes its physical and biological features, so encouraging an optimal environment for plant growth (Fig. 2). The benefits of vermicompost are exhibited in different ways, including the increased germination rates of seeds, the formation of robust root systems, greater plant health, higher yields, and enhanced tolerance to environmental stressors (Gupta et al., 2021). This section addresses the contributions of vermicomposting to plant performance, emphasizing the processes via which it stimulates growth and bolsters resilience.

5.1 Seed Germination and Early Growth

A main advantage of vermicomposting is its beneficial effect on seed germination. Numerous plant nutrients, including micronutrients like zinc, copper, and iron as well as macronutrients like potassium, phosphorus, and nitrogen, are found in vermicompost. Furthermore, it comprises growth hormones and enzymes that promote seed development. These nutrients are available to seedlings throughout their earliest growth phases, facilitating faster and more uniform germination relative to traditional soil or chemical fertilizers (Toor et al., 2023). Research has indicated that using vermicompost in seedbeds or as a soil supplement improves germination rates for a range of crops, including vegetables, grains, and ornamental plants. A study by Joshi et al. (2015) demonstrated that vermicompost markedly enhanced the germination rate and initial growth of tomato seedlings compared to other organic and chemical fertilizers. The inclusion of vital nutrients in the compost accelerates germination and promotes the establishment of strong roots and shoots, so creating a stable foundation for future growth stages.

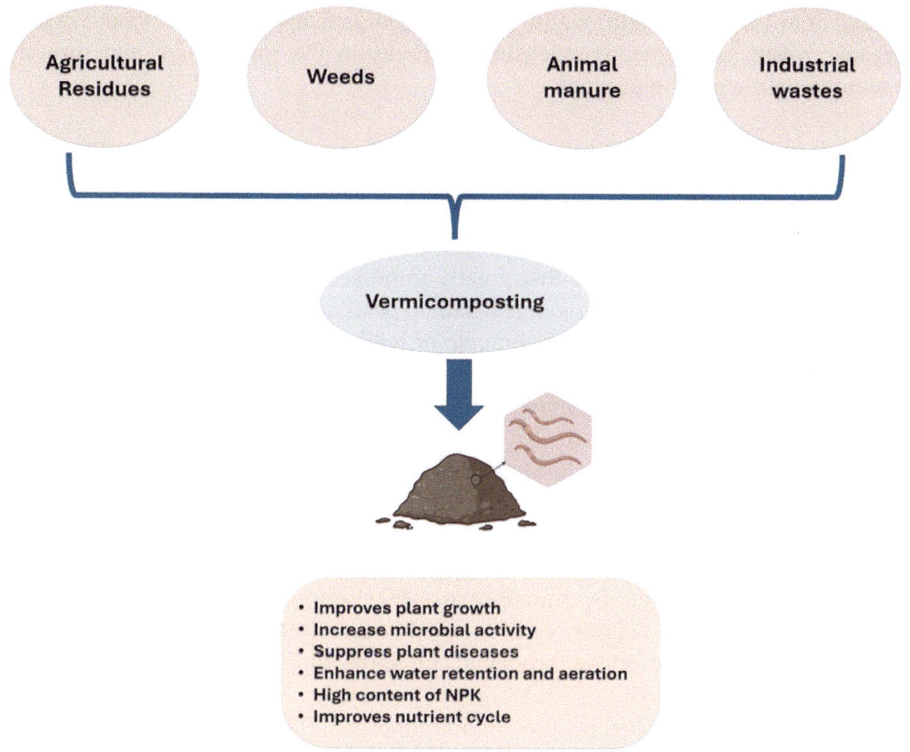

Fig. 2 Organic waste sources that enhances plant growth, water retention, and disease suppression

5.2 *Root Development and Soil Structure*

Vermicomposting substantially enhances soil structure, facilitating root growth. The application of vermicompost promotes soil porosity and aeration, which in turn facilitates enhanced water absorption, a situation that is good for root systems. Additionally, this mechanism supports earthworms in their activities of fragmenting compacted soils, resulting in the construction of simpler routes for the availability of water and minerals for the root zone (Nikmehr et al., 2025). Vermicompost has additionally been shown to improve the toughness and depth of plant roots. Research conducted by Rehman et al. (2023) suggests that crops treated with vermicompost exhibit higher root growth when compared to those cultivated in soil enriched with synthetic fertilizers. A robust root system is connected with greater nutrient absorption and improved drought tolerance, ultimately contributing to overall plant health.

The development of soil's structure results in a rise in microbiological activity, which promotes plant development. Mycorrhizal fungus and nitrogen-fixing bacteria are two examples of these microbes, both of which increase nutrient absorption by plants while also creating symbiotic relationships with the root systems of plants

(Wang et al., 2024b). Additionally, these microorganisms aid in the breakdown of organic waste, consequently releasing nutrients in the rhizosphere and directly encouraging the growth of plants.

5.3 Health and Pest and Disease Resistance of Plants

Vermicomposting is a practice that greatly improves plant nutrition and improves plant health by making plants more resistant to pests and diseases. Beneficial microorganisms that can effectively outcompete or inhibit harmful pathogens proliferate in the soil environment that vermicompost enriches, and the nutrients in vermicompost help to strengthen plant tissues, making them more resilient to disease and stress (Weinmann et al., 2023).

Beneficial microorganisms like bacteria, fungus, and actinomycetes are essential for preventing and mitigation of disease outbreaks. Furthermore, the organic chemicals included in vermicompost have been shown to boost plants' systemic resistance, hence boosting their immunological responses to pathogens (Brar et al., 2024). A study carried out by Kumar et al. (2021) implies that plants cultivated in soil treated with vermicompost demonstrated better disease resistance and general vitality in contrast to those produced in soils supplemented with synthetic fertilizers.

Vermicompost also contains growth hormones and trace elements, such as cytokinins, auxins, and gibberellins, which aid in the development of plants more quickly and become more resilient to stress. In order to maintain healthy plants in the face of environmental stressors, these hormones are crucial for the processes of cell division, elongation, and differentiation (Vijayabharathi et al., 2015). Consequently, plants cultivated on soils supplemented with vermicompost are more robust, healthy, and better able to withstand environmental stresses including drought, temperature fluctuations, and pest infestations.

5.4 Yield Enhancement

The implementation of vermicomposting has various substantial benefits, notably affecting agricultural output. This strategy facilitates the provision of critical nutrients and supports the enhancement of soil structure and root development, all of which are instrumental in enhancing plant yield (Siddique et al., 2024). Numerous research show that using vermicompost can lead to heightened output levels across many crop types, including vegetables, fruits, and grains, by encouraging general plant health and preserving a consistent supply of nutrients. For example, a research conducted by Wang et al. (2017) indicated that the application of vermicompost in tomato growing resulted in significantly improved fruit yields compared to those achieved by traditional fertilizers. Similarly, Saha et al. (2012) revealed that a study focusing on wheat demonstrated how adding vermicompost increased nutrient accessibility and positively impacted grain yield and quality as well as fostering vigorous root growth. Additionally, the increased soil fertility and microbial activity associated with vermicomposting produce an optimal environment for plant growth, hence promoting enhanced yields.

5.5 Resistance to Environmental Stressors

Enhancing the plants' ability to withstand environmental stressors such as drought, salt, and sharp temperature swings, vermicompost application helps to improve plant performance. The improved nutrient availability and superior soil structure found in vermicompost-amended soils are primarily responsible for this greater resilience. Additionally, vermicompost greatly makes the soil more capable of holding onto water, which helps plants cope with water scarcity. This trait makes sure that moisture stays extended periods of time in the soil, which reduces the need for frequent irrigation (Stankovic et al., 2022).

In addition, plants cultivated in soils supplemented with vermicompost demonstrate higher heat tolerance and increased survival rates when subjected to adverse situations. This phenomenon can be accounted for by the substances in vermicompost that promote growth, which activate pathways related with stress response and bolster plant resistance. Consequently, vermicomposting presents a sustainable way to ameliorate the detrimental effects of climate variation and environmental stress on agricultural productivity (Hasanuzzaman et al., 2013).

6 Challenges and Optimization in Vermicomposting

Vermicomposting has received growing attention in recent years as a process for recycling organic waste into high-quality compost, which in turn increases soil health and supports sustainable agriculture practices. Nevertheless, the process meets different problems that require resolution to optimize its effectiveness and scalability (Sharma & Garg, 2018). This section will discuss the key problems associated with vermicomposting and the techniques that may be adopted to optimize the process for both small-scale and large-scale enterprises.

6.1 Challenges in Vermicomposting

Expanding the vermicomposting process to meet the needs of large-scale agricultural operations is a major problem. Large-scale operations require strong infrastructure, even while small-scale vermiculture systems, like home worm bins, may offer simple management techniques. Large-scale earthworm farms need to be designed to maintain ideal environmental conditions, including temperature, moisture, and aeration, all of which are critical for effective decomposition (Wang et al., 2022). Additionally, handling large amounts of organic waste requires careful preparation when it comes to vermicompost collecting, processing, and storage.

6.2 Optimization Strategy

In response to the issues associated with scalability, there is a continuing attempt to build automated systems and mechanized processing processes. The adoption of automatic aeration and temperature control systems serves to maintain appropriate environmental conditions. Additionally, the testing of continuous flow-through systems, which permit the continual entry of organic waste into the processing environment, aims to boost operating efficiency. Furthermore, the deployment of modular systems provides flexible extension, hence enabling the management of trash on a greater scale (Karakus & Durresi, 2017).

The health and productivity of contribute significantly to the efficacy of the vermicomposting process. Several factors, including the species of earthworms employed, their feeding preferences, and their capacity to adapt to varied environmental conditions, considerably affect the overall efficiency of vermicomposting. Furthermore, earthworms are prone to several illnesses, parasites, and environmental stressors, including severe temperatures and moisture imbalances (Toor et al., 2024). For instance, exposure to high-temperature circumstances can create stress in earthworms, resulting in reduced activity or mortality. Such consequences can hamper the efficacy of the composting process.

6.3 Optimization Strategy

The decision-making process of appropriate species of earthworms, specifically *Eisenia fetida*, usually referred to as red wiggler worms, is crucial in boosting the success of vermiculture activities. Furthermore, it is vital to maintain appropriate environmental conditions, which include regulating temperature, managing moisture levels, and maintaining adequate aeration. Such methods are crucial for alleviating stress and promoting the health of earthworms. In addition, researchers are researching the application of probiotics and microbial inoculants to strengthen the resistance of earthworms in the face of illnesses (Ning et al., 2020).

One of the biggest challenges in vermicomposting is the composition of organic waste. Different kinds of organic materials break down at different rates, and an imbalance in the carbon-to-nitrogen (C:N) ratio can negatively affect vermicomposting efficiency. For example, materials with high nitrogen content, like fresh manure or food scraps, can upset this balance and produce unpleasant odors, while materials with too much carbon, like dry leaves or straw, can slow down the decomposition process and reduce the amount of nutrients in the finished compost product (Ratnasari et al., 2023).

6.4 Optimization Strategy

For vermicomposting to be effective, a certain carbon-to-nitrogen (C:N) ratio must be maintained. For most worm species, an ideal C:N ratio between 20:1 and 30:1 is generally considered adequate. Furthermore, effective preprocessing techniques for organic waste, particularly shredding and grinding, can considerably facilitate the breakdown process by increasing the amount of surface area available for earthworm and microbiological activities. Furthermore, the regular monitoring of moisture levels and aeration is required to ensure that the vermicomposting process stays efficient and devoid of unpleasant odors (Ravindran & Mnkeni, 2016).

The assurance of constant quality in vermicompost involves considerable hurdles, particularly when managing substantial quantities of organic waste. Disparities in feedstock, variations in processing conditions, and oscillations in earthworm activity contribute to discrepancies in both the nutrient composition and texture of the end product. Such fluctuation in the quality of vermicompost may discourage farmers and gardeners from depending on it as a trustworthy soil amendment, thereby constraining its commercial practicality (Singha & Deka, 2024).

6.5 Optimization Strategy

Standardizing input materials, processing conditions, and monitoring procedures is necessary to ensure effective quality control. The application of quality control measures, such as routinely evaluating the final product's nutrient content, pH levels, and microbial activity, helps to maintain consistency. Furthermore, investigations on how various feedstocks affect the final vermicompost quality are also ongoing; studies have shown that specific combinations of organic waste may produce compost of higher quality (Alauddin et al., 2023).

7 Conclusion

Vermicomposting is acknowledged as a sustainable and successful method of increasing soil fertility and digesting organic waste. This process converts organic waste into nutrient-rich compost by utilizing earthworms' natural decomposition skills. It offers significant advantages for agricultural and environmental health. As mentioned below, vermicomposting contributes to the construction of soil structure, the facilitation of nutrient cycling, the improvement of plant development, and the encouragement of adaptability to environmental challenges. These attributes made it a crucial practice in sustainable agriculture, particularly in view of the rising demand for ecologically friendly practices and organic farming around the world. Nevertheless, although the benefits linked with vermicomposting, certain problems must be solved to permit its optimal implementation, particularly on a larger scale. Concerns such as the scaling of operations, the preservation of earthworm health and productivity, the management of diverse kinds of organic waste, and the assurance of uniform quality in vermicompost demand rigorous control. Addressing these challenges is crucial to properly grasp the advantages of vermicomposting. The adoption of technical advances, the deployment of automated systems, and the design of new management practices are crucial stages in addressing these difficulties, thereby optimizing the process for both small- and large-scale applications. The continual rise of research and development implies a high promise for vermicomposting. Innovations in the selection of worm species, advancements in waste processing techniques, and the integration of vermiculture with other environmental friendly practices are expected to improve the efficacy and economical feasibility of the process. In addition, the growing need for environmentally friendly and organic farming methods is projected to create new chances for the broad deployment of vermicomposting across different agricultural systems. In summary, vermicomposting acts as an outstanding strategy for raising soil health, improving plant production, and supporting sustainable agriculture practices. With further optimization and innovation, vermicomposting is poised to become a significant component of a more resilient and sustainable farming system.

References

Alauddin, M., Khan, F. I., Amyotte, P. R., & Rayner Brown, K. (2023). Fundamentals of inherently safer design. In *Methods in chemical process safety* (Vol. 7, pp. 23–81). Elsevier. https://doi.org/10.1016/BS.MCPS.2023.05.002

Alshehrei, F., & Ameen, F. (2021). Vermicomposting: A management tool to mitigate solid waste. *Saudi Journal of Biological Sciences, 28*(6), 3284–3293. https://doi.org/10.1016/J.SJBS.2021.02.072

Assunção, A. G. L., Cakmak, I., Clemens, S., González-Guerrero, M., Nawrocki, A., & Thomine, S. (2022). Micronutrient homeostasis in plants for more sustainable agriculture and healthier human nutrition. *Journal of Experimental Botany, 73*(6), 1789–1799. https://doi.org/10.1093/JXB/ERAC014

Atiyeh, R. M., Subler, S., Edwards, C. A., Bachman, G., Metzger, J. D., & Shuster, W. (2000). Effects of vermicomposts and composts on plant growth in horticultural container media and soil. *Pedobiologia, 44*(5), 579–590. https://doi.org/10.1078/S0031-4056(04)70073-6

Bastida, F., Eldridge, D. J., García, C., Kenny Png, G., Bardgett, R. D., & Delgado-Baquerizo, M. (2021). Soil microbial diversity–biomass relationships are driven by soil carbon content across global biomes. *The ISME Journal, 15*(7), 2081–2091. https://doi.org/10.1038/S41396-021-00906-0

Bera, T., Samui, S., Dey, A., & Ankireddypalli, J. (2024). Soil carbon sequestration in the context of climate change. *Advances in Global Change Research, 78*, 63–106. https://doi.org/10.1007/978-981-99-7935-6_3

Bhatia, T., & Sindhu, S. S. (2024). Sustainable management of organic agricultural wastes: Contributions in nutrients availability, pollution mitigation and crop production. *Discover Agriculture, 2*(1), 1–42. https://doi.org/10.1007/S44279-024-00147-7
Blouin, M., Barrere, J., Meyer, N., Lartigue, S., Barot, S., & Mathieu, J. (2019). Vermicompost significantly affects plant growth. A meta-analysis. *Agronomy for Sustainable Development, 39*(4), 1–15. https://doi.org/10.1007/S13593-019-0579-X/FIGURES/9
Brar, B., Bala, K., Saharan, B. S., Sadh, P. K., & Duhan, J. S. (2024). Bio-boosting agriculture: Harnessing the potential of fungi-bacteria-plant synergies for crop improvement. *Discover Plants, 1*(1), 1–24. https://doi.org/10.1007/S44372-024-00023-0
Chatterjee, R., Debnath, A., & Mishra, S. (2020). Vermicompost and soil health. In *Soil health* (pp. 69–88). Springer Nature. https://doi.org/10.1007/978-3-030-44364-1_4
Cruz, G. S. J., Hermes, P. H., Miriam, S. V., & López, A. M. (2024). Benefits of vermicompost in agriculture and factors affecting its nutrient content. *Journal of Soil Science and Plant Nutrition, 24*(3), 4898–4917. https://doi.org/10.1007/S42729-024-01880-0
Edwards, C. A., & Arancon, N. Q. (2022). The role of earthworms in organic matter and nutrient cycles. In *Biology and ecology of earthworms* (pp. 233–274). Springer. https://doi.org/10.1007/978-0-387-74943-3_8
Farzadfar, S., Knight, J. D., & Congreves, K. A. (2021). Soil organic nitrogen: An overlooked but potentially significant contribution to crop nutrition. *Plant and Soil, 462*(1), 7–23. https://doi.org/10.1007/S11104-021-04860-W
Gaur, V. K., Sharma, P., Sirohi, R., Awasthi, M. K., Dussap, C. G., & Pandey, A. (2020). Assessing the impact of industrial waste on environment and mitigation strategies: A comprehensive review. *Journal of Hazardous Materials, 398*, 123019. https://doi.org/10.1016/J.JHAZMAT.2020.123019
Gill, S., Iqbal, M., Jabbar, A., Amin, B. A. Z., Gill, A. N., Ilyas, M. F., Tauqeer, H. M., & Virk, Z. A. (2023). Vermicompost for sustainable future: Nature-based solution for environmental degradation, climate change, and food security. *Climate-Resilient Agriculture, 1*, 549–569. https://doi.org/10.1007/978-3-031-37424-1_25
Gupta, S., Kulkarni, M. G., White, J. F., Stirk, W. A., Papenfus, H. B., Doležal, K., Ördög, V., Norrie, J., Critchley, A. T., & Van Staden, J. (2021). Categories of various plant biostimulants—Mode of application and shelf-life. In *Biostimulants for crops from seed germination to plant development: A practical approach* (pp. 1–60). Elsevier. https://doi.org/10.1016/B978-0-12-823048-0.00018-6
Hasanuzzaman, M., Nahar, K., Alam, M. M., Roychowdhury, R., & Fujita, M. (2013). Physiological, biochemical, and molecular mechanisms of heat stress tolerance in plants. *International Journal of Molecular Sciences, 14*(5), 9643–9684. https://doi.org/10.3390/IJMS14059643
Ingrao, C., Strippoli, R., Lagioia, G., & Huisingh, D. (2023). Water scarcity in agriculture: An overview of causes, impacts and approaches for reducing the risks. *Heliyon, 9*(8), e18507. https://doi.org/10.1016/J.HELIYON.2023.E18507
Jacoby, R., Peukert, M., Succurro, A., Koprivova, A., & Kopriva, S. (2017). The role of soil microorganisms in plant mineral nutrition—Current knowledge and future directions. *Frontiers in Plant Science, 8*, 292271. https://doi.org/10.3389/FPLS.2017.01617/PDF
Joshi, R., Singh, J., & Vig, A. P. (2015). Vermicompost as an effective organic fertilizer and biocontrol agent: Effect on growth, yield and quality of plants. *Reviews in Environmental Science and Biotechnology, 14*(1), 137–159. https://doi.org/10.1007/S11157-014-9347-1/METRICS
Karakus, M., & Durresi, A. (2017). A survey: Control plane scalability issues and approaches in software-defined networking (SDN). *Computer Networks, 112*, 279–293. https://doi.org/10.1016/J.COMNET.2016.11.017
Karimi, S., Raza, T., & Mechri, M. (2024). Composting and vermitechnology in organic waste management. In *Environmental engineering and waste management* (pp. 449–470). Springer. https://doi.org/10.1007/978-3-031-58441-1_16
Khaldoon, S., Lalung, J., Maheer, U., Kamaruddin, M. A., Yhaya, M. F., Alsolami, E. S., Alorfi, H. S., Hussein, M. A., & Rafatullah, M. (2022). A review on the role of earthworms in plastics degradation: Issues and challenges. *Polymers, 14*(21), 4770. https://doi.org/10.3390/POLYM14214770

Kumar, S., Gauraha, A. K., Sharma, S., & Bhaskar, S. K. (2021). An Economic Analysis of Production and Marketing of Vermicompost in Balod District of Chhattisgarh. *International Journal of Current Microbiology and Applied Sciences, 10*(7), 532–538. https://doi.org/10.20546/ijcmas.2021.1007.058

Lim, S. L., Wu, T. Y., Lim, P. N., & Shak, K. P. Y. (2015). The use of vermicompost in organic farming: Overview, effects on soil and economics. *Journal of the Science of Food and Agriculture, 95*(6), 1143–1156. https://doi.org/10.1002/JSFA.6849

Lirikum, Kakati, L. N., Thyug, L., & Mozhui, L. (2022). Vermicomposting: An eco-friendly approach for waste management and nutrient enhancement. *Tropical Ecology, 63*(3), 325–337. https://doi.org/10.1007/S42965-021-00212-Y

Liu, Z., Huang, Y., Xu, L., Li, J., Zhang, L., Zhang, Y., Wang, J., & Chen, Y. (2025). Remediation of soda saline-alkali soil using vermicompost: The remediation mechanisms and enhanced improvement by maize straw. *Journal of Soils and Sediments, 25*(1), 41–58. https://doi.org/10.1007/S11368-024-03953-0/METRICS

Maharjan, K. K., Noppradit, P., & Techato, K. (2022). Suitability of vermicomposting for different varieties of organic waste: A systematic literature review (2012–2021). *Organic Agriculture, 12*(4), 581–602. https://doi.org/10.1007/S13165-022-00413-2/METRICS

Malhotra, H., Vandana, Sharma, S., & Pandey, R. (2018). Phosphorus nutrition: Plant growth in response to deficiency and excess. In *Plant nutrients and abiotic stress tolerance* (pp. 171–190). Springer Singapore. https://doi.org/10.1007/978-981-10-9044-8_7

Manzoor, A., Naveed, M. S., Ali, R. M. A., Naseer, M. A., Ul-Hussan, M., Saqib, M., Hussain, S., & Farooq, M. (2024). Vermicompost: A potential organic fertilizer for sustainable vegetable cultivation. *Scientia Horticulturae, 336*, 113443. https://doi.org/10.1016/J.SCIENTA.2024.113443

Minhas, B., Katoch, P., Singha, A., Minhas, N., & Kaushik, N. (2024). Microbial metabolites and their role to improve soil health. In *Advancements in microbial biotechnology for soil health* (Microorganisms for sustainability) (Vol. 50, pp. 67–95). Springer. https://doi.org/10.1007/978-981-99-9482-3_5

Mishra, P., & Dash, D. (2022). Contributing effects of vermicompost on soil health and farmers' socioeconomic sustainability. In *Trends of applied microbiology for sustainable economy* (pp. 737–757). Elsevier. https://doi.org/10.1016/B978-0-323-91595-3.00005-7

Mohite, D. D., Chavan, S. S., Jadhav, V. S., Kanase, T., Kadam, M. A., & Singh, A. S. (2024). Vermicomposting: A holistic approach for sustainable crop production, nutrient-rich bio fertilizer, and environmental restoration. *Discover Sustainability, 5*(1), 1–13. https://doi.org/10.1007/S43621-024-00245-Y/TABLES/1

Nair, P. K. R., Kumar, B. M., & Nair, V. D. (2021). Soil organic matter (SOM) and nutrient cycling. In *An introduction to agroforestry: Four decades of scientific developments* (pp. 383–411). Springer. https://doi.org/10.1007/978-3-030-75358-0_16

Nazli, F., Najm-ul-Seher, Khan, M. Y., Jamil, M., Nadeem, S. M., & Ahmad, M. (2020). Soil microbes and plant health. In *Plant disease management strategies for sustainable agriculture through traditional and modern approaches* (pp. 111–135). Springer. https://doi.org/10.1007/978-3-030-35955-3_6

Nikmehr, S., Etesami, H., & Rizwan, M. (2025). Improving soil fertility and moisture with organic amendments in arid regions. In *Sustainable agriculture under drought stress: Integrated soil, water and nutrient management* (pp. 333–364). Elsevier. https://doi.org/10.1016/B978-0-443-23956-4.00021-1

Ning, Y., Zhou, H., Wang, S., Li, X., Liu, D., Xu, W., & Zhou, D. (2020). Temporal dynamics of earthworm (Eisenia fetida) microbial communities after cadmium stress based on a compound mathematical model. *Environmental Science and Pollution Research, 27*(14), 16326–16338. https://doi.org/10.1007/S11356-020-08122-7

Pahalvi, H. N., Rafiya, L., Rashid, S., Nisar, B., & Kamili, A. N. (2021). Chemical fertilizers and their impact on soil health. In *Microbiota and biofertilizers, Vol 2: Ecofriendly tools for reclamation of degraded soil environs* (pp. 1–20). Springer. https://doi.org/10.1007/978-3-030-61010-4_1

Pathma, J., & Sakthivel, N. (2012). Microbial diversity of vermicompost bacteria that exhibit useful agricultural traits and waste management potential. *Springerplus, 1*(1), 1–19. https://doi.org/10.1186/2193-1801-1-26/TABLES/2

Ratnasari, A., Syafiuddin, A., Mehmood, M. A., & Boopathy, R. (2023). A review of the vermicomposting process of organic and inorganic waste in soils: Additives effects, bioconversion process, and recommendations. *Bioresource Technology Reports, 21*, 101332. https://doi.org/10.1016/J.BITEB.2023.101332

Ravi, R. K., Anusuya, S., Balachandar, M., & Muthukumar, T. (2019). Microbial interactions in soil formation and nutrient cycling. In *Mycorrhizosphere and pedogenesis* (pp. 363–382). Springer. https://doi.org/10.1007/978-981-13-6480-8_21

Ravindran, B., & Mnkeni, P. N. S. (2016). Bio-optimization of the carbon-to-nitrogen ratio for efficient vermicomposting of chicken manure and waste paper using Eisenia fetida. *Environmental Science and Pollution Research, 23*(17), 16965–16976. https://doi.org/10.1007/S11356-016-6873-0/METRICS

Raza, S. T., Rong, L., Rene, E. R., Ali, Z., Iqbal, H., Sahito, Z. A., & Chen, Z. (2024). Effects of vermicompost preparation and application on waste recycling, NH3, and N2O emissions: A systematic review on vermicomposting. *Environmental Technology & Innovation, 35*, 103722. https://doi.org/10.1016/J.ETI.2024.103722

Razzaghi, F., Obour, P. B., & Arthur, E. (2020). Does biochar improve soil water retention? A systematic review and meta-analysis. *Geoderma, 361*, 114055. https://doi.org/10.1016/J.GEODERMA.2019.114055

Reddy, P. P. (2016). Integrated nutrient management. In *Sustainable intensification of crop production* (pp. 193–206). Springer. https://doi.org/10.1007/978-981-10-2702-4_13

Rehman, A., Farooq, M., Lee, D. J., & Siddique, K. H. M. (2022). Sustainable agricultural practices for food security and ecosystem services. *Environmental Science and Pollution Research, 29*(56), 84076–84095. https://doi.org/10.1007/S11356-022-23635-Z

Rehman, S. U., De Castro, F., Aprile, A., Benedetti, M., & Fanizzi, F. P. (2023). Vermicompost: Enhancing Plant Growth and Combating Abiotic and Biotic Stress. In Agronomy (Vol. 13, Issue 4). https://doi.org/10.3390/agronomy13041134

Saha, S., Dutta, D., Ray, D. P., & Karmakar, R. (2012). Vermicompost and soil quality. In *Farming for food and water security* (pp. 243–264). Springer. https://doi.org/10.1007/978-94-007-4500-1_10

Saha, J. K., Selladurai, R., Coumar, M. V., Dotaniya, M. L., Kundu, S., & Patra, A. K. (2017). Soil and its role in the ecosystem. In *Soil pollution—An emerging threat to agriculture* (pp. 11–36). Springer. https://doi.org/10.1007/978-981-10-4274-4_2

Saleem, S., Mushtaq, N. U., Rasool, A., Shah, W. H., Tahir, I., & Rehman, R. U. (2023). Plant nutrition and soil fertility: Physiological and molecular avenues for crop improvement. In *Sustainable plant nutrition: Molecular interventions and advancements for crop improvement* (pp. 23–49). Elsevier. https://doi.org/10.1016/B978-0-443-18675-2.00009-2

Sardans, J., & Peñuelas, J. (2021). Potassium control of plant functions: Ecological and agricultural implications. *Plants, 10*(2), 419. https://doi.org/10.3390/PLANTS10020419

Sharma, K., & Garg, V. K. (2018). Vermicomposting: A green technology for organic waste management. In *Waste to wealth* (Energy, environment, and sustainability) (pp. 199–235). Springer. https://doi.org/10.1007/978-981-10-7431-8_10

Sharma, K., & Garg, V. K. (2023). Vermicomposting technology for organic waste management. In *Current developments in biotechnology and bioengineering: Advances in composting and vermicomposting technology* (pp. 29–56). Elsevier. https://doi.org/10.1016/B978-0-323-91874-9.00009-7

Shrivastav, P., Prasad, M., Singh, T. B., Yadav, A., Goyal, D., Ali, A., & Dantu, P. K. (2020). Role of nutrients in plant growth and development. In *Contaminants in agriculture: Sources, impacts and management* (pp. 43–59). Springer. https://doi.org/10.1007/978-3-030-41552-5_2

Shukla, B. K., Bharti, G., Sharma, P. K., Sharma, M., Rawat, S., Maurya, N., Srivastava, R., & Srivastav, Y. (2024). Sustainable construction practices with recycled and waste materials for a circular economy. *Asian Journal of Civil Engineering, 25*(7), 5255–5276. https://doi.org/10.1007/S42107-024-01111-Y

Siddique, K. H. M., Bolan, N., Rehman, A., & Farooq, M. (2024). Enhancing crop productivity for recarbonizing soil. *Soil and Tillage Research, 235*, 105863. https://doi.org/10.1016/J.STILL.2023.105863

Singh, A., Karmegam, N., Singh, G. S., Bhadauria, T., Chang, S. W., Awasthi, M. K., Sudhakar, S., Arunachalam, K. D., Biruntha, M., & Ravindran, B. (2020). Earthworms and vermicompost: An eco-friendly approach for repaying nature's debt. *Environmental Geochemistry and Health, 42*(6), 1617–1642. https://doi.org/10.1007/S10653-019-00510-4/METRICS

Singha, W. J., & Deka, H. (2024). Instrumental characterization of matured vermicompost produced from organic waste. In *Earthworm technology in organic waste management: Recent trends and advances* (pp. 231–255). Elsevier. https://doi.org/10.1016/B978-0-443-16050-9.00017-7

Sinha, D., & Tandon, P. K. (2020). An overview of nitrogen, phosphorus and potassium: Key players of nutrition process in plants. In *Sustainable solutions for elemental deficiency and excess in crop plants* (pp. 85–117). Springer. https://doi.org/10.1007/978-981-15-8636-1_5

Stankovic, S., Baptista, P., Carillo, P., & Bandurska, H. (2022). Drought stress responses: Coping strategy and resistance. *Plants, 11*(7), 922. https://doi.org/10.3390/PLANTS11070922

Toor, M. D., Kizilkaya, R., Ullah, I., Koleva, L., Basit, A., & Mohamed, H. I. (2023). Potential role of vermicompost in abiotic stress tolerance of crop plants: A review. *Journal of Soil Science and Plant Nutrition, 23*(4), 4765–4787. https://doi.org/10.1007/S42729-023-01476-0

Toor, M. D., Basit, A., Okorie, B., Nath, D., Din, M. M. U., Kumar Verma, P., Sajjad, S., Ullah, I., Yousef, H. N., & Mohamed, H. I. (2024). Earthworms as catalysts for climate-resilient agriculture: Enhancing food security and water management in the face of climate change. *Water, Air, & Soil Pollution, 235*(12), 1–35. https://doi.org/10.1007/S11270-024-07576-6

Trevisan, S., Francioso, O., Quaggiotti, S., & Nardi, S. (2010). Humic substances biological activity at the plant-soil interface. *Plant Signaling & Behavior, 5*(6), 635–643. https://doi.org/10.4161/PSB.5.6.11211

Ullah, M. S., Malekian, R., Randhawa, G. S., Gill, Y. S., Singh, S., Esau, T. J., Zaman, Q. U., Afzaal, H., Du, D. L., & Farooque, A. A. (2024). The potential of biochar incorporation into agricultural soils to promote sustainable agriculture: Insights from soil health, crop productivity, greenhouse gas emission mitigation and feasibility perspectives—A critical review. *Reviews in Environmental Science and Bio/Technology, 23*(4), 1105–1130. https://doi.org/10.1007/S11157-024-09712-4

Uz, I., & Tavali, I. E. (2014). Short-term effect of vermicompost application on biological properties of an alkaline soil with high lime content from Mediterranean region of Turkey. *The Scientific World Journal, 2014*(1), 395282. https://doi.org/10.1155/2014/395282

Vijayabharathi, R., Sathya, A., & Gopalakrishnan, S. (2015). Plant growth-promoting microbes from herbal vermicompost. In *Plant-growth-promoting rhizobacteria (PGPR) and medicinal plants* (pp. 71–88). Springer. https://doi.org/10.1007/978-3-319-13401-7_4

Walia, S. S., & Kaur, T. (2024a). Beneficial role of vermicompost: Nutrient content in vermicompost and success stories. In *Earthworms and vermicomposting* (pp. 135–146). Springer Singapore. https://doi.org/10.1007/978-981-99-8953-9_13

Walia, S. S., & Kaur, T. (2024b). Harvesting of vermicompost. In *Earthworms and vermicomposting* (pp. 109–116). Springer Singapore. https://doi.org/10.1007/978-981-99-8953-9_9

Walia, S. S., & Kaur, T. (2024c). Preparation of vermicompost. In *Earthworms and vermicomposting* (pp. 73–87). Springer Singapore. https://doi.org/10.1007/978-981-99-8953-9_7

Walia, S. S., & Kaur, T. (2024d). Role of earthworms in vermicomposting. In *Earthworms and vermicomposting* (pp. 55–60). Springer Singapore. https://doi.org/10.1007/978-981-99-8953-9_5

Wang, X. X., Zhao, F., Zhang, G., Zhang, Y., & Yang, L. (2017). Vermicompost improves tomato yield and quality and the biochemical properties of soils with different tomato planting history in a greenhouse study. *Frontiers in Plant Science, 8*, 279043. https://doi.org/10.3389/FPLS.2017.01978/BIBTEX

Wang, L. K., Wang, M.-H. S., Hung, Y.-T., Li, K. H., Aziz, H. A., Yusoff, M. S., & Palaniandy, P. (2022). Vermicomposting process for treating agricultural and food wastes. In *Waste treatment in the biotechnology, agricultural and food industries* (pp. 173–203). Springer. https://doi.org/10.1007/978-3-031-03591-3_3

Wang, F., Zhang, Y., Su, Y., Wu, D., & Xie, B. (2024a). Pollutant control and nutrient recovery of organic solid waste by earthworms: Mechanism and agricultural benefits of vermicomposting. *Journal of Environmental Chemical Engineering, 12*(3), 112610. https://doi.org/10.1016/J.JECE.2024.112610

Wang, X., Chi, Y., & Song, S. (2024b). Important soil microbiota's effects on plants and soils: A comprehensive 30-year systematic literature review. *Frontiers in Microbiology, 15*, 1347745. https://doi.org/10.3389/FMICB.2024.1347745/PDF

Weinmann, M., Bradáčová, K., & Nikolic, M. (2023). Relationship between mineral nutrition, plant diseases, and pests. In *Marschner's mineral nutrition of plants* (pp. 445–476). Elsevier. https://doi.org/10.1016/B978-0-12-819773-8.00009-5

Yatoo, A. M., Rasool, S., Ali, S., Majid, S., Rehman, M. U., Ali, M. N., Eachkoti, R., Rasool, S., Rashid, S. M., & Farooq, S. (2020). Vermicomposting: An eco-friendly approach for recycling/management of organic wastes. In *Bioremediation and biotechnology: Sustainable approaches to pollution degradation* (pp. 167–187). Springer. https://doi.org/10.1007/978-3-030-35691-0_8

Zhang, F., & Cheng, W. (2022). The mechanism of bacterial resistance and potential bacteriostatic strategies. *Antibiotics, 11*(9), 1215. https://doi.org/10.3390/ANTIBIOTICS11091215

Index

S. Rajpoot, A. P. Das (eds.), *Innovative Technologies for Waste Management*, Sustainable Environmental Waste Management Strategies,
https://doi.org/10.1007/978-3-031-97184-6

MIX
Papier aus verantwortungsvollen Quellen
Paper from responsible sources
FSC® C105338

FSC
www.fsc.org

If you have any concerns about our products,
you can contact us on
ProductSafety@springernature.com

In case Publisher is established outside the EU,
the EU authorized representative is:
Springer Nature Customer Service Center GmbH
Europaplatz 3, 69115 Heidelberg, Germany

Printed by Libri Plureos GmbH
in Hamburg, Germany